FIFTH EDITION

FUNDAMENTALS OF
PHYSICS

FIFTH EDITION

FUNDAMENTALS OF
PHYSICS

PART 1

DAVID HALLIDAY
University of Pittsburgh

ROBERT RESNICK
Rensselaer Polytechnic Institute

JEARL WALKER
Cleveland State University

JOHN WILEY & SONS, INC.

New York • Chichester • Brisbane • Toronto • Singapore

ACQUISITIONS EDITOR Stuart Johnson
DEVELOPMENTAL EDITOR Rachel Nelson
SENIOR PRODUCTION SUPERVISOR Cathy Ronda
PRODUCTION ASSISTANT Raymond Alvarez
MARKETING MANAGER Catherine Faduska
ASSISTANT MARKETING MANAGER Ethan Goodman
DESIGNER Dawn L. Stanley
MANUFACTURING MANAGER Mark Cirillo
PHOTO EDITOR Hilary Newman
COVER PHOTO William Warren/Westlight
ILLUSTRATION EDITOR Edward Starr
ILLUSTRATION Radiant/Precision Graphics

This book was set in Times Roman by Progressive Information Technologies, and printed and bound by Von Hofmann Press. The cover was printed by Phoenix Color Corp.

The paper on this book was manufactured by a mill whose forest management programs include sustained yield harvesting of its timberlands. Sustained yield harvesting principles ensure that the number of trees cut each year does not exceed the amount of new growth.

ISBN 0-471-14561-0

Printed in the United States of America

10 9 8 7 6 5

Hello There!

You are about to begin your first college level physics course. You may have heard from friends and fellow students that physics is a difficult course, especially if you don't plan to go on to a career in the hard sciences. But that doesn't mean it has to be difficult for you. The key to success in this course is to have a good understanding of each chapter before moving on to the next. When learned a little bit at a time, physics is straightforward and simple. Here are some ideas that can make this text and this class work for you:

- Read through the **Sample Problems** and solutions carefully. These problems are similar to many of the end-of-chapter exercises, so reading them will help you solve homework problems. In addition, they offer a look at how an experienced physicist would approach solving the problem.

- Try to answer the **Checkpoint** questions as you read through the chapter. Most of these can be answered by thinking through the problem, but it helps if you have some scratch paper and a pencil nearby to work out some of the harder questions. Hold the answer page at the back of the book with your thumb (or a tab or bookmark) so you can refer to it easily. The end-of-chapter **Questions** are very similar— you can use them to quiz yourself after you've read the whole chapter.

- Use the **Review & Summary** sections at the end of each chapter as the first place to look for formulas you might need to solve homework problems. These sections are also helpful in making study sheets for exams.

- The biggest tip I can give you is pretty obvious. Do your homework! Understanding the homework problems is the best way to master the material and do well in the course. Doing a lot of homework problems is also the best way to review for exams. And by all means, consult your classmates whenever you are stuck. Working in groups will make your studying more effective.

The study of basic physics is required for degrees in Engineering, Physics, Biology, Chemistry, Medicine, and many other sciences because the fundamentals of physics are the framework on which every other science is built. Therefore, a solid grasp of basic physical principles will help you understand upper-level science courses and make your study of these courses easier.

I took introductory physics because it was a prerequisite for my B.S. in Mechanical Engineering. Even though engineering and pure physics are worlds apart, I find myself using this book as a reference almost every day. I urge you to keep it after you have finished your course. The knowledge you will gain from this book and your introductory physics course is the foundation for all other sciences. This is the primary reason to take this class seriously and be successful in it.

Best of luck!

Josh Kane

Josh Kane

Preface

For four editions, *Fundamentals of Physics* has been successful in preparing physics students for careers in science and engineering. The first three editions were coauthored by the highly regarded team of David Halliday and Robert Resnick, who developed a groundbreaking text replete with conceptual structure and applications. In the fourth edition, the insights provided by new coauthor, Jearl Walker, took the text into the 1990s and met the challenge of guiding students through a time of tremendous advances and a ferment of activity in the science of physics. Now, in the fifth edition, we have expanded on the conventional strengths of the earlier editions and enhanced the applications that help students forge a bridge between concepts and reasoning. We not only *tell* students how physics works, we *show* them, and we give them the opportunity to show us what they have learned by testing their understanding of the concepts and applying them to real-world scenarios. Concept checkpoints, problem solving tactics, sample problems, electronic computations, exercises and problems—all of these skill-building signposts have been developed to help students establish a connection between conceptual theories and application. The students reading this text today are the scientists and engineers of tomorrow. It is our hope that the fifth edition of *Fundamentals of Physics* will help prepare these students for future endeavors by contributing to the enhancement of physics education.

CHANGES IN THE FIFTH EDITION

Although we have retained the basic framework of the fourth edition of *Fundamentals of Physics,* we have made extensive changes in portions of the book. Each chapter and element has been scrutinized to ensure clarity, currency, and accuracy, reflecting the needs of today's science and engineering students.

Content Changes

Mindful that textbooks have grown large and that they tend to increase in length from edition to edition, we have reduced the length of the fifth edition by combining several chapters and pruning their contents. In doing so, six chapters have been rewritten completely, while the remaining chapters have been carefully edited and revised, often ex-

tensively, to enhance their clarity, incorporating ideas and suggestions from dozens of reviewers.

- *Chapters 7 and 8 on energy* (and sections of later chapters dealing with energy) have been rewritten to provide a more careful treatment of energy, work, and the work–kinetic energy theorem. As the same time, the text material and problems at the end of each chapter still allow the instructor to present the more traditional treatment of these subjects.

- *Temperature, heat, and the first law of thermodynamics* have been condensed from two chapters to one chapter (*Chapter 19*).

- *Chapter 21 on entropy* now includes a statistical mechanical presentation of entropy that is tied to the traditional thermodynamical presentation.

- *Chapters on Faraday's law and inductance* have been combined into one new chapter (*Chapter 31*).

- *Treatment of Maxwell's equations* has been streamlined and moved up earlier into the chapter on magnetism and matter (*Chapter 32*).

- *Coverage of electromagnetic oscillations and alternating currents* has been combined into one chapter (*Chapter 33*).

- *Chapters 39, 40, and 41 on quantum mechanics* have been rewritten to modernize the subject. They now include experimental and theoretical results of the last few years. In addition, quantum physics and special relativity are introduced in some of the early chapters in short sections that can be covered quickly. These early sections lay some of the groundwork for the ''modern physics'' topics that appear later in the extended version of the text and add an element of suspense about the subject.

New Pedagogy

In the interest of addressing the needs of science and engineering students, we have added a number of new pedagogical features intended to help students forge a bridge between concepts and reasoning and to marry theory with practice. These new features are designed to help students test their understanding of the material. They were also developed to help students prepare to apply the information to exam questions and real-world scenarios.

- To provide opportunities for students to check their understanding of the physics concepts they have just read, we have placed **Checkpoint** questions within the chapter presentations. Nearly 300 Checkpoints have been added to help guide the student away from common errors and misconceptions. All of the Checkpoints require decision making and reasoning on the part of the student (rather than computations requiring calculators) and focus on the key points of the physics that students need to understand in order to tackle the exercises and problems at the end of each chapter. Answers to all of the Checkpoints are found in the back of the book, sometimes with extra guidance to the student.

- Continuing our focus on the key points of the physics, we have included additional **Checkpoint-type questions** in the Questions section at the end of each chapter. These new questions require decision making and reasoning on the part of the student; they ask the student to organize the physics concepts rather than just plug numbers into equations. Answers to the odd-numbered questions are now provided in the back of the book.

- To encourage the use of computer math packages and graphing calculators, we have added an **Electronic Computation problem section** to the Exercises and Problems sections of many of the chapters.

These new features are just a few of the pedagogical elements available to enhance the student's study of physics. A number of tried-and-true features of the previous edition have been retained and refined in the fifth edition, as described below.

CHAPTER FEATURES

The pedagogical elements that have been retained from previous editions have been carefully planned and crafted to motivate students and guide their reasoning process.

- *Puzzlers* Each chapter opens with an intriguing photograph and a ''puzzler'' that is designed to motivate the student to read the chapter. The answer to each puzzler is provided within the chapter, but it is not identified as such to ensure that the student reads the entire chapter.

- *Sample Problems* Throughout each chapter, sample problems provide a bridge from the concepts of the chapter to the exercises and problems at the end of the chapter. Many of the nearly 400 sample problems featured in the text have been replaced with new ones that more sharply focus on the common difficulties students experience in solving the exercises and problems. We have been especially mindful of the mathematical difficulties students face. The sample problems also provide

an opportunity for the student to see how a physicist thinks through a problem.

- *Problem Solving Tactics* To help further bridge concepts and applications and to add focus to the key physics concepts, we have refined and expanded the number of problem solving tactics that are placed within the chapters, particularly in the earlier chapters. These tactics provide guidance to the students about how to organize the physics concepts, how to tackle mathematical requirements in the exercises and problems, and how to prepare for exams.

- *Illustrations* Because the illustrations in a physics textbook are so important to an understanding of the concepts, we have altered nearly 30 percent of the illustrations to improve their clarity. We have also removed some of the less effective illustrations and added many new ones.

- *Review & Summary* A review and summary section is found at the end of each chapter, providing a quick review of the key definitions and physics concepts *without* being a replacement for reading the chapter.

- *Questions* Approximately 700 thought-provoking questions emphasizing the conceptual aspects of physics appear at the ends of the chapters. Many of these questions relate back to the checkpoints found throughout the chapters, requiring decision making and reasoning on the part of the student. Answers to the odd-numbered questions are provided in the back of the book.

- *Exercises & Problems* There are approximately 3400 end-of-chapter exercises and problems in the text, arranged in order of difficulty, starting with the exercises (labeled ''E''), followed by the problems (labeled ''P''). Particularly challenging problems are identified with an asterisk (*). Those exercises and problems that have been retained from previous editions have been edited for greater clarity; many have been replaced. Answers to the odd-numbered exercises and problems are provided in the back of the book.

VERSIONS OF THE TEXT

The fifth edition of *Fundamentals of Physics* is available in a number of different versions, to accommodate the individual needs of instructors and students alike. The Regular Edition consists of Chapters 1 through 38 (ISBN 0-471-10558-9). The Extended Edition contains seven additional chapters on quantum physics and cosmology (Chapters 1–45) (ISBN 0-471-10559-7). Both editions are available as single, hardcover books, or in the alternative versions listed on page ix:

- Volume 1—Chapters 1–21 (Mechanics/Thermodynamics), cloth, 0-471-15662-0
- Volume 2—Chapters 22–45 (E&M and Modern Physics), cloth, 0-471-15663-9
- Part 1—Chapters 1–12, paperback, 0-471-14561-0
- Part 2—Chapters 13–21, paperback, 0-471-14854-7
- Part 3—Chapters 22–33, paperback, 0-471-14855-5
- Part 4—Chapters 34–38, paperback, 0-471-14856-3
- Part 5—Chapters 39–45, paperback, 0-471-15719-8

The Extended edition of the text is also available on CD ROM.

SUPPLEMENTS

The fifth edition of *Fundamentals of Physics* is supplemented by a comprehensive ancillary package carefully developed to help teachers teach and students learn.

Instructor's Supplements

- *Instructor's Manual* by J. RICHARD CHRISTMAN, U.S. Coast Guard Academy. This manual contains lecture notes outlining the most important topics of each chapter, as well as demonstration experiments, and laboratory and computer exercises; film and video sources are also included. Separate sections contain articles that have appeared recently in the *American Journal of Physics* and *The Physics Teacher*.
- *Instructor's Solutions Manual* by JERRY J. SHI, Pasadena City College. This manual provides worked-out solutions for all the exercises and problems found at the end of each chapter within the text. *This supplement is available only to instructors.*
- *Solutions Disk.* An electronic version of the Instructor's Solutions Manual, for instructors only, available in TeX for Macintosh and Windows™.
- *Test Bank* by J. RICHARD CHRISTMAN, U.S. Coast Guard Academy. More than 2200 multiple-choice questions are included in the Test Bank for *Fundamentals of Physics*.

- *Computerized Test Bank.* IBM and Macintosh versions of the entire Test Bank are available with full editing features to help you customize tests.
- *Animated Illustrations.* Approximately 85 text illustrations are animated for enhanced lecture demonstrations.
- *Transparencies.* More than 200 four-color illustrations from the text are provided in a form suitable for projection in the classroom.

Student's Supplements

- *A Student's Companion* by J. RICHARD CHRISTMAN, U.S. Coast Guard Academy. Much more than a traditional study guide, this student manual is designed to be used in close conjunction with the text. The Student's Companion is divided into four parts, each of which corresponds to a major section of the text, beginning with an overview "chapter." These overviews are designed to help students understand how the important topics are integrated and how the text is organized. For each chapter of the text, the corresponding Companion chapter offers: Basic Concepts, Problem Solving, Notes, Mathematical Skills, and Computer Projects and Notes.
- *Solutions Manual* by J. RICHARD CHRISTMAN, U.S. Coast Guard Academy and EDWARD DERRINGH, Wentworth Institute. This manual provides students with complete worked-out solutions to 30 percent of the exercises and problems found at the end of each chapter within the text.
- **Interactive Learningware** by JAMES TANNER, Georgia Institute of Technology, with the assistance of GARY LEWIS, Kennesaw State College. This software contains 200 problems from the end-of-chapter exercises and problems, presented in an interactive format, providing detailed feedback for the student. Problems from Chapter 1 to 21 are included in Part 1, from Chapters 22 to 38 in Part 2. The accompanying workbooks allow the student to keep a record of the worked-out problems. The Learningware is available in IBM 3.5″ and Macintosh formats.
- **CD Physics.** The entire Extended Version of the text (Chapters 1–45) is available on CD ROM, along with the student solutions manual, study guide, animated illustrations, and Interactive Learningware.

Acknowledgments

A textbook contains far more contributions to the elucidation of a subject than those made by the authors alone. J. Richard Christman, of the U.S. Coast Guard Academy, has once again created many fine supplements for us; his knowledge of our book and his recommendations to students and faculty are invaluable. James Tanner, of Georgia Institute of Technology, and Gary Lewis, of Kennesaw State College, have provided us with innovative software, closely tied to the text's exercises and problems. J. Richard Christman, of the U.S. Coast Guard Academy, and Glen Terrell, of the University of Texas at Arlington, contributed problems to the Electronic Computation sections of the text. Jerry Shi, of Pasadena City College, performed the Herculean task of working out solutions for every one of the Exercises and Problems in the text. We thank John Merrill, of Brigham Young University, and Edward Derringh, of the Wentworth Institute of Technology for their many contributions in the past. We also thank George W. Hukle of Oxnard, California, for his check of the answers at the back of the book.

At John Wiley, publishers, we have been fortunate to receive strong coordination and support from our former editor, Cliff Mills. Cliff guided our efforts and encouraged us along the way. When Cliff moved on to other responsibilities at Wiley, we were ably guided to completion by his successor, Stuart Johnson. Rachel Nelson has coordinated the developmental editing and multilayered preproduction process. Catherine Faduska, our senior marketing manager, and Ethan Goodman, assistant marketing manager, have been tireless in their efforts on behalf of this edition. Jennifer Bruer has built a fine supporting package of ancillary materials. Monica Stipanov and Julia Salsbury managed the review and administrative duties admirably.

We thank Lucille Buonocore, our able production manager, and Cathy Ronda, our production editor, for pulling all the pieces together and guiding us through the complex production process. We also thank Dawn Stanley, for her design; Brenda Griffing, for her copy editing; Edward Starr, for managing the line art program; Lilian Brady, for her proofreading; and all other members of the production team.

Stella Kupferburg and her team of photo researchers, particularly Hilary Newman and Pat Cadley, were inspired in their search for unusual and interesting photographs that communicate physics principles beautifully. We thank Boris Starosta and Irene Nunes for their careful development of a full-color line art program, for which they scrutinized and suggested revisions of every piece. We also owe a debt of gratitude for the line art to the late John Balbalis, whose careful hand and understanding of physics can still be seen in every diagram.

We especially thank Edward Millman for his developmental work on the manuscript. With us, he has read every word, asking many questions from the point of view of a student. Many of his questions and suggested changes have added to the clarity of this volume. Irene Nunes added a final, valuable developmental check in the last stages of the book.

We owe a particular debt of gratitude to the numerous students who used the fourth edition of *Fundamentals of Physics* and took the time to fill out the response cards and return them to us. As the ultimate consumers of this text, students are extremely important to us. By sharing their opinions with us, your students help us ensure that we are providing the best possible product and the most value for their textbook dollars. We encourage the users of this book to contact us with their thoughts and concerns so that we can continue to improve this text in the years to come. In particular, we owe a special debt of gratitude to the students who participated in a final focus group at Union College in Schenecdaty, New York: Matthew Glogowski, Josh Kane, Lauren Papa, Phil Tavernier, Suzanne Weldon, and Rebecca Willis.

Finally, our external reviewers have been outstanding and we acknowledge here our debt to each member of that team:

MARIS A. ABOLINS
Michigan State University

BARBARA ANDERECK
Ohio Wesleyan University

ALBERT BARTLETT
University of Colorado

MICHAEL E. BROWNE
University of Idaho

TIMOTHY J. BURNS
Leeward Community College

JOSEPH BUSCHI
Manhattan College

PHILIP A. CASABELLA
Rensselaer Polytechnic Institute

RANDALL CATON
Christopher Newport College

J. RICHARD CHRISTMAN
U.S. Coast Guard Academy

ROGER CLAPP
University of South Florida

W. R. CONKIE
Queen's University

PETER CROOKER
University of Hawaii at Manoa

WILLIAM P. CRUMMETT
*Montana College of Mineral
Science and Technology*

EUGENE DUNNAM
University of Florida

ROBERT ENDORF
University of Cincinnati

F. PAUL ESPOSITO
University of Cincinnati

JERRY FINKELSTEIN
San Jose State University

ALEXANDER FIRESTONE
Iowa State University

ALEXANDER GARDNER
Howard University

ANDREW L. GARDNER
Brigham Young University

JOHN GIENIEC
*Central Missouri State
University*

JOHN B. GRUBER
San Jose State University

ANN HANKS
American River College

SAMUEL HARRIS
Purdue University

EMILY HAUGHT
Georgia Institute of Technology

LAURENT HODGES
Iowa State University

JOHN HUBISZ
North Carolina State University

JOEY HUSTON
Michigan State University

DARRELL HUWE
Ohio University

CLAUDE KACSER
University of Maryland

LEONARD KLEINMAN
University of Texas at Austin

EARL KOLLER
Stevens Institute of Technology

ARTHUR Z. KOVACS
Rochester Institute of Technology

KENNETH KRANE
Oregon State University

SOL KRASNER
University of Illinois at Chicago

PETER LOLY
University of Manitoba

ROBERT R. MARCHINI
Memphis State University

DAVID MARKOWITZ
University of Connecticut

HOWARD C. MCALLISTER
University of Hawaii at Manoa

W. SCOTT MCCULLOUGH
Oklahoma State University

JAMES H. MCGUIRE
Tulane University

DAVID M. MCKINSTRY
Eastern Washington University

JOE P. MEYER
Georgia Institute of Technology

ROY MIDDLETON
University of Pennsylvania

IRVIN A. MILLER
Drexel University

EUGENE MOSCA
United States Naval Academy

MICHAEL O'SHEA
Kansas State University

PATRICK PAPIN
San Diego State University

GEORGE PARKER
North Carolina State University

ROBERT PELCOVITS
Brown University

OREN P. QUIST
South Dakota State University

JONATHAN REICHART
SUNY—Buffalo

MANUEL SCHWARTZ
University of Louisville

DARRELL SEELEY
Milwaukee School of Engineering

BRUCE ARNE SHERWOOD
Carnegie Mellon University

JOHN SPANGLER
St. Norbert College

ROSS L. SPENCER
Brigham Young University

HAROLD STOKES
Brigham Young University

JAY D. STRIEB
Villanova University

DAVID TOOT
Alfred University

J. S. TURNER
University of Texas at Austin

T. S. VENKATARAMAN
Drexel University

GIANFRANCO VIDALI
Syracuse University

FRED WANG
Prairie View A&M

ROBERT C. WEBB
Texas A&M University

GEORGE WILLIAMS
University of Utah

DAVID WOLFE
University of New Mexico

We hope that our words here reveal at least some of the wonder of physics, the fundamental clockwork of the universe. And, hopefully, those words might also reveal some of our awe of that clockwork.

DAVID HALLIDAY
6563 NE Windermere Road
Seattle, WA 98105

ROBERT RESNICK
Rensselaer Polytechnic Institute
Troy, NY 12181

JEARL WALKER
Cleveland State University
Cleveland, OH 44115

HOW TO USE THIS BOOK:

You are about to begin what could be one the most exciting course that you will undertake in college. It offers you the opportunity to learn what makes our world "tick" and to gain insight into the role physics plays in our everyday lives. This knowledge will not come without some effort, however, and this book has been carefully designed and written with an awareness of the kinds of difficulties and challenges you may face. Therefore, before you begin, we have provided a visual overview of some of the key features of the book that will aid in your studies.

Chapter Opening Puzzlers

Each chapter opens with an intriguing example of physics in action. By presenting high-interest applications of each chapters concepts, the puzzlers are intended to peak your interest and motivate you to read the chapter.

In 1977, Kitty O'Neil set a dragster record by reaching 392.54 mi/h in a sizzling time of 3.72 s. In 1958, Eli Beeding Jr. rode a rocket sled from a standstill to a speed of 72.5 mi/h in an elapsed time of 0.04 s (less than an eye blink). How can we compare these two rides to see which was more exciting (or more frightening)—by final speeds, by elapsed times, or by some other quantity?

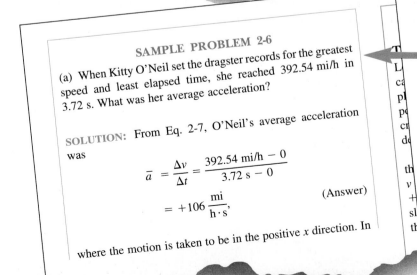

SAMPLE PROBLEM 2-6

(a) When Kitty O'Neil set the dragster records for the greatest speed and least elapsed time, she reached 392.54 mi/h in 3.72 s. What was her average acceleration?

SOLUTION: From Eq. 2-7, O'Neil's average acceleration was

$$\bar{a} = \frac{\Delta v}{\Delta t} = \frac{392.54 \text{ mi/h} - 0}{3.72 \text{ s} - 0}$$

$$= +106 \frac{\text{mi}}{\text{h} \cdot \text{s}}, \qquad \text{(Answer)}$$

where the motion is taken to be in the positive x direction. In

Answers to Puzzlers

All chapter-opening puzzlers are answered later in the chapter, either in text discussion or in a sample problem.

If the car
r, the bob
oninertial

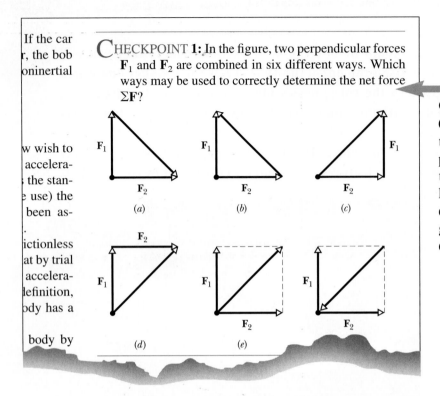

CHECKPOINT 1: In the figure, two perpendicular forces $\mathbf{F}_1$ and $\mathbf{F}_2$ are combined in six different ways. Which ways may be used to correctly determine the net force $\Sigma\mathbf{F}$?

w wish to
accelera-
the stan-
use) the
been as-

ictionless
at by trial
accelera-
lefinition,
ody has a

body by

Checkpoints
Checkpoints appear throughout the text, focusing on the key points of physics you will need to tackle the exercises and problems found at the end of each chapter. These checkpoints help guide you away from common errors and misconceptions.

Checkpoint Questions
Checkpoint-type questions at the end of each chapter ask you to organize the physics concepts rather than plug numbers into equations. Answers to the odd-numbered questions are provided in the back of the book.

ey puck in

$\mathbf{v} = -2t\mathbf{i}$

ponents of
ion vector
nd and t is
-2 and 3?

cal projec-
t identical
n the same
nal speeds

(c)

peed (a) a

$4.9\mathbf{j}$ (x is
). Has the

9. Figure 4-25 shows three paths for a kicked football. Ignoring the effects of air on the flight, rank the paths according to (a) time of flight, (b) initial vertical velocity component, (c) initial horizontal velocity component, and (d) initial speed. Place the greatest first in each part.

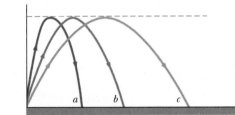

FIGURE 4-25   Question 9.

10. Figure 4-26 shows the velocity and acceleration of a particle at a particular instant in three situations. In which situation, and at that instant, is (a) the speed increasing, (b) the speed decreasing, (c) the speed not changing, (d) $\mathbf{v} \cdot \mathbf{a}$ positive, (e) $\mathbf{v} \cdot \mathbf{a}$ negative, and (f) $\mathbf{v} \cdot \mathbf{a} = 0$?

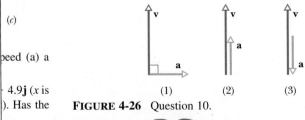

FIGURE 4-26   Question 10.

Sample Problems

The sample problems offer you the opportunity to work through the physics concepts just presented. Often built around real-world applications, they are closely coordinated with the end-of-chapter Questions, Exercises, and Problems.

SAMPLE PROBLEM 4-1

The position vector for a particle is initially

$$\mathbf{r}_1 = -3\mathbf{i} + 2\mathbf{j} + 5\mathbf{k}$$

and then later is

$$\mathbf{r}_2 = 9\mathbf{i} + 2\mathbf{j} + 8\mathbf{k}$$

(see Fig. 4-2). What is the displacement from $\mathbf{r}_1$ to $\mathbf{r}_2$?

SOLUTION: Recall from Chapter 3 that we add (or subtract) two vectors in unit-vector notation by combining the components, axis by axis. So Eq. 4-2 becomes

$$\mathbf{\Delta r} = (9\mathbf{i} + 2\mathbf{j} + 8\mathbf{k}) - (-3\mathbf{i} + 2\mathbf{j} + 5\mathbf{k})$$
$$= 12\mathbf{i} + 3\mathbf{k}. \qquad \text{(Answer)}$$

The displacement vector is parallel to the xz plane, because it lacks any y component, a fact that is easier to pick out in the numerical result than in Fig. 4-2.

Sample Problem 4-1. The displacement $\mathbf{\Delta r} = $
d of $\mathbf{r}_1$ to the head of $\mathbf{r}_2$.

PROBLEM SOLVING TACTICS

TACTIC 1: *Reading Force Problems*
Read the problem statement several times until you have a good mental picture of what the situation is, what data are given, and what is requested. In Sample Problem 5-1, for example, you should tell yourself: "Someone is pushing a sled. Its speed changes, so acceleration is involved. The motion is along a straight line. A force is given in one part and asked for in the other, and so the situation looks like Newton's second law applied to one-dimensional motion."

If you know what the problem is about but don't know what to do next, put the problem aside and reread the text. If you are hazy about Newton's second law, reread that section. Study the sample problems. The one-dimensional-motion parts of Sample Problem 5-1 and the constant acceleration should send you back to Chapter 2 and especially to Table 2-1, which displays all the equations you are likely to need.

TACTIC 2: *Draw Two Types of Figures*
You may need two figures. One is a rough sketch of the actual real-world situation. When you draw the forces on it, place the tail of each force vector either on the boundary of or within the body feeling that force. The other figure is a free-body diagram in which the forces on a *single* body are drawn, with the body represented with a dot or a sketch. Place the tail of each force vector on the dot or sketch.

TACTIC 3: *What I   em?*
If you are

Problem Solving Tactics

Careful attention has been paid to helping you develop your problem-solving skills. Problem-solving tactics are closely related to the sample problems and can be found throughout the text, though most fall within the first half. The tactics are designed to help you work through assigned homework problems and prepare for exams. Collectively, they represent the stock in trade of experienced problem solvers and practicing scientists and engineers.

Review and Summary

Review & Summary sections at the end of each chapter review the most important concepts and equations.

REVIEW & SUMMARY

Conservative Forces

A force is a **conservative force** if the net work it does on a particle moving along a closed path from an initial point and then back to that point is zero. Or, equivalently, it is conservative if its work on a particle moving between two points does not depend on the path taken by the particle. The gravitational force (weight) and the spring force are conservative forces; the kinetic frictional force is a **nonconservative force.**

Potential Energy

A **potential energy** is energy that is associated with the configuration of a system in which a conservative force acts. When the conservative force does work W on a particle within the system, the change ΔU in the potential energy of the system is

$$\Delta U = -W. \qquad (8\text{-}1)$$

If the particle moves from point x_i to point x_f, the change in the potential energy of the system is

$$\Delta U = -\int_{x_i}^{x_f} F(x)\, dx. \qquad (8\text{-}6)$$

Gravitational Potential Energy

The potential energy associated with a system consisting of the Earth and a nearby particle is the **gravitational potential energy.** If the particle moves from height y_i to height y_f, the change in gravitational potential energy of the particle–Earth system is

$$\Delta U = mg(y_f - y_i) = mg\,\Delta y. \qquad (8\text{-}7)$$

If the **reference position** of the particle is set as $y_i = 0$ and the corresponding gravitational potential energy of the system is set as $U_i = 0$, then the gravitational potential energy U when the particle is at any position y is

$$U = mgy. \qquad (8\text{-}9)$$

in which the subscripts refer to different instants during an transfer process. This conservation can also be written as

$$\Delta E = \Delta K + \Delta U = 0.$$

Potential Energy Curves

If we know the **potential energy function** $U(x)$ for a sy which a force F acts on a particle, we can find the force

$$F(x) = -\frac{dU(x)}{dx}.$$

If $U(x)$ is given on a graph, then at any value of x, the for the negative of the slope of the curve there and the kinetic of the particle is given by

$$K(x) = E - U(x),$$

where E is the mechanical energy of the system. A **turnin** is a point x where the particle reverses its motion (there, The particle is in **equilibrium** at points where the slope $U(x)$ curve is zero (there, $F(x) = 0$).

Work by Nonconservative Forces

If a nonconservative applied force $\mathbf{F}$ does work on particle part of a system having a potential energy, then the wo done on the system by $\mathbf{F}$ is equal to the change ΔE in the m ical energy of the system:

$$W_{app} = \Delta K + \Delta U = \Delta E. \qquad (8\text{-}24$$

If a kinetic frictional force $\mathbf{f}_k$ does work on an obj change ΔE in the total mechanical energy of the object a system containing it is given by

$$\Delta E = -f_k d,$$

in which d is the displacement of the object during the wo

Exercises and Problems

A hallmark of this text, nearly 3400 end-of-chapter exercises and problems are arranged in order of difficulty, starting with the exercises (labeled "E"), followed by the problems (labeled "P"). Particularly difficult problems are identified with an asterisk (*). Answers to all the odd-numbered exercises and problems are provided in the back of the book. New electronic computation problems, which require the use of math packages and graphing calculators, have been added to many of the chapters.

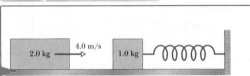

FIGURE 10-44 Problem 56.

57P. Two 22.7 kg ice sleds are placed a short distance apart, one directly behind the other, as shown in Fig. 10-45. A 3.63 kg cat, standing on one sled, jumps across to the other and immediately back to the first. Both jumps are made at a speed of 3.05 m/s relative to the ice. Find the final speeds of the two sleds.

FIGURE 10-45 Problem 57.

58P. The bumper of a 1200 kg car is designed so that it can just absorb all the energy when the car runs head-on into a solid wall at 5.00 km/h. The car is involved in a collision in which it runs at 70.0 km/h into the rear of a 900 kg car moving at 60.0 km/h in the same direction. The 900 kg car is accelerated to 70.0 km/h as a result of the collision. (a) What is the speed of the 1200 kg car immediately after impact? (b) What is the ratio of the kinetic energy absorbed in the collision to that which can be absorbed by the bumper of the 1200 kg car?

59P. A railroad freight car weighing 32 tons and traveling at 5.0

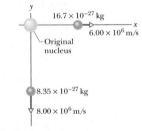

FIGURE 10-46 Exercise 62.

63E. In a game of pool, the cue ball strikes anothe at rest. After the collision, the cue ball moves at 3.5 line making an angle of 22.0° with its original dir tion, and the second ball has a speed of 2.00 m/s angle between the direction of motion of the secon original direction of motion of the cue ball and (b speed of the cue ball. (c) Is kinetic energy conserv

64E. Two vehicles A and B are traveling west and tively, toward the same intersection, where they co together. Before the collision, A (total weight 2700 with a speed of 40 mi/h and B (total weight 3600 lb of 60 mi/h. Find the magnitude and direction of the v (interlocked) vehicles immediately after the collisi

65E. In a game of billiards, the cue ball is given a V and strikes the pack of 15 stationary balls. All engage in nume   and ball–cushion co time lat   ne accident) all

Brief Contents

Contents

CHAPTER 5

FORCE AND MOTION—I *81*

Can a man pull two railroad passenger cars with his teeth?

CHAPTER 6

FORCE AND MOTION—II *108*

Why do cats sometimes survive long falls better than shorter ones?

CHAPTER 7

KINETIC ENERGY AND WORK *130*

How much work is required in lifting great weights?

CHAPTER 8

POTENTIAL ENERGY AND CONSERVATION OF ENERGY *155*

How far will a bungee-cord jumper fall?

CHAPTER 9

SYSTEMS OF PARTICLES *186*

How does a ballerina seemingly "turn off" gravity?

CHAPTER 10

COLLISIONS   *214*

*Is a board or a concrete block
easier to break in karate?*

CHAPTER 11

ROTATION   *238*

*What advantages does physics offer
in judo throws?*

CHAPTER 12

ROLLING, TORQUE, AND
ANGULAR MOMENTUM   *268*

*Why is a quadruple somersault
so difficult in trapeze acts?*

APPENDICES   *A1*

ANSWERS TO CHECKPOINTS AND
ODD-NUMBERED QUESTIONS, EXERCISES
AND PROBLEMS   *AN1*

INDEX   *I1*

FIFTH EDITION

FUNDAMENTALS OF
PHYSICS

You can watch the Sun set and disappear over a calm ocean once while you lie on a beach, and once again if you stand up. Surprisingly, by measuring the time between the two sunsets, you can approximate the radius of the Earth. How can such a simple observation be used to measure the Earth?

1-1 MEASURING THINGS

Physics is based on measurement. We discover physics by learning how to measure the quantities that are involved in physics. Among these quantities are length, time, mass, temperature, pressure, and electrical resistance.

To describe a physical quantity we first define a **unit,** that is, a measure of the quantity that is defined to be exactly 1.0. Then we define a **standard,** that is, a reference to which all other examples of the quantity are compared. For example, the unit of length is the meter, and, as you will see, the standard for the meter is defined to be the distance traveled by light in vacuum during a certain fraction of a second. We are free to define a unit and its standard in any way we care to. The important thing is to do so in such a way that scientists around the world will agree that our definitions are both sensible and practical.

Once we have set up a standard, say, for length, we must work out procedures by which any length whatever, be it the radius of a hydrogen atom, the wheelbase of a skateboard, or the distance to a star, can be expressed in terms of the standard. Rulers, which approximate our length standard, give us one such procedure for measuring length. But many of our comparisons must be indirect. You cannot use a ruler, for example, to measure either the radius of an atom or the distance to a star.

There are so many physical quantities that it is a problem to organize them. Fortunately, they are not all independent. Speed, for example, is the ratio of a length to a time. So what we do is pick out—by international agreement—a small number of physical quantities, such as length and time, and assign standards to them alone. We then define all other physical quantities in terms of these *base quantities* and their standards. Speed, for example, is defined in terms of the base quantities length and time and the associated base standards.

The base standards must be both accessible and invariable. If we define the length standard as the distance between one's nose and the index finger on an outstretched arm, we certainly have an accessible standard—but it will, of course, vary from person to person. The demand for precision in science and engineering pushes us to aim first for invariability. We then exert great effort to make duplicates of the base standards that are accessible to those who need them.

1-2 THE INTERNATIONAL SYSTEM OF UNITS

In 1971, the 14th General Conference on Weights and Measures picked seven quantities as base quantities, thereby forming the basis of the International System of Units, abbreviated SI from its French name and popularly known as the *metric system.* Table 1-1 shows the units for the three base quantities—length, mass, and time—that we use in the early chapters of this book. The units for the quantities were chosen to be on a ''human scale.''

Many SI *derived units* are defined in terms of these base units. For example, the SI unit for power, called the **watt** (abbreviated as W), is defined in terms of the base units for mass, length, and time. Thus, as you will see in Chapter 7,

$$1 \text{ watt} = 1 \text{ W} = 1 \text{ kg} \cdot \text{m}^2/\text{s}^3. \tag{1-1}$$

To express the very large and very small quantities that we often run into in physics, we use the so-called *scientific notation,* which employs powers of 10. In this notation,

$$3{,}560{,}000{,}000 \text{ m} = 3.56 \times 10^9 \text{ m} \tag{1-2}$$

and

$$0.000\ 000\ 492 \text{ s} = 4.92 \times 10^{-7} \text{ s}. \tag{1-3}$$

Since the advent of computers, scientific notation sometimes takes on an even briefer look, as in 3.56 E9 m and 4.92 E −7 s, where E stands for ''exponent of ten.'' It is briefer still on some calculators, where E is replaced with an empty space.

As a further convenience when dealing with very large or very small measurements, we use the prefixes listed in Table 1-2. As you can see, each prefix represents a certain power of 10, as a factor. Attaching a prefix to an SI unit has the effect of multiplying by the associated factor. Thus we can express a particular electric power as

$$1.27 \times 10^9 \text{ watts} = 1.27 \text{ gigawatts} = 1.27 \text{ GW} \tag{1-4}$$

or a particular time interval as

$$2.35 \times 10^{-9} \text{ s} = 2.35 \text{ nanoseconds} = 2.35 \text{ ns}. \tag{1-5}$$

Some prefixes, as used in milliliter, centimeter, kilogram, and megabyte, are already familiar to you.

1-3 CHANGING UNITS

We often need to change the units in which a physical quantity is expressed. We do so by a method called *chain-link conversion.* In this method, we multiply the original measurement by a **conversion factor** (a ratio of units that is equal to unity). For example, because 1 min and 60 s are identical time intervals, we can write

$$\frac{1 \text{ min}}{60 \text{ s}} = 1 \quad \text{and} \quad \frac{60 \text{ s}}{1 \text{ min}} = 1.$$

This is *not* the same as writing $\frac{1}{60} = 1$ or $60 = 1$; the *number* and its *unit* must be treated together.

2

TABLE 1-1 SOME SI BASE UNITS

QUANTITY	UNIT NAME	UNIT SYMBOL
Length	meter	m
Time	second	s
Mass	kilogram	kg

Because multiplying any quantity by unity leaves it unchanged, we can introduce such conversion factors wherever we find them useful. In chain-link conversion, we use the factors to cancel the unwanted units. For example, to convert 2 min to seconds, we have

$$2 \text{ min} = (2 \text{ min})(1) = (2 \text{ min})\left(\frac{60 \text{ s}}{1 \text{ min}}\right)$$
$$= 120 \text{ s}. \tag{1-6}$$

If you introduce the conversion factor in a way that the units do *not* cancel, invert the factor and try again. Units obey the same algebraic rules as variables and numbers.

Appendix D and the inside back cover give conversion factors between SI and other systems of units. The United States is the only major country (in fact, almost the only country) that has not officially adopted the International System of Units.

SAMPLE PROBLEM 1-1

The research submersible ALVIN is diving at a speed of 36.5 fathoms per minute.

(a) Express this speed in meters per second. A *fathom* (fath) is precisely 6 ft.

SOLUTION: To find the speed in meters per second, we write

$$36.5 \frac{\text{fath}}{\text{min}} = \left(36.5 \frac{\text{fath}}{\text{min}}\right)\left(\frac{1 \text{ min}}{60 \text{ s}}\right)\left(\frac{6 \text{ ft}}{1 \text{ fath}}\right)\left(\frac{1 \text{ m}}{3.28 \text{ ft}}\right)$$
$$= 1.11 \text{ m/s}. \tag{Answer}$$

(b) What is this speed in miles per hour?

SOLUTION: As above, we have

$$36.5 \frac{\text{fath}}{\text{min}} = \left(36.5 \frac{\text{fath}}{\text{min}}\right)\left(\frac{60 \text{ min}}{1 \text{ h}}\right)\left(\frac{6 \text{ ft}}{1 \text{ fath}}\right)\left(\frac{1 \text{ mi}}{5280 \text{ ft}}\right)$$
$$= 2.49 \text{ mi/h}. \tag{Answer}$$

(c) What is this speed in light-years per year?

SOLUTION: A light-year (ly) is the distance that light travels in 1 year, 9.46×10^{12} km.

We start from the result of (a) above:

$$1.11 \frac{\text{m}}{\text{s}} = \left(1.11 \frac{\text{m}}{\text{s}}\right)\left(\frac{1 \text{ ly}}{9.46 \times 10^{12} \text{ km}}\right)$$
$$\times \left(\frac{1 \text{ km}}{1000 \text{ m}}\right)\left(\frac{3.16 \times 10^7 \text{ s}}{1 \text{ y}}\right)$$
$$= 3.71 \times 10^{-9} \text{ ly/y}. \tag{Answer}$$

We can write this in the even more unlikely form of 3.71 nly/y, where "nly" is an abbreviation for nanolight-year.

SAMPLE PROBLEM 1-2

How many square meters are in an area of 6.0 km²?

SOLUTION: Each kilometer in the original measure must be converted. The surest way is to separate them:

TABLE 1-2 PREFIXES FOR SI UNITS^a

FACTOR	PREFIX	SYMBOL	FACTOR	PREFIX	SYMBOL
10^{24}	yotta-	Y	10^{-24}	yocto-	y
10^{21}	zetta-	Z	10^{-21}	zepto-	z
10^{18}	exa-	E	10^{-18}	atto-	a
10^{15}	peta-	P	10^{-15}	femto-	f
10^{12}	tera-	T	**10^{-12}**	**pico-**	**p**
10^9	**giga-**	**G**	**10^{-9}**	**nano-**	**n**
10^6	**mega-**	**M**	**10^{-6}**	**micro-**	**μ**
10^3	**kilo-**	**k**	**10^{-3}**	**milli-**	**m**
10^2	hecto-	h	**10^{-2}**	**centi-**	**c**
10^1	deka-	da	10^{-1}	deci-	d

^aThe most commonly used prefixes are shown in bold type.

$$6.0 \text{ km}^2 = 6.0 \text{ (km)(km)}$$
$$= 6.0 \text{ (km)(km)} \left(\frac{1000 \text{ m}}{1 \text{ km}} \right) \left(\frac{1000 \text{ m}}{1 \text{ km}} \right)$$
$$= 6.0 \times 10^6 \text{ m}^2.$$

PROBLEM SOLVING TACTICS

TACTIC 1: *Significant Figures and Decimal Places*

If you calculated the answer to (a) in Sample Problem 1-1 without your calculator automatically rounding it off, the number 1.112804878 might have appeared in the display. The precision implied by this number is meaningless. We have properly rounded the number to 1.11 m/s, which is the precision that is justified by the precision of the original data: The given speed of 36.5 fath/min consists of three digits, called **significant figures.** Any fourth digit that might appear to the right of the 5 is not known, and so the result of the conversion is not reliable beyond three digits, or three significant figures.

In general, no final result should have more significant figures than the original data from which it was derived.

If multiple steps of calculation are involved, you should retain more significant figures than the original data have. However, when you come to the final result, you should round off according to the original data with the least significant figures. (The answers to sample problems in this book are usually presented with the symbol = instead of ≈ even if rounding off is involved.)

When a number such as 3.15 or 3.15 × 10³ is provided in a problem, the number of significant figures is apparent. But how about the number 3000? Is it known to only one significant figure (it could be written as 3 × 10³)? Or is it known to as many as four significant figures (it could be written as 3.000 × 10³)? In this book, we assume that all the zeros in such provided numbers as 3000 are significant, but you had better not make that assumption elsewhere.

Don't confuse *significant figures* with *decimal places*. Consider the lengths 35.6 mm, 3.56 m, and 0.00356 m. They all have three significant figures but, in sequence, they have one, two, and five decimal places.

1-4 LENGTH

In 1792, the newborn Republic of France established a new system of weights and measures. Its cornerstone was the meter, defined to be one ten-millionth of the distance from the north pole to the equator. Eventually, for practical reasons, this Earth standard was abandoned and the meter came to be defined as the distance between two fine lines engraved near the ends of a platinum–iridium bar, the **standard meter bar,** which was kept at the International Bureau of Weights and Measures near Paris. Accurate copies of the bar were sent to standardizing laboratories throughout the world. These **secondary standards** were used to produce other, still more accessible standards so that ultimately every measuring device derived its authority from the standard meter bar through a complicated chain of comparisons.

In 1959, the yard was legally defined to be

$$1 \text{ yard} = 0.9144 \text{ meter (exactly)}, \tag{1-7}$$

which is equivalent to

$$1 \text{ inch} = 2.54 \text{ centimeters (exactly)}. \tag{1-8}$$

Table 1-3 lists some interesting lengths. One corresponds to a virus; an example is shown in Fig. 1-1.

Eventually, modern science and technology required a more precise standard than the distance between two fine scratches on a metal bar. In 1960, a new standard for the meter, based on the wavelength of light, was adopted. Specifically, the meter was redefined to be 1,650,763.73 wavelengths of a particular orange-red light emitted by atoms of krypton-86 in a gas discharge tube.* This awkward number of wavelengths was chosen so that the new standard would be as consistent as possible with the old meter-bar standard.

TABLE 1-3 SOME APPROXIMATE LENGTHS

LENGTH	METERS
Distance to the farthest observed quasar (1996)	2×10^{26}
Distance to the Andromeda galaxy	2×10^{22}
Distance to the nearest star (Proxima Centauri)	4×10^{16}
Distance to the farthest planet (Pluto)	6×10^{12}
Radius of the Earth	6×10^6
Height of Mt. Everest	9×10^3
Thickness of this page	1×10^{-4}
Wavelength of light	5×10^{-7}
Length of a typical virus	1×10^{-8}
Radius of a hydrogen atom	5×10^{-11}
Radius of a proton	1×10^{-15}

*The number 86 in the notation krypton-86 identifies a particular one of the five stable isotopes (or types) of this element. An equivalent notation is 86Kr. The number is called the *mass number* of the isotope in question.

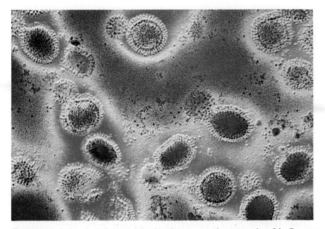

FIGURE 1-1 A ''false-color'' electron micrograph of influenza virus particles. Lipoproteins (colored yellow) taken from the host surround the core (colored green) of the virus. Together they are less than 50 nm in diameter.

The krypton-86 atoms of the atomic length standard are available everywhere, are identical, and emit light of precisely the same wavelength. As Philip Morrison of MIT has pointed out, every atom is a storehouse of natural standards, more secure than the International Bureau of Weights and Measures.

By 1983, the demand for higher precision had reached such a point that even the krypton-86 standard could not meet it, and in that year a bold step was taken. The meter was redefined as the distance traveled by a light wave in a specified time interval. In the words of the 17th General Conference on Weights and Measures:

The meter is the length of the path traveled by light in vacuum during a time interval of 1/299,792,458 of a second.

This number was chosen so that the speed of light c could be exactly

$$c = 299,792,458 \text{ m/s}.$$

Measurements of the speed of light had become extremely precise, so it made sense to adopt the speed of light as a defined quantity and to use it to redefine the meter.

SAMPLE PROBLEM 1-3

Both 100 yd and 100 m are used as distances for dashes in track meets.

(a) Which is longer?

SOLUTION: From Eq. 1-7, 100 yd is equal to 91.44 m, so that 100 m is longer than 100 yd.

(b) By how many meters is it longer?

SOLUTION: We represent the difference as ΔL, where Δ is the capital Greek letter delta. Thus

$$\Delta L = 100 \text{ m} - 100 \text{ yd}$$
$$= 100 \text{ m} - 91.44 \text{ m} = 8.56 \text{ m}. \quad \text{(Answer)}$$

(c) By how many feet is it longer?

SOLUTION: We can express this difference by the method used in Sample Problem 1-1,

$$\Delta L = (8.56 \text{ m})\left(\frac{3.28 \text{ ft}}{1 \text{ m}}\right) = 28.1 \text{ ft}. \quad \text{(Answer)}$$

1-5 TIME

Time has two aspects. For civil and some scientific purposes, we want to know the time of day (see Fig. 1-2) so that we can order events in sequence. And in much scientific work, we want to know how long an event lasts. Thus any time standard must be able to answer two questions: ''*When* did it happen?'' and ''What was its *duration*?'' Table 1-4 shows some time intervals.

Any phenomenon that repeats itself is a possible time standard. The rotation of the Earth, which determines the length of the day, has been used in this way for centuries. A quartz clock, in which a quartz ring continuously vibrates, can be calibrated against the rotating Earth via astronomical observations and used to measure time intervals in the laboratory. However, the calibration cannot be carried out

FIGURE 1-2 When the metric system was proposed in 1792, the hour was redefined to provide a 10-hour day. The idea did not catch on. The maker of this 10-hour watch wisely provided a small dial that kept conventional 12-hour time. Do the two dials indicate the same time?

TABLE 1-4 SOME APPROXIMATE TIME INTERVALS

TIME INTERVAL	SECONDS
Lifetime of the proton (predicted)	1×10^{39}
Age of the universe	5×10^{17}
Age of the pyramid of Cheops	1×10^{11}
Human life expectancy (U.S.)	2×10^{9}
Length of a day	9×10^{4}
Time between human heartbeats	8×10^{-1}
Lifetime of the muon	2×10^{-6}
Shortest laboratory light pulse (1989)	6×10^{-15}
Lifetime of most unstable particle	1×10^{-23}
The Planck time[a]	1×10^{-43}

[a]This is the earliest time after the ''Big Bang'' at which the laws of physics as we know them can be applied.

with the accuracy called for by modern scientific and engineering technology.

To meet the need for a better time standard, atomic clocks have been developed. Figure 1-3 shows such a clock, based on a characteristic frequency of the isotope cesium-133, at the National Institute of Standards and Technology (NIST). It forms the basis in this country for Coordinated Universal Time (UTC), for which time signals are available by shortwave radio (stations WWV and WWVH) and by telephone. (To set a clock extremely accurately at your particular location, you would have to account for the travel time that is required for these signals to reach you.)

Figure 1-4 shows variations in the length of one day on Earth over a 4-year period, as determined by comparison with a cesium clock. Because of the seasonal variation displayed by Fig. 1-4, we suspect the rotating Earth when there is a difference between Earth and atom as timekeepers. The variation is probably due to tidal effects caused by the Moon and to large-scale atmospheric winds.

The 13th General Conference on Weights and Measures in 1967 adopted a standard second based on the cesium clock:

One second is the time taken by 9,192,631,770 vibrations of the light (of a specified wavelength) emitted by a cesium-133 atom.

In principle, two cesium clocks would have to run for 6000 years before their readings would differ by more than 1 s. Even such accuracy pales in comparison to that of clocks currently being developed; their precision may be as fine as 1 part in 10^{18}, that is, 1 s in 1×10^{18} s (about 3×10^{10} y).

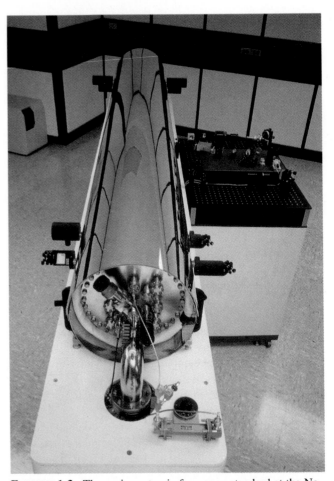

FIGURE 1-3 The cesium atomic frequency standard at the National Institute of Standards and Technology in Boulder, Colorado. It is the primary standard for the unit of time in the United States. Dial (303) 499-7111 and set your watch by it. Call (900) 410-8463 or http://tycho.usno.navy.mil/time.html for Naval Observatory time signals.

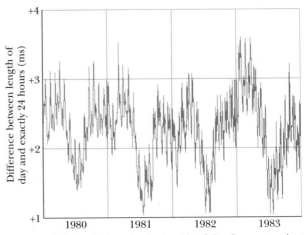

FIGURE 1-4 Variations in the length of the day over a 4-year period. Note that the entire vertical scale amounts to only 3 ms (= 0.003 s).

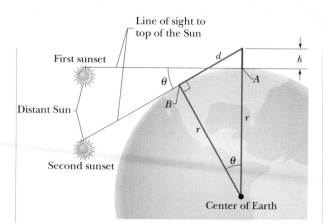

SAMPLE PROBLEM 1-4*

Suppose that you watch the Sun set over a calm ocean while lying on the beach, starting a stopwatch just as the top of the Sun disappears. You then stand, elevating your eyes by a height $h = 1.70$ m, and stop the watch when the top of the Sun again disappears. If the elapsed time on the watch is $t = 11.1$ s, what is the radius r of the Earth?

SOLUTION: As shown in Fig. 1-5, your line of sight to the top of the Sun, as the top first disappears, is tangent to the Earth's surface at your location, at point A. The figure also shows that your line of sight to the top of the Sun as the top again disappears is tangent to the Earth's surface at point B. Let d represent the distance between point B and the location of your eyes when you are standing, and draw radii r as shown in Fig. 1-5. From the Pythagorean theorem, we then have

$$d^2 + r^2 = (r + h)^2 = r^2 + 2rh + h^2,$$

or

$$d^2 = 2rh + h^2. \tag{1-9}$$

Because the height h is so much smaller than the Earth's radius r, the term h^2 is negligible compared to the term $2rh$, and we can rewrite Eq. 1-9 as

$$d^2 = 2rh. \tag{1-10}$$

In Fig. 1-5, the angle between the two tangent points A and B is θ, which is also the angle through which the Sun moves about the Earth during the measured time $t = 11.1$ s. During a full day, which is approximately 24 h, the Sun moves through an angle of $360°$ about the Earth. This allows us to write

$$\frac{\theta}{360°} = \frac{t}{24 \text{ h}},$$

which, with $t = 11.1$ s, gives us

$$\theta = \frac{(360°)(11.1 \text{ s})}{(24 \text{ h})(60 \text{ min/h})(60 \text{ s/min})} = 0.04625°.$$

From Fig. 1-5 we see that $d = r \tan \theta$. Substituting this for d in Eq. 1-10 gives us

$$r^2 \tan^2\theta = 2rh$$

or

$$r = \frac{2h}{\tan^2\theta}.$$

Substituting $\theta = 0.04625°$ and $h = 1.70$ m, we find

$$r = \frac{(2)(1.70 \text{ m})}{\tan^2 0.04625°} = 5.22 \times 10^6 \text{ m}, \quad \text{(Answer)}$$

which is within 20% of the accepted value (6.37×10^6 m) for the (mean) radius of the Earth.

*Adapted from "Doubling Your Sunsets, or How Anyone Can Measure the Earth's Size with a Wristwatch and Meter Stick," by Dennis Rawlins, *American Journal of Physics,* Feb. 1979, Vol. 47, pp. 126–128. This technique works best at the equator.

FIGURE 1-5 Sample Problem 1-4. Your line of sight to the top of the setting Sun rotates through the angle θ when you move from a prone position at point A, elevating your eyes by a distance h. (Angle θ and distance h are exaggerated here for clarity.)

1-6 MASS

The Standard Kilogram

The SI standard of mass is a platinum–iridium cylinder (Fig. 1-6) kept at the International Bureau of Weights and Measures near Paris and assigned, by international agreement, a mass of 1 kilogram. Accurate copies have been sent to standardizing laboratories in other countries, and the masses of other bodies can be determined by balancing them against a copy. Table 1-5 shows some masses expressed in kilograms.

The U.S. copy of the standard kilogram is housed in a vault at NIST. It is removed, no more than once a year, for the purpose of checking duplicate copies that are used elsewhere. Since 1889, it has been taken to France twice for recomparison with the primary standard. The procedure of comparison with the standard kilogram will someday be

FIGURE 1-6 The international 1 kg standard of mass, a cylinder 39 cm in height and in diameter.

TABLE 1-5 **SOME APPROXIMATE MASSES**

OBJECT	KILOGRAMS
Known universe	1×10^{53}
Our galaxy	2×10^{41}
Sun	2×10^{30}
Moon	7×10^{22}
Asteroid Eros	5×10^{15}
Small mountain	1×10^{12}
Ocean liner	7×10^{7}
Elephant	5×10^{3}
Grape	3×10^{-3}
Speck of dust	7×10^{-10}
Penicillin molecule	5×10^{-17}
Uranium atom	4×10^{-25}
Proton	2×10^{-27}
Electron	9×10^{-31}

replaced with a more reliable and accessible procedure involving a base unit of mass that is the mass of an atom.

A Second Mass Standard

The masses of atoms can be compared with each other more precisely than they can be compared with the standard kilogram. For this reason, we have a second mass standard. It is the carbon-12 atom, which, by international agreement, has been assigned a mass of 12 **atomic mass units** (u). The relation between the two units is

$$1 \text{ u} = 1.6605402 \times 10^{-27} \text{ kg}, \qquad (1\text{-}11)$$

with an uncertainty of ± 10 in the last two decimal places. Scientists can, with reasonable precision, experimentally determine the masses of other atoms relative to the mass of carbon-12. What we presently lack is a reliable means of extending that precision to more common units of mass, such as a kilogram.

REVIEW & SUMMARY

Measurement in Physics

Physics is based on measurement of physical quantities. Certain physical quantities have been chosen as **base quantities** (such as length, time, and mass); each has been defined in terms of a **standard** and given a **unit** measure (such as meter, second, and kilogram). Other physical quantities (such as speed) are defined in terms of the base quantities and their standards and units.

SI Units

The unit system emphasized in this book is the International System of Units (SI). The three physical quantities displayed in Table 1-1 are used in the early chapters of this book. Standards, which must be both accessible and invariable, have been established for these base quantities by international agreement. These standards are used in all physical measurement, for both the base quantities and the quantities derived from them. Scientific notation and the prefixes of Table 1-2 can be used to simplify measurement notation in many cases.

Changing Units

Conversion of units from one system to another (for example, from miles per hour to kilometers per second) may be performed by using *chain-link conversions* in which the units are treated as algebraic quantities and the original data are multiplied successively, by *conversion factors* written as unity, until the desired units are obtained.

Length

The unit of length—the meter—is defined as the distance traveled by light during a precisely specified time interval. The yard, together with its multiples and submultiples, is legally defined in this country in terms of the meter.

Time

The unit of time—the second—was formerly defined in terms of the rotation of the Earth. It is now defined in terms of the vibrations of the light emitted by an atomic (cesium-133) source. Accurate time signals are sent worldwide by radio signals keyed to atomic clocks in standardizing laboratories.

Mass

The unit of mass—the kilogram—is defined in terms of a particular platinum–iridium prototype kept near Paris, France. For measurements on an atomic scale, the atomic mass unit, defined in terms of the atom carbon-12, is usually used.

EXERCISES & PROBLEMS

SECTION 1-2 The International System of Units

1E. Use the prefixes in Table 1-2 to express (a) 10^6 phones; (b) 10^{-6} phone; (c) 10^1 cards; (d) 10^9 lows; (e) 10^{12} bulls; (f) 10^{-1} mate; (g) 10^{-2} pede; (h) 10^{-9} Nannette; (i) 10^{-12} boo; (j) 10^{-18} boy; (k) 2×10^2 withits; (l) 2×10^3 mockingbirds. Now that you have the idea, invent a few similar expressions. (See, in this connection, p. 61 of *A Random Walk in Science*, compiled by R.

L. Weber; Crane, Russak & Co., New York, 1974.)

2E. Some of the prefixes of the SI units have crept into everyday language. (a) What is the weekly equivalent of an annual salary of $36 K (= 36 kilobucks)? (b) A lottery awards 10 megabucks as the top prize, payable over 20 years. How much is in each monthly check? (c) The hard disk of a computer has a capacity of 30 MB (= 30 megabytes). At 8 bytes per word, how many words can it store? [In computerese, *mega* means 1,048,576 (= 2^{20}), not 1,000,000.]

SECTION 1-4 Length

3E. A space shuttle orbits the Earth at an altitude of 300 km. What is this altitude in (a) miles and (b) millimeters?

4E. What is your height in meters?

5E. The micrometer (1 μm) is often called the *micron*. (a) How many microns make up 1.0 km? (b) What fraction of a centimeter equals 1.0 μm? (c) How many microns are in 1.0 yd?

6E. The Earth is approximately a sphere of radius 6.37 × 10^6 m. (a) What is its circumference in kilometers? (b) What is its surface area in square kilometers? (c) What is its volume in cubic kilometers?

7E. Calculate the number of kilometers in 20.0 mi using only the following conversion factors: 1 mi = 5280 ft, 1 ft = 12 in., 1 in. = 2.54 cm, 1 m = 100 cm, and 1 km = 1000 m.

8E. Give the relation between (a) a square yard and a square foot; (b) a square inch and a square centimeter; (c) a square mile and a square kilometer; and (d) a cubic meter and a cubic centimeter.

9P. A unit of area, often used in measuring land areas, is the *hectare,* defined as 10^4 m². An open-pit coal mine consumes 75 hectares of land, down to a depth of 26 m, each year. What volume of earth, in cubic kilometers, is removed in this time?

10P. The *cord* is a volume of cut wood equal to a stack 8 ft long, 4 ft wide, and 4 ft high. How many cords of wood are in 1.0 m³?

11P. A room is 20 ft, 2 in. long and 12 ft, 5 in. wide. What is the floor area in (a) square feet and (b) square meters? If the ceiling is 12 ft, $2\frac{1}{2}$ in. above the floor, what is the volume of the room in (c) cubic feet and (d) cubic meters?

12P. Antarctica is roughly semicircular, with a radius of 2000 km (Fig. 1-7). The average thickness of its ice cover is 3000 m. How many cubic centimeters of ice does Antarctica contain? (Ignore the curvature of the Earth.)

FIGURE 1-7 Problem 12.

13P. A typical sugar cube has an edge length of 1 cm. If you had a cubical box that contained a mole of sugar cubes, what would its edge length be? (One mole = 6.02 × 10^{23} units.)

14P. Hydraulic engineers often use, as a unit of volume of water, the *acre-foot,* defined as the volume of water that will cover 1 acre of land to a depth of 1 ft. A severe thunderstorm dumps 2.0 in. of rain in 30 min on a town of area 26 km². What volume of water, in acre-feet, fell on the town?

15P. A certain brand of house paint claims a coverage of 460 ft²/gal. (a) Express this quantity in square meters per liter. (b) Express this quantity in SI base units (see Appendixes A and D). (c) What is the inverse of the original quantity, and what is its physical significance?

16P. Astronomical distances are so large compared to terrestrial ones that much larger units of length are needed. An *astronomical unit* (AU) is equal to the average distance from the Earth to the Sun, about 92.9 × 10^6 mi. A *parsec* (pc) is the distance at which 1 AU would subtend an angle of exactly 1 second of arc (Fig. 1-8). A *light-year* (ly) is the distance that light, traveling through a vacuum with a speed of 186,000 mi/s, would cover in 1.0 year. (a) Express the distance from the Earth to the Sun in parsecs and in light-years. (b) Express 1 ly and 1 pc in miles. Although ''light-year'' appears frequently in popular writing, the parsec is preferred by astronomers.

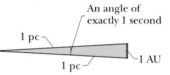

FIGURE 1-8 Problem 16.

17P. During a total solar eclipse, your view of the Sun is almost exactly replaced by your view of the moon. Assuming that the distance from you to the Sun is about 400 times the distance from you to the moon, (a) find the ratio of the Sun's diameter to the moon's diameter. (b) What is the ratio of their volumes? (c) Hold up a dime (or another small coin) so that it would just eclipse the full moon, and measure the angle it subtends at the eye. From this experimental result and the given distance between the moon and the Earth (= 3.8 × 10^5 km), estimate the diameter of the moon.

18P*. The standard kilogram (see Fig. 1-6) is in the shape of a circular cylinder with its height equal to its diameter. Show that, for a circular cylinder of fixed volume, this equality gives the smallest surface area, thus minimizing the effects of surface contamination and wear.

SECTION 1-5 Time

19E. Express the speed of light, 3.0 × 10^8 m/s, in (a) feet per nanosecond and (b) millimeters per picosecond.

20E. Enrico Fermi once pointed out that a standard lecture period (50 min) is close to 1 microcentury. How long is a microcentury in minutes, and what is the percent difference from Fermi's approximation?

21E. How many seconds are in 1 year (= 365.25 days)?

22E. A certain pendulum clock (with a 12-hour dial) happens to gain 1.0 min/day. After setting the clock to the correct time, how long must one wait until it again indicates the correct time?

23E. What is the age of the universe (see Table 1-4) in days?

24E. (a) A unit of time sometimes used in microscopic physics is the *shake*. One shake equals 10^{-8} s. Are there more shakes in a second than there are seconds in a year? (b) Humans have existed for about 10^6 years, whereas the universe is about 10^{10} years old. If the age of the universe is taken to be 1 "day," for how many "seconds" have humans existed?

25E. The maximum speeds of various animals are roughly as follows, in miles per hour: (a) snail, 3.0×10^{-2}; (b) spider, 1.2; (c) human, 23; and (d) cheetah, 70. Convert these data to meters per second. (All four calculations involve the identical conversion factor. You might want to determine that factor first and store it in the memory of your calculator, to use when needed.)

26P. An astronomical unit (AU) is the average distance of the Earth from the Sun, approximately 1.50×10^8 km. The speed of light is about 3.0×10^8 m/s. Express the speed of light in terms of astronomical units per minute.

27P. Until 1883, every city and town in the United States kept its own local time. Today, travelers reset their watches only when the time change equals 1 h. How far, on the average, must you travel in degrees of longitude until your watch must be reset by 1.0 h? *Hint:* The Earth rotates 360° in about 24 h.

28P. Assuming the length of the day uniformly increases by 0.001 s per century, calculate the cumulative effect on the measure of time over 20 centuries. Such slowing of the Earth's rotation is indicated by observations of the occurrences of solar eclipses during this period.

29P. On two *different* tracks, the winners of the mile race ran their races in 3 min, 58.05 s and 3 min, 58.20 s. In order to conclude that the runner with the shorter time was indeed faster, what is the maximum error, in feet, that can be permitted in laying out the mile distances?

30P. Five clocks are being tested in a laboratory. Exactly at noon, as determined by the WWV time signal, on successive days of a week the clocks read as in the following table. Rank the five clocks according to their relative value as good timekeepers, best to worst. Justify your choice.

CLOCK	SUN.	MON.	TUES.	WED.	THURS.	FRI.	SAT.
A	12:36:40	12:36:56	12:37:12	12:37:27	12:37:44	12:37:59	12:38:14
B	11:59:59	12:00:02	11:59:57	12:00:07	12:00:02	11:59:56	12:00:03
C	15:50:45	15:51:43	15:52:41	15:53:39	15:54:37	15:55:35	15:56:33
D	12:03:59	12:02:52	12:01:45	12:00:38	11:59:31	11:58:24	11:57:17
E	12:03:59	12:02:49	12:01:54	12:01:52	12:01:32	12:01:22	12:01:12

31P*. The time it takes the Moon to return to a given position as seen against the background of the fixed stars is called a *sidereal* month. The time interval between identical phases of the Moon is called a *lunar* month. The lunar month is longer than the sidereal month. Why, and by how much?

SECTION 1-6 Mass

32E. Using conversions and data in the chapter, determine the number of hydrogen atoms required to obtain 1.0 kg of hydrogen. A hydrogen atom has a mass of 1.0 u.

33E. One molecule of water (H_2O) contains two atoms of hydrogen and one atom of oxygen. A hydrogen atom has a mass of 1.0 u and an atom of oxygen has a mass of 16 u, approximately. (a) What is the mass in kilograms of one molecule of water? (b) How many molecules of water are in the world's oceans, which have an estimated total mass of 1.4×10^{21} kg?

34E. The Earth has a mass of 5.98×10^{24} kg. The average mass of the atoms that make up the Earth is 40 u. How many atoms are in the Earth?

35P. What mass of water fell on the town in Problem 14 during the thunderstorm? One cubic meter of water has a mass of 10^3 kg.

36P. A person on a diet might lose 2.3 kg (about 5 lb) per week. Express the mass loss rate in milligrams per second.

37P. (a) Assuming that the density (mass/volume) of water is exactly 1 g/cm³, express the density of water in kilograms per cubic meter (kg/m³). (b) Suppose that it takes 10 h to drain a container of 5700 m³ of water. What is the "mass flow rate," in kilograms per second, of water from the container?

38P. Grains of fine California beach sand are approximately spheres with an average radius of 50 μm; they are made of silicon dioxide, 1 m³ of which has a mass of 2600 kg. What mass of sand grains would have a total surface area equal to the surface area of a cube 1 m on an edge?

39P. The density of iron is 7.87 g/cm³, and the mass of an iron atom is 9.27×10^{-26} kg. If the atoms are spherical and tightly packed, (a) what is the volume of an iron atom and (b) what is the distance between the centers of adjacent atoms?

2
Motion Along a Straight Line

In 1977, Kitty O'Neil set a dragster record by reaching 392.54 mi/h in a sizzling time of 3.72 s. In 1958, Eli Beeding Jr. rode a rocket sled from a standstill to a speed of 72.5 mi/h in an elapsed time of 0.04 s (less than an eye blink). How can we compare these two rides to see which was more exciting (or more frightening)—by final speeds, by elapsed times, or by some other quantity?

2-1 MOTION

The world, and everything in it, moves. Even seemingly stationary things, such as a roadway, move with the Earth's rotation, the Earth's orbit around the Sun, the Sun's orbit around the center of the Milky Way galaxy, and that galaxy's migration relative to other galaxies. The classification and comparison of motions (called **kinematics**) is often challenging. What exactly do you measure, and how do you compare?

Before we attempt an answer, we shall examine some general properties of motion that is restricted in three ways.

1. The motion is along a straight line only. The line may be vertical (that of a falling stone), horizontal (that of a car on a level highway), or slanted, but it must be straight.

2. The cause of the motion will not be specified until Chapter 5. In this chapter you study only the motion itself. Does the object speed up, slow down, stop, or reverse direction; and, if the motion does change, how is time involved in the change?

3. The moving object is either a **particle** (by which we mean a pointlike object such as an electron) or an object that moves like a particle (such that every portion moves in the same direction and at the same rate). A pig slipping down a straight playground slide might be considered to be moving like a particle; however, a rotating merry-go-round would not, because different points around its rim move in different directions.

2-2 POSITION AND DISPLACEMENT

To locate an object means to find its position relative to some reference point, often the **origin** (or zero point) of an axis such as the x axis in Fig. 2-1. The **positive direction** of the axis is in the direction of increasing numbers (coordinates), which is toward the right as Fig. 2-1 is drawn. The opposite direction is the **negative direction.**

For example, a particle might be located at $x = 5$ m, which means that it is 5 m in the positive direction from the origin. If it were at $x = -5$ m, it would be just as far from the origin but in the opposite direction. On the axis, a coor-

dinate of -5 m is less than one of -1 m, and both coordinates are less than a coordinate of $+5$ m.

A change from one position x_1 to another position x_2 is called a **displacement** Δx, where

$$\Delta x = x_2 - x_1. \tag{2-1}$$

(The symbol Δ, which represents a change in a quantity, means the final value of that quantity less the initial value.) When numbers are inserted for the position values x_1 and x_2, a displacement in the positive direction (toward the right in Fig. 2-1) always comes out positive, and one in the opposite direction (left in the figure) negative. For example, if the particle moves from $x_1 = 5$ m to $x_2 = 12$ m, then $\Delta x = (12 \text{ m}) - (5 \text{ m}) = +7$ m. The positive result indicates that the motion is in the positive direction. If the particle then returns to $x = 5$ m, the displacement for the full trip is zero. The actual number of meters covered for the full trip is immaterial; displacement involves only the original and final positions.

If we ignore the sign of a displacement (and thus the direction), we are left with the **magnitude** of the displacement Δx, which is always positive.

Displacement is an example of a **vector quantity,** which is a quantity that has both a direction and a magnitude. We explore vectors more fully in Chapter 3 (in fact, some of you may have already read that chapter), but here all we need is the idea that displacement has two features: (1) its magnitude is the distance (such as the number of meters) between the original and final positions, and (2) its direction on an axis, from an original position to a final position, is represented by a plus sign or a minus sign.

What follows is the first of many checkpoints that you will see in this book. Each consists of one or more questions whose answers require some reasoning or a mental calculation, and each gives you a quick check of your understanding. The answers are listed in the back of the book.

CHECKPOINT **1:** Here are three pairs of initial and final positions, respectively, along an x axis. Which pairs give a negative displacement: (a) -3 m, $+5$ m; (b) -3 m, -7 m; (c) 7 m, -3 m?

2-3 AVERAGE VELOCITY AND AVERAGE SPEED

A compact way to describe position is with a graph of position x plotted as a function of time t—a graph of $x(t)$. As a simple example, Fig. 2-2 shows $x(t)$ for a stationary jack rabbit (which we treat as a particle) at $x = -2$ m.

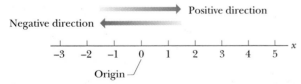

FIGURE 2-1 Position is determined on an axis that is marked in units of length and that extends indefinitely in opposite directions.

Figure 2-3a, also for a rabbit, is more interesting, because it involves motion. The rabbit is apparently first noticed at $t = 0$ when it is at the position $x = -5$ m. It moves toward $x = 0$, passes through that point at $t = 3$ s, and then moves on to increasingly larger positive values of x.

Figure 2-3b depicts the actual straight-line motion of the rabbit and is something like what you would see. The graph in Fig. 2-3a is more abstract and quite unlike what you would see, but it is richer in information. It also reveals how fast the rabbit moves. Actually, several quantities are associated with the phrase "how fast." One of them is the **average velocity** $\bar{v}$, which is the ratio of the displacement Δx that occurs during a particular time interval Δt to that interval:*

$$\bar{v} = \frac{\Delta x}{\Delta t} = \frac{x_2 - x_1}{t_2 - t_1}. \qquad (2\text{-}2)$$

On a graph of x versus t, $\bar{v}$ is the **slope** of the straight line that connects two particular points on the $x(t)$ curve: one is the point that corresponds to x_2 and t_2, and the other is the point that corresponds to x_1 and t_1. Like displacement, $\bar{v}$ has both magnitude and direction. (Average velocity is another vector quantity.) Its magnitude is the magnitude of the line's slope. A positive $\bar{v}$ (and slope) tells us that the line slants upward toward the right; a negative $\bar{v}$ (and slope), that the line slants upward to the left. The average velocity $\bar{v}$ always has the same sign as the displacement Δx because Δt in Eq. 2-2 is always positive.

Figure 2-4 shows how to find $\bar{v}$ for the rabbit of Fig. 2-3 for the time interval $t = 1$ s to $t = 4$ s. The average velocity during that time interval is $\bar{v} = 6$ m/3 s $= +2$ m/s; it is computed as the slope of the straight line that connects the point on the curve at the beginning of the interval and the point on the curve at the end of the interval.

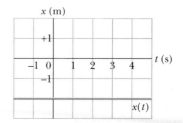

FIGURE 2-2 The graph of $x(t)$ for a jack rabbit that is stationary at $x = -2$ m. The value of x is -2 m for all times t.

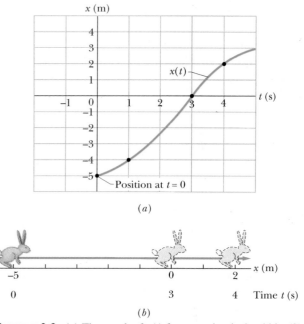

(a)

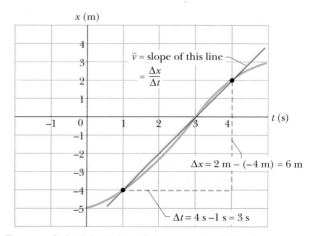

(b)

FIGURE 2-3 (a) The graph of $x(t)$ for a moving jack rabbit. (b) The path associated with the graph. The scale below the x axis shows the times at which the rabbit reaches various x values.

SAMPLE PROBLEM 2-1

You drive a beat-up pickup truck down a straight road for 5.2 mi at 43 mi/h, at which point you run out of fuel. You walk 1.2 mi farther, to the nearest gas station, in 27 min (= 0.450 h). What is your average velocity from the time you started your truck to the time you arrived at the station? Find the answer both numerically and graphically.

SOLUTION: To calculate $\bar{v}$ we need your displacement Δx, from start to finish, and the elapsed time Δt. Assume, for convenience, that your starting point is at the origin of an x axis (so $x_1 = 0$) and that you move in the positive direction.

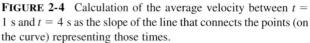

FIGURE 2-4 Calculation of the average velocity between $t = 1$ s and $t = 4$ s as the slope of the line that connects the points (on the curve) representing those times.

---

*In this book, a bar over a symbol means an average value of the quantity that the symbol represents.

You end up at $x_2 = 5.2$ mi + 1.2 mi = +6.4 mi, and so, $\Delta x = x_2 - x_1 = +6.4$ mi. To get the driving time, we rearrange Eq. 2-2 and insert the data about the driving:

$$\Delta t = \frac{\Delta x}{\bar{v}} = \frac{5.2 \text{ mi}}{43 \text{ mi/h}} = 0.121 \text{ h},$$

or about 7.3 min. So the total time, start to finish, is

$$\Delta t = 0.121 \text{ h} + 0.450 \text{ h} = 0.571 \text{ h}.$$

Finally, we insert Δx and Δt into Eq. 2-2:

$$\bar{v} = \frac{\Delta x}{\Delta t} = \frac{6.4 \text{ mi}}{0.571 \text{ h}} \approx +11 \text{ mi/h}. \quad \text{(Answer)}$$

To find $\bar{v}$ graphically, we must first plot $x(t)$, as in Fig. 2-5 where the start and finish points on the curve are the origin and P, respectively. Your average velocity is the slope of the straight line connecting those points. The dashed lines show that the slope also gives $\bar{v} = 6.4$ mi/0.57 h = +11 mi/h.

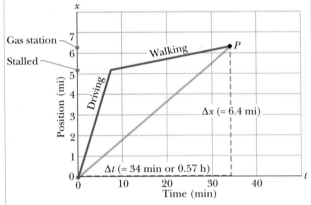

FIGURE 2-5 Sample Problem 2-1. The lines marked "Driving" and "Walking" are the position–time plots for the driver–walker in Sample Problem 2-1. The slope of the straight line joining the origin and point P is the average velocity for the trip.

SAMPLE PROBLEM 2-2

Suppose that you next carry the fuel back to the truck, making the return trip in 35 min. What is your average velocity for the full journey, from the start of your driving to your arrival back at the truck with the fuel?

SOLUTION: As previously, we must find your displacement Δx from start to finish and then divide it by the time interval Δt between start and finish. In this problem, however, the finish is back at the truck. You started at $x_1 = 0$. Back at the truck you are at position $x_2 = 5.2$ mi. And so Δx is 5.2 − 0 = 5.2 mi. The total time Δt you take in going from start to finish is

$$\Delta t = \frac{5.2 \text{ mi}}{43 \text{ mi/h}} + 27 \text{ min} + 35 \text{ min}$$

$$= 0.121 \text{ h} + 0.450 \text{ h} + 0.583 \text{ h} = 1.15 \text{ h}.$$

So $\quad \bar{v} = \dfrac{\Delta x}{\Delta t} = \dfrac{5.2 \text{ mi}}{1.15 \text{ h}} \approx +4.5$ mi/h.   (Answer)

This is slower than the average velocity computed in Sample Problem 2-1 because here the displacement is smaller and the time interval longer.

CHECKPOINT 2: In Sample Problems 2-1 and 2-2, suppose that after refueling the truck, you drive back to x_1 at 52 mi/h. What is your average velocity for the entire trip?

Average speed $\bar{s}$ is a different way of describing "how fast" a particle moves. Whereas the average velocity involves the particle's displacement Δx, the average speed involves the total distance covered (for example, the number of meters run), independent of direction. That is,

$$\bar{s} = \frac{\text{total distance}}{\Delta t}. \quad (2\text{-}3)$$

Average speed also differs from average velocity in that average speed does *not* include direction and thus lacks any algebraic sign. Sometimes $\bar{s}$ is the same (except for the absence of a sign) as $\bar{v}$. But, as demonstrated in Sample Problem 2-3, when an object doubles back on its path, the results can be quite different.

SAMPLE PROBLEM 2-3

In Sample Problem 2-2, what is your average speed?

SOLUTION: From the beginning of your drive to your return to the truck with the fuel, you covered a total of 5.2 mi + 1.2 mi + 1.2 mi = 7.6 mi, taking 1.15 h; so

$$\bar{s} = \frac{7.6 \text{ mi}}{1.15 \text{ h}} \approx 6.6 \text{ mi/h}. \quad \text{(Answer)}$$

PROBLEM SOLVING TACTICS

TACTIC 1: *Do You Understand the Problem?*

For beginning problem solvers, no difficulty is more common than simply not understanding the problem. The best test of understanding is this: Can you explain the problem, in your own words, to a friend? Give it a try.

Write down the given data, with units, using the symbols of the chapter. (In Sample Problems 2-1 and 2-2, the given data allow you to find your net displacement Δx and the corresponding time interval Δt.) Identify the unknown and its symbol. (In these sample problems, the unknown is your average velocity, symbol $\bar{v}$.) Then find the connection between the unknown and the data. (The connection is Eq. 2-2, the definition of average velocity.)

TACTIC 2: *Are the Units OK?*

Be sure to use a consistent set of units when putting numbers into the equations. In Sample Problems 2-1 and 2-2, which involve a truck, the logical units in terms of the given data are miles for distances, hours for time intervals, and miles per hour for velocities. You may need to make conversions.

TACTIC 3: *Is Your Answer Reasonable?*

Look at your answer and ask yourself whether it makes sense. Is it far too large or far too small? Is the sign correct? Are the units appropriate? In Sample Problem 2-1, for example, the correct answer is 11 mi/h. If you find 0.00011 mi/h, -11 mi/h, 11 mi/s, or 11,000 mi/h, you should realize at once that you have done something wrong. The error may lie in your method, in your algebra, or in your arithmetic.

TACTIC 4: *Reading a Graph*

Figures 2-2, 2-3*a*, 2-4, and 2-5 are graphs that you should be able to read easily. In each graph, the variable on the horizontal axis is the time t, the direction of increasing time being to the right. In each, the variable on the vertical axis is the position x of the moving particle with respect to the origin, the direction of increasing x being upward.

Always note the units (seconds or minutes; meters, kilometers, or miles) in which the variables are expressed, and note whether the variables are positive or negative.

2-4 INSTANTANEOUS VELOCITY AND SPEED

You have now seen two ways to describe how fast something moves: average velocity and average speed, both of which are measured over a time interval Δt. But the phrase "how fast" more commonly refers to how fast a particle is moving at a given instant—and that is its **instantaneous velocity** (or simply **velocity**) v.

The velocity at any instant is obtained from the average velocity by shrinking the time interval Δt closer and closer to 0. As Δt dwindles, the average velocity approaches a limiting value, which is the velocity at that instant:

$$v = \lim_{\Delta t \to 0} \frac{\Delta x}{\Delta t} = \frac{dx}{dt}. \qquad (2\text{-}4)$$

Velocity is another vector quantity and thus has an associated direction.

In the language of calculus, the instantaneous velocity is the rate at which a particle's position x is changing with time at a given instant. According to Eq. 2-4, the velocity of a particle at any instant is the slope of its position-time curve at the point representing that instant.

Speed is the magnitude of velocity; that is, speed is velocity that has been stripped of any indication of direction, either in words or via an algebraic sign.* A velocity of $+5$ m/s and one of -5 m/s both have an associated speed of 5 m/s. The speedometer in a car measures the speed, not the velocity, because it cannot ascertain anything about the direction of motion.

SAMPLE PROBLEM 2-4

Figure 2-6*a* is an $x(t)$ plot for an elevator cab that is initially stationary, then moves upward (which we take to be the positive direction), and then stops. Plot v as a function of time.

SOLUTION: The slope, and so also the velocity, is zero in the intervals containing points a and d, when the cab is stationary. During the interval bc the slope is constant and nonzero; so the cab moves with a constant velocity. We calculate the slope of $x(t)$ as

$$\frac{\Delta x}{\Delta t} = v = \frac{24 \text{ m} - 4.0 \text{ m}}{8.0 \text{ s} - 3.0 \text{ s}} = +4.0 \text{ m/s}.$$

The plus sign indicates that the cab is moving in the positive x direction. These values ($v = 0$ and $v = 4$ m/s) are plotted in Fig. 2-6*b*. In addition, as the cab initially begins to move and then later slows to a stop, v might vary as indicated in the intervals 1 s to 3 s and 8 s to 9 s. (Figure 2-6*c* is considered in Section 2-5.)

Given a $v(t)$ graph such as Fig. 2-6*b*, we could "work backward" to produce the shape of the associated $x(t)$ graph (Fig. 2-6*a*). However, we would not know the actual values for x at various times, because the $v(t)$ graph indicates only *changes* in x. To find such a change in x during any interval, we must, in the language of calculus, calculate the area "under the curve" on the $v(t)$ graph for the same interval. For example, during the interval 3 s to 8 s in which the cab has a velocity of 4.0 m/s, the change in x is given by the "area" under the $v(t)$ curve:

$$\text{area} = (4.0 \text{ m/s})(8.0 \text{ s} - 3.0 \text{ s}) = +20 \text{ m}.$$

(This area is positive because the $v(t)$ curve is above the t axis.) Figure 2-6*a* shows that x does indeed increase by 20 m in the interval. But Fig. 2-6*b* does not tell us the *values* of x at the beginning and end of the interval. For that we need additional information.

---

*Speed and average speed can be quite different, so you must be careful solving problems that involve either quantity.

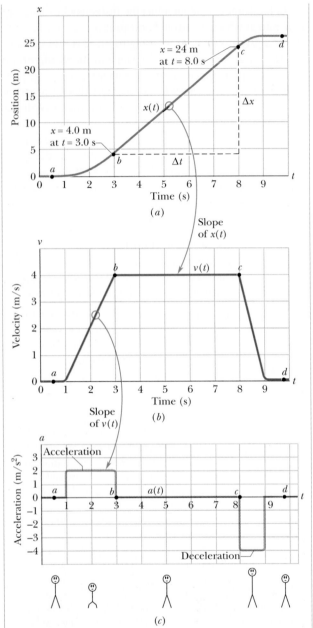

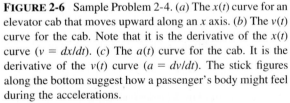

FIGURE 2-6 Sample Problem 2-4. (a) The $x(t)$ curve for an elevator cab that moves upward along an x axis. (b) The $v(t)$ curve for the cab. Note that it is the derivative of the $x(t)$ curve ($v = dx/dt$). (c) The $a(t)$ curve for the cab. It is the derivative of the $v(t)$ curve ($a = dv/dt$). The stick figures along the bottom suggest how a passenger's body might feel during the accelerations.

SAMPLE PROBLEM 2-5

The position of a particle moving on the x axis is given by

$$x = 7.8 + 9.2t - 2.1t^3. \qquad (2\text{-}5)$$

What is its velocity at $t = 3.5$ s? Is the velocity constant, or is it continuously changing?

SOLUTION: For simplicity, the units have been omitted but you can insert them if you like by changing the coefficients to 7.8 m, 9.2 m/s, and -2.1 m/s³. To find the velocity, we use Eq. 2-4 with the right side of Eq. 2-5 substituted for x:

$$v = \frac{dx}{dt} = \frac{d}{dt}(7.8 + 9.2t - 2.1t^3),$$

which becomes

$$v = 0 + 9.2 - (3)(2.1)t^2 = 9.2 - 6.3t^2. \qquad (2\text{-}6)$$

At $t = 3.5$ s,

$$v = 9.2 - (6.3)(3.5)^2 = -68 \text{ m/s}. \qquad \text{(Answer)}$$

At $t = 3.5$ s, the particle is moving toward decreasing x (note the minus sign) with a speed of 68 m/s. Since the quantity t appears in Eq. 2-6, the velocity v depends on t and so is continuously changing.

C**HECKPOINT 3:** The following equations give the position $x(t)$ of a particle in four situations (in each equation, x is in meters, t is in seconds, and $t > 0$): (1) $x = 3t - 2$; (2) $x = -4t^2 - 2$; (3) $x = 2/t^2$; and (4) $x = -2$. (a) In which situation is the velocity v of the particle constant? (b) In which is v in the negative x direction? (c) In which is the particle slowing?

PROBLEM SOLVING TACTICS

TACTIC 5: *Derivatives and Slopes*

Every derivative can be interpreted as the slope of a curve at a point. In Sample Problem 2-4, for example, the velocity of the cab at any instant (a derivative; see Eq. 2-4) is the slope of the $x(t)$ curve of Fig. 2-6a at that instant. Here's how you can find a slope at a point (and thus a derivative) graphically.

Figure 2-7 shows an $x(t)$ plot for a moving particle. To find the velocity of the particle at $t = 1$ s, put a dot on the curve at the point that represents $t = 1$ s. Then draw a line tangent to the curve through the dot (*tangent* means *touching;* the tangent line touches the curve at a single point, the dot), judging carefully by eye. Then construct a right triangle ABC with sides parallel to the axes. (Although the slope is the same no matter what the size of this triangle, the larger the triangle, the more precise will be your measurement of the slope.) Find Δx and Δt, using the vertical and horizontal scales. The slope (derivative) is the quotient $\Delta x/\Delta t$. In Fig. 2-7,

$$\text{slope} = \frac{\Delta x}{\Delta t} = \frac{5.5 \text{ m} - 2.3 \text{ m}}{1.8 \text{ s} - 0.3 \text{ s}} = \frac{3.2 \text{ m}}{1.5 \text{ s}} = +2.1 \text{ m/s}.$$

As Eq. 2-4 tells you, this slope is the velocity of the particle at $t = 1$ s. If you change the scale on either axis of Fig. 2-7, the appearance of the curve and the angle θ will change, but the value you find for the velocity at $t = 1$ s will not.

If you have a mathematical expression for the function $x(t)$, as in Sample Problem 2-5, you can find the derivative dx/dt by the methods of calculus and avoid this graphical method.

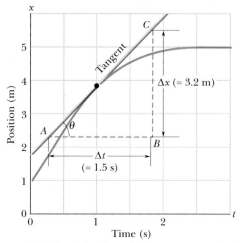

FIGURE 2-7 The derivative of a curve at any point is the slope of its tangent line at that point. At $t = 1.0$ s, the slope of the tangent line (and thus dx/dt, the instantaneous velocity) is $\Delta x/\Delta t = +2.1$ m/s.

2-5 ACCELERATION

When a particle's velocity changes, the particle is said to undergo **acceleration** (or to accelerate). The **average acceleration** $\bar{a}$ over an interval Δt is computed as

$$\bar{a} = \frac{v_2 - v_1}{t_2 - t_1} = \frac{\Delta v}{\Delta t}. \qquad (2\text{-}7)$$

The **instantaneous acceleration** (or simply **acceleration**) is the derivative of the velocity:

$$a = \frac{dv}{dt}. \qquad (2\text{-}8)$$

In words, the acceleration of a particle at any instant is the rate at which its velocity is changing at that instant. According to Eq. 2-8, the acceleration at any point is the slope of the curve of $v(t)$ at that point.

We can combine Eq. 2-8 with Eq. 2-4 to write

$$a = \frac{dv}{dt} = \frac{d}{dt}\left(\frac{dx}{dt}\right) = \frac{d^2x}{dt^2}. \qquad (2\text{-}9)$$

In words, the acceleration of a particle at any instant is the second derivative of its position $x(t)$ with respect to time.

A common unit of acceleration is the meter per second per second: m/(s · s) or m/s². You will see other units in the problems, but they will each be in the form of distance/(time · time) or distance/time². Acceleration has both magnitude and direction (it is yet another vector quantity). The algebraic sign represents the direction on an axis just as it does for displacement and velocity.

Figure 2-6c is a plot of the acceleration of the elevator cab discussed in Sample Problem 2-4. Compare this $a(t)$ curve with the $v(t)$ curve—each point on the $a(t)$ curve is the derivative (slope) of the corresponding point on the $v(t)$ curve. When v is constant (at either 0 or 4 m/s), the derivative is zero and so also is the acceleration. When the cab first begins to move, the $v(t)$ curve has a positive derivative (the slope is positive), which means that $a(t)$ is positive. When the cab slows to a stop, the derivative and slope of the $v(t)$ curve are negative; that is, $a(t)$ is negative.

Next compare the slopes of the $v(t)$ curve during the two acceleration periods. The slope associated with the cab's stopping (commonly called ''deceleration'') is steeper, because the cab stops in half the time it took to get up to speed. The steeper slope means that the magnitude of the deceleration is larger than that of the acceleration, as indicated in Fig. 2-6c.

The sensations you would feel while riding in the cab of Fig. 2-6 are indicated by the sketched figures. When the car first accelerates, you feel as though you are pressed downward; when later the cab is braked to a stop, you seem to be stretched upward. In between, you feel nothing special. Your body reacts to accelerations (it is an accelerometer) but not to velocities (it is not a speedometer). When you are in a car traveling at 60 mi/h or an airplane traveling at 600 mi/h, you have no bodily awareness of the motion. But if the car or plane quickly changes velocity, you may become keenly aware of the change, perhaps even frightened by it. Part of the thrill of an amusement park ride is due to the quick changes of velocity that you undergo. A more extreme example is shown in the photographs of Fig. 2-8, which were taken while a rocket sled was rapidly accelerated and then rapidly braked to a stop.

Large accelerations are sometimes expressed in terms of ''g'' units, with

$$1g = 9.8 \text{ m/s}^2 \qquad (g \text{ unit}). \qquad (2\text{-}10)$$

(As we shall discuss in Section 2-8, g is the magnitude of the acceleration of a falling object near the Earth's surface.) On a roller coaster, you have brief accelerations up to $3g$, which is $(3)(9.8 \text{ m/s}^2)$ or about 29 m/s².

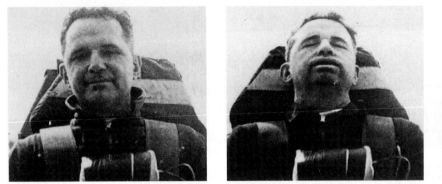

FIGURE 2-8 Colonel J. P. Stapp in a rocket sled as it is brought up to high speed (acceleration out of the page) and then very rapidly braked (acceleration into the page).

SAMPLE PROBLEM 2-6

(a) When Kitty O'Neil set the dragster records for the greatest speed and least elapsed time, she reached 392.54 mi/h in 3.72 s. What was her average acceleration?

SOLUTION: From Eq. 2-7, O'Neil's average acceleration was

$$\bar{a} = \frac{\Delta v}{\Delta t} = \frac{392.54 \text{ mi/h} - 0}{3.72 \text{ s} - 0}$$

$$. = +106 \frac{\text{mi}}{\text{h} \cdot \text{s}}, \qquad \text{(Answer)}$$

where the motion is taken to be in the positive x direction. In more conventional units, her acceleration was 47.1 m/s^2, which is 4.8g.

(b) What was the average acceleration when Eli Beeding Jr. reached 72.5 mi/h in 0.04 s on a rocket sled?

SOLUTION: Again from Eq. 2-7,

$$\bar{a} = \frac{\Delta v}{\Delta t} = \frac{72.5 \text{ mi/h} - 0}{0.04 \text{ s} - 0}$$

$$= +1.8 \times 10^3 \frac{\text{mi}}{\text{h} \cdot \text{s}} \approx +800 \text{ m/s}^2, \qquad \text{(Answer)}$$

or about 80g.

Recall our question in the chapter opener, where O'Neil and Beeding were introduced. How can we tell who had the more exciting ride? Should we compare final speeds, elapsed times, or some other quantity? You now can answer that question. Because the human body senses acceleration rather than speed, you should compare accelerations, and so Beeding wins out, even though his final speed was considerably slower than O'Neil's. In fact, Beeding's acceleration could have been lethal, had it continued for much longer.

PROBLEM SOLVING TACTICS

TACTIC 6: *An Acceleration's Sign*

Look again at the algebraic sign for the accelerations that are calculated in Sample Problem 2-6. In many common examples of acceleration, the sign has its nonscientific meaning: positive acceleration means that the speed of an object is increasing, and negative acceleration means that the speed is decreasing (the object is decelerating).

Such meanings cannot be interpreted without some thought, however. For example, if a car with an initial velocity $v = -27$ m/s (= -60 mi/h) is braked to a stop in 5.0 s, $\bar{a} = +5.4$ m/s^2. The acceleration is *positive,* but the car has slowed. The reason is the difference in signs: the direction of the acceleration is opposite that of the velocity.

Here then is the proper way to interpret the signs:

> If the signs of the velocity and acceleration of a particle are the same, the speed of the particle increases. If the signs are opposite, the speed decreases.

The interpretation will have more meaning in Chapter 4, where we explore the vector nature of velocity and acceleration.

CHECKPOINT 4: A wombat moves along an x axis. What is the sign of its acceleration if it is moving (a) in the positive direction with increasing speed, (b) in the positive direction with decreasing speed, (c) in the negative direction with increasing speed, and (d) in the negative direction with decreasing speed?

FIGURE 2-8 *Continued*

SAMPLE PROBLEM 2-7

A particle's position is given by

$$x = 4 - 27t + t^3,$$

where the units of the coefficients are meters, meters per second, and meters per second cubed, respectively, and the x axis is shown in Fig. 2-1.

(a) Find $v(t)$ and $a(t)$.

SOLUTION: To get $v(t)$, we differentiate $x(t)$ with respect to t:

$$v = -27 + 3t^2. \qquad \text{(Answer)}$$

To get $a(t)$, we differentiate $v(t)$ with respect to t:

$$a = +6t. \qquad \text{(Answer)}$$

(b) Is there ever a time when $v = 0$?

SOLUTION: Setting $v(t) = 0$ yields

$$0 = -27 + 3t^2,$$

which has the solution $t = \pm 3$ s. (Answer)

(c) Describe the particle's motion for $t \geq 0$.

SOLUTION: To answer, we examine the expressions for $x(t)$, $v(t)$, and $a(t)$.

At $t = 0$ the particle is at $x = +4$ m, is moving leftward with a velocity of -27 m/s, and is not accelerating.

For $0 < t < 3$ s, the particle continues to move to the left, but at decreasing speed, because it is now accelerating to the right. (Check $v(t)$ and $a(t)$ for, say, $t = 2$ s.) The rate of the acceleration is increasing.

At $t = 3$ s, the particle stops momentarily ($v = 0$) and is as far to the left as it will ever get ($x = -50$ m). It continues to accelerate to the right at an increasing rate.

For $t > 3$ s, its acceleration to the right continues to increase, and its velocity, which is now also to the right, increases rapidly. (Note that now the signs of v and a match.) The particle moves continuously to the right.

2-6 CONSTANT ACCELERATION: A SPECIAL CASE

In many common types of motion, the acceleration is either constant or approximately so. For example, you might accelerate a car at an approximately constant rate when a traffic light turns from red to green. (Graphs of your position, velocity, and acceleration would resemble those in Fig. 2-9.) If you later had to brake the car to a stop, the deceleration during the braking might also be approximately constant.

Such cases are so ubiquitous that a special set of equations has been derived for dealing with them. One approach to the derivation of the equations is given in this section. A second approach is given in the next section.

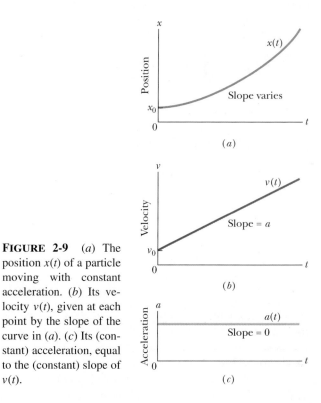

FIGURE 2-9 (*a*) The position $x(t)$ of a particle moving with constant acceleration. (*b*) Its velocity $v(t)$, given at each point by the slope of the curve in (*a*). (*c*) Its (constant) acceleration, equal to the (constant) slope of $v(t)$.

19

Throughout both sections and later when you work on the homework problems, keep in mind that *the equations are valid only for constant acceleration (or situations in which you can approximate the acceleration as being constant).*

When the acceleration is constant, the average acceleration and instantaneous acceleration are equal and we can write Eq. 2-7, with some changes in notation, as

$$a = \frac{v - v_0}{t - 0}.$$

Here v_0 is the velocity at time $t = 0$, and v is the velocity at any later time t. We can recast this equation as

$$v = v_0 + at. \qquad (2\text{-}11)$$

As a check, note that this equation reduces to $v = v_0$ for $t = 0$, as it must. As a further check, take the derivative of Eq. 2-11. Doing so yields $dv/dt = a$, which is the definition of a. Figure 2-9*b* shows a plot of Eq. 2-11, the $v(t)$ function.

In similar manner we can rewrite Eq. 2-2 (with a few changes in notation) as

$$\bar{v} = \frac{x - x_0}{t - 0}$$

and then as

$$x = x_0 + \bar{v}t, \qquad (2\text{-}12)$$

in which x_0 is the position of the particle at $t = 0$, and $\bar{v}$ is the average velocity between $t = 0$ and a later time t.

If you plot v against t using Eq. 2-11, a straight line results. Under these conditions, the *average* velocity over any time interval (say, $t = 0$ to a later time t) is the average of the velocity at the beginning of the interval ($= v_0$) and the velocity at the end of the interval ($= v$). For the interval $t = 0$ to the later time t then, the average velocity is

$$\bar{v} = \tfrac{1}{2}(v_0 + v). \qquad (2\text{-}13)$$

Substituting the right side of Eq. 2-11 for v yields, after a little rearrangement,

$$\bar{v} = v_0 + \tfrac{1}{2}at. \qquad (2\text{-}14)$$

Finally, substituting Eq. 2-14 into Eq. 2-12 yields

$$x - x_0 = v_0 t + \tfrac{1}{2}at^2. \qquad (2\text{-}15)$$

As a check, note that putting $t = 0$ yields $x = x_0$, as it must. As a further check, taking the derivative of Eq. 2-15 yields Eq. 2-11, again as it must. Figure 2-9*a* is a plot of Eq. 2-15.

Five quantities can possibly be involved in any given problem regarding constant acceleration, namely, $x - x_0$,

v, t, a, and v_0. Usually, one of these quantities is *not* involved in the problem, *either as a given or as an unknown*. We are then presented with three of the remaining quantities and asked to find the fourth.

Equations 2-11 and 2-15 each contain four of these quantities, but not the same four. In Eq. 2-11, the "missing ingredient" is the displacement, $x - x_0$. In Eq. 2-15, it is the velocity v. These two equations can also be combined in three ways to yield three additional equations, each of which involves a different "missing ingredient." First, we can eliminate t to obtain

$$v^2 = v_0^2 + 2a(x - x_0). \qquad (2\text{-}16)$$

This equation is useful if we do not know t and are not required to find it. Second, we can eliminate the acceleration a between Eqs. 2-11 and 2-15 to produce an equation in which a does not appear:

$$x - x_0 = \tfrac{1}{2}(v_0 + v)t. \qquad (2\text{-}17)$$

Finally, we can eliminate v_0, obtaining

$$x - x_0 = vt - \tfrac{1}{2}at^2. \qquad (2\text{-}18)$$

Note the subtle difference between this equation and Eq. 2-15. One involves the initial velocity v_0; the other involves the velocity v at time t.

Table 2-1 lists Eqs. 2-11, 2-15, 2-16, 2-17, and 2-18 and shows which one of the five possible quantities is missing from each. To solve a constant acceleration problem, you must decide which of the five quantities is *not* involved in the problem, either as a given or as an unknown. Select the correct equation from Table 2-1 and substitute for the three given quantities to find the unknown. Or, instead of using the table, you might find it easier to remember only Eqs. 2-11 and 2-15, and solve them as simultaneous equations when needed.

TABLE 2-1   EQUATIONS FOR MOTION WITH CONSTANT ACCELERATION[a]

EQUATION NUMBER	EQUATION	MISSING QUANTITY
2-11	$v = v_0 + at$	$x - x_0$
2-15	$x - x_0 = v_0 t + \tfrac{1}{2}at^2$	v
2-16	$v^2 = v_0^2 + 2a(x - x_0)$	t
2-17	$x - x_0 = \tfrac{1}{2}(v_0 + v)t$	a
2-18	$x - x_0 = vt - \tfrac{1}{2}at^2$	v_0

[a]Make sure that the acceleration is indeed constant before using the equations in this table. Note that if you differentiate Eq. 2-15 you get Eq. 2-11. The other three equations are found by eliminating one or another of the variables between Eqs. 2-11 and 2-15.

CHECKPOINT **5:** The following equations give the position $x(t)$ of a particle in four situations: (1) $x = 3t - 4$; (2) $x = -5t^3 + 4t^2 + 6$; (3) $x = 2/t^2 - 4/t$; (4) $x = 5t^2 - 3$. To which of these situations do the equations of Table 2-1 apply?

SAMPLE PROBLEM 2-8

Spotting a police car, you brake a Porsche from 75 km/h to 45 km/h over a displacement of 88 m.

(a) What is the acceleration, assumed to be constant?

SOLUTION: In this problem we are given v_0, v, and $x - x_0$ and need to find a. The time is not involved, being neither given nor required. Table 2-1 then leads us to Eq. 2-16. Solving this equation for a yields

$$a = \frac{v^2 - v_0^2}{2(x - x_0)} = \frac{(45 \text{ km/h})^2 - (75 \text{ km/h})^2}{(2)(0.088 \text{ km})}$$

$$= -2.05 \times 10^4 \text{ km/h}^2 \approx -1.6 \text{ m/s}^2. \quad \text{(Answer)}$$

(In converting hours squared to seconds squared in the last step, we must convert *both* the hour units.) Note that the velocities are positive and the acceleration is negative, which is consistent with a slowing of the car.

(b) What is the elapsed time?

SOLUTION: Now time is involved in the problem, but the acceleration is not. Table 2-1 suggests Eq. 2-17. Solving that equation for t, we obtain

$$t = \frac{2(x - x_0)}{v_0 + v} = \frac{(2)(0.088 \text{ km})}{(75 + 45) \text{ km/h}}$$

$$= 1.5 \times 10^{-3} \text{ h} = 5.4 \text{ s}. \quad \text{(Answer)}$$

(c) If you continue to slow down with the acceleration calculated in (a), how much time will elapse in bringing the car to rest from 75 km/h?

SOLUTION: The quantity not given or asked for here is the displacement, $x - x_0$. Table 2-1 then suggests that we use Eq. 2-11. Solving for t gives

$$t = \frac{v - v_0}{a} = \frac{0 - (75 \text{ km/h})}{(-2.05 \times 10^4 \text{ km/h}^2)}$$

$$= 3.7 \times 10^{-3} \text{ h} = 13 \text{ s}. \quad \text{(Answer)}$$

(d) In (c), what distance will be covered?

SOLUTION: From Eq. 2-15, we have, for the displacement of the car,

$$x - x_0 = v_0 t + \tfrac{1}{2}at^2$$

$$= (75 \text{ km/h})(3.7 \times 10^{-3} \text{ h})$$

$$+ \tfrac{1}{2}(-2.05 \times 10^4 \text{ km/h}^2)(3.7 \times 10^{-3} \text{ h})^2$$

$$= 0.137 \text{ km} \approx 140 \text{ m}. \quad \text{(Answer)}$$

(Neglecting the sign on the acceleration would give a wrong result—you should always be alert to signs.)

(e) Suppose that later, using the acceleration calculated in (a) but a different initial velocity, you bring your car to rest after traversing 200 m. What is the total braking time?

SOLUTION: The missing quantity here is the initial velocity, so we use Eq. 2-18. Setting v (the final velocity) equal to zero (''rest'') and solving this equation for t, we obtain

$$t = \left(-\frac{(2)(x - x_0)}{a}\right)^{1/2} = \left(-\frac{(2)(200 \text{ m})}{-1.6 \text{ m/s}^2}\right)^{1/2}$$

$$= 16 \text{ s}. \quad \text{(Answer)}$$

PROBLEM SOLVING TACTICS

TACTIC 7: *Check the Dimensions*

The dimension of a velocity is L/T, that is, length L divided by time T. The dimension of an acceleration is L/T^2; and so on. In any equation, the dimensions of all terms must be the same. If you are in doubt about an equation, check its dimensions.

To check the dimensions of Eq. 2-15 ($x - x_0 = v_0 t + \tfrac{1}{2}at^2$), we note that every term must be a length, because that is the dimension of x and of x_0. The dimension of the term $v_0 t$ is $(L/T)(T)$, which is L. The dimension of $\tfrac{1}{2}at^2$ is $(L/T^2)(T^2)$, which is also L. This equation checks out. A pure number such as $\tfrac{1}{2}$ or π has no dimension.

2-7 ANOTHER LOOK AT CONSTANT ACCELERATION*

The first two equations in Table 2-1 are the basic equations from which the others are derived. Those two can be obtained by integration of the acceleration with the condition that a is constant. The definition of a (in Eq. 2-8) is

$$a = \frac{dv}{dt},$$

which can be rewritten as

$$dv = a \, dt.$$

To take the *indefinite integral* (or *antiderivative*) of both sides, we write

$$\int dv = \int a \, dt.$$

---

*This section is intended for students who have had integral calculus.

Since acceleration a is a constant, it can be taken outside the integration. Then we obtain

$$\int dv = a \int dt$$

or
$$v = at + C. \tag{2-19}$$

To evaluate the constant of integration C, we let $t = 0$, at which time $v = v_0$. Substituting these values into Eq. 2-19 (which must hold for all values of t, including $t = 0$) yields

$$v_0 = (a)(0) + C = C.$$

Substituting this into Eq. 2-19 gives us Eq. 2-11.

To derive the other basic equation in Table 2-1, we rewrite the definition of velocity (Eq. 2-4) as

$$dx = v \, dt$$

and then take the indefinite integral of both sides to obtain

$$\int dx = \int v \, dt.$$

There is no reason to believe that v is constant, so we cannot move it outside the integration. But we can substitute for v with Eq. 2-11:

$$\int dx = \int (v_0 + at) dt.$$

Since v_0 is a constant, this can be rewritten as

$$\int dx = v_0 \int dt + a \int t \, dt.$$

Integration now yields

$$x = v_0 t + \tfrac{1}{2}at^2 + C', \tag{2-20}$$

where C' is another constant of integration. At time $t = 0$, we have $x = x_0$. Substituting these values in Eq. 2-20 yields $x_0 = C'$. Replacing C' with x_0 in Eq. 2-20 gives us Eq. 2-15.

2-8 FREE-FALL ACCELERATION

If you tossed an object either up or down and could somehow eliminate the effects of air on its flight, you would find that the object accelerates downward at a certain rate. That rate is called the **free-fall acceleration** g. The acceleration is independent of the object's characteristics, such as mass, density, or shape; it is the same for all objects.

Two examples of free-fall acceleration are shown in Fig. 2-10, which is a series of stroboscopic photos of a feather and an apple. As these objects descend, they accelerate downward—both at the same rate g. Their speeds

FIGURE 2-10 A feather and an apple, undergoing free fall in a vacuum, move downward at the same acceleration g. The acceleration causes the increase in distance between images during the fall.

increase together. The value of g varies slightly with latitude and with elevation. At sea level in the midlatitudes the value is 9.8 m/s² (or 32 ft/s²), which is what you should use for the problems in this chapter.

The equations of motion in Table 2-1 for constant acceleration apply to free fall near the Earth's surface. That is, they apply to an object in vertical flight, either up or down, when the effects of the air can be neglected. However, we can make them simpler to use with two minor changes. (1) The directions of motion are now along the vertical y axis instead of the x axis, with the positive direction of y upward. (This change will reduce confusion in later chapters when combined horizontal and vertical motions are examined.) (2) The free-fall acceleration is now negative, that is, downward on the y axis, and so we replace a with $-g$ in the equations.

With these small changes, Eqs. 2-11 and 2-15 to 2-18 of Table 2-1 become, for free fall, Eqs. 2-21 to 2-25 in Table 2-2. Note carefully that:

The free-fall acceleration near the Earth's surface is $a = -g = -9.8$ m/s², and the *magnitude* of the acceleration is $g = 9.8$ m/s². Do not substitute -9.8 m/s² for g.

TABLE 2-2 **EQUATIONS FOR FREE FALL**

EQUATION NUMBER	EQUATION	MISSING QUANTITY
2-21	$v = v_0 - gt$	$y - y_0$
2-22	$y - y_0 = v_0 t - \frac{1}{2}gt^2$	v
2-23	$v^2 = v_0^2 - 2g(y - y_0)$	t
2-24	$y - y_0 = \frac{1}{2}(v_0 + v)t$	g
2-25	$y - y_0 = vt + \frac{1}{2}gt^2$	v_0

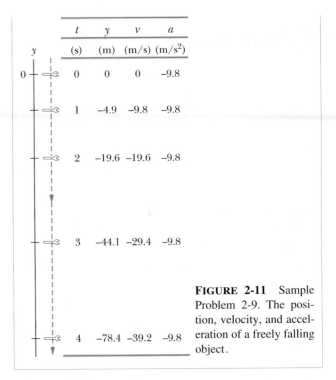

t (s)	y (m)	v (m/s)	a (m/s^2)
0	0	0	−9.8
1	−4.9	−9.8	−9.8
2	−19.6	−19.6	−9.8
3	−44.1	−29.4	−9.8
4	−78.4	−39.2	−9.8

FIGURE 2-11 Sample Problem 2-9. The position, velocity, and acceleration of a freely falling object.

Suppose that you toss a tomato directly upward with an initial velocity v_0 and then catch it when it returns to the level of release. During its *free-fall flight* (just after its release and just before it is caught), the equations of Table 2-2 apply to its motion. The acceleration is always $a = -g = -9.8$ m/s^2 and thus downward. The velocity, however, changes, as indicated by Eqs. 2-21 and 2-23: during the ascent, the magnitude of the (positive) velocity decreases, until it is momentarily zero. Because the tomato has then stopped, it is at its maximum height. During the descent, the magnitude of the (negative) velocity increases.

SAMPLE PROBLEM 2-9

A worker drops a wrench down the elevator shaft of a tall building.

(a) Where is the wrench 1.5 s later?

SOLUTION: Here we know the time t and the acceleration magnitude g, and we can assume the wrench is initially at rest ($v_0 = 0$). We want the displacement. So the missing quantity is the velocity v, which is neither given nor required. This suggests Eq. 2-22 of Table 2-2. Choose the release point of the wrench to be at the origin of the y axis. Setting $y_0 = 0$, $v_0 = 0$, and $t = 1.5$ s in Eq. 2-22 gives

$$y = (0)(1.5 \text{ s}) - \tfrac{1}{2}(9.8 \text{ m/s}^2)(1.5 \text{ s})^2$$
$$= -11 \text{ m}. \qquad \text{(Answer)}$$

The minus sign means that after falling 1.5 s the wrench is below its release point, which we certainly expect.

(b) How fast is the wrench falling just then?

SOLUTION: The wrench's velocity is, by Eq. 2-21,

$$v = v_0 - gt = 0 - (9.8 \text{ m/s}^2)(1.5 \text{ s})$$
$$= -15 \text{ m/s}. \qquad \text{(Answer)}$$

Here the minus sign means that the wrench is falling downward. Again, no great surprise. Figure 2-11 displays the important features of the motion up to $t = 4$ s.

SAMPLE PROBLEM 2-10

In 1939, Joe Sprinz of the San Francisco Baseball Club attempted to break the record for catching a baseball dropped from the greatest height. Members of the Cleveland Indians had set the record the preceding year when they caught baseballs dropped about 700 ft from atop a building. Sprinz used a blimp at 800 ft. Ignore the effects of air on the ball and assume that the ball falls 800 ft.

(a) Find its time of fall.

SOLUTION: Mentally erect a vertical y axis with its origin at the point of the ball's release in the blimp, which means that $y_0 = 0$. The initial velocity v_0 is zero. The missing ingredient is v, and so Eq. 2-22 is required:

$$y - y_0 = v_0 t - \tfrac{1}{2}gt^2$$
$$-800 \text{ ft} = 0t - \tfrac{1}{2}(32 \text{ ft/s}^2)t^2$$
$$16t^2 = 800$$
$$t = 7.1 \text{ s}. \qquad \text{(Answer)}$$

When taking a square root, we have the option of attaching a plus or minus sign to the square root. Here we choose the plus sign, since the ball reaches the ground *after* it is released.

(b) What is the velocity of the ball just before it is caught?

SOLUTION: To get the velocity from the original data, rather than from the result of (a), we use Eq. 2-23:

$$v^2 = v_0^2 - 2g(y - y_0)$$
$$= 0 - (2)(32 \text{ ft/s}^2)(-800 \text{ ft})$$
$$= 5.12 \times 10^4 \text{ ft}^2/\text{s}^2$$
$$v = -226 \text{ ft/s} \ (\approx -154 \text{ mi/h}). \qquad \text{(Answer)}$$

Since the ball is moving downward, we choose the minus sign in our option of signs in taking the square root.

Neglecting the effects of air is actually unwarranted in such a fall. If you included them, you would find that the fall time was longer and the final speed was smaller than the values calculated above. Still, the speed must have been considerable, because when Sprinz finally managed to get a ball in his glove (on his fifth attempt) the impact slammed the glove and hand into his face, fracturing the upper jaw in 12 places, breaking five teeth, and knocking him unconscious. And he dropped the ball.

SAMPLE PROBLEM 2-11

A pitcher tosses a baseball straight up, with an initial speed of 12 m/s. See Fig. 2-12.

(a) How long does the ball take to reach its highest point?

SOLUTION: The ball is at its highest point when its velocity v becomes zero. From Eq. 2-21, we have

$$t = \frac{v_0 - v}{g} = \frac{12 \text{ m/s} - 0}{9.8 \text{ m/s}^2} = 1.2 \text{ s}. \qquad \text{(Answer)}$$

(b) How high does the ball rise above its release point?

SOLUTION: We take the release point of the ball to be the origin of the y axis. Putting $y_0 = 0$ in Eq. 2-23 and solving for y, we obtain

$$y = \frac{v_0^2 - v^2}{2g} = \frac{(12 \text{ m/s})^2 - (0)^2}{(2)(9.8 \text{ m/s}^2)}$$
$$= 7.3 \text{ m}. \qquad \text{(Answer)}$$

If we wanted to take advantage of the fact that we also know the time of flight, having found it in (a), we could also calculate the height of rise from Eq. 2-25. Check it out.

(c) How long will it take for the ball to reach a point 5.0 m above its release point?

SOLUTION: Inspection of Eqs. 2-21 to 2-25 suggests that we try Eq. 2-22. With $y_0 = 0$, we have

$$y = v_0 t - \tfrac{1}{2}gt^2,$$

so $$5.0 \text{ m} = (12 \text{ m/s})t - (\tfrac{1}{2})(9.8 \text{ m/s}^2)t^2.$$

If we temporarily omit the units (having noted that they are consistent), we can rewrite this as

$$4.9t^2 - 12t + 5.0 = 0.$$

Solving this quadratic equation for t yields*

$$t = 0.53 \text{ s} \quad \text{and} \quad t = 1.9 \text{ s}. \qquad \text{(Answer)}$$

There are two such times! This is not really surprising because the ball passes twice through $y = 5.0$ m, once on the way up and once on the way down.

We can check our findings because the time at which the ball reaches its maximum height should lie halfway between these two times, or at

$$t = \tfrac{1}{2}(0.53 \text{ s} + 1.9 \text{ s}) = 1.2 \text{ s}.$$

This is exactly what we found in (a) for the time to reach maximum height.

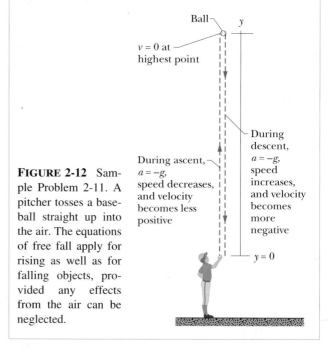

FIGURE 2-12   Sample Problem 2-11. A pitcher tosses a baseball straight up into the air. The equations of free fall apply for rising as well as for falling objects, provided any effects from the air can be neglected.

Ball

y

$v = 0$ at highest point

During ascent, $a = -g$, speed decreases, and velocity becomes less positive

During descent, $a = -g$, speed increases, and velocity becomes more negative

$y = 0$

CHECKPOINT 6: In Sample Problem 2-11, what is the sign of the ball's displacement (a) during its ascent and (b) during its descent? (c) What is the ball's acceleration at its highest point?

PROBLEM SOLVING TACTICS

TACTIC 8: *Meanings of Minus Signs*

In Sample Problems 2-9, 2-10, and 2-11, many answers emerged automatically with minus signs. It is important to know what these signs mean. For falling-body problems, we established a vertical axis (the y axis) and we chose—quite arbitrarily—its upward direction to be positive.

---

*See Appendix E for a general solution to quadratic equations.

We then chose the origin of the *y* axis (that is, the *y* = 0 position) to suit the problem. In Sample Problem 2-9, the origin was the worker's hand; in Sample Problem 2-10 it was at the blimp; in Sample Problem 2-11 it was the pitcher's hand. A negative value of *y* means that the body is below the chosen origin.

A negative velocity means that the body is moving in the direction of decreasing *y*, that is, downward. This is true no matter where the body is located.

We have taken the acceleration to be negative (= −9.8 m/s²) in all problems dealing with falling bodies. A negative acceleration means that, as time goes on, the velocity of the body becomes either less positive or more negative. This is true no matter where the body is located and no matter how fast or in what direction it is moving. In Sample Problem 2-11, the acceleration of the ball is negative throughout its flight, whether the ball is rising or falling.

TACTIC 9: *Unexpected Answers*

Mathematics often generates answers that you might not have thought of as possibilities, as in Sample Problem 2-11c. If you get more answers than you expect, do not discard out of hand the ones that do not seem to fit. Examine them carefully for physical meaning; it is often there.

If time is your variable, even a negative value can mean something; negative time simply refers to time before *t* = 0, the (arbitrary) time at which you decided to start your stopwatch.

2-9 THE PARTICLES OF PHYSICS

As we progress through the book, we plan to step aside occasionally from the familiar world of large, tangible objects and look at nature on a much finer scale. The "particles" that we have dealt with in this chapter, for example, have included pigs, baseballs, and dragsters. In the spirit of our plan, we ask: How small can a particle be? What are the *ultimate* particles of nature? **Particle physics**—for so the field that relates to our inquiry is called—attracts the attention of many of the best of today's physicists.

The realization that matter, on its finest scale, is not continuous but is made up of lumps of material called atoms was the beginning of understanding for physics and chemistry. With modern microscopes, we can "photograph" these atoms, as Fig. 2-13 makes clear. We describe the "lumpiness" of matter by saying that matter is *quantized,* using a word that comes from the Latin word *quantus,* meaning "how much." Quantization is a central feature of nature, and as we go along you will see other physical quantities that are quantized when looked at on a fine enough scale. This pervasiveness of quantization is reflected in the name we give to physics at the atomic level—**quantum physics.**

There is no sharp discontinuity between the quantum world and the world of large-scale objects. The quantum world and the laws that govern it are universal but, as we move from atoms to baseballs and automobiles, the fact of quantization becomes less noticeable and finally totally undetectable. The "graininess" effectively disappears, and the laws of **classical physics** that govern the motions of large objects emerge as special limiting forms of the more general laws of quantum physics.

The Structure of Atoms

An **atom** consists of a central, almost unimaginably compact and dense **nucleus** that is surrounded by one or more lightweight **electrons.** An atom is usually considered to be spherical; so is the nucleus. The radius of a typical atom is on the order of 10^{-10} m; the radius of a nucleus is 100,000 times smaller, about 10^{-15} m. An atom is held together by electrical attraction between the electrons, which are electrically negative, and **protons,** which are electrically positive and reside within the nucleus. The nature of that attraction is explored later in this book, but for now you might realize that were it not present, atoms could not exist, and so neither could you.

The Structure of Nuclei

The simplest nucleus, that of common hydrogen, has a single proton. There are two other, rare, versions of hydrogen: they differ from the common version by the presence

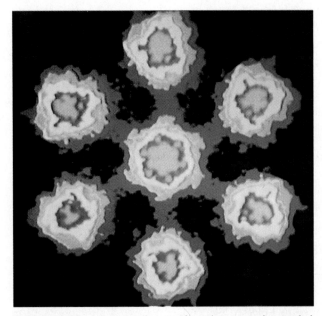

FIGURE 2-13 A hexagonal array of uranium atoms is revealed in this image from a scanning transmission electron microscope. The color has been added by a computer.

of one or two **neutrons** (electrically neutral particles) inside the nucleus. Hydrogen, in any of its versions, is an example of an **element;** each element is distinguished from all the others by the number of protons in the nucleus. When there is only one proton, the element is hydrogen. When, instead, there are six, the element is carbon. The various versions of each element are called **isotopes;** they are distinguished by the number of neutrons.

Roughly speaking, the purpose of the neutrons is to glue together the protons, which, being all electrically positive and closely packed, strongly repel one another. If the neutrons did not provide the glue, the only type of atom that could exist would be common hydrogen; all others would blow apart.

Such instability can be found in many isotopes of common elements, but thankfully not the elements on which your existence depends. For example, 19 of the 21 isotopes† of copper are unstable and undergo transformations to become other elements. The two stable isotopes are the ones used in electronics and other technology.

The Structure of the Particles Within Atoms

The electron is simple but perplexingly so. When detected, it appears to be infinitesimal in size; that is, it has no size and no structure. It is a member of a family of pointlike particles called **leptons;** there are six basic types, each with an **antiparticle** version.

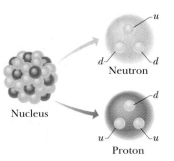

FIGURE 2-14 A representation of the nucleus of an atom, showing the neutrons and protons that make it up. These particles, in turn, are composed of ''up'' (u) and ''down'' (d) quarks.

Protons and neutrons are believed to be different from electrons and the other leptons, because each of the former appears to be a bundle of three simpler particles called **quarks.*** A proton consists of two ''up'' quarks and one ''down'' quark, and a neutron two ''down'' quarks and one ''up'' quark (Fig. 2-14). Other particles that were first thought to be fundamental appear to be similar bundles.

Provocatively, quarks come in six basic types† (each with its antiparticle) just as do leptons. Here physicists wonder: Is there a basic reason for the match in number of types? Or is the match simply coincidence? We do not know.

---

*On a quirk, the word ''quark'' was lifted from James Joyce's *Finnegans Wake*:

> Three quarks for Muster Mark.
> Sure he hasn't got much of a bark
> And sure any he has it's all beside the mark.

†The other types are called charm, strange, top, and bottom (even physicists have their moments).

REVIEW & SUMMARY

Position

The *position x* of a particle on an axis locates the particle with respect to the **origin,** or zero point, of the axis. The position is either positive or negative, according to which side of the origin the particle is on, or zero if the particle is at the origin. The **positive direction** on an axis is the direction of increasing positive numbers; the opposite direction is the **negative direction.**

Displacement

The *displacement* Δx of a particle is the change in its position:

$$\Delta x = x_2 - x_1. \tag{2-1}$$

Displacement is a vector quantity. It is positive if the particle has moved in the positive direction of the x axis, and negative if it has moved in the negative direction.

Average Velocity

When a particle has moved from position x_1 to x_2 during a time interval $\Delta t = t_2 - t_1$, its *average velocity* is

$$\bar{v} = \frac{\Delta x}{\Delta t} = \frac{x_2 - x_1}{t_2 - t_1}. \tag{2-2}$$

The algebraic sign of $\bar{v}$ indicates the direction of motion ($\bar{v}$ is a vector quantity). Average velocity does not depend on the actual distance a particle covers, but instead depends on its original and final positions.

On a graph of x versus t, the average velocity for a time interval Δt is the slope of the straight line connecting the points on the curve that represent the ends of the interval.

Average Speed

The *average speed* $\bar{s}$ of a particle depends on the full distance it covers in a time interval Δt:

$$\bar{s} = \frac{\text{total distance}}{\Delta t}. \tag{2-3}$$

Instantaneous Velocity

If we allow Δt to approach zero in Eq. 2-2, then $\bar{v}$ will approach a limiting value v, the *instantaneous velocity* (or simply **velocity**) of the particle at the time in question; that is,

$$v = \lim_{\Delta t \to 0} \frac{\Delta x}{\Delta t} = \frac{dx}{dt}. \qquad (2\text{-}4)$$

The instantaneous velocity (at a particular time) may be found as the slope (at that particular time) of the graph of x versus t. **Speed** is the magnitude of instantaneous velocity.

Average Acceleration

Average acceleration is the ratio of a change in velocity Δv to the time interval Δt in which the change occurs:

$$\bar{a} = \frac{\Delta v}{\Delta t}. \qquad (2\text{-}7)$$

The algebraic sign indicates the direction of $\bar{a}$.

Instantaneous Acceleration

Instantaneous acceleration (or simply **acceleration**) is the rate of change of velocity with time:

$$a = \frac{dv}{dt}. \qquad (2\text{-}8)$$

The acceleration can also be written as the second derivative of the position $x(t)$ with respect to time:

$$a = \frac{d^2 x}{dt^2}. \qquad (2\text{-}9)$$

On a graph of v versus t, the acceleration a at any time t is the slope of the curve at the point that represents t.

Constant Acceleration

Figure 2-9 shows $x(t)$, $v(t)$, and $a(t)$ for the important case in which a is constant. In this circumstance, the five equations in Table 2-1 describe the motion:

$$v = v_0 + at, \qquad (2\text{-}11)$$

$$x - x_0 = v_0 t + \tfrac{1}{2} a t^2, \qquad (2\text{-}15)$$

$$v^2 = v_0^2 + 2a(x - x_0), \qquad (2\text{-}16)$$

$$x - x_0 = \tfrac{1}{2}(v_0 + v)t, \qquad (2\text{-}17)$$

$$x - x_0 = vt - \tfrac{1}{2} a t^2. \qquad (2\text{-}18)$$

These are *not* valid when the acceleration is not constant.

Free-Fall Acceleration

An important example of straight-line motion with constant acceleration is that of an object rising or falling freely near the Earth's surface. The constant acceleration equations describe this motion, but we make two changes in notation: (1) we refer the motion to the vertical y axis with $+y$ vertically *up*; (2) we replace a with $-g$, where g is the magnitude of the free-fall acceleration. Near the Earth's surface, $g = 9.8$ m/s^2 ($= 32$ ft/s^2). The free-fall equations are those shown as Eqs. 2-21 to 2-25.

The Structure of Atoms

All ordinary matter is composed of **atoms,** which, in a simple model, consist of **electrons** that surround a highly compact central core, the **nucleus. Neutrons** and **protons** reside inside nuclei. Each **element** is distinguished by the number of protons in its nucleus. Variations of an element, differing in the number of neutrons, are **isotopes** of the element.

Quarks and Leptons

Electrons appear to be pointlike particles with no size or internal structure, but protons and neutrons appear to have size and to contain more elementary particles, called **quarks.** There are six basic types of quarks, each with an antiparticle version. Electrons are members of a family of particles, **leptons,** that also come in six basic types, each with an antiparticle version.

QUESTIONS

1. (a) Can an object have zero velocity and still be accelerating? (b) Can an object have constant velocity and still have a varying speed? (c) Can the velocity of an object reverse direction when the object's acceleration is constant? (d) Can an object be increasing in speed as its acceleration decreases?

2. Figure 2-15 is a graph of a particle's position along an x axis versus time. (a) At time $t = 0$, what is the sign of the particle's position? Is the particle's velocity positive, negative, or zero at (b) $t = 1$ s, (c) $t = 2$ s, and (d) $t = 3$ s? (e) How many times does the particle go through the point $x = 0$?

3. At $t = 0$, a particle moving along an x axis is at position $x_0 = -20$ m. The signs of the particle's initial velocity v_0 (at time t_0) and constant acceleration a are, respectively, for four situations: (1) +, +; (2) +, −; (3) −, +; (4) −, −. In which situation will

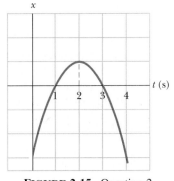

FIGURE 2-15 Question 2.

the particle (a) undergo a momentary stop, (b) definitely pass through the origin (given enough time), and (c) definitely not pass through the origin?

4. Figure 2-16 gives the velocity of a particle moving on an x axis. What are (a) the initial and (b) the final directions of travel? (c) Does the particle stop momentarily? (d) Is the acceleration positive or negative? (e) Is it constant or varying?

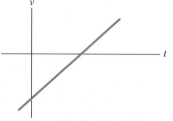

FIGURE 2-16   Question 4.

5. The initial and final velocities, respectively, of a particle in four situations are: (a) 2 m/s, 3 m/s; (b) −2 m/s, 3 m/s; (c) −2 m/s, −3 m/s; (d) 2 m/s, −3 m/s. The magnitude of the particle's constant acceleration is the same in all four situations. Rank the situations according to the magnitude of the particle's displacement, greatest first, during the change from initial to final velocity.

6. The following equations give the velocity $v(t)$ of a particle in four situations: (a) $v = 3$; (b) $v = 4t^2 + 2t − 6$; (c) $v = 3t − 4$; (d) $v = 5t^2 − 3$. To which of these situations do the equations of Table 2-1 apply?

7. Suppose that a hot-air balloonist drops an apple over the side while the balloon is accelerating upward at 4.0 m/s^2 during liftoff. (a) What is the apple's acceleration once it has been released? (b) If the velocity of the balloon is 2 m/s upward at the instant of release, what is the apple's velocity just then?

8. (a) Graph y, v, and a versus t for a cream tangerine that is thrown straight up from a cliff but which, upon falling back down, barely passes by the cliff edge. (b) On the same figures, graph the same quantities for a cream tangerine that is dropped (released with no initial speed) from the cliff edge.

9. You throw a ball straight up from the edge of a cliff, and it lands on the ground below the cliff. If you had, instead, thrown the ball down from the cliff edge with the same speed, would the ball's speed just before landing be larger than, smaller than, or the same as previously? (*Hint:* Consider Eq. 2-23.)

10. The driver of a blue car, moving at a speed of 60 mi/h, suddenly realizes that she is about to rear-end a red car, moving at a speed of 40 mi/h. To avoid a collision, what is the maximum speed the blue car can have just as it reaches the red car? (Warmup for Problem 57)

11. At $t = 0$ and $x = 0$, an initially stationary blue car begins to accelerate at the constant rate of 2.0 m/s^2 in the positive direction of the x axis. At $t = 2$ s, a red car, traveling in an adjacent lane and in the same direction, passes $x = 0$ with a speed of 8.0 m/s and a constant acceleration of 3.0 m/s^2. What pair of simultaneous equations should be solved to find when the red car passes the blue car? (Warmup for Problem 56)

12. Figure 2-17 shows that a particle moving along an x axis undergoes three periods of acceleration. Without written computation, rank the acceleration periods according to the increases they produce in the particle's velocity, greatest first.

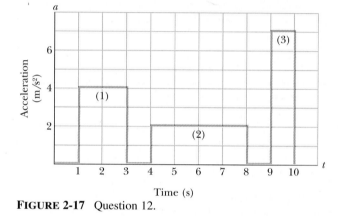

FIGURE 2-17   Question 12.

13. A second ball is dropped from a cliff 1 s after a first ball was dropped. As both fall, does the distance between them increase, decrease, or stay the same?

EXERCISES & PROBLEMS

In several of the problems that follow you are asked to graph position, velocity, and acceleration versus time. Usually a sketch will suffice, appropriately labeled and with straight and curved portions apparent. If you have a computer or programmable calculator, you might use it to produce the graph.

SECTION 2-3   Average Velocity and Average Speed

1E. Carl Lewis ran the 100 m dash in about 10 s, and Bill Rodgers ran the marathon (26 mi, 385 yd) in about 2 h 10 min. (a) What are their average speeds? (b) If Lewis could have maintained his sprint speed during a marathon, how long would he have taken to finish?

2E. During a hard sneeze, your eyes might shut for 0.50 s. If you are driving a car at 90 km/h, how far does it move during that time?

3E. On average, a blink lasts about 100 ms. How far does a MiG-25 "Foxbat" fighter travel during a pilot's blink if the plane's average velocity is 2110 mi/h?

4E. Boston Red Sox pitcher Roger Clemens could routinely throw a fastball at a horizontal speed of 160 km/h. How long did the ball take to reach home plate 18.4 m away?

5E. Figure 2-18 is a plot of the age of ancient seafloor material, in millions of years, against the distance from a particular ocean ridge. Seafloor material extruded from this ridge moves away from it at approximately uniform speed. Find that speed in centimeters per year.

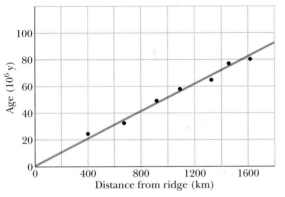

FIGURE 2-18 Exercise 5.

6E. When the legal speed limit for the New York Thruway was increased from 55 mi/h ($= 88.5$ km/h) to 65 mi/h ($= 105$ km/h), how much time was saved by a motorist who drove the 435 mi ($= 700$ km) between the Buffalo entrance and the New York City exit at the legal speed limit?

7E. Using the tables in Appendix D, find the speed of light ($= 3 \times 10^8$ m/s) in miles per hour, feet per second, and light-years per year.

8E. An automobile travels on a straight road for 40 km at 30 km/h. It then continues in the same direction for another 40 km at 60 km/h. (a) What is the average velocity of the car during this 80 km trip? (Assume that it moves in the positive x direction.) (b) What is its average speed? (c) Graph x versus t and indicate how the average velocity is found on the graph.

9P. Compute your average velocity in the following two cases. (a) You walk 240 ft at a speed of 4.0 ft/s and then run 240 ft at a speed of 10 ft/s along a straight track. (b) You walk for 1.0 min at a speed of 4.0 ft/s and then run for 1.0 min at 10 ft/s along a straight track. (c) Graph x versus t for both cases and indicate how the average velocity is found on the graph.

10P. A car travels up a hill at a constant speed of 40 km/h and returns down the hill at a constant speed of 60 km/h. Calculate the average speed for the round trip.

11P. You drive on Interstate 10 from San Antonio to Houston, half the *time* at 35 mi/h ($= 56$ km/h) and the other half at 55 mi/h ($= 89$ km/h). On the way back you travel half the *distance* at 35 mi/h and the other half at 55 mi/h. What is your average speed (a) from San Antonio to Houston, (b) from Houston back to San Antonio, and (c) for the entire trip? (d) What is your average velocity for the entire trip? (e) Graph x versus t for (a), assuming the motion is all in the positive x direction. Indicate how the average velocity can be found on the graph.

12P. The position of an object moving in a straight line is given by $x = 3t - 4t^2 + t^3$, where x is in meters and t in seconds. (a) What is the position of the object at $t = 1, 2, 3,$ and 4 s? (b) What is the object's displacement between $t = 0$ and $t = 4$ s? (c) What is the average velocity for the time interval from $t = 2$ s to $t = 4$ s? (d) Graph x versus t for $0 \le t \le 4$ s and indicate how the answer for (c) can be found on the graph.

13P. The position of a particle moving along the x axis is given in centimeters by $x = 9.75 + 1.50t^3$, where t is in seconds. Consider the time interval $t = 2.00$ s to $t = 3.00$ s and calculate (a) the average velocity; (b) the instantaneous velocity at $t = 2.00$ s; (c) the instantaneous velocity at $t = 3.00$ s; (d) the instantaneous velocity at $t = 2.50$ s; and (e) the instantaneous velocity when the particle is midway between its positions at $t = 2.00$ s and $t = 3.00$ s. (f) Graph x versus t and indicate your answers graphically.

14P. A high-performance jet plane, practicing radar avoidance maneuvers, is in horizontal flight 35 m above the level ground. Suddenly, the plane encounters terrain that slopes gently upward at 4.3°, an amount difficult to detect (see Fig. 2-19). How much time does the pilot have to make a correction to avoid flying into the ground? The speed of the plane is 1300 km/h.

FIGURE 2-19 Problem 14.

15P. Two trains, each having a speed of 30 km/h, are headed at each other on the same straight track. A bird that can fly 60 km/h flies off the front of one train when they are 60 km apart and heads directly for the other train. On reaching the other train it flies directly back to the first train, and so forth. (We have no idea *why* a bird would behave in this way.) (a) How many trips can the bird make from one train to the other before they crash? (b) What is the total distance the bird travels?

SECTION 2-4 Instantaneous Velocity and Speed

16E. (a) If a particle's position is given by $x = 4 - 12t + 3t^2$ (where t is in seconds and x is in meters), what is its velocity at $t = 1$ s? (b) Is it moving toward increasing or decreasing x just then? (c) What is its speed just then? (d) Is the speed larger or smaller at later times? (Try answering the next two questions without further calculation.) (e) Is there ever an instant when the velocity is zero? (f) Is there a time after $t = 3$ s when the particle is moving toward decreasing x?

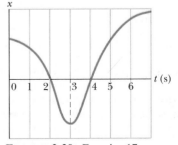

FIGURE 2-20 Exercise 17.

17E. The graph in Fig. 2-20 pertains to an armadillo that scampers left (direction of decreasing *x*) and right along an *x* axis. (a) When, if ever, is the animal to the left of the origin on the axis? When, if ever, is its velocity (b) negative, (c) positive, or (d) zero?

18E. Sketch a graph of *x* versus *t* for a mouse that is constrained in a narrow corridor (the *x* axis) and that scurries in the following sequence: (1) runs leftward (the direction of decreasing *x*) with a constant speed of 1.2 m/s, (2) gradually slows to 0.6 m/s toward the left, (3) gradually speeds up to 2.0 m/s toward the left, (4) gradually slows to a stop and then speeds up to 1.2 m/s toward the right. Where is the curve steepest? The least steep?

19P. How far does the runner whose velocity–time graph is shown in Fig. 2-21 travel in 16 s?

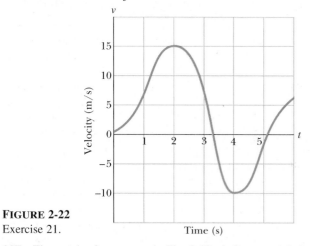

FIGURE 2-21
Problem 19.

SECTION 2-5 Acceleration

20E. A particle had a velocity of 18 m/s and 2.4 s later its velocity was 30 m/s in the opposite direction. (a) What was the magnitude of the average acceleration of the particle during this 2.4 s interval? (b) Graph *v* versus *t* and indicate how the average acceleration can be found on the graph.

21E. An object moves in a straight line as described by the velocity–time graph of Fig. 2-22. Sketch a graph that represents the acceleration of the object as a function of time.

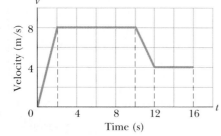

FIGURE 2-22
Exercise 21.

22E. The graph of *x* versus *t* in Fig. 2-23*a* is for a particle in straight-line motion. (a) State, for each of the intervals *AB*, *BC*, *CD*, and *DE*, whether the velocity *v* is positive, negative, or 0 and whether the acceleration *a* is positive, negative, or 0. (Ignore the end points of the intervals.) (b) From the curve, is there any inter-

val over which the acceleration is obviously not constant? (c) If the axes are shifted upward together such that the time axis ends up running along the dashed line, do any of your answers change?

23E. Repeat Exercise 22 for the motion described by the graph of Fig. 2-23*b*.

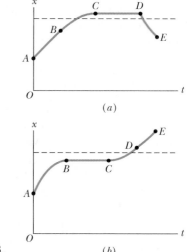

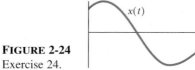

FIGURE 2-23
Exercises 22 and 23.

24E. A particle moves along the *x* axis with *x*(*t*) as shown in Fig. 2-24. Make rough sketches of velocity versus time and acceleration versus time for this motion.

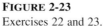

FIGURE 2-24
Exercise 24.

25E. Sketch a graph that is a possible description of position as a function of time for a particle that moves along the *x* axis and, at *t* = 1 s, has (a) zero velocity and positive acceleration; (b) zero velocity and negative acceleration; (c) negative velocity and positive acceleration; (d) negative velocity and negative acceleration. (e) For which of these situations is the speed of the particle increasing at *t* = 1 s?

26E. Consider the two quantities $(dx/dt)^2$ and d^2x/dt^2. (a) Are these two equivalent expressions for the same thing? (b) What are the SI units of these two quantities?

27E. A particle moves along the *x* axis according to the equation $x = 50t + 10t^2$, where *x* is in meters and *t* is in seconds. Calculate (a) the average velocity of the particle during the first 3.0 s of its motion, (b) the instantaneous velocity of the particle at *t* = 3.0 s, and (c) the instantaneous acceleration of the particle at *t* = 3.0 s. (d) Graph *x* versus *t* and indicate how the answer to (a) can be obtained from the plot. (e) Indicate the answer to (b) on the graph. (f) Plot *v* versus *t* and indicate on it the answer to (c).

28E. (a) If the position of a particle is given by $x = 20t - 5t^3$, where *x* is in meters and *t* is in seconds, when, if ever, is the particle's velocity zero? (b) When is its acceleration *a* zero? (c) When is *a* negative? Positive? (d) Graph *x*(*t*), *v*(*t*), and *a*(*t*).

29P. A man stands still from $t = 0$ to $t = 5.00$ min; from $t = 5.00$ min to $t = 10.0$ min he walks briskly in a straight line at a constant speed of 2.20 m/s. What are his average velocity and average acceleration during the time intervals (a) 2.00 min to 8.00 min and (b) 3.00 min to 9.00 min? (c) Sketch x versus t and v versus t, and indicate how the answers to (a) and (b) can be obtained from the graphs.

30P. If the position of an object is given by $x = 2.0t^3$, where x is measured in meters and t in seconds, find (a) the average velocity and (b) the average acceleration between $t = 1.0$ s and $t = 2.0$ s. Then find (c) the instantaneous velocities and (d) the instantaneous accelerations at $t = 1.0$ s and $t = 2.0$ s. (e) Compare the average and instantaneous quantities and in each case explain why the larger one is larger. (f) Graph x versus t and v versus t, and indicate on the graphs your answers to (a) through (d).

31P. In an arcade video game, a spot is programmed to move across the screen according to $x = 9.00t - 0.750t^3$, where x is distance in centimeters measured from the left edge of the screen and t is time in seconds. When the spot reaches a screen edge, at either $x = 0$ or $x = 15.0$ cm, t is reset to 0 and the spot starts moving again according to $x(t)$. (a) At what time after starting is the spot instantaneously at rest? (b) Where does this occur? (c) What is its acceleration when this occurs? (d) In what direction is it moving just prior to coming to rest? (e) Just after? (f) When does it first reach an edge of the screen after $t = 0$?

32P. The position of a particle moving along the x axis depends on the time according to the equation $x = et^2 - bt^3$, where x is in feet and t in seconds. (a) What dimensions and units must e and b have? For the following, let their numerical values be 3.0 and 1.0, respectively. (b) At what time does the particle reach its maximum positive x position? (c) What distance does the particle cover in the first 4.0 s? (d) What is its displacement from $t = 0$ to $t = 4.0$ s? (e) What is its velocity at $t = 1.0, 2.0, 3.0,$ and 4.0 s? (f) What is its acceleration at these times?

SECTION 2-6 Constant Acceleration: A Special Case

33E. The head of a rattlesnake can accelerate 50 m/s² in striking a victim. If a car could do as well, how long would it take to reach a speed of 100 km/h from rest?

34E. An object has a constant acceleration of +3.2 m/s². At a certain instant its velocity is +9.6 m/s. What is its velocity (a) 2.5 s earlier and (b) 2.5 s later?

35E. An automobile increases its speed uniformly from 25 km/h to 55 km/h in 0.50 min. A bicycle rider uniformly speeds up to 30 km/h from rest in 0.50 min. Calculate their accelerations.

36E. Suppose a rocketship in deep space moves with constant acceleration equal to 9.8 m/s², which gives the illusion of normal gravity during the flight. (a) If it starts from rest, how long will it take to acquire a speed one-tenth that of light, which travels at 3.0×10^8 m/s? (b) How far will it travel in so doing?

37E. A jumbo jet must reach a speed of 360 km/h ($= 225$ mi/h) on the runway for takeoff. What is the least constant acceleration needed for takeoff from a 1.80 km runway?

38E. A muon (an elementary particle) enters an electric field with a speed of 5.00×10^6 m/s, whereupon the field slows it at the rate of 1.25×10^{14} m/s². (a) How far does the muon take to stop? (b) Graph x versus t and v versus t for the muon.

39E. An electron with initial velocity $v_0 = 1.50 \times 10^5$ m/s enters a region 1.0 cm long where it is electrically accelerated (Fig. 2-25). It emerges with velocity $v = 5.70 \times 10^6$ m/s. What was its acceleration, assumed constant? (Such a process occurs in the electron gun in a cathode-ray tube, used in television receivers and oscilloscopes.)

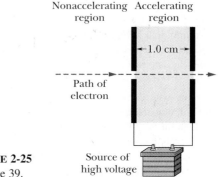

Nonaccelerating region   Accelerating region

←1.0 cm→

Path of electron

FIGURE 2-25
Exercise 39.

Source of high voltage

40E. A car can be braked to a stop from 60 mi/h in 43 m. (a) What is the magnitude of the acceleration in SI units and g units? (Assume that a is constant.) (b) What is the stopping time? If your reaction time T for braking is 400 ms, to how many "reaction times" does the stopping time correspond?

41E. A world's land speed record was set by Colonel John P. Stapp when, on March 19, 1954, he rode a rocket-propelled sled that moved down a track at 1020 km/h. He and the sled were brought to a stop in 1.4 s. (See Fig. 2-8.) What acceleration did he experience? Express your answer in g units.

42E. On a dry road a car with good tires may be able to brake with a deceleration of 4.92 m/s² (assume that it is constant). (a) How long does such a car, initially traveling at 24.6 m/s, take to come to rest? (b) How far does it travel in this time? (c) Graph x versus t and v versus t for the deceleration.

43E. A rocket-driven sled running on a straight, level track is used to investigate the physiological effects of large accelerations on humans. One such sled can attain a speed of 1600 km/h in 1.8 s starting from rest. Find (a) the acceleration (assumed constant) in g units and (b) the distance traveled.

44E. The brakes on your automobile are capable of creating a deceleration of 17 ft/s². (a) If you are going 85 mi/h and suddenly see a state trooper, what is the minimum time in which you can get your car under the 55 mi/h speed limit? (The answer reveals the futility of braking to keep your high speed from being detected with a radar gun.) (b) Graph x versus t and v versus t for such a deceleration.

45E. A motorcycle is moving at 30 m/s when the rider applies the brakes, giving the motorcycle a constant deceleration. During the 3.0 s interval immediately after braking begins, the speed decreases to 15 m/s. What distance does the motorcycle travel from the instant braking begins until it comes to rest?

46P. A hot rod can accelerate from 0 to 60 km/h in 5.4 s. (a) What is its average acceleration, in m/s^2, during this time? (b) How far will it travel during the 5.4 s, assuming its acceleration is constant? (c) How much time would it require to go a distance of 0.25 km if its acceleration could be maintained at the value in (a)?

47P. A train started from rest and moved with constant acceleration. At one time it was traveling 30 m/s, and 160 m farther on it was traveling 50 m/s. Calculate (a) the acceleration, (b) the time required to travel the 160 m mentioned, (c) the time required to attain the speed of 30 m/s, and (d) the distance moved from rest to the time the train had a speed of 30 m/s. (e) Graph x versus t and v versus t for the train, from rest.

48P. A car traveling 56.0 km/h is 24.0 m from a barrier when the driver slams on the brakes. The car hits the barrier 2.00 s later. (a) What was the car's constant deceleration before impact? (b) How fast was the car traveling at impact?

49P. A car moving with constant acceleration covers the distance between two points 60.0 m apart in 6.00 s. Its speed as it passes the second point is 15.0 m/s. (a) What is the speed at the first point? (b) What is the acceleration? (c) At what prior distance from the first point was the car at rest? (d) Graph x versus t and v versus t for the car, from rest.

50P. Two subway stops are separated by 1100 m. If a subway train accelerates at $+1.2$ m/s^2 from rest through the first half of the distance and decelerates at -1.2 m/s^2 through the second half, what are (a) its travel time and (b) its maximum speed? (c) Graph x, v, and a versus t for the trip.

51P. To stop a car, you require first a certain reaction time to begin braking; then the car slows under the constant braking deceleration. Suppose that the total distance moved by your car during these two phases is 186 ft when its initial speed is 50 mi/h, and 80 ft when the initial speed is 30 mi/h. What are (a) your reaction time and (b) the magnitude of the deceleration?

52P. You are driving toward a traffic signal when it turns yellow. Your speed is the legal speed limit of $v_0 = 35$ mi/h; your best deceleration rate is $a = 17$ ft/s^2. Your best reaction time to begin braking is $T = 0.75$ s. To avoid having the front of your car enter the intersection after the light turns red, should you brake to a stop or continue to move at 35 mi/h if the distance to the intersection and the duration of the yellow light are (a) 40 m and 2.8 s, and (b) 32 m and 1.8 s?

53P. When a driver brings a car to a stop by braking as hard as possible, the stopping distance can be regarded as the sum of a "reaction distance," which is initial speed times the driver's reaction time, and a "braking distance," which is the distance covered during braking. The following table gives typical values:

INITIAL SPEED (m/s)	REACTION DISTANCE (m)	BRAKING DISTANCE (m)	STOPPING DISTANCE (m)
10	7.5	5.0	12.5
20	15	20	35
30	22.5	45	67.5

(a) What reaction time is the driver assumed to have? (b) What is the car's stopping distance if the initial speed is 25 m/s?

54P. (a) If the maximum acceleration that is tolerable for passengers in a subway train is 1.34 m/s^2, and subway stations are located 806 m apart, what is the maximum speed a subway train can attain between stations? (b) What is the travel time between stations? (c) If the subway train stops for 20 s at each station, what is the maximum average speed of a subway train, from one start-up to the next? (d) Graph x, v, and a versus t.

55P. An elevator cab in the New York Marquis Marriott has a total run of 624 ft. Its maximum speed is 1000 ft/min. Its acceleration and deceleration both have a magnitude of 4.0 ft/s^2. (a) How far does the cab move while accelerating to full speed from rest? (b) How long does it take to make the nonstop 624 ft run, starting and ending at rest?

56P. At the instant the traffic light turns green, an automobile starts with a constant acceleration a of 2.2 m/s^2. At the same instant a truck, traveling with a constant speed of 9.5 m/s, overtakes and passes the automobile. (a) How far beyond the traffic signal will the automobile overtake the truck? (b) How fast will the car be traveling at that instant?

57P. When a high-speed passenger train traveling at 100 mi/h rounds a bend, the engineer is shocked to see that a locomotive has improperly entered onto the track from a siding 0.42 mi ahead; see Fig. 2-26. The locomotive is moving at 18 mi/h. The engineer of the passenger train immediately applies the brakes. (a) What must be the magnitude of the resulting constant acceleration if a collision is to be just avoided? (b) Assume that the

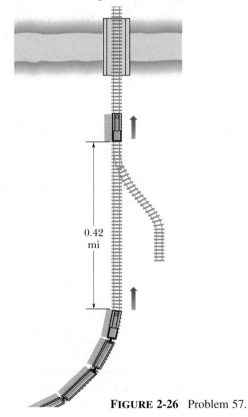

FIGURE 2-26 Problem 57.

0.42 mi

engineer is at $x = 0$ when, at $t = 0$, he first spots the locomotive. Sketch the $x(t)$ curves representing the locomotive and passenger train for the situations in which a collision is just avoided and not quite avoided.

58P. Two trains, one traveling at 72 km/h and the other at 144 km/h, are headed toward one another along a straight, level track. When they are 950 m apart, each engineer sees the other's train and applies the brakes. The brakes decelerate each train at the rate of 1.0 m/s². Is there a collision?

59P. Sketch a $v(t)$ graph that would be associated with the $a(t)$ graph shown in Fig. 2-27.

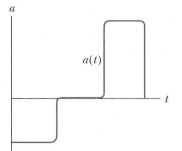

FIGURE 2-27
Problem 59.

SECTION 2-8 Free-Fall Acceleration

60E. At a construction site a pipe wrench strikes the ground with a speed of 24 m/s. (a) From what height was it inadvertently dropped? (b) How long was it falling? (c) Sketch graphs of y, v, and a versus t for the wrench.

61E. (a) With what speed must a ball be thrown vertically from ground level to rise to a maximum height of 50 m? (b) How long will it be in the air? (c) Sketch graphs of y, v, and a versus t for the ball. On the first two graphs, indicate the time at which 50 m is reached.

62E. Raindrops fall to Earth from a cloud 1700 m above the Earth's surface. If they were not slowed by air resistance, how fast would the drops be moving when they struck the ground? Would it be safe to walk outside during a rainstorm?

63E. The single cable supporting an unoccupied construction elevator breaks when the elevator is at rest at the top of a 120-m-high building. (a) With what speed does the elevator strike the ground? (b) How long was it falling? (c) What was its speed when it passed the halfway point on the way down? (d) How long had it been falling when it passed the halfway point?

64E. A hoodlum throws a stone vertically downward with an initial speed of 12.0 m/s from the roof of a building, 30.0 m above the ground. (a) How long does it take the stone to reach the ground? (b) What is the speed of the stone at impact?

65E. The Zero Gravity Research Facility at the NASA Lewis Research Center includes a 145 m drop tower. This is an evacuated vertical tower through which, among other possibilities, a 1 m diameter sphere containing an experimental package can be dropped. (a) How long is the sphere in free fall? (b) What is its speed just as it reaches a catching device at the bottom of the tower? (c) When caught, the sphere experiences an average de-

celeration of 25g as its speed is reduced to zero. Through what distance does it travel during the deceleration?

66E. A model rocket, propelled by burning fuel, takes off vertically. Plot qualitatively (numbers not required) graphs of y, v, and a versus t for the rocket's flight. Indicate when the fuel is exhausted, when the rocket reaches maximum height, and when it returns to the ground.

67E. A rock is dropped from a 100 m high cliff. How long does it take to fall (a) the first 50 m and (b) the second 50 m?

68P. A startled armadillo leaps upward (Fig. 2-28), rising 0.544 m in 0.200 s. (a) What was its initial speed? (b) What is its speed at this height? (c) How much higher does it go?

FIGURE 2-28 Problem 68.

69P. An object falls from a bridge that is 45 m above the water. It falls directly into a model boat, moving with constant velocity, that was 12 m from the point of impact when the object was released. What was the speed of the boat?

70P. A model rocket is fired vertically and ascends with a constant vertical acceleration of 4.00 m/s² for 6.00 s. Its fuel is then exhausted and it continues as a free-fall particle. (a) What is the maximum altitude reached? (b) What is the total time elapsed from takeoff until the rocket strikes the Earth?

71P. A basketball player, standing near the basket to grab a rebound, jumps 76.0 cm vertically. How much (total) time does the player spend (a) in the top 15.0 cm of this jump and (b) in the bottom 15.0 cm? Does this help explain why such players seem to hang in the air at the tops of their jumps?

72P. At the National Physical Laboratory in England, a measurement of the free-fall acceleration g was made by throwing a glass ball straight up in an evacuated tube and letting it return. Let ΔT_L in Fig. 2-29 be the time interval between the two passages of the ball across a certain lower level, ΔT_U the time interval between the two passages across an upper level, and H the distance between the two levels. Show that

$$g = \frac{8H}{\Delta T_L^2 - \Delta T_U^2}.$$

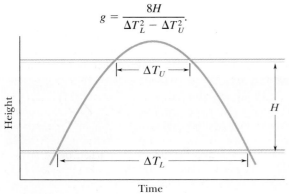

FIGURE 2-29 Problem 72.

73P. A ball of moist clay falls to the ground from a height of 15.0 m. It is in contact with the ground for 20.0 ms before coming to rest. What is the average acceleration of the clay during the time it is in contact with the ground? (Treat the ball as a particle.)

74P. A ball is thrown *down* vertically with an initial *speed* of v_0 from a height of h. (a) What is its speed just before it strikes the ground? (b) How long does the ball take to reach the ground? (c) What would be the answers to (a) and (b) if the ball were thrown *upward* from the same height and with the same initial speed? Before solving any equations, decide if the answers here should be greater than, less than, or the same as in (a) and (b).

75P. Figure 2-30 shows a simple device for measuring your reaction time. It consists of a cardboard strip marked with a scale and two large dots. A friend holds the strip *vertically,* with thumb and forefinger at the dot on the right in Fig. 2-30. You then position your thumb and forefinger at the other dot (on the left in Fig. 2-30), being careful not to touch the strip. Your friend releases the strip, and you try to pinch it as soon as possible after you see it begin to fall. The mark at the place where you pinch the strip gives your reaction time. (a) How far from the lower dot should you place the 50.0 ms mark? (b) How much higher should the marks for 100, 150, 200, and 250 ms be? (For example, should the 100 ms marker be two times as far from the dot as the 50 ms marker? Can you find any pattern in the answers?)

Reaction time (ms)

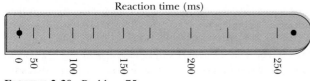

FIGURE 2-30 Problem 75.

76P. A juggler tosses balls vertically a certain distance into the air. How much higher must they be tossed if they are to spend twice as much time in the air?

77P. A stone is thrown vertically upward. On its way up it passes point A with speed v, and point B, 3.00 m higher than A, with speed $\frac{1}{2}v$. Calculate (a) the speed v and (b) the maximum height reached by the stone above point B.

78P. To test the quality of a tennis ball, you drop it onto the floor from a height of 4.00 m. It rebounds to a height of 3.00 m. If the ball was in contact with the floor for 10.0 ms, what was its average acceleration during contact

79P. Water drips from the nozzle of a shower onto the floor 200 cm below. The drops fall at regular (equal) intervals of time, the first drop striking the floor at the instant the fourth drop begins to fall. Find the locations of the second and third drops when the first strikes the floor.

80P. A lead ball is dropped into a lake from a diving board 5.20 m above the water. It hits the water with a certain velocity and then sinks to the bottom with this same constant velocity. It reaches the bottom 4.80 s after it is dropped. (a) How deep is the lake? (b) What is the average velocity of the ball? (c) Suppose that all the water is drained from the lake. The ball is now thrown from the diving board so that it again reaches the bottom in 4.80 s. What is the initial velocity of the ball?

81P. An object falls from height h from rest. If it travels $0.50h$ in the last 1.00 s, find (a) the time and (b) the height of its fall. Explain the physically unacceptable solution of the quadratic equation in t that you obtain.

82P. A woman fell 144 ft from the top of a building, landing on the top of a metal ventilator box, which she crushed to a depth of 18 in. She survived without serious injury. What acceleration (assumed uniform) did she experience during the collision? Express your answer in terms of g units.

83P. A stone is dropped into a river from a bridge 144 ft above the water. Another stone is thrown vertically down 1.00 s after the first is dropped. Both stones strike the water at the same time. (a) What was the initial speed of the second stone? (b) Plot velocity versus time on a graph for each stone, taking zero time as the instant the first stone is released.

84P. A parachutist bails out and freely falls 50 m. Then the parachute opens, and thereafter she decelerates at 2.0 m/s². She reaches the ground with a speed of 3.0 m/s. (a) How long was the parachutist in the air? (b) At what height did the fall begin?

85P. Two objects begin a free fall from rest from the same height 1.0 s apart. How long after the first object begins to fall will the two objects be 10 m apart?

86P. As Fig. 2-31 shows, Clara jumps from a bridge, followed closely by Jim. How long did Jim wait after Clara jumped? Assume that Jim is 170 cm tall and that the jumping-off level is at the top of the figure. Make scale measurements directly on the figure.

87P. A hot-air balloon is ascending at the rate of 12 m/s and is 80 m above the ground when a package is dropped over the side. (a) How long does the package take to reach the ground? (b) With what speed does it hit the ground?

88P. An elevator without a ceiling is ascending with a constant speed of 10 m/s. A boy on the elevator throws a ball directly

FIGURE 2-31 Problem 86.

upward, from a height of 2.0 m above the elevator floor, just as the elevator floor is 28 m above the ground. The initial speed of the ball with respect to the elevator is 20 m/s. (a) What is the maximum height attained by the ball? (b) How long does it take for the ball to return to the elevator floor?

89P. A steel ball is dropped from a building's roof and passes a window, taking 0.125 s to fall from the top to the bottom of the window, a distance of 1.20 m. It then falls to a sidewalk and bounces "perfectly" back past the window, moving from bottom to top in 0.125 s. (The upward flight is a reverse of the fall.) The time spent below the bottom of the window is 2.00 s. How tall is the building?

90P. A drowsy cat spots a flowerpot that sails first up and then down past an open window. The pot was in view for a total of 0.50 s, and the top-to-bottom height of the window is 2.00 m. How high above the window top did the flowerpot go?

ELECTRONIC COMPUTATION

Problems in this section are to be solved with a graphing calculator or with a mathematics package on a computer.

91. A driver is to test the braking of a car equipped with antilock brakes, which can decelerate the car at the maximum possible rate without the tires skidding. The test is made on a dry straight section of a test track. In response to a visual signal, the driver stops the car in the shortest possible distance. The stopping distances for six initial speeds are given in the following table. (a) Find an expression for the stopping distance d in terms of the initial speed v_i, the response time T_R of the driver to the visual signal, and the magnitude a' (assumed constant) of the car's deceleration. Using a least squares fit for the data, find (b) a' and (c) T_R.

v_i (m/s)	5.00	10.0	15.0	20.0	25.0	30.0
d (m)	4.6	12.1	22.3	35.2	51.0	69.5

92. A motorcyclist starts from rest and accelerates along a horizontal straight track. Photogates are attached to six posts evenly spaced every 10.0 m along the track; the first post is at the starting point. The photogates on each post measure the time required by the motorcyclist to reach that post. The following table gives the results for one test. (a) Find an expression for the distance d to each post in terms of the time t to reach that post and the acceleration a' of the motorcyclist (assumed constant). (b) Using the data of the table, graph d versus t^2. (c) Using a linear regression fit of the data, find the acceleration of the motorcyclist.

Post number	1	2	3	4	5	6
Distance traveled (m)	0	10.0	20.0	30.0	40.0	50.0
Time required (s)	0	1.63	2.33	2.83	3.31	3.79

93. The motorcyclist of Problem 92 next starts from rest and accelerates uniformly along another straight test track. A series of six posts equally spaced by a distance $d_0 = 10.0$ m line the track, beginning an unknown distance d from the motorcyclist's starting point. The following table gives the speed v at each post. (a) Find an expression for the square of the speed v_j at post j in terms of the speed v_1 at post 1, the distance d_0, the constant acceleration a', and the post number j. (b) Graph v_j^2 versus $(j - 1)$. Using a linear regression fit of the data, find (c) acceleration a' and (d) distance d.

Post number j	1	2	3	4	5	6
Speed v (m/s)	14.0	18.3	21.7	24.6	27.5	30.0

94. Suppose the coordinate of a particle moving along the x axis is given as a function of time t by

$$x(t) = -32.0 + 24.0t^2 e^{-0.0300t},$$

where x is in meters and t is in seconds. (a) Write expressions for the velocity v and acceleration a as functions of the time. (b) Plot graphs of the coordinate, velocity, and acceleration as functions of time from $t = 0$ to $t = 100$ s. (c) Find the time when the coordinate of the particle is zero; then find the velocity and acceleration at that time. (d) Find the time when the velocity of the particle is zero; then find the coordinate and acceleration at that time.

3
Vectors

For two decades spelunking teams crawled, climbed, and squirmed through 200 km of Mammoth Cave and the Flint Ridge cave system, seeking a connection. The photograph shows Richard Zopf pushing his pack through the Tight Tube, far inside the Flint Ridge system. After 12 hours of ''caving'' along a labyrinthine route, Zopf and six others waded through a stretch of chilling water and found themselves in Mammoth Cave. Their breakthrough established the Mammoth–Flint cave system as being the longest cave in the world. How can their final point be related to their initial point other than in terms of the actual route they covered?

3-1 VECTORS AND SCALARS

A particle confined to a straight line can move in only two directions. We can take its motion to be positive in one of these directions and negative in the other. For a particle moving in three dimensions, however, a plus sign or minus sign is no longer enough to indicate the direction of the motion. Instead, we must use a *vector*.

A **vector** has magnitude as well as direction, and vectors follow certain rules of combination, which we examine below; a **vector quantity** is a quantity that has both a magnitude and a direction. Some physical quantities that can be represented by vectors are displacement, velocity, and acceleration.

Not all physical quantities involve a direction. Temperature, pressure, energy, mass, and time, for example, do not "point" in the spatial sense. We call such quantities **scalars,** and we deal with them by the rules of ordinary algebra. A single value, with a sign (as in $-40°F$), specifies a scalar.

The simplest vector quantity is displacement, or change of position. A vector that represents a displacement is called, reasonably, a **displacement vector**. (Similarly, we have velocity vectors and acceleration vectors.) If a particle changes its position by moving from A to B in Fig. 3-1a, we say that it undergoes a displacement from A to B, which we represent with an arrow pointing from A to B. The arrow specifies the vector graphically. To distinguish vector symbols from other kinds of arrows, we use the outline of a triangle as the arrowhead.

The arrows from A to B, from A' to B', and from A'' to B'' in Fig. 3-1a represent the same *change of position* for the particle and we make no distinction among them. All three arrows have the same magnitude and direction and thus specify identical displacement vectors.

The displacement vector tells us nothing about the actual path that the particle takes. In Fig. 3-1b, for example, all three paths connecting points A and B correspond to the

same displacement vector, that of Fig. 3-1a. Displacement vectors represent only the overall effect of the motion, not the motion itself.

Another way to specify a vector is to give its magnitude and the angles it makes with two reference lines. We do that in Sample Problem 3-1.

SAMPLE PROBLEM 3-1

The 1972 team that connected the Mammoth–Flint cave system went from the Austin Entrance in the Flint Ridge system to Echo River in Mammoth Cave (see Fig. 3-2a), traveling a net 2.6 km westward, 3.9 km southward, and 25 m upward. What was their displacement vector?

SOLUTION: We first take an overhead view (Fig. 3-2b) to find their horizontal displacement d_h. The magnitude of d_h is given by the Pythagorean theorem:

$$d_h = \sqrt{(2.6 \text{ km})^2 + (3.9 \text{ km})^2} = 4.69 \text{ km}.$$

The angle θ of their displacement vector relative to the west is given by

$$\tan \theta = \frac{3.9 \text{ km}}{2.6 \text{ km}} = 1.5,$$

or $\qquad \theta = \tan^{-1} 1.5 = 56°.$

We next take a side view (Fig. 3-2c), to find the magnitude of the overall displacement d,

$$d = \sqrt{(4.69 \text{ km})^2 + (0.025 \text{ km})^2} = 4.69 \text{ km} \approx 4.7 \text{ km},$$

and the angle ϕ,

$$\phi = \tan^{-1} \frac{0.025 \text{ km}}{4.69 \text{ km}} = 0.3°.$$

So the team's displacement vector had a magnitude of 4.7 km and was at an angle of 56° south of west and at an angle of 0.3° upward. The net vertical motion was, of course, insignificant compared to the horizontal motion, but that fact would have been no comfort to the team as they climbed up and down countless times. The route they actually covered was quite different from the displacement vector, which merely points from start to finish.

3-2 ADDING VECTORS: GRAPHICAL METHOD

Suppose that, as in the vector diagram of Fig. 3-3a, a particle moves from A to B and then later from B to C. We can represent its overall displacement (no matter what its actual path) with two successive displacement vectors, AB and BC. The net effect of these two displacements is a single displacement from A to C. We call AC the **vector sum** (or **resultant**) of the vectors AB and BC. This sum is

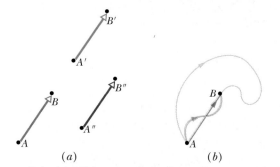

FIGURE 3-1 (*a*) All three arrows represent the same displacement. (*b*) All three paths connecting the two points correspond to the same displacement vector.

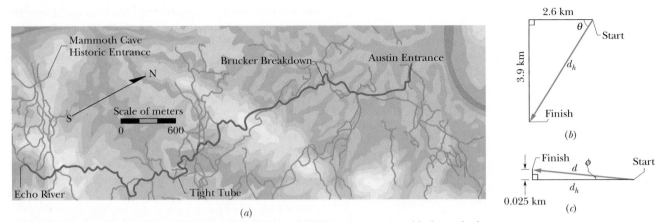

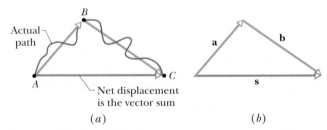

FIGURE 3-2 Sample Problem 3-1. (*a*) Part of the Mammoth–Flint cave system, with the spelunking team's route from the Austin Entrance to Echo River indicated. (*b*) An overhead view of the team's displacement. (*c*) A side view. (Adapted from map by Cave Research Foundation.)

not the usual algebraic sum, and we need more than simple numbers to specify it.

In Fig. 3-3*b*, we redraw the vectors of Fig. 3-3*a* and relabel them in the way that we shall use from now on, namely, with boldface symbols such as **a**, **b**, and **s**. In handwriting, you can place an arrow over the symbol, as in $\vec{a}$. If we want to indicate only the magnitude of the vector (a quantity that is always positive), we shall use the italic symbol, such as *a*, *b*, and *s*. (You can use just a handwritten symbol.) The boldface symbol always implies both properties of the vector, magnitude and direction.

We can represent the relation among the three vectors in Fig. 3-3*b* with the vector equation

$$\mathbf{s} = \mathbf{a} + \mathbf{b}, \qquad (3\text{-}1)$$

which says that the vector **s** is the vector sum of vectors **a** and **b**. The symbol + in Eq. 3-1 and the words ''sum'' and ''add'' have different meanings for vectors than they do in the usual algebra.

Figure 3-3 suggests a procedure for adding two-dimensional vectors **a** and **b** graphically. (1) On a sheet of paper, lay out vector **a** to some convenient scale and at the proper angle. (2) Lay out vector **b** to the same scale, with its tail at the head of vector **a,** again at the proper angle. (3) Construct the vector sum **s** by drawing an arrow from the tail of **a** to the head of **b**. Note that this procedure takes into account both the magnitudes and the directions of the vectors; you can easily generalize it to add more than two vectors.

Vector addition, defined in this way, has two important properties. First, the order of addition does not matter. That is,

$$\mathbf{a} + \mathbf{b} = \mathbf{b} + \mathbf{a} \qquad \text{(commutative law).} \quad (3\text{-}2)$$

Figure 3-4 should convince you that this is the case.

Second, if there are more than two vectors, it does not matter how we group them as we add them. Thus if we want to add vectors **a**, **b**, and **c**, we can add **a** and **b** first and then add their vector sum to **c**. On the other hand, we can add **b** and **c** first, and then add *that* sum to **a**. We get the same result either way. In equation form,

$$(\mathbf{a} + \mathbf{b}) + \mathbf{c} = \mathbf{a} + (\mathbf{b} + \mathbf{c}) \qquad \text{(associative law).} \quad (3\text{-}3)$$

Figure 3-5 should convince you that Eq. 3-3 is correct.

FIGURE 3-3 (*a*) *AC* is the vector sum of the vectors *AB* and *BC*. (*b*) The same vectors relabeled.

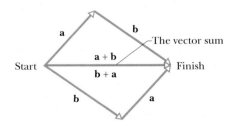

FIGURE 3-4 The two vectors **a** and **b** can be added in either order; see Eq. 3-2.

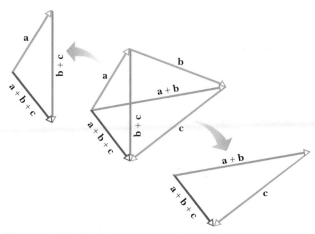

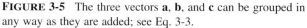

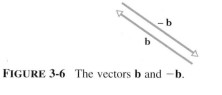

FIGURE 3-5 The three vectors **a**, **b**, and **c** can be grouped in any way as they are added; see Eq. 3-3.

The vector −**b** is a vector with the same magnitude as **b** but the opposite direction (see Fig. 3-6). If you try to add the two vectors in Fig. 3-6, you will see that

$$\mathbf{b} + (-\mathbf{b}) = 0.$$

Adding −**b** has the effect of subtracting **b**. We use this property to define the difference between two vectors: let **d** = **a** − **b**. Then

$$\mathbf{d} = \mathbf{a} - \mathbf{b} = \mathbf{a} + (-\mathbf{b}) \qquad \text{(subtraction).} \quad (3\text{-}4)$$

That is, we find the difference vector **d** by adding the vector −**b** to the vector **a**. Figure 3-7 shows how this is done graphically.

Remember, although we have used displacement vectors as a prototype, the rules for addition and subtraction hold for vectors of all kinds, whether they represent forces, velocities, or anything else. However, as in ordinary arithmetic, it is still true that we can add only vectors of the same kind. We can add two displacements, for example, or

FIGURE 3-6 The vectors **b** and −**b**.

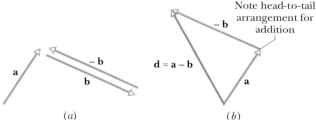

FIGURE 3-7 (*a*) Vectors **a**, **b**, and −**b**. (*b*) To subtract vector **b** from vector **a**, add vector −**b** to vector **a**.

two velocities, but it makes no sense to add a displacement and a velocity. In the world of scalars, that would be like trying to add 21 s and 12 m.

CHECKPOINT **1:** The magnitudes of displacements **a** and **b** are 3 m and 4 m, respectively, and **c** = **a** + **b**. Considering various orientations of **a** and **b**, what is (a) the maximum possible magnitude for **c** and (b) the minimum possible magnitude?

SAMPLE PROBLEM 3-2

In an orienteering class, you have the goal of moving as far (straight-line distance) from base camp as possible by making three straight-line moves. You may use the following displacements in any order: (a) **a**, 2.0 km due east; (b) **b**, 2.0 km 30° north of east; (c) **c**, 1.0 km due west. Or you may substitute either −**b** for **b** or −**c** for **c**. What is the greatest distance you can be from base camp at the end of the third displacement?

SOLUTION: Using a convenient scale, we draw vectors **a**, **b**, **c**, −**b**, and −**c** as in Fig. 3-8*a*. We then mentally slide the vectors over the page, connecting three of them in head-to-tail arrangements to find their vector sum **d**. The tail of the first vector represents base camp. The head of the third vector represents the point at which you stop. The vector sum **d** extends from the tail of the first vector to the head of the third vector. Its magnitude *d* is your distance from base camp.

We find that distance *d* is greatest for a head-to-tail arrangement of vectors **a**, **b**, and −**c**. They can be in any order, because their vector sum is the same for any order. The order shown in Fig. 3-8*b* is for the vector sum

$$\mathbf{d} = \mathbf{b} + \mathbf{a} + (-\mathbf{c}) = \mathbf{b} + \mathbf{a} - \mathbf{c}.$$

Using the scale given in Fig. 3-8*a*, we measure the length *d* of this vector sum, finding

$$d = 4.8 \text{ m.} \qquad \text{(Answer)}$$

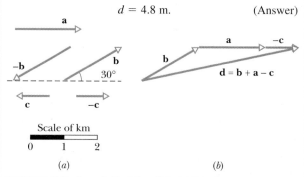

FIGURE 3-8 Sample Problem 3-2. (*a*) Displacement vectors; three are to be used. (*b*) Your distance from base camp is greatest if you undergo displacements **a**, **b**, and −**c**, in any order. One choice of order is shown; it gives vector sum **d** = **b** + **a** − **c**.

3-3 VECTORS AND THEIR COMPONENTS

Adding vectors graphically can be tedious. A neater and easier technique involves algebra but requires that the vectors be placed on a rectangular coordinate system. The x and y axes are usually drawn in the plane of the page, as in Fig. 3-9a. The z axis comes directly out of the page at the origin; we ignore it for now and deal only with two-dimensional vectors.

Vector **a** in Fig. 3-9 is in the xy plane. If we drop perpendicular lines from the ends of **a** to the coordinate axes, the quantities a_x and a_y so formed are called the **components** of the vector **a** in the x and y directions. The process of forming them is called **resolving the vector**. In general, a vector will have three components, although for the case of Fig. 3-9a the component along the z axis is zero. As Fig. 3-9b shows, if you move a vector in such a way that it remains parallel to its original direction at all times, the values of its components remain unchanged.

We can easily find the components of **a** in Fig. 3-9a from the right triangle there:

$$a_x = a \cos \theta \quad \text{and} \quad a_y = a \sin \theta, \qquad (3\text{-}5)$$

where θ is the angle that the vector **a** makes with the direction of increasing x, and a is its magnitude. Figure 3-9c shows that the vector and its x and y components form a right triangle. It also shows how we can reconstruct a vector from its components: we arrange the components head to tail. Then we complete a right triangle with the vector being along the hypotenuse, from the tail of one component to the head of the other component.

Depending on the value of θ, the components of a vector may be positive, negative, or zero. In a figure, we use small, solid triangles as arrowheads to indicate the signs of the components, according to the usual conven-

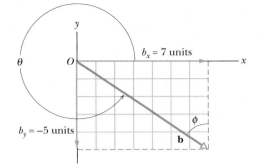

FIGURE 3-10 The components of **b** are positive on the x axis and negative on the y axis.

tion: positive in the direction of increasing coordinate values, and negative in the opposite direction. Figure 3-10 shows a vector **b** for which b_y is negative and b_x is positive.

Once a vector has been resolved into its components, the components themselves can be used in place of the vector. For example, the two numbers a and θ specify the magnitude and direction of the two-dimensional vector **a** in Fig. 3-9a. But instead of a and θ, we can specify the vector with the two other numbers a_x and a_y. Both sets of numbers contain exactly the same information, and we can pass back and forth readily between the two descriptions. To obtain a and θ if we are given a_x and a_y, we note (see Fig. 3-9c) that

$$a = \sqrt{a_x^2 + a_y^2} \quad \text{and} \quad \tan \theta = \frac{a_y}{a_x}. \qquad (3\text{-}6)$$

In solving problems, you may use *either* the a_x, a_y notation *or* the a, θ notation.

CHECKPOINT 2: In the figure, which of the ways indicated for combining the x and y components of vector **a** are proper to determine that vector?

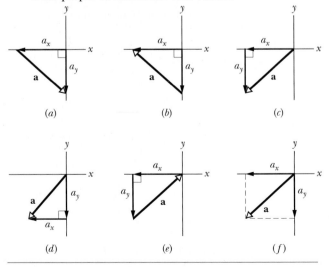

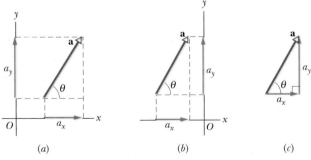

FIGURE 3-9 (a) The components of vector **a**. (b) The components are unchanged if the vector is shifted, as long as the magnitude and orientation are maintained. (c) The components form the legs of a right triangle whose hypotenuse is the magnitude of the vector.

SAMPLE PROBLEM 3-3

A small airplane leaves an airport on an overcast day and is later sighted 215 km away, in a direction making an angle of 22° east of north. How far east and north is the airplane from the airport when sighted?

SOLUTION: On an *xy* coordinate system the situation is as shown in Fig. 3-11, where for convenience the origin of the system has been placed at the airport. The airplane's displacement vector **d** points from the origin to where the airplane is sighted.

To answer the question, we find the components of **d**. With Eq. 3-5 and the angle $\theta = 68°$ (= 90° − 22°), we have

$$d_x = d \cos \theta = (215 \text{ km})(\cos 68°)$$
$$= 81 \text{ km} \qquad \text{(Answer)}$$

and
$$d_y = d \sin \theta = (215 \text{ km})(\sin 68°)$$
$$= 199 \text{ km}. \qquad \text{(Answer)}$$

So the airplane was spotted 199 km north and 81 km east of the airport.

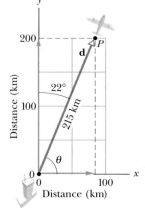

FIGURE 3-11 Sample Problem 3-3. A plane takes off from an airport at the origin and is later sighted at *P*.

PROBLEM SOLVING TACTICS

TACTIC 1: *Angles—Degrees and Radians*

Angles that are measured relative to the positive direction of the *x* axis are positive if they are measured in the counterclockwise direction, and negative if clockwise. For example, 210° and −150° are the same angle. Most calculators (try yours) will accept angles in either form when a trig function is taken.

Angles may be measured in degrees or radians (rad). You can relate the two measures by remembering that one full circle is equivalent to 360° and to 2π rad. So if you needed to convert, say, 40° to radians, you would write

$$40° \frac{2\pi \text{ rad}}{360°} = 0.70 \text{ rad}.$$

Is the answer reasonable? Quickly check by realizing that 40° is one-ninth of a full circle, and with a full circle equivalent to 2π rad, which is about 6.3 rad, the angle should be $\frac{1}{9}(6.3)$, which it is. Or, check by remembering that 1 rad ≈ 57°.

Most calculators are in the degree mode when they are turned on, so that angles must be entered in degrees. However, you may be able to change yours to the radian mode. Check it before your first exam.

TACTIC 2: *Trig Functions*

You need to know the definitions of the common trig functions—sine, cosine, and tangent—because they are part of the language of science and engineering. They are given in Fig. 3-12 in a form that does not depend on how the triangle is labeled.

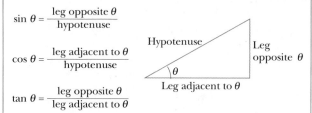

$$\sin \theta = \frac{\text{leg opposite } \theta}{\text{hypotenuse}}$$
$$\cos \theta = \frac{\text{leg adjacent to } \theta}{\text{hypotenuse}}$$
$$\tan \theta = \frac{\text{leg opposite } \theta}{\text{leg adjacent to } \theta}$$

FIGURE 3-12 A triangle used to define the trigonometric functions. See also Appendix E.

You should also be able to sketch how the trig functions vary with angle, as in Fig. 3-13, in order to be able to judge whether a calculator result is reasonable. Even knowing the signs of the functions in the various quadrants can be of help.

TACTIC 3: *Inverse Trig Functions*

The most important of the inverse trig functions are $\sin^{-1}$, $\cos^{-1}$, and $\tan^{-1}$. When they are taken on a calculator, you must consider the reasonableness of the answer you get, because there is usually another possible answer that the calculator does not give. The range of operation for a calculator in taking each inverse trig function is indicated in Fig. 3-13. As an example, $\sin^{-1} 0.5$ has associated angles of 30° (which is displayed by the calculator, since 30° falls within its range of operation) and 150°. To see both angles, draw a horizontal line through 0.5 in Fig. 3-13a and note where it cuts the sine curve.

How do you distinguish a correct answer? As an example, reconsider the calculation of θ in Sample Problem 3-1, where $\tan \theta = 1.5$. Taking $\tan^{-1} 1.5$ on your calculator tells you that $\theta = 56°$, but $\theta = 236°$ (= 180° + 56°) also has a tangent of 1.5. Which is correct? From the physical situation (Fig. 3-2b), 56° is reasonable and 236° is clearly not.

TACTIC 4: *Measuring Vector Angles*

The equations for $\cos \theta$ and $\sin \theta$ in Eq. 3-5 and the equation for $\tan \theta$ in Eq. 3-6 are valid only if the angle is measured relative to the positive direction of the *x* axis. If it is measured relative to some other direction, then the trig functions in Eq. 3-5 may have to be interchanged, and the ratio in Eq. 3-6 may have to be inverted. A safer method is to convert the given angle into one that is measured from the positive direction of the *x* axis, as shown in Sample Problem 3-3.

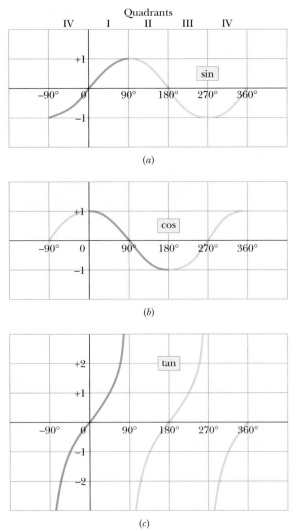

FIGURE 3-13 Three useful curves to remember. A calculator's range of operation for taking *inverse* trig functions is indicated by the darker portions of the colored curves.

3-4 UNIT VECTORS

A **unit vector** is a vector that has a magnitude of exactly 1 and points in a particular direction. It lacks both dimension and unit. Its sole purpose is to point, that is, to specify a direction. The unit vectors in the positive directions of the x, y, and z axes are labeled **i**, **j**, and **k**, as shown in Fig. 3-14.* The arrangement of axes in Fig. 3-14 is said to be a **right-handed coordinate system.** The system remains right-handed if it is rotated rigidly to a new orientation. We use such coordinate systems exclusively in this text.

_____
*To distinguish handwritten unit vectors from other symbols, including other vectors, you might top them with a "hat": $\hat{i}, \hat{j}, \hat{k}$.

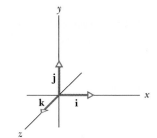

FIGURE 3-14 Unit vectors **i**, **j**, and **k** define the directions of a right-handed rectangular coordinate system.

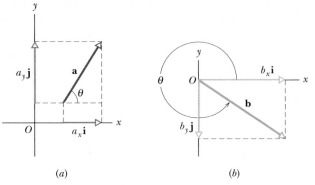

FIGURE 3-15 (a) The vector components of vector **a**. (b) The vector components of vector **b**.

Unit vectors are very useful for expressing other vectors; for example, we can express **a** and **b** of Figs. 3-9 and 3-10 as

$$\mathbf{a} = a_x\mathbf{i} + a_y\mathbf{j} \tag{3-7}$$

and

$$\mathbf{b} = b_x\mathbf{i} + b_y\mathbf{j}. \tag{3-8}$$

These two equations are illustrated in Fig. 3-15. The quantities $a_x\mathbf{i}$ and $a_y\mathbf{j}$ are the **vector components** of **a**, in contrast to a_x and a_y, which are its **scalar components** (or, as before, simply its **components**).

Look back for a moment at Sample Problem 3-1. If you superimpose the coordinate system of Fig. 3-14 at the Austin Entrance of Fig. 3-2a, with **i** eastward, **j** northward, and **k** upward, then the displacement **d** to Echo River is neatly expressed as

$$\mathbf{d} = -(2.6 \text{ km})\mathbf{i} - (3.9 \text{ km})\mathbf{j} + (0.025 \text{ km})\mathbf{k}.$$

3-5 ADDING VECTORS BY COMPONENTS

As we have noted, adding vectors graphically is tedious; it also has limited accuracy and is challenging in three dimensions. Here we find a more direct technique—adding vectors by combining their components, axis by axis.

To start, consider the statement

$$\mathbf{r} = \mathbf{a} + \mathbf{b}, \tag{3-9}$$

which says that the vector **r** is the same as the vector (**a** + **b**). If that is so, then each component of **r** must be the same as the corresponding component of (**a** + **b**):

$$r_x = a_x + b_x, \qquad (3\text{-}10)$$

$$r_y = a_y + b_y, \qquad (3\text{-}11)$$

$$r_z = a_z + b_z. \qquad (3\text{-}12)$$

In other words, two vectors are equal if their corresponding components are equal. Equations 3-10 to 3-12 tell us that to add vectors **a** and **b**, we must (1) resolve the vectors into their scalar components; (2) axis by axis, combine these scalar components to get the components of the sum **r**; and (3) if necessary, combine the components of **r** to get **r** itself. We have a choice in step 3. We can express **r** in unit-vector notation, or we can give the magnitude of **r** and its direction (by stating one angle when we are working in two dimensions or two angles when we are working in three dimensions).

$\mathbb{C}$HECKPOINT **3:** (a) In the figure, what are the signs of the x components of $\mathbf{d}_1$ and $\mathbf{d}_2$? (b) What are the signs of the y components of $\mathbf{d}_1$ and $\mathbf{d}_2$? (c) What are the signs of the x and y components of $\mathbf{d}_1 + \mathbf{d}_2$?

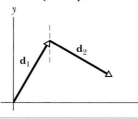

SAMPLE PROBLEM 3-4

In a road rally, you are given the following instructions: from the starting point use available roads to drive 36 km due east to checkpoint "Able," then 42 km due north to "Baker," and then 25 km northwest to "Charlie." (The roadway and checkpoints are shown in Fig. 3-16.)

(a) At "Charlie," what are the magnitude and orientation of your displacement **d** from the starting point?

SOLUTION: Figure 3-16 also shows a convenient orientation for an xy coordinate system, as well as vectors representing the three displacements you have undergone. The scalar components of **d** are

$$d_x = a_x + b_x + c_x = 36 \text{ km} + 0 + (25 \text{ km})(\cos 135°)$$
$$= (36 + 0 - 17.7) \text{ km} = 18.3 \text{ km}$$
and
$$d_y = a_y + b_y + c_y = 0 + 42 \text{ km} + (25 \text{ km})(\sin 135°)$$
$$= (0 + 42 + 17.7) \text{ km} = 59.7 \text{ km}.$$

Next, we use Eq. 3-6 to find the magnitude and direction of **d**:

$$d = \sqrt{d_x^2 + d_y^2} = \sqrt{(18.3 \text{ km})^2 + (59.7 \text{ km})^2}$$
$$= 62 \text{ km} \qquad \text{(Answer)}$$
and

$$\theta = \tan^{-1} \frac{d_y}{d_x} = \tan^{-1} \frac{59.7 \text{ km}}{18.3 \text{ km}} = 73°, \quad \text{(Answer)}$$

where θ is the angle shown in Fig. 3-16.

(b) Write **d** in unit-vector notation.

SOLUTION: We simply write **d** as

$$\mathbf{d} = (x \text{ component})\mathbf{i} + (y \text{ component})\mathbf{j}$$
$$= (18.3 \text{ km})\mathbf{i} + (59.7 \text{ km})\mathbf{j}. \qquad \text{(Answer)}$$

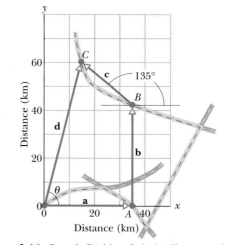

FIGURE 3-16 Sample Problem 3-4. A rally route, showing the origin, checkpoints Able (A), Baker (B), and Charlie (C), and the road network.

SAMPLE PROBLEM 3-5

Here are three vectors, each expressed in unit-vector notation:

$$\mathbf{a} = 4.2\mathbf{i} - 1.6\mathbf{j},$$

$$\mathbf{b} = -1.6\mathbf{i} + 2.9\mathbf{j},$$

$$\mathbf{c} = -3.7\mathbf{j}.$$

All three lie in the xy plane, and none of them has a z component. Find the vector **r** that is the sum of these three vectors. For convenience, the units have been omitted from these vector expressions; you may take them to be meters.

SOLUTION: From Eqs. 3-10 and 3-11 we have

$$r_x = a_x + b_x + c_x = 4.2 - 1.6 + 0 = 2.6$$
and
$$r_y = a_y + b_y + c_y = -1.6 + 2.9 - 3.7 = -2.4.$$

Thus

$$\mathbf{r} = 2.6\mathbf{i} - 2.4\mathbf{j}. \qquad \text{(Answer)}$$

Figure 3-17a shows the three vectors and their sum. Figure 3-17b shows **r** and its vector components.

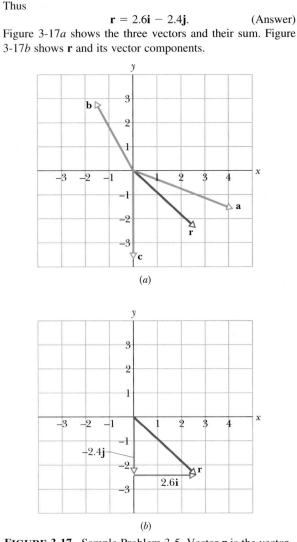

(a)

(b)

FIGURE 3-17   Sample Problem 3-5. Vector **r** is the vector sum of the other three vectors.

3-6 VECTORS AND THE LAWS OF PHYSICS

So far, in every figure that includes a coordinate system, the x and y axes are parallel to the edges of the book page. And so, when a vector **a** is included, its components a_x and a_y are also parallel to the edges (as in Fig. 3-18a). The only reason for that orientation of the axes is that it looks proper: there is no deeper reason. We could, instead, rotate the axes (but not the vector **a**) through an angle ϕ as in Fig. 3-18b, in which case the components would have new values, call them a_x' and a_y'. Since there are an infinite number of choices of ϕ, there are an infinite number of different pairs of components for **a**.

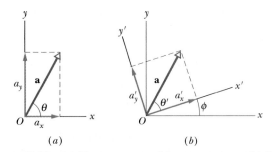

FIGURE 3-18   (a) The vector **a** and its components. (b) The same vector, with the axes of the coordinate system rotated through an angle ϕ.

Which then is the "right" pair of components? The answer is that they are all equally valid because each pair (with its axes) just gives us a different way of describing the same vector **a**; all produce the same magnitude and direction for the same vector. In Fig. 3-18 we have

$$a = \sqrt{a_x^2 + a_y^2} = \sqrt{a_x'^2 + a_y'^2} \qquad (3\text{-}13)$$

and

$$\theta = \theta' + \phi. \qquad (3\text{-}14)$$

The point is that we have great freedom in choosing a coordinate system, because the relations among vectors (including, for example, the vector addition of Eq. 3-1) do not depend on the location of the origin of the coordinate system or on the orientation of the axes. This is also true of the relations of physics; they are all independent of the choice of coordinate system. Add to that the simplicity and richness of the language of vectors and you can see why the laws of physics are almost always presented in that language: one equation, like Eq. 3-9, can represent three (or even more) relations, like Eqs. 3-10, 3-11, and 3-12.

3-7 MULTIPLYING VECTORS*

There are three ways in which vectors can be multiplied. None is exactly like the usual algebraic multiplication.

Multiplying a Vector by a Scalar

If we multiply a vector **a** by a scalar s, we get a new vector. Its magnitude is the product of the magnitude of **a** and the absolute value of s. Its direction is the direction of **a** if s is positive, but the opposite direction if s is negative. To divide **a** by s, we multiply **a** by 1/s.

In either multiplication or division, the scalar may be a pure number or a physical quantity; in the latter case, the

---

*This material will not be employed until later (Chapter 7 for scalar products and Chapter 12 for vector products), and so your instructor may wish to postpone assignment of the section.

physical nature of the product differs from that of the original vector **a**.

The Scalar Product

There are two ways to multiply a vector by a vector: one way produces a scalar and the other produces a new vector. Students commonly confuse the two ways, and so starting now, you should carefully distinguish between them.

The **scalar product** of the vectors **a** and **b** in Fig. 3-19a is written as **a** · **b** and defined to be

$$\mathbf{a} \cdot \mathbf{b} = ab \cos \phi, \qquad (3\text{-}15)$$

where a is the magnitude of **a**, b is the magnitude of **b**, and ϕ is the angle* between **a** and **b**. Note that there are only scalars on the right side of Eq. 3-15 (including the value of $\cos \phi$). Thus **a** · **b** on the left side represents a *scalar* quantity; because of the notation, it is also known as the **dot product** and is spoken as "a dot b."

A dot product can be regarded as the product of two quantities: (1) the magnitude of one of the vectors and (2) the scalar component of the second vector along the direction of the first vector. For example, in Fig. 3-19b, **a** has a scalar component $a \cos \phi$ along the direction of **b**; note that a perpendicular dropped from the head of **a** to **b** determines that component. Similarly, **b** has a scalar component $b \cos \phi$ along the direction of **a**. If ϕ is 0°, the component of one vector along the other is maximum, and so also is

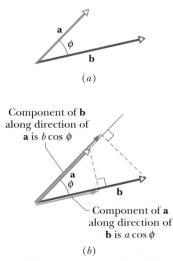

(a)

Component of **b** along direction of **a** is $b \cos \phi$

Component of **a** along direction of **b** is $a \cos \phi$

(b)

FIGURE 3-19 (a) Two vectors **a** and **b**, with an angle ϕ between them. (b) Each vector has a component along the direction of the other vector.

*In Fig. 3-19a, there are actually two angles between the vectors: ϕ and $360° - \phi$. Either can be used in Eq. 3-15, because their cosines are the same.

the dot product. If, instead, ϕ is 90°, the component of one vector along the other is zero, and so is the dot product.

Equation 3-15 can be rewritten as follows to emphasize the components:

$$\mathbf{a} \cdot \mathbf{b} = (a \cos \phi)(b) = (a)(b \cos \phi). \qquad (3\text{-}16)$$

The commutative law applies to a scalar product, so we can write

$$\mathbf{a} \cdot \mathbf{b} = \mathbf{b} \cdot \mathbf{a}.$$

When two vectors are in unit-vector notation, we write their dot product as

$$\mathbf{a} \cdot \mathbf{b} = (a_x \mathbf{i} + a_y \mathbf{j} + a_z \mathbf{k}) \cdot (b_x \mathbf{i} + b_y \mathbf{j} + b_z \mathbf{k}), \qquad (3\text{-}17)$$

for which the distributive law applies: each component of the first vector is to be dotted into each component of the second vector.

C**HECKPOINT 4:** Vectors **C** and **D** have magnitudes of 3 units and 4 units, respectively. What is the angle between **C** and **D** if **C** · **D** equals (a) zero, (b) 12 units, and (c) −12 units?

SAMPLE PROBLEM 3-6

What is the angle ϕ between $\mathbf{a} = 3.0\mathbf{i} - 4.0\mathbf{j}$ and $\mathbf{b} = -2.0\mathbf{i} + 3.0\mathbf{k}$?

SOLUTION: By Eq. 3-15, the dot product is

$$\mathbf{a} \cdot \mathbf{b} = ab \cos \phi = \sqrt{3.0^2 + 4.0^2} \, \sqrt{2.0^2 + 3.0^2} \cos \phi$$
$$= 18.0 \cos \phi. \qquad (3\text{-}18)$$

We next find the dot product with Eq. 3-17:

$$\mathbf{a} \cdot \mathbf{b} = (3.0\mathbf{i} - 4.0\mathbf{j}) \cdot (-2.0\mathbf{i} + 3.0\mathbf{k}).$$

With the distributive law, this yields

$$\mathbf{a} \cdot \mathbf{b} = (3.0\mathbf{i}) \cdot (-2.0\mathbf{i}) + (3.0\mathbf{i}) \cdot (3.0\mathbf{k})$$
$$+ (-4.0\mathbf{j}) \cdot (-2.0\mathbf{i}) + (-4.0\mathbf{j}) \cdot (3.0\mathbf{k}).$$

We next apply Eq. 3-15 to each term. The angle between vectors for the first term is 0°, and for the other three terms it is 90°. We then have

$$\mathbf{a} \cdot \mathbf{b} = -(6.0)(1) + (9.0)(0) + (8.0)(0) - (12)(0)$$
$$= -6.0. \qquad (3\text{-}19)$$

Setting the results of Eqs. 3-18 and 3-19 equal to each other, we find

$$18.0 \cos \phi = -6.0$$

or

$$\phi = \cos^{-1} \frac{-6.0}{18.0} = 109° \approx 110°. \qquad \text{(Answer)}$$

The Vector Product

The **vector product** of **a** and **b**, written **a** × **b**, produces a third vector **c** whose magnitude is

$$c = ab \sin \phi, \qquad (3\text{-}20)$$

where ϕ is the *smaller* of the two angles between **a** and **b**.* Because of the notation, **a** × **b** is also known as the **cross product** and in speech it is "a cross b." If **a** and **b** are parallel or antiparallel, **a** × **b** = 0. The magnitude of **a** × **b**, which can be written as |**a** × **b**|, is maximum when **a** and **b** are perpendicular to each other.

The direction of **c** is perpendicular to the plane that contains **a** and **b**. Figure 3-20a shows how to determine the direction of **c** = **a** × **b** with what is known as the **right-hand rule.** Place the vectors **a** and **b** tail to tail without altering their orientations and imagine a line that is perpendicular to their plane where they meet. Pretend to grasp the line with your *right* hand in such a way that your fingers would sweep **a** into **b** through the smaller angle between them. Your outstretched thumb points in the direction of **c**.

The order of the vector multiplication is important. In Fig. 3-20b, we are determining the direction of **c**′ = **b** × **a**, so the fingers are placed to sweep **b** into **a** through the smaller angle. The thumb ends up in the opposite direction from previously, and so it must be that **c**′ = −**c**; that is,

$$\mathbf{b} \times \mathbf{a} = -(\mathbf{a} \times \mathbf{b}). \qquad (3\text{-}21)$$

In other words, the commutative law does not apply to a vector product.

In unit-vector notation, we write

$$\mathbf{a} \times \mathbf{b} = (a_x\mathbf{i} + a_y\mathbf{j} + a_z\mathbf{k})$$
$$\times \ (b_x\mathbf{i} + b_y\mathbf{j} + b_z\mathbf{k}), \qquad (3\text{-}22)$$

and the distributive law applies: each component of the first vector is to be crossed into each component of the second vector. You should use that law to expand Eq. 3-22; check your result against the expansion given in Appendix E. You can also evaluate a cross product by setting up and evaluating a determinant, as shown in Appendix E.

We can use a cross product of unit vectors to see if any given rectangular coordinate system is a right-handed coordinate system: if **i** × **j** = **k**, it is. Check the system shown in Fig. 3-14.

Cʜᴇᴄᴋᴘᴏɪɴᴛ **5:** Vectors **C** and **D** have magnitudes of 3 units and 4 units, respectively. What is the angle between **C** and **D** if the magnitude of the vector product **C** × **D** equals (a) zero and (b) 12 units?

---

*You must use the smaller of the two angles between the vectors because $\sin \phi$ and $\sin(360° - \phi)$ differ in algebraic sign.

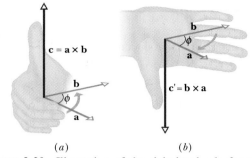

(a)   (b)

FIGURE 3-20 Illustration of the right-hand rule for vector products. (*a*) Swing vector **a** into vector **b** with the fingers of your right hand. Your thumb shows the direction of vector **c** = **a** × **b**. (*b*) Showing that (**a** × **b**) = −(**b** × **a**).

SAMPLE PROBLEM 3-7

Vector **a** lies in the xy plane in Fig. 3-21. It has a magnitude of 18 units and points in a direction 250° from the direction of increasing x. Vector **b** has a magnitude of 12 units and points along the direction of increasing z.

(a) What is the scalar product of these two vectors?

SOLUTION: The angle ϕ between these two vectors is 90° so that, from Eq. 3-15,

$$\mathbf{a} \cdot \mathbf{b} = ab \cos \phi = (18)(12)(\cos 90°) = 0. \quad \text{(Answer)}$$

The scalar product of any two vectors that are at right angles to each other is zero. This is consistent with the fact that neither of these vectors has a component along the direction of the other vector.

(b) What is the vector product **c** = **a** × **b**?

SOLUTION: The magnitude of the vector product is, from Eq. 3-20,

$$ab \sin \phi = (18)(12)(\sin 90°) = 216. \quad \text{(Answer)}$$

The direction of **c** is perpendicular to the plane formed by **a** and **b**. It must then be perpendicular to **b**, which means that it is perpendicular to the z axis. This in turn means that **c** must lie in the xy plane. The right-hand rule of Fig. 3-20 shows that **c** points as shown in Fig. 3-21. Because **c** is also perpendicular to **a**, the direction of **c** makes an angle of 250° − 90° = 160° with the direction of increasing x.

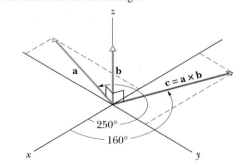

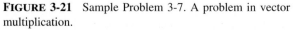

FIGURE 3-21 Sample Problem 3-7. A problem in vector multiplication.

SAMPLE PROBLEM 3-8

If $\mathbf{c} = \mathbf{a} \times \mathbf{b}$, where $\mathbf{a} = 3\mathbf{i} - 4\mathbf{j}$ and $\mathbf{b} = -2\mathbf{i} + 3\mathbf{k}$, what is $\mathbf{c}$?

SOLUTION: From Eq. 3-22 we have

$$\mathbf{c} = (3\mathbf{i} - 4\mathbf{j}) \times (-2\mathbf{i} + 3\mathbf{k}),$$

which, by the distributive law, becomes

$$\mathbf{c} = -(3\mathbf{i} \times 2\mathbf{i}) + (3\mathbf{i} \times 3\mathbf{k}) + (4\mathbf{j} \times 2\mathbf{i}) - (4\mathbf{j} \times 3\mathbf{k}).$$

We next evaluate each term with Eq. 3-20, determining the direction with the right-hand rule. (For the first term, $\phi = 0°$; for the other terms, $\phi = 90°$.) We find

$$\mathbf{c} = 0 - 9\mathbf{j} - 8\mathbf{k} - 12\mathbf{i} = -12\mathbf{i} - 9\mathbf{j} - 8\mathbf{k}. \quad \text{(Answer)}$$

The vector $\mathbf{c}$ is perpendicular to both $\mathbf{a}$ and $\mathbf{b}$, a fact that you can check by showing $\mathbf{c} \cdot \mathbf{a} = 0$ and $\mathbf{c} \cdot \mathbf{b} = 0$; that is, there is no component of $\mathbf{c}$ along the direction of either $\mathbf{a}$ or $\mathbf{b}$.

PROBLEM SOLVING TACTICS

TACTIC 5: *Common Errors with Cross Products*

Several errors are common in finding a cross product. (1) Failure to arrange vectors tail to tail is tempting when an illustration presents them head to tail: you must mentally shift (or better, redraw) one vector to the proper arrangement without changing its orientation. (2) Failing to use the right hand in applying the right-hand rule is easy when the right hand is occupied with a calculator or pencil. (3) Failure to sweep the first vector of the product into the second vector can occur when the orientations of the vectors require an awkward twisting of your hand to apply the right-hand rule. And sometimes that happens when you try to make the sweep mentally rather than actually using your hand. (4) Failure to work with a right-handed coordinate system results when you forget how to draw such a system (see Fig. 3-14).

REVIEW & SUMMARY

Scalars and Vectors

Scalars, such as temperature, have magnitude only. They are specified by a number with a unit (82°F) and obey the rules of arithmetic and ordinary algebra. *Vectors,* such as displacement, have both magnitude and direction (5 m, north) and obey the special rules of vector algebra.

Adding Vectors Geometrically

Two vectors $\mathbf{a}$ and $\mathbf{b}$ may be added geometrically by drawing them to a common scale and placing them head to tail. The vector connecting the tail of the first to the head of the second is the sum vector $\mathbf{s}$, as Fig. 3-3 shows. To subtract $\mathbf{b}$ from $\mathbf{a}$, reverse the direction of $\mathbf{b}$ to get $-\mathbf{b}$; then add $-\mathbf{b}$ to $\mathbf{a}$: see Fig. 3-7. Vector addition is commutative and obeys the associative law.

Components of a Vector

The (scalar) *components* a_x and a_y of any two-dimensional vector $\mathbf{a}$ are found by dropping perpendicular lines from the ends of $\mathbf{a}$ onto the coordinate axes. The components are given by

$$a_x = a \cos \theta \quad \text{and} \quad a_y = a \sin \theta, \quad (3\text{-}5)$$

where θ is the angle from the positive direction of the x axis to the direction of $\mathbf{a}$. The algebraic sign of a component indicates its direction along the associated axis. Given the components, we can reconstruct the vector from

$$a = \sqrt{a_x^2 + a_y^2} \quad \text{and} \quad \tan \theta = \frac{a_y}{a_x}. \quad (3\text{-}6)$$

Unit-Vector Notation

Often we find it useful to introduce *unit vectors* $\mathbf{i}$, $\mathbf{j}$, and $\mathbf{k}$, whose magnitudes are unity and whose directions are those of the x, y,

and z axes, respectively, in a right-handed coordinate system. We can write a vector $\mathbf{a}$ in terms of unit vectors as

$$\mathbf{a} = a_x\mathbf{i} + a_y\mathbf{j} + a_z\mathbf{k}, \quad (3\text{-}7)$$

in which $a_x\mathbf{i}$, $a_y\mathbf{j}$, and $a_z\mathbf{k}$ are the **vector components** and a_x, a_y, and a_z are the **scalar components** of $\mathbf{a}$.

Adding Vectors in Component Form

To add vectors in component form, we use the rules

$$r_x = a_x + b_x; \quad r_y = a_y + b_y; \quad r_z = a_z + b_z. \quad (3\text{-}10 \text{ to } 3\text{-}12)$$

Here $\mathbf{a}$ and $\mathbf{b}$ are the vectors to be added, and $\mathbf{r}$ is the vector sum.

Vectors and Physical Laws

Any physical situation involving vectors can be described using many possible coordinate systems. We usually choose the one that simplifies the work. However, the relation between the vector quantities does not depend on our choice. The laws of physics are also independent of our choice of coordinates.

Scalar Times Vector

The product of a scalar s and a vector $\mathbf{v}$ is a new vector whose magnitude is sv and whose direction is the same as that of $\mathbf{v}$ if s is positive, and opposite that of $\mathbf{v}$ if s is negative. To divide $\mathbf{v}$ by s, multiply $\mathbf{v}$ by $(1/s)$.

The Scalar Product

The scalar (or **dot**) product of two vectors $\mathbf{a}$ and $\mathbf{b}$ is written $\mathbf{a} \cdot \mathbf{b}$ and is the *scalar* quantity given by

$$\mathbf{a} \cdot \mathbf{b} = ab \cos \phi, \quad (3\text{-}15)$$

in which ϕ is the angle between the directions of **a** and **b**. The scalar product may be positive, zero, or negative, depending on the value of ϕ. Figure 3-19b indicates that a scalar product is the product of the magnitude of one vector and the component of the second vector along the direction of the first vector. In unit-vector notation, we have

$$\mathbf{a} \cdot \mathbf{b} = (a_x\mathbf{i} + a_y\mathbf{j} + a_z\mathbf{k}) \cdot (b_x\mathbf{i} + b_y\mathbf{j} + b_z\mathbf{k}), \quad (3\text{-}17)$$

which obeys the distributive law. Note that $\mathbf{a} \cdot \mathbf{b} = \mathbf{b} \cdot \mathbf{a}$.

The Vector Product

The vector (or **cross**) product of two vectors **a** and **b** is written

$\mathbf{a} \times \mathbf{b}$ and is a *vector* **c** whose magnitude c is given by

$$c = ab \sin \phi, \quad (3\text{-}20)$$

in which ϕ is the smaller of the angles between the directions of **a** and **b**. The direction of **c** is perpendicular to the plane defined by **a** and **b** and is given by a right-hand rule, as shown in Fig. 3-20. Note that $\mathbf{a} \times \mathbf{b} = -(\mathbf{b} \times \mathbf{a})$. In unit-vector notation we have

$$\mathbf{a} \times \mathbf{b} = (a_x\mathbf{i} + a_y\mathbf{j} + a_z\mathbf{k}) \times (b_x\mathbf{i} + b_y\mathbf{j} + b_z\mathbf{k}), \quad (3\text{-}22)$$

to which we may apply the distributive law.

QUESTIONS

1. Displacement vector **N** points from coordinates (5 m, 3 m) to coordinates (7 m, 6 m) in the xy plane. Which of the following displacement vectors are equivalent to **N**: vector **A**, which points from (−6 m, −5 m) to (−4 m, −2 m); vector **B**, which points from (−6 m, 1 m) to (−4 m, 4 m); vector **C**, which points from (−8 m, −6 m) to (−10 m, −9 m)?

2. If $\mathbf{d} = \mathbf{a} + \mathbf{b} + (-\mathbf{c})$, does (a) $\mathbf{a} + (-\mathbf{d}) = \mathbf{c} + (-\mathbf{b})$, (b) $\mathbf{a} = (-\mathbf{b}) + \mathbf{d} + \mathbf{c}$, and (c) $\mathbf{c} + (-\mathbf{d}) = \mathbf{a} + \mathbf{b}$?

3. Equation 3-2 shows that the addition of two vectors **a** and **b** is commutative. Does that mean subtraction is commutative, so that $\mathbf{a} - \mathbf{b} = \mathbf{b} - \mathbf{a}$?

4. In a game within a three-dimensional maze, you must move your game piece from *start,* at xyz coordinates (0, 0, 0), to *finish,* at coordinates (−2 cm, 4 cm, −4 cm). The game piece can undergo only the displacements (in centimeters) given below. If, along the way, the game piece lands at coordinates (−5 cm, −1 cm, −1 cm) or (5 cm, 2 cm, −1 cm), you lose the game. Which vectors and in what sequence will get your game piece to *finish?*

$$\mathbf{p} = -7\mathbf{i} + 2\mathbf{j} - 3\mathbf{k} \quad \mathbf{r} = 2\mathbf{i} - 3\mathbf{j} + 2\mathbf{k}$$
$$\mathbf{q} = 2\mathbf{i} - \mathbf{j} + 4\mathbf{k} \quad \mathbf{s} = 3\mathbf{i} + 5\mathbf{j} - 3\mathbf{k}.$$

5. Describe two vectors **a** and **b** such that
(a) $\mathbf{a} + \mathbf{b} = \mathbf{c}$ and $a + b = c$;
(b) $\mathbf{a} + \mathbf{b} = \mathbf{a} - \mathbf{b}$;
(c) $\mathbf{a} + \mathbf{b} = \mathbf{c}$ and $a^2 + b^2 = c^2$.

6. (a) Can a vector have zero magnitude if one of its components is not zero? (b) Can two vectors having different magnitudes be combined to give a vector sum of zero? Can three vectors give a vector sum of zero if they (c) are in the same plane, and (d) are not in the same plane?

7. Can the magnitude of the difference between two vectors ever be greater than (a) the magnitude of one of the vectors, (b) the magnitudes of both vectors, and (c) the magnitude of their sum?

8. Can the sum of the magnitudes of two vectors ever be equal to the magnitude of the sum of the same two vectors? If no, why not? If yes, when?

9. Which of the arrangements of axes in Fig. 3-22 can be labeled

"right-handed coordinate system"? As usual, each axis label indicates the positive side of the axis.

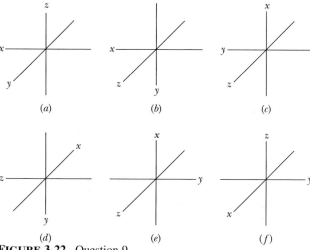

FIGURE 3-22   Question 9.

10. The x and y components of four vectors, **a, b, c,** and **d** are given below. For which vectors will your calculator give you the correct angle θ when you use it to find θ with Eq. 3-6? Answer first by examining Fig. 3-13, and then check your answers with your calculator.

$$a_x = 3 \quad a_y = 3 \quad c_x = -3 \quad c_y = -3$$
$$b_x = -3 \quad b_y = 3 \quad d_x = 3 \quad d_y = -3.$$

11. In Fig. 3-23, what are the signs of the x and y components of (a) **A**, and (b) **D**, where $\mathbf{D} = \mathbf{A} - \mathbf{B}$?

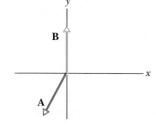

FIGURE 3-23   Question 11.

12. Figure 3-24 shows vector **A** and four choices for vector **B** that differ only in orientation. Rank the choices according to the absolute value of **A·B** that they give, greatest first.

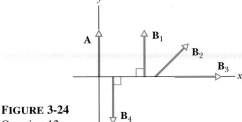

FIGURE 3-24
Question 12.

13. Figure 3-25 shows vector **A** and four other vectors that have the same magnitude but differ in orientation. (a) Which of those other four vectors have the same dot product with **A**? (b) Which have a negative dot product with **A**?

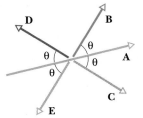

FIGURE 3-25   Question 13.

14. If **A** = 2i + 4j and **B** = 7k, what is **A·B**?

15. If **a·b** = **a·c**, must **b** equal **c**?

16. In Fig. 3-26, what is the direction of (a) **E** × **F**, (b) **F** × **E**, and (c) **G** × **E**, where **G** is in the *xy* plane? (d) Does the answer to (c) change if **G** is shifted parallel to the *z* axis without any change in its direction?

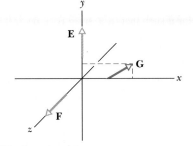

FIGURE 3-26   Question 16.

17. If **A** = 2i + 4j, what is **A** × **B** when (a) **B** = 8i + 16j and (b) **B** = − 8i − 16j? (This question can be answered without computation.)

EXERCISES & PROBLEMS

SECTION 3-2 Adding Vectors: Graphical Method

1E. Consider two displacements, one of magnitude 3 m and another of magnitude 4 m. Show how the displacement vectors may be combined to get a resultant displacement of magnitude (a) 7 m, (b) 1 m, and (c) 5 m.

2E. A woman walks 250 m in the direction 30° east of north, then 175 m directly east. (a) Using graphical methods, find her final displacement from the starting point. (b) Compare the magnitude of her displacement with the distance she walked.

3E. A person walks in the following pattern: 3.1 km north, then 2.4 km west, and finally 5.2 km south. (a) Construct the vector diagram that represents this motion. (b) How far and in what direction would a bird fly in a straight line from the same starting point to the same final point?

4E. A car is driven east for a distance of 50 km, then north for 30 km, and then in a direction 30° east of north for 25 km. Draw the vector diagram and determine the total displacement of the car from its starting point.

5P. Vector **a** has a magnitude of 5.0 units and is directed east. Vector **b** is directed 35° west of north and has a magnitude of 4.0 units. Construct vector diagrams for calculating **a** + **b** and **b** − **a**. Estimate the magnitudes and directions of **a** + **b** and **b** − **a** from your diagrams.

6P. A bank in downtown Boston is robbed (see the map in Fig. 3-27). To elude police, the robbers escape by helicopter, making three successive flights described by the following displacements: 20 miles, 45° south of east; 33 miles, 26° north of west; 16

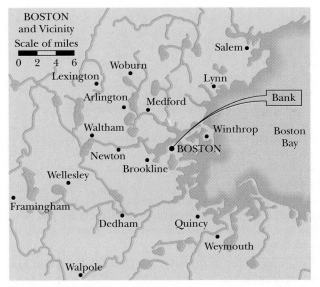

FIGURE 3-27   Problem 6.

miles, 18° east of south. At the end of the third flight they are captured. In what town are they apprehended? (Use the graphical method to add these displacements on the map.)

7P. Three vectors **a**, **b**, and **c**, each having a magnitude of 50 units, lie in the *xy* plane and make angles of 30°, 195°, and 315° with the positive *x* axis, respectively. Find graphically the magnitudes and directions of the vectors (a) **a** + **b** + **c**, (b) **a** − **b** + **c**, and (c) a vector **d** such that (**a** + **b**) − (**c** + **d**) = 0.

SECTION 3-3 Vectors and Their Components

8E. (a) Express the following angles in radians: 20.0°, 50.0°, 100°. (b) Convert the following angles to degrees: 0.330 rad, 2.10 rad, 7.70 rad.

9E. What are the *x* and *y* components of a vector **a** in the *xy* plane if its direction is 250° counterclockwise from the positive *x* axis and its magnitude is 7.3 units?

10E. The *x* component of a certain vector is −25.0 units and the *y* component is +40.0 units. (a) What is the magnitude of the vector? (b) What is the angle between the direction of the vector and the positive direction of *x*?

11E. A displacement vector **r** in the *xy* plane is 15 m long and directed as shown in Fig. 3-28. Determine the *x* and *y* components of the vector.

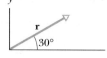

FIGURE 3-28
Exercise 11.

12E. A heavy piece of machinery is raised by sliding it 12.5 m along a plank oriented at 20.0° to the horizontal, as shown in Fig. 3-29. (a) How high above its original position is it raised? (b) How far is it moved horizontally?

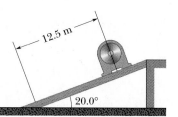

FIGURE 3-29
Exercise 12.

13E. The minute hand of a wall clock measures 10 cm from axis to tip. What is the displacement vector of its tip (a) from a quarter after the hour to half past, (b) in the next half hour, and (c) in the next hour?

14E. A ship sets out to sail to a point 120 km due north. An unexpected storm blows the ship to a point 100 km due east of its starting point. How far, and in what direction, must it now sail to reach its original destination?

15P. A person desires to reach a point that is 3.40 km from her present location and in a direction that is 35.0° north of east. However, she must travel along streets that are oriented either north–south or east–west. What is the minimum distance she could travel to reach her destination?

16P. Rock *faults* are ruptures along which opposite faces of rock have slid past each other. In Fig. 3-30, points *A* and *B* coincided before the rock in the foreground slid down to the right. The net displacement *AB* is along the plane of the fault. The horizontal component of *AB* is the *strike-slip AC*. The component of *AB* that is directly down the plane of the fault is the *dip-slip AD*. (a) What is the net displacement *AB* if the strike-slip is 22.0 m and the dip-slip is 17.0 m? (b) If the plane of the fault is inclined 52.0° to the horizontal, what is the vertical component of *AB*?

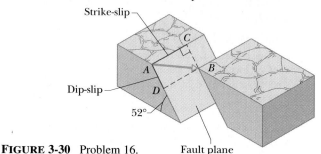

FIGURE 3-30 Problem 16.   Fault plane

17P. A wheel with a radius of 45.0 cm rolls without slipping along a horizontal floor (Fig. 3-31). At time t_1, the dot *P* painted on the rim of the wheel is at the point of contact between the wheel and the floor. At a later time t_2, the wheel has rolled through one-half of a revolution. What is the displacement of *P* during this interval?

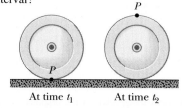

FIGURE 3-31 Problem 17.

18P. A room has dimensions 10.0 ft (height) × 12.0 ft × 14.0 ft. A fly starting at one corner flies around, ending up at the diagonally opposite corner. (a) What is the magnitude of its displacement? (b) Could the length of its path be less than this distance? Greater than this distance? Equal to this distance? (c) Choose a suitable coordinate system and find the components of the displacement vector in that system. (d) If the fly walks rather than flies, what is the length of the shortest path it can take? (This can be answered without calculus.)

SECTION 3-5 Adding Vectors by Components

19E. Find the vector components of the sum **r** of the vector displacements **c** and **d** whose components in meters along three perpendicular directions are $c_x = 7.4$, $c_y = -3.8$, $c_z = -6.1$; $d_x = 4.4$, $d_y = -2.0$, $d_z = 3.3$.

20E. (a) What is the sum in unit-vector notation of the two vectors **a** = 4.0**i** + 3.0**j** and **b** = −13**i** + 7.0**j**? (b) What are the magnitude and direction of **a** + **b**?

21E. Calculate the x and y components, magnitudes, and directions of (a) $\mathbf{a} + \mathbf{b}$ and (b) $\mathbf{b} - \mathbf{a}$ if $\mathbf{a} = 3.0\mathbf{i} + 4.0\mathbf{j}$ and $\mathbf{b} = 5.0\mathbf{i} - 2.0\mathbf{j}$.

22E. Two vectors are given by $\mathbf{a} = 4\mathbf{i} - 3\mathbf{j} + \mathbf{k}$ and $\mathbf{b} = -\mathbf{i} + \mathbf{j} + 4\mathbf{k}$. Find (a) $\mathbf{a} + \mathbf{b}$, (b) $\mathbf{a} - \mathbf{b}$, and (c) a vector $\mathbf{c}$ such that $\mathbf{a} - \mathbf{b} + \mathbf{c} = 0$.

23E. Given two vectors $\mathbf{a} = 4.0\mathbf{i} - 3.0\mathbf{j}$ and $\mathbf{b} = 6.0\mathbf{i} + 8.0\mathbf{j}$, find the magnitudes and directions of (a) $\mathbf{a}$, (b) $\mathbf{b}$, (c) $\mathbf{a} + \mathbf{b}$, (d) $\mathbf{b} - \mathbf{a}$, and (e) $\mathbf{a} - \mathbf{b}$. How do the orientations of the last two compare?

24P. If $\mathbf{a} - \mathbf{b} = 2\mathbf{c}$, $\mathbf{a} + \mathbf{b} = 4\mathbf{c}$, and $\mathbf{c} = 3\mathbf{i} + 4\mathbf{j}$, then what are $\mathbf{a}$ and $\mathbf{b}$?

25P. If vector $\mathbf{B}$ is added to vector $\mathbf{A}$, the result is $6.0\mathbf{i} + 1.0\mathbf{j}$. If $\mathbf{B}$ is subtracted from $\mathbf{A}$, the result is $-4.0\mathbf{i} + 7.0\mathbf{j}$. What is the magnitude of $\mathbf{A}$?

26P. A vector $\mathbf{B}$, when added to the vector $\mathbf{C} = 3.0\mathbf{i} + 4.0\mathbf{j}$, yields a resultant vector that is in the positive y direction and has a magnitude equal to that of $\mathbf{C}$. What is the magnitude of $\mathbf{B}$?

27P. Two vectors $\mathbf{a}$ and $\mathbf{b}$ have equal magnitudes of 10.0 units. They are oriented as shown in Fig. 3-32 and their vector sum is $\mathbf{r}$. Find (a) the x and y components of $\mathbf{r}$, (b) the magnitude of $\mathbf{r}$, and (c) the angle $\mathbf{r}$ makes with the positive x axis.

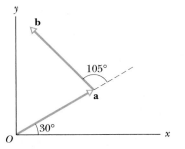

FIGURE 3-32 Problem 27.

28P. A golfer takes three putts to get the ball into the hole. The first putt displaces the ball 12 ft north, the second 6.0 ft southeast, and the third 3.0 ft southwest. What displacement was needed to get the ball into the hole on the first putt?

29P. A radar station detects an airplane approaching directly from the east. At first observation, the range to the plane is 1200 ft at 40° above the horizon. The airplane is tracked for another 123° in the vertical east–west plane, the range at final contact being 2580 ft. See Fig. 3-33. Find the displacement of the airplane during the period of observation.

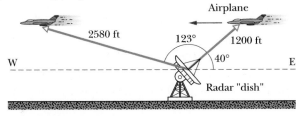

FIGURE 3-33 Problem 29.

30P. (a) A man leaves his front door, walks 1000 m east, 2000 m north, and then takes a penny from his pocket and drops it from a cliff 500 m high. Set up a coordinate system and write

down an expression, using unit vectors, for the displacement of the penny from the house to its landing point. (b) The man then returns to his front door, following a different path. What is his resultant displacement for the round trip?

31P. A particle undergoes three successive displacements in a plane, as follows: 4.00 m southwest, 5.00 m east, and 6.00 m in a direction 60.0° north of east. Choose the y axis pointing north and the x axis pointing east and find (a) the components of each displacement, (b) the components of the resultant displacement, (c) the magnitude and direction of the resultant displacement, and (d) the displacement that would be required to bring the particle back to the starting point.

32P. Prove that two vectors must have equal magnitudes if their sum is perpendicular to their difference.

33P. Two vectors of lengths a and b make an angle θ with each other when placed tail to tail. Prove, by taking components along two perpendicular axes, that the length of their sum is

$$r = \sqrt{a^2 + b^2 + 2ab \cos \theta}.$$

34P. (a) Using unit vectors, write expressions for the body diagonals (the straight lines from one corner to another through the center) of a cube in terms of its edges, which have length a. (b) Determine the angles that the body diagonals make with the adjacent edges. (c) Determine the length of the body diagonals.

SECTION 3-6 Vectors and the Laws of Physics

35E. A vector $\mathbf{a}$ with a magnitude of 17.0 m is directed 56.0° counterclockwise from the $+x$ axis, as shown in Fig. 3-34. (a) What are the components a_x and a_y of the vector? (b) A second coordinate system is inclined by 18.0° with respect to the first. What are the components a_x' and a_y' in this primed coordinate system?

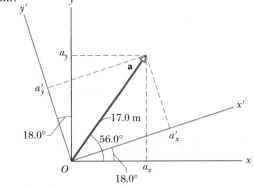

FIGURE 3-34 Exercise 35.

SECTION 3-7 Multiplying Vectors

36E. A vector $\mathbf{d}$ has a magnitude of 2.5 m and points north. What are the magnitudes and directions of the vectors (a) $4.0\mathbf{d}$ and (b) $-3.0\mathbf{d}$?

37E. Consider $\mathbf{a}$ in the positive direction of x, $\mathbf{b}$ in the positive direction of y, and a scalar d. What is the direction of $\mathbf{b}/d$ if d is (a) positive and (b) negative? What is the magnitude of (c) $\mathbf{a} \cdot \mathbf{b}$ and

(d) $\mathbf{a} \cdot \mathbf{b}/d$? What is the direction of (e) $\mathbf{a} \times \mathbf{b}$ and (f) $\mathbf{b} \times \mathbf{a}$? (g) What are the magnitudes of the cross products in (e) and (f)? (h) What are the magnitude and direction of $\mathbf{a} \times \mathbf{b}/d$?

38E. In a right-handed coordinate system show that

$$\mathbf{i} \cdot \mathbf{i} = \mathbf{j} \cdot \mathbf{j} = \mathbf{k} \cdot \mathbf{k} = 1 \quad \text{and} \quad \mathbf{i} \cdot \mathbf{j} = \mathbf{j} \cdot \mathbf{k} = \mathbf{k} \cdot \mathbf{i} = 0.$$

If the coordinate system is rectangular but not right-handed, do the results change?

39E. In a right-handed coordinate system show that

$$\mathbf{i} \times \mathbf{i} = \mathbf{j} \times \mathbf{j} = \mathbf{k} \times \mathbf{k} = 0$$

and $\qquad \mathbf{i} \times \mathbf{j} = \mathbf{k}; \quad \mathbf{k} \times \mathbf{i} = \mathbf{j}; \quad \mathbf{j} \times \mathbf{k} = \mathbf{i}.$

If the coordinate system is rectangular but not right-handed, do the results change?

40E. Show for any vector $\mathbf{a}$ that $\mathbf{a} \cdot \mathbf{a} = a^2$ and that $\mathbf{a} \times \mathbf{a} = 0$.

41E. Find (a) "north cross west," (b) "down dot south," (c) "east cross up," (d) "west dot west," and (e) "south cross south." Let each vector have unit magnitude.

42E. A vector $\mathbf{a}$ of magnitude 10 units and another vector $\mathbf{b}$ of magnitude 6.0 units point in directions differing by 60°. Find (a) the scalar product of the two vectors and (b) the magnitude of the vector product $\mathbf{a} \times \mathbf{b}$.

43E. Two vectors, $\mathbf{r}$ and $\mathbf{s}$, lie in the xy plane. Their magnitudes are 4.50 and 7.30 units, respectively, and their directions are 320° and 85.0°, respectively, as measured counterclockwise from the positive x axis. What are the values of (a) $\mathbf{r} \cdot \mathbf{s}$ and (b) $\mathbf{r} \times \mathbf{s}$?

44E. For the vectors in Fig. 3-35, calculate (a) $\mathbf{a} \cdot \mathbf{b}$, (b) $\mathbf{a} \cdot \mathbf{c}$, and (c) $\mathbf{b} \cdot \mathbf{c}$.

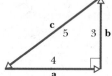

FIGURE 3-35 Exercises 44 and 45.

45E. For the vectors in Fig. 3-35, calculate (a) $\mathbf{a} \times \mathbf{b}$, (b) $\mathbf{a} \times \mathbf{c}$, and (c) $\mathbf{b} \times \mathbf{c}$.

46P. *Scalar Product in Unit-Vector Notation.* Let two vectors be represented in terms of their components as

$$\mathbf{a} = a_x \mathbf{i} + a_y \mathbf{j} + a_z \mathbf{k} \quad \text{and} \quad \mathbf{b} = b_x \mathbf{i} + b_y \mathbf{j} + b_z \mathbf{k}.$$

Show that

$$\mathbf{a} \cdot \mathbf{b} = a_x b_x + a_y b_y + a_z b_z.$$

47P. (a) Determine the components and magnitude of $\mathbf{r} = \mathbf{a} - \mathbf{b} + \mathbf{c}$ if $\mathbf{a} = 5.0\mathbf{i} + 4.0\mathbf{j} - 6.0\mathbf{k}$, $\mathbf{b} = -2.0\mathbf{i} + 2.0\mathbf{j} + 3.0\mathbf{k}$, and $\mathbf{c} = 4.0\mathbf{i} + 3.0\mathbf{j} + 2.0\mathbf{k}$. (b) Calculate the angle between $\mathbf{r}$ and the positive z axis.

48P. Use the definition of scalar product, $\mathbf{a} \cdot \mathbf{b} = ab \cos \theta$, and the fact that $\mathbf{a} \cdot \mathbf{b} = a_x b_x + a_y b_y + a_z b_z$ (see Problem 46) to calculate the angle between the two vectors given by $\mathbf{a} = 3.0\mathbf{i} + 3.0\mathbf{j} + 3.0\mathbf{k}$ and $\mathbf{b} = 2.0\mathbf{i} + 1.0\mathbf{j} + 3.0\mathbf{k}$.

49P. *Vector Product in Unit-Vector Notation.* Show that for the vectors $\mathbf{a}$ and $\mathbf{b}$ of Problem 46,

$$\mathbf{a} \times \mathbf{b} = \mathbf{i}(a_y b_z - a_z b_y) + \mathbf{j}(a_z b_x - a_x b_z) + \mathbf{k}(a_x b_y - a_y b_x).$$

50P. Two vectors are given by $\mathbf{a} = 3.0\mathbf{i} + 5.0\mathbf{j}$ and $\mathbf{b} = 2.0\mathbf{i} + 4.0\mathbf{j}$. Find (a) $\mathbf{a} \times \mathbf{b}$, (b) $\mathbf{a} \cdot \mathbf{b}$, and (c) $(\mathbf{a} + \mathbf{b}) \cdot \mathbf{b}$.

51P. Two vectors $\mathbf{a}$ and $\mathbf{b}$ have the components, in arbitrary units, $a_x = 3.2$, $a_y = 1.6$, $b_x = 0.50$, $b_y = 4.5$. (a) Find the angle between the directions of $\mathbf{a}$ and $\mathbf{b}$. (b) Find the components of a vector $\mathbf{c}$ that is perpendicular to $\mathbf{a}$, is in the xy plane, and has a magnitude of 5.0 units.

52P. Vector $\mathbf{a}$ lies in the yz plane 63° from the $+y$ axis, has a positive z component, and has magnitude 3.20 units. Vector $\mathbf{b}$ lies in the xz plane 48° from the $+x$ axis, has a positive z component, and has magnitude 1.40 units. Find (a) $\mathbf{a} \cdot \mathbf{b}$, (b) $\mathbf{a} \times \mathbf{b}$, and (c) the angle between $\mathbf{a}$ and $\mathbf{b}$.

53P. Three vectors are given by $\mathbf{a} = 3.0\mathbf{i} + 3.0\mathbf{j} - 2.0\mathbf{k}$, $\mathbf{b} = -1.0\mathbf{i} - 4.0\mathbf{j} + 2.0\mathbf{k}$, and $\mathbf{c} = 2.0\mathbf{i} + 2.0\mathbf{j} + 1.0\mathbf{k}$. Find (a) $\mathbf{a} \cdot (\mathbf{b} \times \mathbf{c})$, (b) $\mathbf{a} \cdot (\mathbf{b} + \mathbf{c})$, and (c) $\mathbf{a} \times (\mathbf{b} + \mathbf{c})$.

54P. Find the angles between the body diagonals of a cube with edge length a. See Problem 34.

55P. Show that the area of the triangle contained between the vectors $\mathbf{a}$ and $\mathbf{b}$ and the red line in Fig. 3-36 is $\frac{1}{2}|\mathbf{a} \times \mathbf{b}|$.

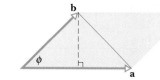

FIGURE 3-36 Problem 55.

56P. (a) Show that $\mathbf{a} \cdot (\mathbf{b} \times \mathbf{a})$ is zero for all vectors $\mathbf{a}$ and $\mathbf{b}$. (b) What is the value of $\mathbf{a} \times (\mathbf{b} \times \mathbf{a})$ if there is an angle ϕ between the directions of $\mathbf{a}$ and $\mathbf{b}$?

57P. Show that $\mathbf{a} \cdot (\mathbf{b} \times \mathbf{c})$ is equal in magnitude to the volume of the parallelepiped formed on the three vectors $\mathbf{a}$, $\mathbf{b}$, and $\mathbf{c}$ as shown in Fig. 3-37.

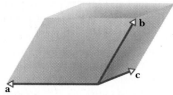

FIGURE 3-37 Problem 57.

58P. The three vectors shown in Fig. 3-38 have magnitudes $a = 3.00$, $b = 4.00$, and $c = 10.0$. (a) Calculate the x and y components of these vectors. (b) Find the numbers p and q such that $\mathbf{c} = p\mathbf{a} + q\mathbf{b}$.

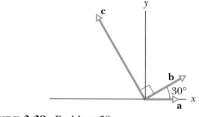

FIGURE 3-38 Problem 58.

In 1922, one of the Zacchinis, a famous family of circus performers, was the first human cannonball to be shot across an arena and into a net. To increase the excitement, the family gradually increased the height and distance of the flight until, in 1939 or 1940, Emanuel Zacchini soared over three Ferris wheels and through a horizontal distance of 225 feet. How could he know where to place the net? And how could he be certain he would clear the Ferris wheels?

4-1 MOVING IN TWO OR THREE DIMENSIONS

This chapter extends the material of the preceding two chapters to two and three dimensions. Many of the ideas of Chapter 2, such as position, velocity, and acceleration, are used here, but they are now a little more complex because of the extra dimensions. To keep the notation manageable, we use the vector algebra of Chapter 3. As you read this chapter, you might want to thumb back to those previous chapters to refresh your memory.

4-2 POSITION AND DISPLACEMENT

One general way of locating a particle-like object is with a **position vector r**, which is a vector that extends from a reference point (usually the origin of a coordinate system) to the object. In the unit-vector notation of Section 3-4, **r** can be written

$$\mathbf{r} = x\mathbf{i} + y\mathbf{j} + z\mathbf{k}, \tag{4-1}$$

where $x\mathbf{i}$, $y\mathbf{j}$, and $z\mathbf{k}$ are the vector components of **r** and the coefficients x, y, and z are its scalar components. (This notation is slightly different from the notation we used in Chapter 3. Take a minute to convince yourself that the two are comparable.)

The coefficients x, y, and z give the object's location along the axes and relative to the origin. That is, the object has the rectangular coordinates (x, y, z). For instance, Fig. 4-1 shows an object P whose position vector is

$$\mathbf{r} = -3\mathbf{i} + 2\mathbf{j} + 5\mathbf{k}$$

and whose rectangular coordinates are $(-3, 2, 5)$. Along the x axis P is 3 units from the origin, in the $-\mathbf{i}$ direction. Along the y axis it is 2 units from the origin, in the $+\mathbf{j}$

direction. And along the z axis it is 5 units from the origin, in the $+\mathbf{k}$ direction.

As an object moves, its position vector changes in such a way that the vector always extends to the object from the origin. If the object has position vector $\mathbf{r}_1$ at time t_1 and position vector $\mathbf{r}_2$ at a later time $t_1 + \Delta t$, then its *displacement* $\Delta\mathbf{r}$ during the time interval Δt is

$$\Delta\mathbf{r} = \mathbf{r}_2 - \mathbf{r}_1. \tag{4-2}$$

Using the unit-vector notation of Eq. 4-1, we can rewrite this displacement as

$$\Delta\mathbf{r} = (x_2\mathbf{i} + y_2\mathbf{j} + z_2\mathbf{k}) - (x_1\mathbf{i} + y_1\mathbf{j} + z_1\mathbf{k})$$

or

$$\Delta\mathbf{r} = (x_2 - x_1)\mathbf{i} + (y_2 - y_1)\mathbf{j} + (z_2 - z_1)\mathbf{k}, \tag{4-3}$$

where coordinates (x_1, y_1, z_1) correspond to position vector $\mathbf{r}_1$ and coordinates (x_2, y_2, z_2) correspond to position vector $\mathbf{r}_2$. We can also rewrite the displacement by substituting Δx for $(x_2 - x_1)$, Δy for $(y_2 - y_1)$, and Δz for $(z_2 - z_1)$.

SAMPLE PROBLEM 4-1

The position vector for a particle is initially

$$\mathbf{r}_1 = -3\mathbf{i} + 2\mathbf{j} + 5\mathbf{k}$$

and then later is

$$\mathbf{r}_2 = 9\mathbf{i} + 2\mathbf{j} + 8\mathbf{k}$$

(see Fig. 4-2). What is the displacement from $\mathbf{r}_1$ to $\mathbf{r}_2$?

SOLUTION: Recall from Chapter 3 that we add (or subtract) two vectors in unit-vector notation by combining the components, axis by axis. So Eq. 4-2 becomes

$$\Delta\mathbf{r} = (9\mathbf{i} + 2\mathbf{j} + 8\mathbf{k}) - (-3\mathbf{i} + 2\mathbf{j} + 5\mathbf{k})$$
$$= 12\mathbf{i} + 3\mathbf{k}. \tag{Answer}$$

The displacement vector is parallel to the xz plane, because it lacks any y component, a fact that is easier to pick out in the numerical result than in Fig. 4-2.

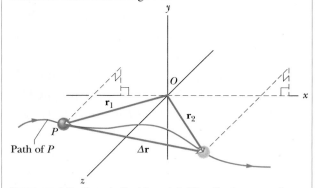

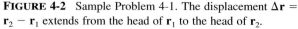

FIGURE 4-2   Sample Problem 4-1. The displacement $\Delta\mathbf{r} = \mathbf{r}_2 - \mathbf{r}_1$ extends from the head of $\mathbf{r}_1$ to the head of $\mathbf{r}_2$.

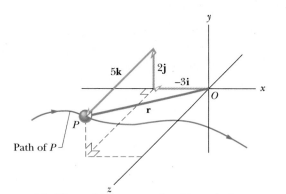

FIGURE 4-1   The position vector **r** for object P is the vector sum of its vector components.

CHECKPOINT **1:** (a) If a bat flies from *xyz* coordinates (−2 m, 4 m, −3 m) to coordinates (6 m, −2 m, −3 m), what is its displacement $\Delta \mathbf{r}$ in unit-vector notation? (b) Is $\Delta \mathbf{r}$ parallel to one of the three coordinate planes? If so, which plane?

4-3 VELOCITY AND AVERAGE VELOCITY

If a particle moves through a displacement $\Delta \mathbf{r}$ in a time interval Δt, then its *average velocity* is

$$\overline{\mathbf{v}} = \frac{\Delta \mathbf{r}}{\Delta t}, \tag{4-4}$$

which can be written in expanded form as

$$\overline{\mathbf{v}} = \frac{\Delta x \mathbf{i} + \Delta y \mathbf{j} + \Delta z \mathbf{k}}{\Delta t}$$

$$= \frac{\Delta x}{\Delta t} \mathbf{i} + \frac{\Delta y}{\Delta t} \mathbf{j} + \frac{\Delta z}{\Delta t} \mathbf{k}. \tag{4-5}$$

The (instantaneous) *velocity* $\mathbf{v}$ is the value that $\overline{\mathbf{v}}$ approaches in the limit as we shrink Δt to 0. It can be written as the derivative

$$\mathbf{v} = \frac{d\mathbf{r}}{dt}. \tag{4-6}$$

Substituting for $\mathbf{r}$ from Eq. 4-1 yields

$$\mathbf{v} = \frac{d}{dt}(x\mathbf{i} + y\mathbf{j} + z\mathbf{k}) = \frac{dx}{dt}\mathbf{i} + \frac{dy}{dt}\mathbf{j} + \frac{dz}{dt}\mathbf{k},$$

which can be rewritten as

$$\mathbf{v} = v_x\mathbf{i} + v_y\mathbf{j} + v_z\mathbf{k}, \tag{4-7}$$

where

$$v_x = \frac{dx}{dt}, \quad v_y = \frac{dy}{dt}, \quad \text{and} \quad v_z = \frac{dz}{dt} \tag{4-8}$$

are the scalar components of $\mathbf{v}$.

Figure 4-3 shows the path of a particle P that is restricted to the *xy* plane. As the particle travels to the right along the curve, its position vector sweeps to the right. At t_1 the position vector is $\mathbf{r}_1$, and at an arbitrary later time $t_1 + \Delta t$ the position vector is $\mathbf{r}_2$. The particle's displacement during Δt is $\Delta \mathbf{r}$. The particle's average velocity $\overline{\mathbf{v}}$ during Δt is, by Eq. 4-4, in the same direction as $\Delta \mathbf{r}$.

Three things happen as we shrink interval Δt toward zero: (1) the vector $\mathbf{r}_2$ in Fig. 4-3 moves toward $\mathbf{r}_1$ so that

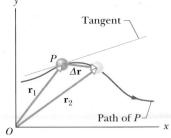

FIGURE 4-3 The position of particle P along its path, both at time t_1 and at a later time $t_1 + \Delta t$. The vector $\Delta \mathbf{r}$ is the displacement of the particle during Δt. The tangent to the particle's path at t_1 is shown.

$\Delta \mathbf{r}$ shrinks toward zero; (2) the direction of $\Delta \mathbf{r}$ (and thus the direction of $\overline{\mathbf{v}}$) approaches the direction of the tangent line in Fig. 4-3; and (3) the average velocity $\overline{\mathbf{v}}$ approaches the instantaneous velocity $\mathbf{v}$.

In the limit as $\Delta t \to 0$, we have $\overline{\mathbf{v}} \to \mathbf{v}$ and, most important here, $\overline{\mathbf{v}}$ takes on the direction of the tangent line. Hence $\mathbf{v}$ has that direction as well. That is:

> The instantaneous velocity $\mathbf{v}$ of a particle is always tangent to the path of the particle.

This is shown in Fig. 4-4, where both $\mathbf{v}$ and its scalar x and y components are included. The result is the same in three dimensions: $\mathbf{v}$ is always tangent to the particle's path.

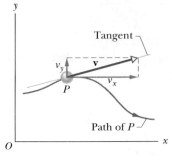

FIGURE 4-4 The velocity $\mathbf{v}$ of particle P along with the scalar components of $\mathbf{v}$. Note that $\mathbf{v}$ lies along the tangent to the path.

CHECKPOINT **2:** The figure shows a circular path taken by a particle. If the instantaneous velocity of the particle is $\mathbf{v} = (2 \text{ m/s})\mathbf{i} - (2 \text{ m/s})\mathbf{j}$, through which quadrant is the particle moving when it is traveling (a) clockwise and (b) counterclockwise around the circle?

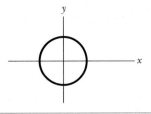

4-4 ACCELERATION AND AVERAGE ACCELERATION

When a particle's velocity changes from $\mathbf{v}_1$ to $\mathbf{v}_2$ in a time period Δt, its average acceleration $\bar{\mathbf{a}}$ during Δt is

$$\bar{\mathbf{a}} = \frac{\mathbf{v}_2 - \mathbf{v}_1}{\Delta t} = \frac{\Delta \mathbf{v}}{\Delta t}. \qquad (4\text{-}9)$$

If we shrink Δt to 0, then in the limit $\bar{\mathbf{a}}$ approaches the (instantaneous) *acceleration* $\mathbf{a}$; that is,

$$\mathbf{a} = \frac{d\mathbf{v}}{dt}. \qquad (4\text{-}10)$$

If the velocity changes in *either* magnitude *or* direction (or both), there is an acceleration.

Substituting $\mathbf{v}$ from Eq. 4-7 into Eq. 4-10 yields

$$\mathbf{a} = \frac{d}{dt}(v_x\mathbf{i} + v_y\mathbf{j} + v_z\mathbf{k})$$

$$= \frac{dv_x}{dt}\mathbf{i} + \frac{dv_y}{dt}\mathbf{j} + \frac{dv_z}{dt}\mathbf{k}$$

or

$$\mathbf{a} = a_x\mathbf{i} + a_y\mathbf{j} + a_z\mathbf{k}, \qquad (4\text{-}11)$$

in which the three scalar components of the acceleration vector are given by

$$a_x = \frac{dv_x}{dt}, \quad a_y = \frac{dv_y}{dt}, \quad \text{and} \quad a_z = \frac{dv_z}{dt}. \qquad (4\text{-}12)$$

Figure 4-5 shows an acceleration vector $\mathbf{a}$ and its scalar components for a particle P moving in two dimensions.

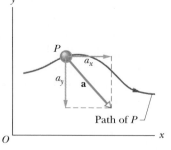

FIGURE 4-5 The acceleration $\mathbf{a}$ of particle P along with the scalar components of $\mathbf{a}$.

SAMPLE PROBLEM 4-2

A rabbit runs across a parking lot on which a set of coordinate axes has, strangely enough, been drawn. The rabbit's path is such that the components of its position with respect to an origin of coordinates are given as functions of time by

$$x = -0.31t^2 + 7.2t + 28$$

and

$$y = 0.22t^2 - 9.1t + 30,$$

with t in seconds and x and y in meters. The rabbit's position vector $\mathbf{r}$ is

$$\mathbf{r}(t) = x(t)\mathbf{i} + y(t)\mathbf{j}.$$

(a) What are the magnitude and direction of the rabbit's position vector at $t = 15$ s?

SOLUTION: At $t = 15$ s, the components of $\mathbf{r}$ are

$$x = (-0.31)(15)^2 + (7.2)(15) + 28 = 66 \text{ m}$$

and

$$y = (0.22)(15)^2 - (9.1)(15) + 30 = -57 \text{ m}.$$

The components and $\mathbf{r}$ itself are shown in Fig. 4-6a.
The magnitude of $\mathbf{r}$ is

$$r = \sqrt{x^2 + y^2} = \sqrt{(66 \text{ m})^2 + (-57 \text{ m})^2}$$
$$= 87 \text{ m}. \qquad \text{(Answer)}$$

The angle θ between $\mathbf{r}$ and the direction of increasing x is

$$\theta = \tan^{-1}\frac{y}{x} = \tan^{-1}\left(\frac{-57 \text{ m}}{66 \text{ m}}\right) = -41°. \quad \text{(Answer)}$$

(Although $\theta = 139°$ has the same tangent as $-41°$, study of the signs of the components of $\mathbf{r}$ rules out 139°.)

(b) Also calculate the position of the rabbit at $t = 0, 5, 10, 20,$ and 25 s and sketch the rabbit's path.

SOLUTION: Proceeding as in (a) leads to the following values of r and θ.

t (s)	x (m)	y (m)	r (m)	θ
0	28	30	41	$+47°$
5	56	-10	57	$-10°$
10	69	-39	79	$-29°$
15	66	-57	87	$-41°$
20	48	-64	80	$-53°$
25	14	-60	62	$-77°$

Figure 4-6b shows a plot of the rabbit's path, drawn with the x and y values.

SAMPLE PROBLEM 4-3

In Sample Problem 4-2, find the magnitude and direction of the rabbit's velocity vector at $t = 15$ s.

SOLUTION: The velocity component along the x axis (see Eq. 4-8) is

$$v_x = \frac{dx}{dt} = \frac{d}{dt}(-0.31t^2 + 7.2t + 28) = -0.62t + 7.2.$$

At $t = 15$ s, this becomes

$$v_x = (-0.62)(15) + 7.2 = -2.1 \text{ m/s}.$$

Similarly,

$$v_y = \frac{dy}{dt} = \frac{d}{dt}(0.22t^2 - 9.1t + 30) = 0.44t - 9.1.$$

At $t = 15$ s, this becomes

$$v_y = (0.44)(15) - 9.1 = -2.5 \text{ m/s}.$$

The vector **v** and its components are shown in Fig. 4-6c.

The magnitude and direction of **v** are

$$v = \sqrt{v_x^2 + v_y^2} = \sqrt{(-2.1 \text{ m/s})^2 + (-2.5 \text{ m/s})^2}$$
$$= 3.3 \text{ m/s} \qquad \text{(Answer)}$$

and

$$\theta = \tan^{-1}\frac{v_y}{v_x} = \tan^{-1}\left(\frac{-2.5 \text{ m/s}}{-2.1 \text{ m/s}}\right)$$
$$= \tan^{-1} 1.19 = -130°. \qquad \text{(Answer)}$$

(Although 50° has the same tangent, inspection of the signs of the velocity components indicates that the desired angle is in the third quadrant, given by 50° − 180° = −130°.) The velocity vector in Fig. 4-6c is tangent to the path of the rabbit and points in the direction in which the rabbit is running at $t = 15$ s.

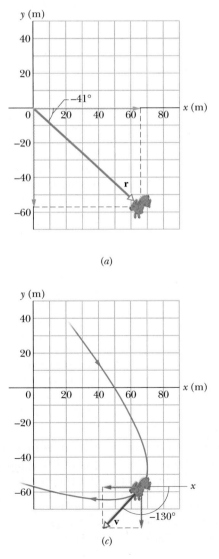

FIGURE 4-6 Sample Problems 4-2, 4-3, and 4-4. (a) The vector **r** and its components at $t = 15$ s. The magnitude of **r** is 87 m. (b) The path of a rabbit across a parking lot, showing the rabbit's position at the indicated times. (c) The velocity **v** of the rabbit at $t = 15$ s. Note that **v** is tangent to the path at the position of the rabbit at $t = 15$ s. (d) The acceleration **a** of the rabbit at $t = 15$ s. The rabbit happens to have this same acceleration for all points of its path.

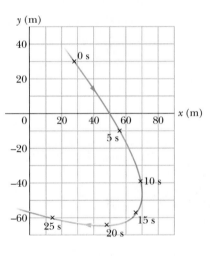

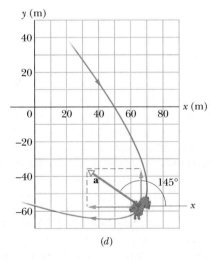

SAMPLE PROBLEM 4-4

In Sample Problem 4-2, determine the magnitude and direction of the rabbit's acceleration vector **a** at $t = 15$ s.

SOLUTION: The acceleration components (see Eq. 4-12) are

$$a_x = \frac{dv_x}{dt} = \frac{d}{dt}(-0.62t + 7.2) = -0.62 \text{ m/s}^2$$

and

$$a_y = \frac{dv_y}{dt} = \frac{d}{dt}(0.44t - 9.1) = 0.44 \text{ m/s}^2.$$

We see that the acceleration does not vary with time; it is a constant. In fact, we have differentiated the time variable completely away. The components of **a**, along with the vector itself, are shown in Fig. 4-6d.

The magnitude and direction of **a** are

$$a = \sqrt{a_x^2 + a_y^2} = \sqrt{(-0.62 \text{ m/s}^2)^2 + (0.44 \text{ m/s}^2)^2}$$

$$= 0.76 \text{ m/s}^2 \qquad \text{(Answer)}$$

and

$$\theta = \tan^{-1}\frac{a_y}{a_x} = \tan^{-1}\left(\frac{0.44 \text{ m/s}^2}{-0.62 \text{ m/s}^2}\right)$$

$$= 145°. \qquad \text{(Answer)}$$

The acceleration vector has the same magnitude and direction for all parts of the rabbit's path. Perhaps a strong southeast wind was blowing across the parking lot, causing the rabbit's acceleration toward the northwest.

CHECKPOINT 3: Here are four descriptions of the position (in meters) of a hockey puck as it moves in the xy plane:
(1) $x = -3t^2 + 4t - 2$ and $y = 6t^2 - 4t$
(2) $x = -3t^3 - 4t$ and $y = -5t^2 + 6$
(3) $\mathbf{r} = 2t^2\mathbf{i} - (4t + 3)\mathbf{j}$
(4) $\mathbf{r} = (4t^3 - 2t)\mathbf{i} + 3\mathbf{j}$
For each description, determine whether the x and y components of the puck's acceleration are constant, and whether the acceleration **a** is constant.

SAMPLE PROBLEM 4-5

A particle with velocity $\mathbf{v}_0 = -2.0\mathbf{i} + 4.0\mathbf{j}$ (in meters per second) at $t = 0$ undergoes a constant acceleration **a** of magnitude $a = 3.0$ m/s^2 at an angle $\theta = 130°$ from the positive direction of the x axis. What is the particle's velocity **v** at $t = 2.0$ s, in unit-vector notation and as a magnitude and direction (with respect to the positive direction of the x axis)?

SOLUTION: Since **a** is constant, Eq. 2-11 ($v = v_0 + at$) applies; it should, however, be used separately to find v_x and v_y

(the x and y components of velocity **v**) because they vary independently of each other. So we write

$$v_x = v_{0x} + a_x t \quad \text{and} \quad v_y = v_{0y} + a_y t.$$

Here v_{0x} (= -2.0 m/s) and v_{0y} (= 4.0 m/s) are the x and y components of $\mathbf{v}_0$, and a_x and a_y are the x and y components of **a**. To find a_x and a_y, we resolve **a** with Eqs. 3-5:

$$a_x = a\cos\theta = (3.0 \text{ m/s}^2)(\cos 130°) = -1.93 \text{ m/s}^2,$$

$$a_y = a\sin\theta = (3.0 \text{ m/s}^2)(\sin 130°) = +2.30 \text{ m/s}^2.$$

When these values are inserted into the equations for v_x and v_y, we find that

$$v_x = -2.0 \text{ m/s} + (-1.93 \text{ m/s}^2)(2.0 \text{ s}) = -5.9 \text{ m/s},$$

$$v_y = 4.0 \text{ m/s} + (2.30 \text{ m/s}^2)(2.0 \text{ s}) = 8.6 \text{ m/s}.$$

So at $t = 2.0$ s, we have, from Eq. 4-7,

$$\mathbf{v} = (-5.9 \text{ m/s})\mathbf{i} + (8.6 \text{ m/s})\mathbf{j}. \qquad \text{(Answer)}$$

The magnitude of **v** is

$$v = \sqrt{(-5.9 \text{ m/s})^2 + (8.6 \text{ m/s})^2}$$

$$= 10 \text{ m/s}. \qquad \text{(Answer)}$$

The angle of **v** is

$$\theta = \tan^{-1}\left(\frac{8.6 \text{ m/s}}{-5.9 \text{ m/s}}\right) = 124° \approx 120°. \qquad \text{(Answer)}$$

Check the last line with your calculator. Does 124° appear on the display, or does $-55.5°$ appear? Now sketch the vector **v** with its components to see which angle is reasonable. To see why some calculators give a mathematically possible but unreasonable result here, reread Tactic 3 in Chapter 3.

PROBLEM SOLVING TACTICS

TACTIC 1: *Trig Functions and Angles*

In Sample Problem 4-3, we computed $\theta = \tan^{-1} 1.19$ and had to find θ. Your calculator might tell you $\theta = 50°$. However, Fig. 3-13c shows that $\theta = 230°$ (= 50° + 180°) has the same tangent. Inspection of the signs of the velocity components v_x and v_y in Fig. 4-6c tells us that this latter angle is the correct one. Some calculators can choose the correct angle for you.

There is still another decision to make. We can stick with 230° or we can relabel it as $-130°$. They are exactly the same angle, as is pointed out in Tactic 1 of Chapter 3. Here, we chose $\theta = -130°$.

TACTIC 2: *Drawing Vectors*

The vectors in Fig. 4-6 were oriented in the following way. (1) Choose the point at which you wish the tail of the vector to be. (2) From that point, draw a line in the direction of increasing x. (3) Using a protractor, mark off the appropriate angle θ, counterclockwise from this line if θ is positive or clockwise if θ is negative.

The vector **r** in Fig. 4-6a should be drawn to the same scale as the two axes because it is a length. The velocity vector **v** in Fig. 4-6c and the acceleration vector **a** in Fig. 4-6d, however, have no established scale and you may make them as long or as short as you wish.

It makes no sense to ask whether, for example, a velocity vector should be longer or shorter than a displacement vector. They are different physical quantities, expressed in different units, and they have no common scale.

CHECKPOINT **4:** If a marble's position is given by $\mathbf{r} = (4t^3 - 2t)\mathbf{i} + 3\mathbf{j}$, with **r** in meters and t in seconds, what must be the units of the coefficients 4, −2, and 3?

4-5 PROJECTILE MOTION

We next consider a particle that moves in a vertical plane during free fall, with its only acceleration being that of the free-fall acceleration **g**, which is downward. Such a **projectile** might be a golf ball (as in Fig. 4-7), a baseball, or any of a variety of other objects. Throughout, we shall assume that the air has no effect on the motion of the projectile.

Figure 4-8, which is analyzed in the next section, shows the path followed by a projectile under such ideal

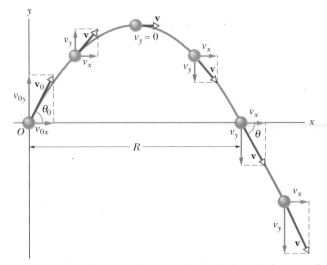

FIGURE 4-8 The path of a projectile that is launched at $x_0 = 0$ and $y_0 = 0$, with an initial velocity $\mathbf{v}_0$. The initial velocity and the velocities at various points along its path are shown, along with their components. Note that the horizontal velocity component remains constant but the vertical velocity component changes continuously. The *range R* is the horizontal distance the projectile has traveled *when it returns to its launch height*.

conditions. The projectile is launched with some initial velocity $\mathbf{v}_0$, which can be written

$$\mathbf{v}_0 = v_{0x}\mathbf{i} + v_{0y}\mathbf{j}. \tag{4-13}$$

The components v_{0x} and v_{0y} can then be found if we know the angle θ_0 between $\mathbf{v}_0$ and the positive x direction:

$$v_{0x} = v_0 \cos \theta_0 \quad \text{and} \quad v_{0y} = v_0 \sin \theta_0. \tag{4-14}$$

During its two-dimensional motion, the projectile's position vector **r** and velocity vector **v** change continuously, but its acceleration vector **a** is constant and *always* directed vertically downward. (The projectile has *no* horizontal acceleration.) As shown in Fig. 4-9, the angle between the direction of the velocity vector and that of the acceleration vector is not constant but varies during the motion.

Projectile motion, like that in Figs. 4-7 through 4-9, looks complicated, but we have the following simplifying feature (known from experiment):

The horizontal motion and the vertical motion are in dependent of each other; that is, neither motion affects the other.

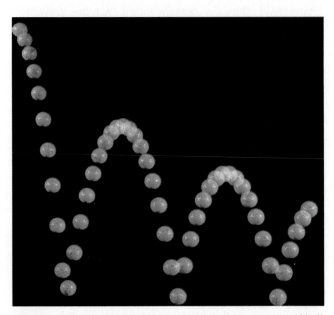

FIGURE 4-7 A stroboscopic photograph of an orange golf ball bouncing off a hard surface. Between impacts, the ball has projectile motion.

This feature allows us to break up a problem involving two-dimensional motion into two separate and easier one-dimensional problems, one for the horizontal motion and

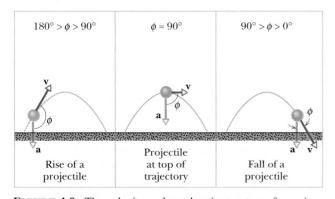

$180° > \phi > 90°$	$\phi = 90°$	$90° > \phi > 0°$
Rise of a projectile	Projectile at top of trajectory	Fall of a projectile

FIGURE 4-9 The velocity and acceleration vectors of a projectile for various motions. Note that the acceleration and velocity vectors do not have any fixed directional relation to each other.

one for the vertical motion. Here are two experiments that show that the horizontal motion and the vertical motion are independent.

Two Golf Balls

Figure 4-10 is a stroboscopic photograph of two golf balls, one simply dropped and the other fired horizontally by a spring mechanism. They have the same vertical motion, each ball falling through the same vertical distance in the

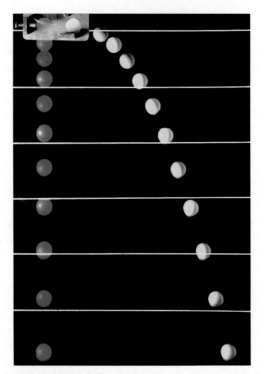

FIGURE 4-10 One ball is released from rest at the same instant that another ball is shot horizontally to the right. Their vertical motions are identical.

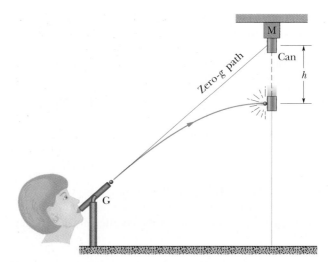

FIGURE 4-11 The projectile ball always hits the falling can. Each falls a distance h from where it would be were there no free-fall acceleration.

same interval of time. *The fact that one ball is moving horizontally while it is falling has no effect on its vertical motion.* To push the experiment to a limit, we say that if you were to fire a rifle horizontally and at the same time drop a bullet, in the absence of air resistance the two bullets would reach level ground at the same time.

A Great Student Rouser

Figure 4-11 shows a demonstration that has enlivened many a physics lecture. It involves a blow gun G, using a ball as a projectile. The target is a can suspended from a magnet M, and the tube of the blow gun is aimed directly at the can. The experiment is arranged so that the magnet releases the can just as the ball leaves the blow gun.

If g (the magnitude of the free-fall acceleration) were zero, the ball would follow the straight line shown in Fig. 4-11 and the can would float in place after the magnet released it. The ball would certainly hit the can.

However, g is *not* zero. The ball *still* hits the can! As Fig. 4-11 shows, during the time of flight of the ball, both ball and can fall the same distance h from their zero-g locations. The harder the demonstrator blows, the greater the ball's initial speed, the shorter the time of flight, and the smaller the value of h.

4-6 PROJECTILE MOTION ANALYZED

Now we are ready to analyze projectile motion, horizontally and vertically:

FIGURE 4-12 The vertical component of this skateboarder's velocity is changing, but not the horizontal component, which matches the skateboard's velocity. As a result, the skateboard stays underneath him, allowing him to land on it.

The Horizontal Motion

Because there is *no acceleration* in the horizontal direction, the horizontal component v_{0x} of the projectile's initial velocity remains unchanged throughout the motion, as demonstrated in Fig. 4-12. The horizontal displacement $x - x_0$ from an initial position x_0 is given by Eq. 2-15 with $a = 0$, which we write as

$$x - x_0 = v_{0x}t.$$

Because $v_{0x} = v_0 \cos \theta_0$, this becomes

$$x - x_0 = (v_0 \cos \theta_0)t. \qquad (4\text{-}15)$$

The Vertical Motion

The vertical motion is the motion we discussed in Section 2-8 for a particle in free fall. Equations 2-21 to 2-25 apply. Equation 2-22, for example, becomes

$$y - y_0 = v_{0y}t - \tfrac{1}{2}gt^2$$
$$= (v_0 \sin \theta_0)t - \tfrac{1}{2}gt^2, \qquad (4\text{-}16)$$

where the initial vertical velocity component v_{0y} is replaced with the equivalent $v_0 \sin \theta_0$. Equations 2-21 and 2-23 are also useful in analyzing projectile motion.

Adapted to our purpose, they are

$$v_y = v_0 \sin \theta_0 - gt \qquad (4\text{-}17)$$

and

$$v_y^2 = (v_0 \sin \theta_0)^2 - 2g(y - y_0). \qquad (4\text{-}18)$$

As is illustrated in Fig. 4-8 and Eq. 4-17, the vertical velocity component behaves just as for a ball thrown vertically upward. It is directed upward initially, its magnitude steadily decreasing to zero, *which marks the maximum height of the path*. The vertical component then reverses direction, and its magnitude becomes larger with time.

The Equation of the Path

We can find the equation of the path (the **trajectory**) of the projectile by eliminating t between Eqs. 4-15 and 4-16. Solving Eq. 4-15 for t and substituting into Eq. 4-16, we obtain, after a little rearrangement,

$$y = (\tan \theta_0)x - \frac{gx^2}{2(v_0 \cos \theta_0)^2} \qquad \text{(trajectory)}. \qquad (4\text{-}19)$$

This is the equation of the path shown in Fig. 4-8. In deriving it, for simplicity we let $x_0 = 0$ and $y_0 = 0$ in Eqs. 4-15 and 4-16, respectively. Because g, θ_0, and v_0 are constants, Eq. 4-19 is of the form $y = ax + bx^2$, in which a and b are constants. This is the equation of a parabola, so the path is *parabolic*.

The Horizontal Range

The *horizontal range R* of the projectile, as Fig. 4-8 shows, is the *horizontal* distance the projectile has traveled when it returns to its initial (launch) height. To find range R, let us put $x - x_0 = R$ in Eq. 4-15 and $y - y_0 = 0$ in Eq. 4-16, obtaining

$$x - x_0 = (v_0 \cos \theta_0)t = R$$

and

$$y - y_0 = (v_0 \sin \theta_0)t - \tfrac{1}{2}gt^2 = 0.$$

Eliminating t between these two equations yields

$$R = \frac{2v_0^2}{g} \sin \theta_0 \cos \theta_0.$$

Using the identity $\sin 2\theta_0 = 2 \sin \theta_0 \cos \theta_0$ (see Appendix E), we obtain

$$R = \frac{v_0^2}{g} \sin 2\theta_0. \qquad (4\text{-}20)$$

Note that R has its maximum value when $\sin 2\theta_0 = 1$, which corresponds to $2\theta_0 = 90°$ or $\theta_0 = 45°$.

The horizontal range R is maximum for a launch angle of 45°.

The Effects of the Air

We have assumed that the air through which the projectile moves has no effect on its motion, a reasonable assumption at low speeds. However, for greater speeds, the disagreement between our calculations and the actual motion of the projectile can be large because the air resists (or opposes) the motion. Figure 4-13, for example, shows two paths for a fly ball that leaves the bat at an angle of 60° with the horizontal and an initial speed of 100 mi/h (data adapted from "The Trajectory of a Fly Ball," by Peter J. Brancazio, *The Physics Teacher,* January 1985). Path I (the baseball player's fly ball) is a calculated path that approximates normal conditions of play, in air. Path II (the physics professor's fly ball) is the path that the ball would follow in a vacuum. Table 4-1 gives some data for the two cases. We shall discuss details of the effect of the air on motion in Chapter 6.

CHECKPOINT **5:** A fly ball is hit to the outfield. During its flight (ignore the effects of air), what happens to its (a) horizontal and (b) vertical components of velocity? What are the (c) horizontal and (d) vertical components of its acceleration during its ascent and its descent, and at the topmost point of its flight?

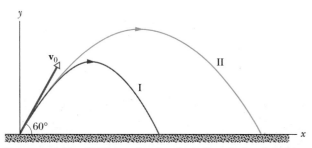

FIGURE 4-13   (I) The path of a fly ball, calculated (using a computer) by taking air resistance into account. (II) The path the ball would follow in a vacuum, calculated by the methods of this chapter. Also see Table 4-1.

TABLE 4-1 TWO FLY BALLS[a]

	PATH I (AIR)	PATH II (VACUUM)
Range	323 ft	581 ft
Maximum height	174 ft	252 ft
Time of flight	6.6 s	7.9 s

[a]See Fig. 4-13. The launch angle is 60° and the launch speed is 100 mi/h.

SAMPLE PROBLEM 4-6

A rescue plane is flying at a constant elevation of 1200 m with a speed of 430 km/h toward a point directly over a person struggling in the water (see Fig. 4-14). At what angle of sight ϕ should the pilot release a rescue capsule if it is to strike (very close to) the person in the water?

SOLUTION: The initial velocity of the capsule is the same as the velocity of the plane at the moment of release. That is, the initial capsule velocity $\mathbf{v}_0$ is horizontal and has a magnitude of 430 km/h. We know the vertical distance the capsule falls, so we can find its time of flight with Eq. 4-16,

$$y - y_0 = (v_0 \sin \theta_0)t - \tfrac{1}{2}gt^2.$$

Putting $y - y_0 = -1200$ m (we use the minus sign because the person is below the origin) and $\theta_0 = 0$, we obtain

$$-1200 \text{ m} = 0 - \tfrac{1}{2}(9.8 \text{ m/s}^2)t^2.$$

Solving for t yields

$$t = \sqrt{\frac{(2)(1200 \text{ m})}{9.8 \text{ m/s}^2}} = 15.65 \text{ s}.$$

The horizontal distance covered by the capsule (and by the plane) in that time is given by Eq. 4-15:

$$\begin{aligned} x - x_0 &= (v_0 \cos \theta_0)t \\ &= (430 \text{ km/h})(\cos 0°)(15.65 \text{ s})(1 \text{ h}/3600 \text{ s}) \\ &= 1.869 \text{ km} = 1869 \text{ m}. \end{aligned}$$

If $x_0 = 0$, then $x = 1869$ m. The angle of sight is then (see Fig. 4-14)

$$\phi = \tan^{-1}\frac{x}{h} = \tan^{-1}\frac{1869 \text{ m}}{1200 \text{ m}} = 57°. \quad \text{(Answer)}$$

Because the plane and the capsule have the same horizontal velocity, the plane remains vertically over the capsule while the capsule is in flight.

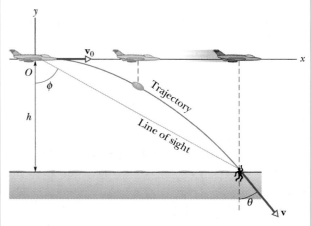

FIGURE 4-14   Sample Problem 4-6. A plane drops a rescue capsule and then continues in level flight. While the capsule is falling, its horizontal velocity component remains equal to the velocity of the plane. The capsule hits the water with velocity **v**, at angle θ from the vertical.

SAMPLE PROBLEM 4-7

A movie stuntman is to run across a rooftop and jump horizontally off it, to land on the roof of the next building (Fig. 4-15). Before he attempts the jump, he wisely asks you to determine whether it is possible. Can he make the jump if his maximum rooftop speed is 4.5 m/s?

SOLUTION: The fall of 4.8 m will take a time t, which can be obtained from Eq. 4-16. Letting $y - y_0 = -4.8$ m (note the sign) and $\theta_0 = 0$, you rearrange Eq. 4-16 to obtain

$$t = \sqrt{-\frac{2(y - y_0)}{g}} = \sqrt{-\frac{(2)(-4.8 \text{ m})}{9.8 \text{ m/s}^2}}$$
$$= 0.990 \text{ s.}$$

You now ask: "How far would he move horizontally in this time?" The answer, from Eq. 4-15, is

$$x - x_0 = (v_0 \cos \theta_0)t$$
$$= (4.5 \text{ m/s})(\cos 0°)(0.990 \text{ s}) = 4.5 \text{ m.}$$

To reach the next building, the stuntman has to move 6.2 m horizontally. Your advice: "Don't jump."

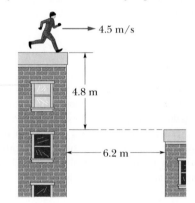

FIGURE 4-15 Sample Problem 4-7. Can the stuntman make the jump?

SAMPLE PROBLEM 4-8

Figure 4-16 shows a pirate ship, moored 560 m from a fort defending the harbor entrance of an island. The harbor defense cannon, located at sea level, has a muzzle velocity of 82 m/s.

(a) To what angle must the cannon be elevated to hit the pirate ship?

SOLUTION: Solving Eq. 4-20 for $2\theta_0$ yields

$$2\theta_0 = \sin^{-1}\frac{gR}{v_0^2} = \sin^{-1}\frac{(9.8 \text{ m/s}^2)(560 \text{ m})}{(82 \text{ m/s})^2}$$
$$= \sin^{-1} 0.816.$$

There are two angles whose sine is 0.816, namely, 54.7° and 125.3°. Thus we find

$$\theta_0 = \tfrac{1}{2}(54.7°) \approx 27° \qquad \text{(Answer)}$$

and $\qquad \theta_0 = \tfrac{1}{2}(125.3°) \approx 63°. \qquad$ (Answer)

The commandant of the fort can elevate the cannon to either of these two angles and (if only there were no intervening air!) hit the pirate ship.

(b) What are the times of flight for the two elevation angles calculated above?

SOLUTION: Solving Eq. 4-15 for t gives, for $\theta_0 = 27°$,

$$t = \frac{x - x_0}{v_0 \cos \theta_0} = \frac{560 \text{ m}}{(82 \text{ m/s})(\cos 27°)}$$
$$= 7.7 \text{ s.} \qquad \text{(Answer)}$$

Repeating the calculation for $\theta_0 = 63°$ yields $t = 15$ s. It is reasonable that the time of flight for the higher elevation angle should be larger.

(c) How far should the pirate ship be from the fort if it is to be beyond range of the cannon?

SOLUTION: We have seen that maximum range corresponds to an elevation angle θ_0 of 45°. Thus, from Eq. 4-20 with $\theta_0 = 45°$,

$$R = \frac{v_0^2}{g} \sin 2\theta_0 = \frac{(82 \text{ m/s})^2}{9.8 \text{ m/s}^2} \sin(2 \times 45°)$$
$$= 690 \text{ m.} \qquad \text{(Answer)}$$

As the pirate ship sails away, the two elevation angles at which the ship can be hit draw closer together, eventually merging at $\theta_0 = 45°$ when the ship is 690 m away. Beyond that distance the ship is safe.

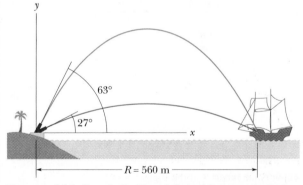
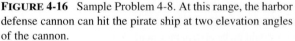

FIGURE 4-16 Sample Problem 4-8. At this range, the harbor defense cannon can hit the pirate ship at two elevation angles of the cannon.

SAMPLE PROBLEM 4-9

Figure 4-17 illustrates the flight of Emanuel Zacchini over three Ferris wheels, located as shown and each 18 m high. Zacchini is launched with speed $v_0 = 26.5$ m/s, at an angle $\theta_0 = 53°$ from the horizontal and with an initial height of 3.0 m above the ground. The net in which he is to land is at the same height.

(a) Does he clear the first Ferris wheel?

SOLUTION: We place the origin on the cannon muzzle so that $x_0 = 0$ and $y_0 = 0$. To find his height y when $x = 23$ m, we use Eq. 4-19:

$$y = (\tan \theta_0)x - \frac{gx^2}{2(v_0 \cos \theta_0)^2}$$

$$= (\tan 53°)(23 \text{ m}) - \frac{(9.8 \text{ m/s}^2)(23 \text{ m})^2}{2(26.5 \text{ m/s})^2(\cos 53°)^2}$$

$$= 20.3 \text{ m.} \qquad \text{(Answer)}$$

Since he begins 3.0 m off the ground, he clears the Ferris wheel by about 5.3 m.

(b) If he reaches his maximum height when he is over the middle Ferris wheel, what is his clearance above it?

SOLUTION: At maximum height, $v_y = 0$ and Eq. 4-18 becomes

$$v_y^2 = (v_0 \sin \theta_0)^2 - 2gy = 0.$$

Solving for y gives

$$y = \frac{(v_0 \sin \theta_0)^2}{2g} = \frac{(26.5 \text{ m/s})^2(\sin 53°)^2}{(2)(9.8 \text{ m/s}^2)} = 22.9 \text{ m,}$$

which means that he clears the middle Ferris wheel by 7.9 m.

(c) What is his time of flight t?

SOLUTION: Of the several ways to find t, we could use Eq. 4-16 and the fact that $y = 0$ when he lands. Then we have

$$y = (v_0 \sin \theta_0)t - \tfrac{1}{2}gt^2 = 0,$$

or $\qquad t = \dfrac{2v_0 \sin \theta_0}{g} = \dfrac{(2)(26.5 \text{ m/s})(\sin 53°)}{9.8 \text{ m/s}^2}$

$$= 4.3 \text{ s.} \qquad \text{(Answer)}$$

(d) How far from the cannon should the center of the net be positioned?

SOLUTION: One way to answer is with Eq. 4-15, with $x_0 = 0$:

$$x = (v_0 \cos \theta_0)t$$
$$= (26.5 \text{ m/s})(\cos 53°)(4.3 \text{ s})$$
$$= 69 \text{ m,} \qquad \text{(Answer)}$$

which is the range R of the flight.

We can now answer the questions that opened this chapter: How could Zacchini know where to place the net? And how could he be certain he would clear the Ferris wheels? He (or someone) did the calculations as we have here. Although he could not take into account the complicated effects of the air on his flight, Zacchini knew that the air would slow him and thus decrease his range from the calculated value. So he used a wide net and biased it toward the cannon. He was then relatively safe whether the effects of the air in a particular flight happened to slow him considerably or very little. Still, the variability of this factor of air effects must have played on his imagination before each flight.

The circus performer still faced a subtle danger: even for shorter flights, his propulsion in the muzzle of the cannon was

so severe that he underwent a momentary blackout. If he landed during the blackout, he could break his neck. So, he had trained himself to awake quickly. Indeed, not waking up in time presents the only real danger to a human cannonball in a circus of today, where a flight is much shorter and safer than Zacchini's.

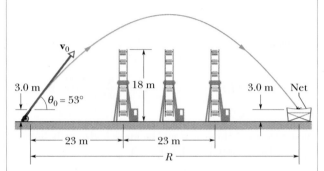

FIGURE 14-17   Sample Problem 4-9. The flight of a human cannonball over three Ferris wheels, and the desired placement of the net.

PROBLEM SOLVING TACTICS

TACTIC 3: *Numbers Versus Algebra*

One way to avoid numerical rounding errors is to solve problems algebraically, substituting numbers only in the final step. That is easy to do in Sample Problems 4-6 to 4-9, and that is the way experienced problem solvers operate. In these early chapters, however, we prefer to solve most problems in pieces, to give you a firmer numerical grasp of what is going on. Later we shall stick more to the algebra.

4-7 UNIFORM CIRCULAR MOTION

A particle is in **uniform circular motion** if it travels around a circle or circular arc at constant speed. Although the speed does not vary, *the particle is accelerating.* That fact may be surprising because we usually think of acceleration as an increase or decrease in speed. But actually **v** is a vector, not a scalar. If **v** changes, even only in direction, there is an acceleration, and that is what happens in uniform circular motion.

We use Fig. 4-18 to find the magnitude and direction of the acceleration. That figure represents a particle in uniform circular motion with speed v in a circle of radius r. Velocity vectors are drawn for two points, p and q, that are symmetric with respect to the y axis. These vectors, $\mathbf{v}_p$ and $\mathbf{v}_q$, have the same magnitude v but—because they point in different directions—they are different vectors. Their x and y components are

$$v_{px} = +v \cos \theta, \qquad v_{py} = +v \sin \theta$$

and $\qquad v_{qx} = +v \cos \theta, \qquad v_{qy} = -v \sin \theta.$

The time required for the particle to move from p to q at constant speed v is

$$\Delta t = \frac{\text{arc}(pq)}{v} = \frac{r(2\theta)}{v}, \qquad (4\text{-}21)$$

in which $\text{arc}(pq)$ is the length of the arc from p to q.

We now have enough information to calculate the components of the average acceleration $\bar{\mathbf{a}}$ experienced by the particle as it moves from p to q in Fig. 4-18. For the x component, we have

$$\bar{a}_x = \frac{v_{qx} - v_{px}}{\Delta t} = \frac{v \cos \theta - v \cos \theta}{\Delta t} = 0.$$

This result is not surprising because it is clear from symmetry in Fig. 4-18 that the x component of velocity has the same value at q and at p.

For the y component of the average acceleration, we find, making use of Eq. 4-21,

$$\bar{a}_y = \frac{v_{qy} - v_{py}}{\Delta t} = \frac{-v \sin \theta - v \sin \theta}{\Delta t}$$

$$= -\frac{2v \sin \theta}{2r\theta/v} = -\left(\frac{v^2}{r}\right)\left(\frac{\sin \theta}{\theta}\right).$$

The minus sign tells us that this acceleration component points vertically downward in Fig. 4-18.

Now let us allow the angle θ in Fig. 4-18 to shrink, approaching zero as a limit. This requires that points p and q in that figure approach their midpoint, shown as point P at the top of the circle. The average acceleration $\bar{\mathbf{a}}$, whose components we have just found, then approaches the instantaneous acceleration $\mathbf{a}$ at point P.

The *direction* of this instantaneous acceleration vector at point P in Fig. 4-18 is downward, toward the center of the circle at O, because the direction of the average acceleration does not change as θ becomes smaller. To find the *magnitude a* of the instantaneous acceleration vector, we need only the mathematical fact that as θ shrinks, the ratio $(\sin \theta)/\theta$ approaches unity. From the relation given above for $\bar{a}_y$, we then have

$$a = \frac{v^2}{r} \qquad \text{(centripetal acceleration).} \qquad (4\text{-}22)$$

We conclude:

When a particle moves at constant speed v in a circle (or a circular arc) of radius r, the acceleration of the particle is directed toward the center of the circle and has a constant magnitude v^2/r.

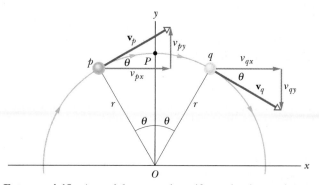

FIGURE 4-18 A particle moves in uniform circular motion at constant speed v in a circle of radius r. Its velocities $\mathbf{v}_p$ and $\mathbf{v}_q$ at points p and q, equidistant from the y axis, are shown, along with its velocity components at those points. The instantaneous acceleration of the particle at any point is directed toward the center of the circle and has a magnitude v^2/r.

In addition, during this acceleration at constant speed, the particle travels the circumference of the circle (a distance of $2\pi r$) in time

$$T = \frac{2\pi r}{v} \qquad \text{(period).} \qquad (4\text{-}23)$$

T is called the *period of revolution*, or simply the *period*, of the motion. It is, in general, the time for a particle to go around a closed path exactly once.

Figure 4-19 shows the relation between the velocity and acceleration vectors at various stages during uniform circular motion. Both vectors have constant magnitude as the motion progresses, but their directions change continuously. The velocity is always tangent to the circle in the direction of motion; the acceleration is always directed radially inward. Because of this, the acceleration associated with uniform circular motion is called a **centripetal** (meaning "center seeking") **acceleration**.

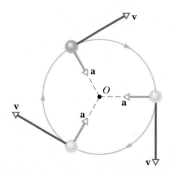

FIGURE 4-19 Velocity and acceleration vectors for a particle in uniform circular motion. Both have constant magnitude but vary continuously in direction.

The acceleration resulting from a change in the direction of a velocity is just as real as one resulting from a change in the magnitude of a velocity. In Fig. 2-8, for example, we saw Colonel John P. Stapp while his rocket sled was braked to rest. He was experiencing a velocity that was constant in direction but changing rapidly in magnitude. On the other hand, an astronaut whirling in a human centrifuge experiences a velocity that is constant in magnitude but changing rapidly in direction. The accelerations that these two people feel are indistinguishable.

CHECKPOINT **6:** An object moves in a circular path in the xy plane, with the center at the origin. When the object is at $x = -2$ m, its velocity is $-(4$ m/s$)\mathbf{j}$. Give the object's (a) velocity and (b) centripetal acceleration when it is at $y = 2$ m.

SAMPLE PROBLEM 4-10

''Top gun'' pilots have long worried about taking a turn too tightly. As a pilot's body undergoes centripetal acceleration, with the head toward the center of curvature, the blood pressure in the brain decreases, leading to loss of brain function.

There are several warning signs to signal a pilot to ease up: when the centripetal acceleration is $2g$ or $3g$, the pilot feels heavy. At about $4g$, the pilot's vision switches to black and white and narrows down to ''tunnel vision.'' If that acceleration is sustained or increased, vision ceases and soon after, the pilot is unconscious—a condition known as g-LOC for ''g-induced loss of consciousness.''

What is the centripetal acceleration, in g units, of a pilot flying an F-22 at speed $v = 1600$ mi/h (716 m/s) through a circular arc with radius of curvature $r = 3.60$ mi (5.80 km)?

SOLUTION: Substituting the given data into Eq. 4-22, we have

$$a = \frac{v^2}{r} = \frac{(716 \text{ m/s})^2}{5800 \text{ m}}$$
$$= 88.39 \text{ m/s}^2 = 9.0g. \qquad \text{(Answer)}$$

If an unwary pilot caught in a dogfight puts the aircraft into such a tight turn, the pilot goes into g-LOC almost immediately, with no warning signs to signal the danger.

SAMPLE PROBLEM 4-11

A satellite is in circular Earth orbit, at an altitude $h = 200$ km above the Earth's surface. There the free-fall acceleration g is 9.20 m/s². What is the orbital speed v of the satellite?

SOLUTION: We have uniform circular motion around the

Earth. The satellite's centripetal acceleration is then the free-fall acceleration g. So we can find v from Eq. 4-22, with $a = g$ and with $r = R_E + h$, where R_E is the Earth's radius (see inside front cover or Appendix C):

$$g = \frac{v^2}{R_E + h}.$$

Solving for v gives

$$v = \sqrt{g(R_E + h)}$$
$$= \sqrt{(9.20 \text{ m/s}^2)(6.37 \times 10^6 \text{ m} + 200 \times 10^3 \text{ m})}$$
$$= 7770 \text{ m/s} = 7.77 \text{ km/s}. \qquad \text{(Answer)}$$

You can show that this is equivalent to 17,400 mi/h and that the satellite would take 1.47 h to complete one orbital revolution; that is, the period T of the motion is 1.47 h.

4-8 RELATIVE MOTION IN ONE DIMENSION

Suppose you see a duck flying north at, say, 20 mi/h. To another duck flying alongside, the first duck is at rest. In other words, the velocity of a particle depends on the **reference frame** of whoever is doing the measuring. For our purposes, a reference frame is the physical object to which you attach your coordinate system.

The reference frame that seems most natural to us in our daily comings and goings is the ground beneath our feet. When a traffic officer tells you that you have been driving at 70 mi/h, the unspoken qualification, ''in a coordinate system attached to the ground,'' is always taken for granted by each of you.

If you are traveling in an airplane or a spaceship, the reference frame of the Earth may not be the most convenient one. You are free to choose any reference frame that you wish. Once having made it, however, you must always be aware of your choice and be careful to make all your measurements with respect to that reference frame.

Suppose that Alex (frame A) is parked by the side of a highway, watching car P (the ''particle'') speed past. Barbara (frame B), driving along the highway at constant speed, is also watching car P. Suppose that, as in Fig. 4-20, they both measure the position of the car at a given moment. From the figure we see that

$$x_{PA} = x_{PB} + x_{BA}. \qquad (4\text{-}24)$$

The terms in Eq. 4-24 are scalars and may be of either sign. The equation is read: ''The coordinate of P as measured by A is equal to the coordinate of P as measured by B plus the coordinate of B as measured by A.'' Note how this reading is supported by the sequence of the subscripts.

Taking the time derivative of Eq. 4-24, we obtain

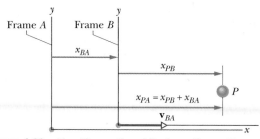

FIGURE 4-20 Alex (frame A) and Barbara (frame B) watch car P. All motion is along the common x axis of the two frames. The vector $\mathbf{v}_{BA}$ shows the relative separation velocity of the two frames. The three position measurements shown are all made at the same instant of time.

$$\frac{d}{dt}(x_{PA}) = \frac{d}{dt}(x_{PB}) + \frac{d}{dt}(x_{BA}),$$

or (because $v = dx/dt$)

$$v_{PA} = v_{PB} + v_{BA}. \qquad (4\text{-}25)$$

This scalar equation is the relation between the velocities of the same object (car P) as measured in the two frames; those measured velocities are different. In words, Eq. 4-25 says: "The velocity of P as measured by A *is equal to* the velocity of P as measured by B *plus* the velocity of B as measured by A." The term v_{BA} is the separation velocity of frame B with respect to frame A; see Fig. 4-20.

We consider only frames that move at constant velocity; such frames are called **inertial reference frames.** In our example, this means that Barbara (frame B) will drive always at constant speed with respect to Alex (frame A). Car P (the moving particle), however, may speed up, slow down, come to rest, or reverse direction.

The time derivative of the velocity equation (Eq. 4-25) yields the acceleration equation,

$$a_{PA} = a_{PB}. \qquad (4\text{-}26)$$

Note that because v_{BA} is a constant, its time derivative is zero. Equation 4-26 tells us:

Observers on different inertial frames of reference (their velocity of separation is constant) will measure the same acceleration for a moving particle.

CHECKPOINT 7: The table gives velocities (km/h) for Barbara and car P of Fig. 4-20 for three situations. For each, what is the missing value and how is the distance between Barbara and car P changing?

SITUATION	v_{BA}	v_{PA}	v_{PB}
1	+50	+50	
2	+30		+40
3		+60	−20

SAMPLE PROBLEM 4-12

Alex, parked by the side of an east–west road, is watching car P, which is moving in a westerly direction. Barbara, driving east at a speed $v_{BA} = 52$ km/h, watches the same car. Take the easterly direction as positive.

(a) If Alex measures a speed of 78 km/h for car P, what velocity will Barbara measure?

SOLUTION: Equation 4-25 may be rearranged to yield

$$v_{PB} = v_{PA} - v_{BA}.$$

We have $v_{PA} = -78$ km/h, the minus sign telling us that car P is moving west, in the negative direction. We also have $v_{BA} = 52$ km/h, so that

$$v_{PB} = (-78 \text{ km/h}) - (52 \text{ km/h})$$
$$= -130 \text{ km/h}. \qquad \text{(Answer)}$$

If car P were connected to Barbara's car by a string wound up on a spool, the string would be unwinding at this speed as the two cars separated.

(b) If Alex sees car P brake to a halt in 10 s, what acceleration (assumed constant) will he measure for it?

SOLUTION: From Eq. 2-11 ($v = v_0 + at$) we have

$$a = \frac{v - v_0}{t} = \frac{0 - (-78 \text{ km/h})}{10 \text{ s}}$$
$$= \left(\frac{78 \text{ km/h}}{10 \text{ s}}\right)\left(\frac{1 \text{ m/s}}{3.6 \text{ km/h}}\right)$$
$$= 2.2 \text{ m/s}^2. \qquad \text{(Answer)}$$

(c) What acceleration would Barbara measure for the braking car?

SOLUTION: Barbara sees the initial speed of the car as -130 km/h, as we calculated in (a). Although the car has braked to rest, it is at rest only in Alex's reference frame. To Barbara, car P is not at rest at all but is receding from her at 52 km/h so that its final velocity in her reference frame is -52 km/h. Thus, from the relation $v = v_0 + at$,

$$a = \frac{v - v_0}{t} = \frac{(-52 \text{ km/h}) - (-130 \text{ km/h})}{10 \text{ s}}$$
$$= 2.2 \text{ m/s}^2. \qquad \text{(Answer)}$$

This is exactly the same acceleration that Alex measured, which reassures us that we made no mistakes.

4-9 RELATIVE MOTION IN TWO DIMENSIONS

Now we move from the scalar world of relative motion in one dimension to the vector world of relative motion in two (and, by extension, in three) dimensions. Figure 4-21 shows reference frames A and B, now two-dimensional. Our two observers are again watching a moving particle P. We once more assume that the two frames are separating at a constant velocity $\mathbf{v}_{BA}$ (the frames are inertial) and, for simplicity, we further assume that their x and y axes remain parallel to each other as they do so.

Let observers on frames A and B each measure the position of particle P at a certain instant. From the vector triangle in Fig. 4-21, we have the vector equation

$$\mathbf{r}_{PA} = \mathbf{r}_{PB} + \mathbf{r}_{BA}. \qquad (4\text{-}27)$$

This relation is the vector equivalent of the scalar Eq. 4-24.

If we take the time derivative of Eq. 4-27, we find a connection between the (vector) velocities of the particle as measured by our two observers; namely,

$$\mathbf{v}_{PA} = \mathbf{v}_{PB} + \mathbf{v}_{BA}. \qquad (4\text{-}28)$$

This is the vector equivalent of the scalar Eq. 4-25. Note that the order of the subscripts is the same as in that equation, and that again $\mathbf{v}_{BA}$ is the constant relative velocity of frame B as observed by the observer on frame A.

If we take the time derivative of Eq. 4-28, we obtain a connection between the two measured accelerations; namely,

$$\mathbf{a}_{PA} = \mathbf{a}_{PB}. \qquad (4\text{-}29)$$

It remains true for motion in three dimensions that all observers on inertial reference frames will measure the same acceleration for a moving particle.

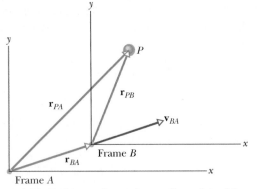

FIGURE 4-21 Reference frames in two dimensions. The vectors $\mathbf{r}_{PA}$ and $\mathbf{r}_{PB}$ show the positions of particle P in frames A and B, respectively. Vector $\mathbf{r}_{BA}$ shows the position of frame B with respect to frame A. Vector $\mathbf{v}_{BA}$ shows the (constant) separation velocity of the two frames.

SAMPLE PROBLEM 4-13

A bat detects an insect (lunch) while the two are flying with velocities $\mathbf{v}_{BG}$ and $\mathbf{v}_{IG}$, respectively, measured with respect to the ground. See Fig. 4-22a. What is the velocity $\mathbf{v}_{IB}$ of the insect with respect to the bat, in unit-vector notation?

SOLUTION: From Fig. 4-22a, the velocities of the insect and the bat, relative to the ground, are given by

$$\mathbf{v}_{IG} = (5.0 \text{ m/s})(\cos 50°)\mathbf{i} + (5.0 \text{ m/s})(\sin 50°)\mathbf{j}$$

and

$$\mathbf{v}_{BG} = (4.0 \text{ m/s})(\cos 150°)\mathbf{i} + (4.0 \text{ m/s})(\sin 150°)\mathbf{j},$$

where in each term, the angle is relative to the positive direction of the x axis. Now here is the key idea: the velocity $\mathbf{v}_{IB}$ of the *insect relative to the bat* is given by the vector sum of the velocity $\mathbf{v}_{IG}$ of the *insect relative to the ground* and the velocity $\mathbf{v}_{GB}$ of the *ground relative to the bat*; that is,

$$\mathbf{v}_{IB} = \mathbf{v}_{IG} + \mathbf{v}_{GB},$$

as shown in Fig. 4-22b. (Note that on the right-hand side of the equation, the two inner subscripts are the same. Also note that the two outer ones are the same as those on the left-hand side and appear in the same order.)

By definition the vector $\mathbf{v}_{GB}$ is in the direction opposite that of the vector $\mathbf{v}_{BG}$. Hence $\mathbf{v}_{GB} = -\mathbf{v}_{BG}$, and so

$$\mathbf{v}_{IB} = \mathbf{v}_{IG} + (-\mathbf{v}_{BG}).$$

Substituting the unit-vector expressions for $\mathbf{v}_{IG}$ and $\mathbf{v}_{BG}$ into this expression, we find

$$\begin{aligned}
\mathbf{v}_{IB} &= (5.0 \text{ m/s})(\cos 50°)\mathbf{i} + (5.0 \text{ m/s})(\sin 50°)\mathbf{j} \\
&\quad - (4.0 \text{ m/s})(\cos 150°)\mathbf{i} - (4.0 \text{ m/s})(\sin 150°)\mathbf{j} \\
&= 3.21\mathbf{i} + 3.83\mathbf{j} + 3.46\mathbf{i} - 2.0\mathbf{j} \\
&\approx (6.7 \text{ m/s})\mathbf{i} + (1.8 \text{ m/s})\mathbf{j}. \qquad \text{(Answer)}
\end{aligned}$$

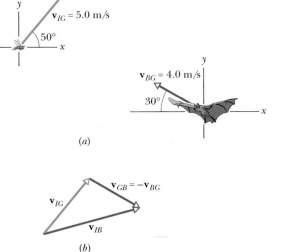

FIGURE 4-22 Sample Problem 4-13. (*a*) A bat detects an insect. (*b*) Velocity vectors of insect and bat.

SAMPLE PROBLEM 4-14

The compass in a plane indicates that the plane is headed (pointed) due east; its airspeed indicator reads 215 km/h. (Airspeed is speed relative to the air.) A steady wind of 65.0 km/h is blowing due north.

(a) What is the velocity of the plane with respect to the ground?

SOLUTION: The moving ''particle'' in this problem is the plane P. There are two reference frames, the ground (G) and the air mass (M). By a simple change of notation, we can rewrite Eq. 4-28 as

$$\mathbf{v}_{PG} = \mathbf{v}_{PM} + \mathbf{v}_{MG}. \qquad (4\text{-}30)$$

Figure 4-23a shows these vectors, which form a right triangle. The terms in Eq. 4-30 are, in sequence, the velocity of the plane with respect to the ground, the velocity of the plane with respect to the air, and the velocity of the air with respect to the ground (that is, the wind velocity). Note that the orientation of the plane in Fig. 4-23a is consistent with a due east reading on its compass. The plane is pointed due east but may not actually be moving in that direction.

The magnitude of the plane's velocity relative to the ground is found from the triangle in Fig. 4-23a:

$$\begin{aligned} v_{PG} &= \sqrt{v_{PM}^2 + v_{MG}^2} \\ &= \sqrt{(215 \text{ km/h})^2 + (65.0 \text{ km/h})^2} \\ &= 225 \text{ km/h}. \qquad \text{(Answer)} \end{aligned}$$

The angle α in Fig. 4-23a follows from

$$\begin{aligned} \alpha &= \tan^{-1}\frac{v_{MG}}{v_{PM}} = \tan^{-1}\frac{65.0 \text{ km/h}}{215 \text{ km/h}} \\ &= 16.8°. \qquad \text{(Answer)} \end{aligned}$$

Thus, with respect to the ground, the plane is flying at 225 km/h in a direction 16.8° north of east. Note that its speed relative to the ground (the ''ground speed'') is greater than its airspeed.

(b) If the pilot wishes to fly due east, what must be the heading? That is, what must the compass read?

SOLUTION: In this case the pilot must head into the wind somewhat, so that the wind velocity will help ensure that the velocity of the plane with respect to the ground points east. The wind velocity is unchanged from the preceding situation (a), and the vector diagram representing the new situation is as shown in Fig. 4-23b. Note that the three vectors still form a right triangle, as they did in Fig. 4-23a, and Eq. 4-30 still holds.

The pilot's ground speed is now, from Fig. 4-23b,

$$\begin{aligned} v_{PG} &= \sqrt{v_{PM}^2 - v_{MG}^2} \\ &= \sqrt{(215 \text{ km/h})^2 - (65.0 \text{ km/h})^2} = 205 \text{ km/h}. \end{aligned}$$

As the orientation of the plane in Fig. 4-23b indicates, the pilot must head into the wind by an angle θ given by

$$\theta = \sin^{-1}\frac{v_{MG}}{v_{PM}} = \sin^{-1}\frac{65.0 \text{ km/h}}{215 \text{ km/h}} = 17.6°. \quad \text{(Answer)}$$

Note that the ground speed is now less than the airspeed because the plane is headed into the wind.

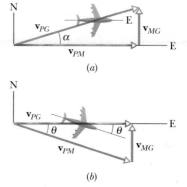

(a)

(b)

FIGURE 4-23 Sample Problem 4-14. (a) An airplane, heading due east, is blown to the north. (b) To travel due east, the plane must head into the wind.

4-10 RELATIVE MOTION AT HIGH SPEEDS (OPTIONAL)

An orbiting satellite has a speed of 17,000 mi/h. Before you call that a high speed, you must answer this question: ''High compared to what?'' Nature has given us a standard; it is the speed of light c, where

$$c = 299{,}792{,}458 \text{ m/s} \qquad \text{(speed of light)}. \quad (4\text{-}31)$$

As we shall examine later, no entity—be it particle or wave—can move faster than the speed of light, no matter what reference frame is used for observation. By this standard, all tangible large-scale objects—no matter how high their speeds seem by ordinary standards—are very slow indeed. The speed of the orbiting satellite, for example, is only 0.0025% of the speed of light. Subatomic particles such as electrons or protons, however, can acquire speeds very close to (but never equal to or greater than) the speed of light. Experiment shows, for example, that an electron accelerated through 10 million volts acquires a speed of 0.9988c; if you accelerate it through 20 million volts, its speed increases, but only to 0.9997c. The speed of light is a barrier that objects can approach but never reach. (Alas, the faster-than-light speeds utilized in science fiction, such as during ''warp drive'' in *Star Trek*, when the speed is $c2^n$, where n is the warp number, are only fictional.)

How can we be sure that the kinematics we have examined so far, which was developed by studying very slow, ordinary objects, also holds for very fast objects, such as high-speed electrons or protons? The answer, which we can find only from experiment, is that the kinematics for slow objects does *not* hold at speeds that approach the speed of light. Einstein's **theory of special relativity,** however, agrees with experiment at *all* speeds.

At "slow" speeds—all speeds that can be acquired by ordinary objects—the kinematic equations of Einstein's theory reduce to those of the kinematics we have examined. The failure of the "slow kinematics" is gradual, its predictions agreeing less and less well with experiment as the speed increases. Here is an example: Eq. 4-25,

$$v_{PA} = v_{PB} + v_{BA} \qquad \text{(slow speeds),}$$

relates the speed of particle P as seen by an observer in frame B to that seen by an observer in frame A. The corresponding equation in Einstein's theory is

$$v_{PA} = \frac{v_{PB} + v_{BA}}{1 + v_{PB}v_{BA}/c^2} \qquad \text{(all speeds).} \quad (4\text{-}32)$$

If $v_{PB} \ll c$ and $v_{BA} \ll c$ (which is always the case for ordinary objects), then the denominator in Eq. 4-32 is very close to unity and Eq. 4-32 reduces to Eq. 4-25, as we know it must.

The speed of light c is the central constant of Einstein's theory, and every relativistic equation contains it. A way to test the validity of such equations is to allow c to become infinitely large. Under those conditions, *all* speeds would be "slow" and "slow kinematics" would never fail. Putting $c \rightarrow \infty$ in Eq. 4-32 does indeed reduce that equation to Eq. 4-25.

SAMPLE PROBLEM 4-15

(Slow speeds)   For the case where $v_{PB} = v_{BA} = 0.0001c$ ($= 67,000$ mi/h!), what do Eqs. 4-25 and 4-32 predict for v_{PA}?

SOLUTION: From Eq. 4-25,

$$v_{PA} = v_{PB} + v_{BA} = 0.0001c + 0.0001c$$
$$= 0.0002c. \qquad \text{(Answer)}$$

From Eq. 4-32,

$$v_{PA} = \frac{v_{PB} + v_{BA}}{1 + v_{PB}v_{BA}/c^2} = \frac{0.0001c + 0.0001c}{1 + (0.0001c)^2/c^2}$$
$$= \frac{0.0002c}{1.00000001} \approx 0.0002c. \qquad \text{(Answer)}$$

Conclusion: At any speed acquired by ordinary tangible objects, Eqs. 4-25 and 4-32 yield essentially the same answer. For such speeds, we can use Eq. 4-25 ("slow kinematics") without a second thought.

SAMPLE PROBLEM 4-16

(High speeds)   For $v_{PB} = v_{BA} = 0.65c$, what do Eqs. 4-25 and 4-32 predict for v_{PA}?

SOLUTION: From Eq. 4-25,

$$v_{PA} = v_{PB} + v_{BA} = 0.65c + 0.65c$$
$$= 1.30c. \qquad \text{(Answer)}$$

From Eq. 4-32,

$$v_{PA} = \frac{v_{PB} + v_{BA}}{1 + v_{PB}v_{BA}/c^2} = \frac{0.65c + 0.65c}{1 + (0.65c)(0.65c)/c^2}$$
$$= \frac{1.30c}{1.423} = 0.91c. \qquad \text{(Answer)}$$

Conclusion: At high speeds, "slow kinematics" and special relativity predict very different results. "Slow kinematics" involves no upper limit on speed and can easily (as in this case) predict a speed greater than the speed of light. Special relativity, on the other hand, *never* predicts a speed that exceeds c, no matter how high the combining speeds. Experiment agrees with special relativity on all counts.

REVIEW & SUMMARY

Position Vector

The location of a particle relative to the origin of a coordinate system is given by a *position vector* **r**, which in unit-vector notation is

$$\mathbf{r} = x\mathbf{i} + y\mathbf{j} + z\mathbf{k}. \qquad (4\text{-}1)$$

Here $x\mathbf{i}$, $y\mathbf{j}$, and $z\mathbf{k}$ are the *vector components* of position vector **r**, and x, y, and z are its *scalar components*. A position vector is described by a magnitude and one or two angles for orientation, or by its vector or scalar components.

Displacement

If a particle moves so that its position vector changes from $\mathbf{r}_1$ to $\mathbf{r}_2$, then the particle's *displacement* $\Delta\mathbf{r}$ is

$$\Delta\mathbf{r} = \mathbf{r}_2 - \mathbf{r}_1. \qquad (4\text{-}2)$$

The displacement can also be written as

$$\Delta\mathbf{r} = (x_2 - x_1)\mathbf{i} + (y_2 - y_1)\mathbf{j} + (z_2 - z_1)\mathbf{k}, \qquad (4\text{-}3)$$

where coordinates (x_1, y_1, z_1) correspond to position vector $\mathbf{r}_1$ and coordinates (x_2, y_2, z_2) correspond to position vector $\mathbf{r}_2$.

Average Velocity

If a particle undergoes a displacement $\Delta\mathbf{r}$ in time Δt, its *average velocity* $\overline{\mathbf{v}}$ for that time interval is

$$\overline{\mathbf{v}} = \frac{\Delta\mathbf{r}}{\Delta t}. \tag{4-4}$$

Velocity

As Δt in Eq. 4-4 is shrunk to 0, $\overline{\mathbf{v}}$ reaches a limit called the (instantaneous) *velocity:*

$$\mathbf{v} = \frac{d\mathbf{r}}{dt}, \tag{4-6}$$

which can be rewritten in unit-vector notation as

$$\mathbf{v} = v_x\mathbf{i} + v_y\mathbf{j} + v_z\mathbf{k}, \tag{4-7}$$

where $v_x = dx/dt$, $v_y = dy/dt$, and $v_z = dz/dt$. When the position of a moving particle is plotted on a coordinate system, $\mathbf{v}$ is always tangent to the curve representing the particle's path.

Average Acceleration

If a particle's velocity changes from $\mathbf{v}_1$ to $\mathbf{v}_2$ in time interval Δt, its *average acceleration* during Δt is

$$\overline{\mathbf{a}} = \frac{\mathbf{v}_2 - \mathbf{v}_1}{\Delta t} = \frac{\Delta\mathbf{v}}{\Delta t}. \tag{4-9}$$

Acceleration

As Δt in Eq. 4-9 is shrunk to 0, $\overline{\mathbf{a}}$ reaches a limiting value called the (instantaneous) *acceleration,* which is

$$\mathbf{a} = \frac{d\mathbf{v}}{dt}. \tag{4-10}$$

In unit-vector notation,

$$\mathbf{a} = a_x\mathbf{i} + a_y\mathbf{j} + a_z\mathbf{k}, \tag{4-11}$$

where $a_x = dv_x/dt$, $a_y = dv_y/dt$, and $a_z = dv_z/dt$.

When $\mathbf{a}$ is constant, the components of $\mathbf{a}$, $\mathbf{v}$, and $\mathbf{r}$ along any axis can be treated as in the one-dimensional motion of Chapter 2.

Projectile Motion

Projectile motion is the motion of a particle that is launched with an initial velocity $\mathbf{v}_0$ and then has the free-fall acceleration $\mathbf{g}$. If $\mathbf{v}_0$ is expressed as a magnitude (the speed v_0) and an angle θ_0, the equations of motion along the horizontal x axis and vertical y axis are

$$x - x_0 = (v_0 \cos \theta_0)t, \tag{4-15}$$

$$y - y_0 = (v_0 \sin \theta_0)t - \tfrac{1}{2}gt^2, \tag{4-16}$$

$$v_y = v_0 \sin \theta_0 - gt, \tag{4-17}$$

$$v_y^2 = (v_0 \sin \theta_0)^2 - 2g(y - y_0). \tag{4-18}$$

The path of a particle in projectile motion is parabolic and is given by

$$y = (\tan \theta_0)x - \frac{gx^2}{2(v_0 \cos \theta_0)^2}, \tag{4-19}$$

where the origin has been chosen so that x_0 and y_0 are both zero. The particle's **range** R, which is the horizontal distance from the launch point to the point at which the particle returns to the launch height, is

$$R = \frac{v_0^2}{g} \sin 2\theta_0. \tag{4-20}$$

Uniform Circular Motion

If a particle travels along a circle or circular arc with radius r at constant speed v, it is in *uniform circular motion* and has an acceleration $\mathbf{a}$ of magnitude

$$a = \frac{v^2}{r}. \tag{4-22}$$

The direction of $\mathbf{a}$ is toward the center of the circle or circular arc, and $\mathbf{a}$ is said to be centripetal. The time for the particle to complete a circle is

$$T = \frac{2\pi r}{v}. \tag{4-23}$$

T is called the *period of revolution,* or simply the *period,* of the motion.

Relative Motion

When two frames of reference A and B are moving relative to each other at constant velocity, they are said to be **inertial reference frames.** The velocity of a moving particle as measured by an observer in frame A, in general, differs from that measured from frame B. The two measured velocities are related by

$$\mathbf{v}_{PA} = \mathbf{v}_{PB} + \mathbf{v}_{BA}, \tag{4-28}$$

in which $\mathbf{v}_{BA}$ is the velocity of B with respect to A. Both observers measure the same acceleration for the particle; that is,

$$\mathbf{a}_{PA} = \mathbf{a}_{PB}. \tag{4-29}$$

If speeds near the speed of light are involved, Eq. 4-25 ($v_{PA} = v_{PB} + v_{BA}$) must be replaced with an equation derived using the **special theory of relativity.** For one-dimensional motion, the correct result is

$$v_{PA} = \frac{v_{PB} + v_{BA}}{1 + v_{PB}v_{BA}/c^2}, \tag{4-32}$$

which becomes identical to Eq. 4-25 if the speeds are all negligible compared to the speed of light c.

QUESTIONS

1. Here are four descriptions for the velocity of a hockey puck in the xy plane, all in meters per second:

(1) $v_x = -3t^2 + 4t - 2$ and $v_y = 6t - 4$
(2) $v_x = -3$ and $v_y = -5t^2 + 6$
(3) $\mathbf{v} = 2t^2\mathbf{i} - (4t + 3)\mathbf{j}$
(4) $\mathbf{v} = -2t\mathbf{i} + 3\mathbf{j}$

For each description, determine whether the x and y components of the acceleration are constant, and whether the acceleration vector **a** is constant. In description 4, if **v** is in meters per second and t is in seconds, what must be the units of the coefficients -2 and 3?

2. Figure 4-24 shows three situations in which identical projectiles are launched from the ground (at the same levels) at identical initial speeds and angles. The projectiles do not land on the same terrain, however. Rank the situations according to the final speeds of the projectiles just before they land, greatest first.

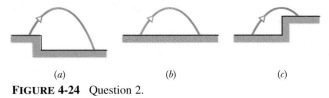

(a) (b) (c)

FIGURE 4-24 Question 2.

3. At what point in the path of a projectile is the speed (a) a minimum and (b) a maximum?

4. At a certain instant, a fly ball has velocity $\mathbf{v} = 25\mathbf{i} - 4.9\mathbf{j}$ (x is horizontal, y is upward, and **v** is in meters per second). Has the ball passed the highest point of its trajectory?

5. You are to launch a rocket, from just above the ground, with one of the following initial velocity vectors: (1) $\mathbf{v}_0 = 20\mathbf{i} + 70\mathbf{j}$, (2) $\mathbf{v}_0 = -20\mathbf{i} + 70\mathbf{j}$, (3) $\mathbf{v}_0 = 20\mathbf{i} - 70\mathbf{j}$, (4) $\mathbf{v}_0 = -20\mathbf{i} - 70\mathbf{j}$. In your coordinate system, x runs along level ground and y increases upward. (a) Rank the vectors according to the launch speed of the projectile, greatest first. (b) Rank the vectors according to the time of flight of the projectile, greatest first.

6. A snowball is thrown from ground level (by someone in a hole) with initial speed v_0 at an angle of 45° relative to the (level) ground, on which the snowball later lands. If the launch angle is increased, do (a) the range and (b) the flight time increase, decrease, or stay the same?

7. A mud ball is launched 2 m above the ground with initial velocity $\mathbf{v}_0 = (2\mathbf{i} + 4\mathbf{j})$ m/s. What is its velocity just before it lands on a surface that is 2 m above the ground?

8. An airplane flying horizontally at a constant speed of 350 km/h over level ground releases a bundle of food supplies. Ignore the effect of the air on the bundle. What are the bundle's initial (a) vertical and (b) horizontal components of velocity? (c) What is its horizontal component of velocity just before hitting the ground? (d) If the airplane's speed were, instead, 450 km/h, would the time of fall be larger, smaller, or the same?

9. Figure 4-25 shows three paths for a kicked football. Ignoring the effects of air on the flight, rank the paths according to (a) time of flight, (b) initial vertical velocity component, (c) initial horizontal velocity component, and (d) initial speed. Place the greatest first in each part.

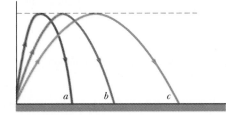

FIGURE 4-25 Question 9.

10. Figure 4-26 shows the velocity and acceleration of a particle at a particular instant in three situations. In which situation, and at that instant, is (a) the speed increasing, (b) the speed decreasing, (c) the speed not changing, (d) $\mathbf{v} \cdot \mathbf{a}$ positive, (e) $\mathbf{v} \cdot \mathbf{a}$ negative, and (f) $\mathbf{v} \cdot \mathbf{a} = 0$?

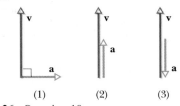

(1) (2) (3)

FIGURE 4-26 Question 10.

11. You are driving directly behind a pickup truck, going at the same speed as the truck. A crate falls from the bed of the truck to the road. (a) Will your car hit the crate before the crate hits the road if you neither brake nor swerve? (b) During the fall, is the horizontal speed of the crate more than, less than, or the same as that of the truck?

12. (a) Is it possible to be accelerating while traveling at constant speed? Is it possible to round a curve with (b) zero acceleration and (c) a constant magnitude of acceleration?

13. While riding in a moving car, you toss an egg directly upward. Does the egg tend to land behind you, in front of you, or back in your hands if the car is (a) traveling at a constant speed, (b) increasing in speed, and (c) decreasing in speed?

14. A passenger in an elevator drops a coin while the elevator descends at constant speed. According to (a) the passenger and (b) someone waiting at the next floor, is the acceleration of the coin equal to, less than, or more than g?

15. A pickpocket standing on the rear observation platform of a train moving with constant velocity drops a wallet over the rear guardrail. Describe the path of the wallet as seen by (a) the pickpocket, (b) her accomplice waiting alongside the track at the drop site, and (c) a police officer on another train moving at constant speed in the direction opposite that of the pickpocket's train, on a parallel track.

16. When the Germans shelled Paris from 70 mi away with the WWI long-range artillery piece nicknamed ''Big Bertha,'' the shells were fired at an angle greater than 45°; the Germans had discovered that a greater angle gave their gun a greater range, possibly even twice as long as that with a 45° angle. Does that result mean that the density of the air at high altitudes increases with altitude or decreases with altitude?

17. Figure 4-27 shows one of four star cruisers that are in a race. As each cruiser passes the starting line, a shuttle craft leaves the cruiser and races toward the finish line. You are the judge and are stationary relative to the start and finish lines. The speeds v_c of the four cruisers relative to you and the speeds v_s of the shuttles relative to their corresponding cruisers are, respectively: (1)

0.70c, 0.40c; (2) 0.40c, 0.70c; (3) 0.20c, 0.90c; (4) 0.50c, 0.60c. Without written calculation, determine (a) which shuttle wins the race and (b) which is last.

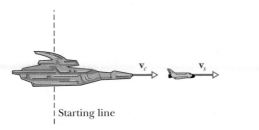

FIGURE 4-27   Question 17.

EXERCISES & PROBLEMS

SECTION 4-2 Position and Displacement

1E. A watermelon has the following coordinates: $x = -5.0$ m, $y = 8.0$ m, and $z = 0$ m. Find its position vector (a) in unit-vector notation and (b) in terms of the vector's magnitude and orientation. (c) Sketch the vector on a right-handed coordinate system. If the watermelon is moved to the xyz coordinates (3.00 m, 0 m, 0 m), what is its displacement (d) in unit-vector notation and (e) in terms of a magnitude and an orientation?

2E. The position vector for an electron is $\mathbf{r} = 5.0\mathbf{i} - 3.0\mathbf{j} + 2.0\mathbf{k}$, in meters. (a) Find the magnitude of $\mathbf{r}$. (b) Sketch the vector on a right-handed coordinate system.

3E. The position vector for a proton is initially $\mathbf{r} = 5.0\mathbf{i} - 6.0\mathbf{j} + 2.0\mathbf{k}$ and then later is $\mathbf{r} = -2.0\mathbf{i} + 6.0\mathbf{j} + 2.0\mathbf{k}$, all in meters. (a) What is the proton's displacement vector, and (b) to what plane is it parallel?

4E. A positron undergoes a displacement $\Delta\mathbf{r} = 2.0\mathbf{i} - 3.0\mathbf{j} + 6.0\mathbf{k}$, ending with the position vector $\mathbf{r} = 3.0\mathbf{j} - 4.0\mathbf{k}$, in meters. What was the positron's former position vector?

SECTION 4-3 Velocity and Average Velocity

5E. A plane flies 300 mi east from city A to city B in 45.0 min and then 600 mi south from city B to city C in 1.50 h. (a) What displacement vector represents the total trip? What are (b) the average velocity (a vector) and (c) the average speed for the trip?

6E. A train moving at a constant speed of 60.0 km/h moves east for 40.0 min, then in a direction 50.0° east of north for 20.0 min, and finally west for 50.0 min. What is the average velocity of the train during this run?

7E. In 3.50 h, a balloon drifts 21.5 km north, 9.70 km east, and 2.88 km in upward elevation from its release point on the ground. Find (a) the magnitude of its average velocity and (b) the angle its average velocity makes with the horizontal.

8E. An ion's position vector is initially $\mathbf{r} = 5.0\mathbf{i} - 6.0\mathbf{j} + 2.0\mathbf{k}$,

and 10 s later it is $\mathbf{r} = -2.0\mathbf{i} + 8.0\mathbf{j} - 2.0\mathbf{k}$, all in meters. What was its average velocity during the 10 s?

9E. The position of an electron is given by $\mathbf{r} = 3.0t\mathbf{i} - 4.0t^2\mathbf{j} + 2.0\mathbf{k}$ (where t is in seconds and the coefficients have the proper units for $\mathbf{r}$ to be in meters). (a) What is $\mathbf{v}(t)$ for the electron? (b) In unit-vector notation, what is $\mathbf{v}$ at $t = 2.0$ s? (c) What are the magnitude and direction of $\mathbf{v}$ just then?

SECTION 4-4 Acceleration and Average Acceleration

10E. A proton initially has $\mathbf{v} = 4.0\mathbf{i} - 2.0\mathbf{j} + 3.0\mathbf{k}$ and then 4.0 s later has $\mathbf{v} = -2.0\mathbf{i} - 2.0\mathbf{j} + 5.0\mathbf{k}$ (in meters per second). (a) In unit-vector notation, what is the average acceleration $\bar{\mathbf{a}}$ over the 4.0 s? (b) What are the magnitude and orientation of $\bar{\mathbf{a}}$?

11E. A particle moves so that its position as a function of time in SI units is $\mathbf{r} = \mathbf{i} + 4t^2\mathbf{j} + t\mathbf{k}$. Write expressions for (a) its velocity and (b) its acceleration as functions of time.

12E. The position $\mathbf{r}$ of a particle moving in an xy plane is given by $\mathbf{r} = (2.00t^3 - 5.00t)\mathbf{i} + (6.00 - 7.00t^4)\mathbf{j}$. Here $\mathbf{r}$ is in meters and t in seconds. Calculate (a) $\mathbf{r}$, (b) $\mathbf{v}$, and (c) $\mathbf{a}$ when $t = 2.00$ s. (d) What is the orientation of a line that is tangent to the particle's path at $t = 2.00$ s?

13E. An ice boat sails across the surface of a frozen lake with constant acceleration produced by the wind. At a certain instant the boat's velocity is $6.30\mathbf{i} - 8.42\mathbf{j}$ in meters per second. Three seconds later, because of a wind shift, the boat is instantaneously at rest. What is its average acceleration during this 3 s interval?

14P. A particle starts from the origin at $t = 0$ with a velocity of $8.0\mathbf{j}$ m/s and moves in the xy plane with a constant acceleration of $(4.0\mathbf{i} + 2.0\mathbf{j})$ m/s². (a) At the instant the x coordinate of the particle is 29 m, what is its y coordinate? (b) What is the speed of the particle at this time?

15P. A particle leaves the origin with an initial velocity $\mathbf{v} =$

3.00**i**, in meters per second. It experiences a constant acceleration **a** = −1.00**i** − 0.500**j**, in meters per second squared. (a) What is the velocity of the particle when it reaches its maximum x coordinate? (b) Where is the particle at this time?

16P. The velocity **v** of a particle moving in the xy plane is given by **v** = $(6.0t − 4.0t^2)$**i** + 8.0**j**. Here **v** is in meters per second and t (> 0) is in seconds. (a) What is the acceleration when t = 3.0 s? (b) When (if ever) is the acceleration zero? (c) When (if ever) is the velocity zero? (d) When (if ever) does the speed equal 10 m/s?

17P. A particle A moves along the line y = 30 m with a constant velocity **v** (v = 3.0 m/s) directed parallel to the positive x axis (Fig. 4-28). A second particle B starts at the origin with zero speed and constant acceleration **a** (a = 0.40 m/s²) at the same instant that particle A passes the y axis. What angle θ between **a** and the positive y axis would result in a collision between these two particles? (If your computation involves an equation with a term such as t^4, substitute u = t^2 and then consider solving the resulting quadratic equation to get u.)

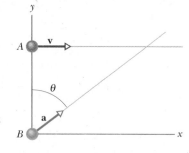

FIGURE 4-28
Problem 17.

SECTION 4-6 Projectile Motion Analyzed

In some of these problems, exclusion of the effects of the air is unwarranted but helps simplify the calculations.

18E. A dart is thrown horizontally toward the bull's-eye, point P on the dart board of Fig. 4-29, with an initial speed of 10 m/s. It hits at point Q on the rim, vertically below P, 0.19 s later. (a) What is the distance PQ? (b) How far away from the dart board did the dart thrower stand?

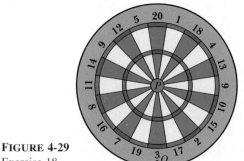

FIGURE 4-29
Exercise 18.

19E. A rifle is aimed horizontally at a target 100 ft away. The bullet hits the target 0.75 in. below the aiming point. (a) What is the bullet's time of flight? (b) What is its muzzle velocity?

20E. Electrons, like all other forms of matter, can undergo free fall. (a) If an electron is projected horizontally with a speed of 3.0 × 10⁶ m/s, how far will it fall in traversing 1.0 m of horizontal distance? (b) Does the answer increase or decrease if the initial speed is increased?

21E. In a cathode-ray tube, a beam of electrons is projected horizontally with a speed of 1.0 × 10⁹ cm/s into the region between a pair of horizontal plates 2.0 cm square. An electric field between the plates causes a constant downward acceleration of the electrons of magnitude 1.0 × 10¹⁷ cm/s². Find (a) the time required for an electron to pass through the plates, (b) the vertical displacement of the beam in passing through the plates (it does not run into a plate), and (c) the velocity of the beam as it emerges from the plates.

22E. A ball rolls horizontally off the edge of a tabletop that is 4.0 ft high. It strikes the floor at a point 5.0 ft horizontally away from the edge of the table. (a) How long was the ball in the air? (b) What was its speed at the instant it left the table?

23E. A projectile is fired horizontally from a gun that is 45.0 m above flat ground. The muzzle velocity is 250 m/s. (a) How long does the projectile remain in the air? (b) At what horizontal distance from the firing point does it strike the ground? (c) What is the magnitude of the vertical component of its velocity as it strikes the ground?

24E. A baseball leaves a pitcher's hand horizontally at a speed of 100 mi/h. The distance to the batter is 60 ft. (a) How long does it take for the ball to travel the first 30 ft horizontally? The second 30 ft? (b) How far does the ball fall under gravity during the first 30 ft of its horizontal travel? (c) During the second 30 ft? (d) Why aren't the quantities in (b) and (c) equal? (Ignore the effect of air resistance.)

25E. A projectile is launched with an initial speed of 30 m/s at an angle of 60° above the horizontal. Calculate the magnitude and direction of its velocity (a) 2.0 s and (b) 5.0 s after launch.

26E. A stone is catapulted rightward with an initial velocity of 20.0 m/s at an angle of 40.0° above level ground. Find its horizontal and vertical displacements (a) 1.10 s, (b) 1.80 s, and (c) 5.00 s after launch.

27E. You throw a ball from a cliff with an initial velocity of 15.0 m/s at an angle of 20.0° below the horizontal. Find (a) its horizontal displacement and (b) its vertical displacement 2.30 s later.

28E. You throw a ball with a speed of 25.0 m/s at an angle of 40.0° above the horizontal directly toward a wall (Fig. 4-30). The wall is 22.0 m from the release point of the ball. (a) How long does the ball take to reach the wall? (b) How far above the release point does the ball hit the wall? (c) What are the horizontal and vertical components of its velocity as it hits the wall? (d) When it hits, has it passed the highest point on its trajectory?

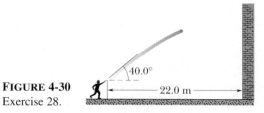

FIGURE 4-30
Exercise 28.

29E. (a) Prove that for a projectile fired from level ground at an angle θ_0 above the horizontal, the ratio of the maximum height H to the range R is given by $H/R = \frac{1}{4} \tan \theta_0$. See Fig. 4-31. (b) For what angle θ_0 does $H = R$?

30E. A projectile is fired from level ground at an angle θ_0 above the horizontal. (a) Show that the elevation angle ϕ of the highest point as seen from the launch point is related to θ_0, the elevation angle of projection, by $\tan \phi = \frac{1}{2} \tan \theta_0$. See Fig. 4-31 and Exercise 29. (b) Calculate ϕ for $\theta_0 = 45°$.

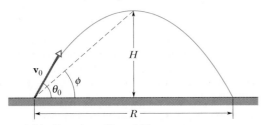

FIGURE 4-31 Exercises 29 and 30.

31E. A stone is projected at a cliff of height h with an initial speed of 42.0 m/s directed 60.0° above the horizontal, as shown in Fig. 4-32. The stone strikes at A, 5.50 s after launching. Find (a) the height h of the cliff, (b) the speed of the stone just before impact at A, and (c) the maximum height H reached above the ground.

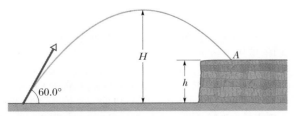

FIGURE 4-32 Exercise 31.

32P. The launching speed of a certain projectile is five times the speed it has at its maximum height. Calculate the elevation angle at launching.

33P. In Sample Problem 4-6, find (a) the speed of the capsule when it hits the water and (b) the angle θ shown in Fig. 4-14.

34P. In the 1991 World Track and Field Championships in Tokyo, Mike Powell (Fig. 4-33) jumped 8.95 m, breaking the 23-year long-jump record set by Bob Beamon by a full 5 cm. Assume that Powell's speed on takeoff was 9.5 m/s (about equal to that of a sprinter) and that $g = 9.80$ in Tokyo. How close did Powell come to the maximum possible range in the absence of air resistance?

35P. A rifle with a muzzle velocity of 1500 ft/s shoots a bullet at a target 150 ft away. How high above the target must the rifle barrel be pointed to ensure that the bullet will hit the target?

36P. Show that the maximum height reached by a projectile is $y_{max} = (v_0 \sin \theta_0)^2/2g$.

37P. A ball is shot from the ground into the air. At a height of 9.1 m, the velocity is observed to be $\mathbf{v} = 7.6\mathbf{i} + 6.1\mathbf{j}$ in meters per second ($\mathbf{i}$ horizontal, $\mathbf{j}$ upward). (a) To what maximum height will the ball rise? (b) What will be the total horizontal distance

FIGURE 4-33 Problem 34. Mike Powell's jump.

traveled by the ball? (c) What is the velocity of the ball (magnitude and direction) the instant before it hits the ground?

38P. In a detective story, a body is found 15 ft from the base of a building and 80 ft below an open window. Would you guess the death to be accidental. Explain your answer.

39P. In Galileo's *Two New Sciences*, the author states that "for elevations [angles of projection] which exceed or fall short of 45° by equal amounts, the ranges are equal. . . ." Prove this statement. (See Fig. 4-34.)

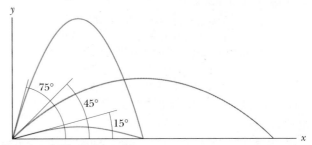

FIGURE 4-34 Problem 39.

40P. The range of a projectile depends not only on v_0 and θ_0 but also on the value g of the free-fall acceleration, which varies from place to place. In 1936, Jesse Owens established a world's running broad jump record of 8.09 m at the Olympic Games at Berlin ($g = 9.8128$ m/s²). Assuming the same values of v_0 and θ_0, by how much would his record have differed if he had competed instead in 1956 at Melbourne ($g = 9.7999$ m/s²)?

41P. A third baseman wishes to throw to first base, 127 ft distant. His best throwing speed is 85 mi/h. (a) If he throws the ball horizontally 3.0 ft above the ground, how far from first base will it hit the ground? (b) At what upward angle must the third baseman throw the ball if the first baseman is to catch it 3.0 ft above the ground? (c) What will be the time of flight in that case?

42P. During volcanic eruptions, chunks of solid rock can be blasted out of the volcano; these projectiles are called *volcanic bombs*. Figure 4-35 shows a cross section of Mt. Fuji, in Japan. (a) At what initial speed would a bomb have to be ejected, at 35° to the horizontal, from the vent at A in order to fall at the foot of the volcano at B? Ignore, for the moment, the effects of air on the bomb's travel. (b) What would be the time of flight? (c) Would the effect of the air increase or decrease your answer in (a)?

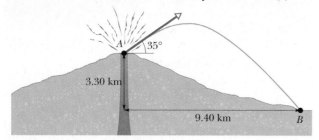

FIGURE 4-35   Problem 42.

43P. At what initial speed must the basketball player throw the ball, at 55° above the horizontal, to make the foul shot, as shown in Fig. 4-36?

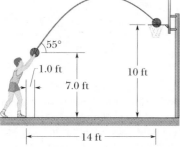

FIGURE 4-36   Problem 43.

44P. A football player punts the football so that it will have a ''hang time'' (time of flight) of 4.5 s and land 50 yd away. If the ball leaves the player's foot 5.0 ft above the ground, what initial velocity (magnitude and direction) must the ball have?

45P. A golfer tees off from the top of a rise, giving the golf ball an initial velocity of 43 m/s at an angle of 30° above the horizontal. The ball strikes the fairway a horizontal distance of 180 m from the tee. Assume the fairway is level. (a) How high is the rise above the fairway? (b) What is the speed of the ball as it strikes the fairway?

46P. A projectile is fired with an initial speed $v_0 = 30.0$ m/s from the level ground at a target on the ground a distance $R = 20.0$ m away (Fig. 4-37). Find the two projection angles.

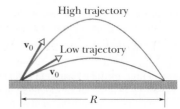

FIGURE 4-37   Problem 46.

47P. What is the maximum vertical height to which a baseball player can throw a ball if his maximum throwing *range* is 60 m?

48P. A certain airplane has a speed of 180 mi/h and is diving at an angle of 30.0° below the horizontal when a radar decoy is released. (See Fig. 4-38.) The horizontal distance between the release point and the point where the decoy strikes the ground is 2300 ft. (a) How high was the plane when the decoy was released? (b) How long was the decoy in the air?

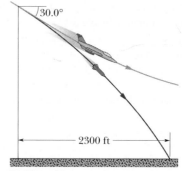

FIGURE 4-38   Problem 48.

49P. A football is kicked off with an initial speed of 64 ft/s at a projection angle of 45°. A receiver 60 yd away in the direction of the kick starts running to meet the ball at that instant. What must be his average speed if he is to catch the ball just before it hits the ground? Neglect air resistance.

50P. A ball rolls horizontally off the top of a stairway with a speed of 5.0 ft/s. The steps are 8.0 in. high and 8.0 in. wide. Which step will the ball hit first?

51P. An airplane, diving at an angle of 53.0° with the vertical, releases a projectile at an altitude of 730 m. The projectile hits the ground 5.00 s after being released. (a) What is the speed of the aircraft? (b) How far did the projectile travel horizontally during its flight? (c) What were the horizontal and vertical components of its velocity just before striking the ground?

52P. A ball is thrown horizontally from a height of 20 m and hits the ground with a speed that is three times its initial speed. What was the initial speed?

53P. (a) During a tennis match, a player serves at 23.6 m/s, the ball leaving the racquet horizontally 2.37 m above the court surface. By how much does the ball clear the net, which is 12 m away and 0.90 m high? (b) Suppose the player serves the ball as before except that the ball leaves the racquet at 5.00° below the horizontal. Does the ball clear the net now?

54P. In Sample Problem 4-8, suppose that a second identical harbor defense cannon is emplaced 30 m above sea level, rather that at sea level. How much longer is the horizontal distance from launch to impact of the second cannon than that of the first, which was found to be 690 m, if the elevation angle of fire is 45°?

55P. A batter hits a pitched ball whose center is 4.0 ft above the ground so that its angle of projection is 45° and its *range* is 350 ft. The ball will be a home run if it clears a 24 ft high fence that is 320 ft from home plate. Will the ball clear the fence? If so, by how much?

56P*. A football kicker can give the ball an initial speed of 25 m/s. Within what two elevation angles must he kick the ball to score a field goal from a point 50 m in front of goalposts whose horizontal bar is 3.44 m above the ground? (You might want to use $\sin^2 \theta + \cos^2 \theta = 1$ to get a relation between $\tan^2 \theta$ and $1/\cos^2 \theta$, and then solve the resulting quadratic equation.)

SECTION 4-7 Uniform Circular Motion

57E. In one model of the hydrogen atom, an electron orbits a proton in a circle of radius 5.28×10^{-11} m with a speed of 2.18×10^6 m/s. (a) What is the acceleration of the electron in this model? (b) What is the period of the motion?

58E. (a) What is the acceleration of a sprinter running at 10 m/s when rounding a bend with a turn radius of 25 m? (b) In what direction does the acceleration vector point?

59E. A magnetic field can force a charged particle to move in a circular path. Suppose that an electron experiences a radial acceleration of 3.0×10^{14} m/s² in a particular magnetic field. (a) What is the speed of the electron if the radius of its circular path is 15 cm? (b) What is the period of the motion?

60E. A sprinter runs at 9.2 m/s around a circular track with a centripetal acceleration of 3.8 m/s². (a) What is the track radius? (b) What is the period of the motion?

61E. An Earth satellite moves in a circular orbit 640 km above the Earth's surface. The period of the motion is 98.0 min. (a) What is the speed of the satellite? (b) What is the free-fall acceleration at the orbit height?

62E. Suppose a space probe can withstand the stresses of a 20g acceleration. (a) What is the minimum turning radius of such a craft moving at a speed of one-tenth the speed of light? (b) How long would it take to complete a 90° turn at this speed?

63E. A rotating fan completes 1200 revolutions every minute. Consider a point on the tip of a blade, at a radius of 0.15 m. (a) Through what distance does the point move in one revolution? (b) What is the speed of the point? (c) What is its acceleration? (d) What is the period of the motion?

64E. The fast train known as the TGV (Train à Grande Vitesse) that runs south from Paris, France, has a scheduled average speed of 216 km/h. (a) If the train goes around a curve at that speed and the acceleration experienced by the passengers is to be limited to 0.050g, what is the smallest radius of curvature for the track that can be tolerated? (b) If there is a curve with a 1.00 km radius, to what speed must the train be slowed to keep the acceleration below the limit?

65E. When a large star becomes a *supernova*, its core may be compressed so tightly that it becomes a *neutron star*, with a radius of about 20 km (about the size of the San Francisco area). If a neutron star rotates once every second, (a) what is the speed of a particle on the star's equator and (b) what is the particle's centripetal acceleration in meters per second squared and in g units? (c) If the neutron star rotates even faster, what happens to the answers to (a) and (b)?

66E. An astronaut is rotated in a horizontal centrifuge at a radius of 5.0 m. (a) What is the astronaut's speed if the centripetal acceleration is 7.0g? (b) How many revolutions per minute are required to produce this acceleration? (c) What is the period of the motion?

67P. (a) What is the centripetal acceleration of an object on the Earth's equator owing to the rotation of the Earth? (b) What would the period of rotation of the Earth have to be for objects on the equator to have a centripetal acceleration equal to 9.8 m/s²?

68P. A carnival Ferris wheel has a 15 m radius and completes five turns about its horizontal axis every minute. (a) What is the period of the motion? (b) What is the centripetal acceleration of a passenger at the highest point? (c) What is the centripetal acceleration at the lowest point?

69P. Calculate the acceleration of a person at latitude 40° owing to the rotation of the Earth. (See Fig. 4-39.)

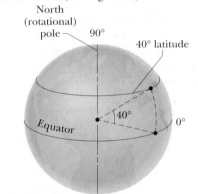

FIGURE 4-39 Problem 69.

70P. A particle P travels with constant speed on a circle of radius $r = 3.00$ m (Fig. 4-40) and completes one revolution in 20.0 s. The particle passes through O at $t = 0$. Find the magnitude and direction of each of the following vectors. (a) With respect to O, find the particle's position vector at $t = 5.00$ s, 7.50 s, and 10.0 s. For the 5.00 s interval from the end of the fifth second to the end of the tenth second, find the particle's (b) displacement and (c) average velocity. Find its (d) velocity and (e) acceleration at the beginning and end of that 5.00 s interval.

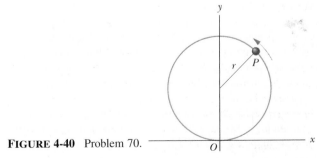

FIGURE 4-40 Problem 70.

71P. A boy whirls a stone in a horizontal circle 2.0 m above the ground by means of a string 1.5 m long. The string breaks, and the stone flies off horizontally and strikes the ground 10 m away. What was the centripetal acceleration of the stone while in circular motion?

SECTION 4-8   Relative Motion in One Dimension

72E. A boat is traveling upstream at 14 km/h with respect to the water of a river. The water itself is flowing at 9 km/h with respect to the ground. (a) What is the velocity of the boat with respect to the ground? (b) A child on the boat walks from front to rear at 6 km/h with respect to the boat. What is the child's velocity with respect to the ground?

73E. A person walks up a stalled 15 m long escalator in 90 s. When standing on the same escalator, now moving, the person is carried up in 60 s. How much time would it take that person to walk up the moving escalator? Does the answer depend on the length of the escalator?

74E. A transcontinental flight of 2700 mi is scheduled to take 50 min longer westward than eastward. The airspeed of the airplane is 600 mi/h, and the jet stream it will fly through is presumed to be moving either due east or due west. What assumptions about the jet stream wind velocity are made in preparing the schedule?

75E. A cameraman on a pickup truck is traveling westward at 40 mi/h while he videotapes a cheetah that is moving westward 30 mi/h faster than the truck. Suddenly, the cheetah stops, turns, and then runs at 60 mi/h eastward, as measured by a suddenly nervous crew member who stands alongside the cheetah's path. The change in the animal's velocity took 2.0 s. What was its acceleration from the perspective of the cameraman? From the perspective of the nervous crew member?

76P. The airport terminal in Geneva, Switzerland, has a "moving sidewalk" to speed passengers through a long corridor. Peter does not use the moving sidewalk; he takes 150 s to walk through the corridor. Paul, who simply stands on the moving sidewalk, covers the same distance in 70 s. Mary boards the sidewalk and walks along it. How long does Mary take to move through the corridor? Assume that Peter and Mary walk at the same speed.

SECTION 4-9   Relative Motion in Two Dimensions

77E. In rugby (Fig. 4-41) a player can legally pass the ball to a teammate as long as the pass is not "forward" (it must not have a velocity component parallel to the length of the field and directed toward the other team's goal). Suppose a player runs parallel to the field's length with a speed of 4.0 m/s while he passes the ball with a speed of 6.0 m/s relative to himself. What is the smallest angle from the forward direction that keeps the pass legal?

78E. Two highways intersect as shown in Fig. 4-42. At the instant shown, a police car P is 800 m from the intersection and moving at 80 km/h. Motorist M is 600 m from the intersection and moving at 60 km/h. (a) In unit-vector notation, what is the velocity of the motorist with respect to the police car? (b) For the

FIGURE 4-41   Exercise 77.

instant shown in Fig. 4-42, how does the direction of the velocity found in (a) compare to the line of sight between the two cars? (c) If the cars maintain their velocities, do the answers to (a) and (b) change as the cars move nearer the intersection?

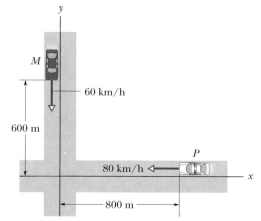

FIGURE 4-42   Exercise 78.

79E. Snow is falling vertically at a constant speed of 8.0 m/s. At what angle from the vertical do the snowflakes appear to be falling as viewed by the driver of a car traveling on a straight road with a speed of 50 km/h?

80E. In a large department store, a shopper is standing on the "up" escalator, which is traveling at an angle of 40° above the horizontal and at a speed of 0.75 m/s. He passes his daughter, who is standing on the identical, adjacent "down" escalator. (See Fig. 4-43.) Find the velocity of the shopper relative to his daughter in unit-vector notation.

81P. A helicopter is flying in a straight line over a level field at a constant speed of 6.2 m/s and at a constant altitude of 9.5 m. A package is ejected horizontally from the helicopter with an initial velocity of 12 m/s relative to the helicopter, and in a direction opposite the helicopter's motion. (a) Find the initial speed of the package relative to the ground. (b) What is the horizontal distance between the helicopter and the package at the instant the package strikes the ground? (c) What angle does the velocity vector of the

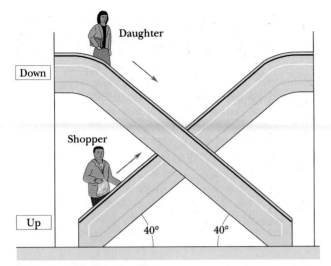

FIGURE 4-43 Exercise 80.

package make with the ground at the instant before impact, as seen from the ground?

82P. A train travels due south at 30 m/s (relative to the ground) in a rain that is blown toward the south by the wind. The path of each raindrop makes an angle of 70° with the vertical, as measured by an observer stationary on the Earth. An observer on the train, however, sees the drops fall perfectly vertically. Determine the speed of the raindrops relative to the Earth.

83P. A light plane attains an airspeed of 500 km/h. The pilot sets out for a destination 800 km to the north but discovers that the plane must be headed 20.0° east of north to fly there directly. The plane arrives in 2.00 h. What was the wind velocity vector?

84P. Two ships, A and B, leave port at the same time. Ship A travels northwest at 24 knots and ship B travels at 28 knots in a direction 40° west of south. (1 knot = 1 nautical mile per hour; see Appendix D.) (a) What are the magnitude and direction of the velocity of ship A relative to B? (b) After what time will they be 160 nautical miles apart? (c) What will be the bearing (direction) of B relative to A at that time?

85P. The New Hampshire State Police use aircraft to enforce highway speed limits. Suppose that one of the airplanes has a speed of 135 mi/h in still air. It is flying straight north so that it is at all times directly above a north–south highway. A ground observer tells the pilot by radio that a 70.0 mi/h wind is blowing, but neglects to give the wind direction. The pilot observes that in spite of the wind the plane can travel 135 mi along the highway in 1.00 h. In other words, the ground speed is the same as if there were no wind. (a) What is the direction of the wind? (b) What is the heading of the plane, that is, the angle between an axis along its length and the highway?

86P. A wooden boxcar is moving along a straight railroad track at speed v_1. A sniper fires a bullet (initial speed v_2) at it from a high-powered rifle. The bullet passes through both walls of the car, its entrance and exit holes being exactly opposite each other as viewed from within the car. From what direction, relative to the track, was the bullet fired? Assume that the bullet was not

deflected upon entering the car, but that its speed decreased by 20%. Take $v_1 = 85$ km/h and $v_2 = 650$ m/s. (Why don't you need to know the width of the boxcar?)

87P. A woman can row a boat at 4.0 mi/h in still water. (a) If she is crossing a river where the current is 2.0 mi/h, in what direction must her boat be headed if she wants to reach a point directly opposite her starting point? (b) If the river is 4.0 mi wide, how long will it take her to cross the river? (c) Suppose that instead of crossing the river she rows 2.0 mi *down* the river and then back to her starting point. How long will she take? (d) How long will she take to row 2.0 mi *up* the river and then back to her starting point? (e) In what direction should she head the boat if she wants to cross in the shortest possible time, and what is that time?

SECTION 4-10 Relative Motion at High Speeds

88E. As we head toward the center of our galaxy, we detect a burst of light traveling past us at speed c relative to us, toward that center. An observer on a second ship, traveling at a speed of $0.98c$ relative to us, also detects the light. What is the speed of the light as measured by the observer on the second ship if that ship is traveling (a) in the same direction as us and (b) in the opposite direction?

89E. An electron moves at speed $0.42c$ with respect to observer B. Observer B moves at speed $0.63c$ with respect to observer A, in the same direction as the electron. What does observer A measure for the speed of the electron?

90P. While traveling in a ship toward the star Betelgeuse, we detect a burst of protons traveling past us toward that star, with a velocity of $0.9800c$ relative to us. Detectors on a second ship, also traveling toward Betelgeuse along our line of travel, measure the velocity of the protons to be $-0.9800c$ relative to them. What is the relative speed between us and the second ship?

91P. Galaxy Alpha is observed to be receding from us with a speed of $0.35c$. Galaxy Beta, located in precisely the opposite direction, is also found to be receding from us at this same speed. What recessional speed would an observer on Galaxy Alpha find (a) for our galaxy and (b) for Galaxy Beta?

ELECTRONIC COMPUTATION

92. If the launch site of a projectile is above the landing place, a launch angle of 45° may not produce the greatest horizontal distance. Suppose a shot putter releases the shot from a point that is a distance h above a horizontal playing field. The initial speed of the shot is v_0 and the launch angle is θ. (a) Show that the horizontal distance from the putter's feet to the landing point is given by

$$d = \frac{v_0 \cos \theta}{g} [v_0 \sin \theta + \sqrt{v_0^2 \sin^2 \theta + 2gh}].$$

(b) Set up a program to calculate d for a series of values of θ, given v_0 and h. (c) Take $v_0 = 9.0$ m/s and $h = 2.1$ m and find to the nearest half degree the launch angle that produces the greatest horizontal distance. Also find that distance. (d) Does the result depend on the initial speed of the shot? Try $v_0 = 5.0$ m/s and $v_0 = 15$ m/s. (The last value is much greater than the speed generated by a champion shot putter.)

93. Immediately after launch, all projectiles move away from the launch site but later some move closer before moving further away again. It depends on the launch angle. Set up a program to compute the distance from the launch site to the projectile every 0.5 s from the time of launch to a few seconds after it is below the launch site. Take the initial speed to be 100 m/s and run the program for launch angles from 5° to 90°, with an interval of 5°. For each launch angle, search the list of distances to find when the projectile is moving away from and when it is moving toward the launch site. Classify the motions according to whether or not the projectile moves toward the launch site during any time interval and, if it does, estimate the interval. (See the article "Projectiles: Are They Coming or Going?" by James S. Walker in *The Physics Teacher*, May 1995.)

94. A batter strikes a baseball at a point 1.00 m above home plate, giving the ball an initial velocity of magnitude v_0, at angle θ above the horizontal. The ball is to barely clear a 2.40 m tall fence at a distance $R = 110$ m from home plate. (a) Find the minimum and maximum values of θ that allow this clearance for $v_0 = 35.0$ m/s. (*Hint:* rather than explicitly solving for θ in the equations of motion, find two expressions of θ, plot them, and then find their intersections.) (b) Find the minimum value of v_0 that allows the clearance for $\theta = 40.0°$. (c) Now find the minimum value of v_0 considering all values of θ and give the corresponding angle. (d) Repeat (c) for Fenway Park, with a 12.2 m

tall "wall" at 96.0 m down the foul line from home plate.

95. A golfer chips balls toward a vertical wall 20.0 m straight ahead, trying to hit a 30.0 cm diameter red circle painted on the wall. The target is centered about a point 1.20 m above the point where the wall intersects the horizontal ground. On one try, the ball leaves the ground with a speed of 15.0 m/s and at an angle of 35.0° above the horizontal. (a) How long does the ball take to reach the wall? (b) Does the ball hit the red circle? (c) What is the speed of the ball just before it hits? (d) Has the ball passed the highest point of its trajectory when it hits?

96. At one instant a bicyclist is 40.0 m due east of a park's flagpole, going due south with a speed of 10.0 m/s. Then 30.0 s later, the cyclist is 40.0 m due north of the flagpole, going due east with a speed of 10.0 m/s. Find (a) the displacement, (b) the average velocity, and (c) the average acceleration of the cyclist during the 30.0 s interval. (d) Compute $(\mathbf{v}_f - \mathbf{v}_i)/2$, where $\mathbf{v}_f$ and $\mathbf{v}_i$ are the velocities at the end and at the beginning of the 30.0 s interval, respectively.

97. At one instant a butterfly has the position vector $\mathbf{D}_i = (2.00 \text{ m})\mathbf{i} + (3.00 \text{ m})\mathbf{j} + (1.00 \text{ m})\mathbf{k}$ relative to the base of a birdbath. Then 40 s later, the butterfly has the position vector $\mathbf{D}_f = (3.00 \text{ m})\mathbf{i} + (1.00 \text{ m})\mathbf{j} + (2.00 \text{ m})\mathbf{k}$. Find (a) the displacement (in unit-vector notation), (b) the magnitude of the displacement, (c) the average velocity, and (d) the average speed of the butterfly during the 40.0 s interval.

On April 4, 1974, John Massis of Belgium managed to move two passenger cars belonging to New York's Long Island Railroad. He did so by clamping his teeth down on a bit that was attached to the cars with a rope and then leaning backward while pressing his feet against the railway ties. The cars weighed about 80 tons. Did Massis have to pull with superhuman force to accelerate them?

5-1 WHAT CAUSES AN ACCELERATION?

If you see the velocity of a particle-like body change in either magnitude or direction, you know that something must have *caused* that change (that acceleration). Indeed, out of common experience, you know that the change in velocity must be due to an interaction between the body and something in its surroundings. For example, if you see a hockey puck that is sliding across an ice rink suddenly stop or suddenly change direction, you will suspect that the puck bumped into a slight hill on the ice surface.

An interaction that causes an acceleration of a body is called a **force**, which is, loosely speaking, a push or pull. For example, the bump on the hockey puck by the hill is a push on the puck, causing an acceleration. The relationship between a force and the acceleration it causes was first understood by Isaac Newton (1642–1727) and is the subject of this chapter. The study of that relationship, as Newton presented it, is called *Newtonian mechanics*. We will focus on its three primary laws of motion.

Newtonian mechanics does not apply to all situations. As we discussed in Chapter 4, if the speeds of the interacting bodies are an appreciable fraction of the speed of light, we must replace Newtonian mechanics with Einstein's special theory of relativity, which holds at any speed, including those near the speed of light. If the interacting bodies are on the scale of atomic structure (for example, they might be electrons within an atom), we must replace Newtonian mechanics with quantum mechanics. Physicists now view Newtonian mechanics as a special case of these two more comprehensive theories. Still, it is a very important special case because it applies to the motion of objects ranging in size from the very small (almost on the scale of atomic structure) to astronomical (objects such as galaxies and clusters of galaxies). Let us now examine the first law of motion in Newtonian mechanics.

5-2 NEWTON'S FIRST LAW

Before Newton formulated his mechanics, it was thought that some influence, a "force," was needed to keep a body moving at constant velocity. Similarly, a body was thought to be in its "natural state" when it was at rest. For it to move with constant velocity, it seemingly had to be propelled in some way, by a push or a pull. Otherwise, it would "naturally" stop moving.

These ideas were reasonable. If you send a book sliding across a wooden floor, it does indeed slow and then stop. If you want to make it move across the floor with constant velocity, you have to continuously pull or push it.

Slide the book over the ice of a skating rink, however,

and it goes a lot farther. You can imagine longer and more slippery surfaces, over which the book would slide farther and farther. In the limit you can think of a long, extremely slippery surface (said to be a **frictionless surface**), over which the book would hardly slow. (We can in fact come close to this situation in the laboratory, by sending a book sliding over a horizontal air table, across which it moves on a film of air.)

We are led to conclude that you do *not* need a force to keep a body moving with constant velocity. And that leads us to the first of Newton's three laws of motion:

> **Newton's First Law:** Consider a body on which no force acts. If the body is at rest, it will remain at rest. If the body is moving with constant velocity, it will continue to do so.

This law fits in nicely with what we discussed in Section 4-8 about reference frames: a body on which no force acts may be stationary in one frame and moving at constant velocity with respect to another. (*Rest* and *moving with constant velocity* are not all that different.)

Newton's first law can be interpreted as a statement about reference frames, in that it defines the kinds of reference frames in which the laws of Newtonian mechanics hold: frames with constant separation velocity. From this point of view the first law is expressed as follows:

> **Newton's First Law:** If no force acts on a body, we can always find a reference frame in which that body has no acceleration.

Newton's first law is sometimes called the *law of inertia,* and the reference frames that it defines are called *inertial reference frames* or just *inertial frames.*

Figure 5-1 shows how you can test a particular frame to see whether it is an inertial frame. With the railroad car at rest, mark the position of the stationary pendulum bob on the table. With the car in motion, the bob remains over the mark *only* if the car is moving in a straight line at

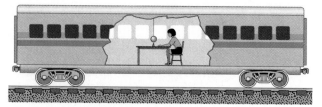

FIGURE 5-1 Testing a railroad car to see whether it is an inertial reference frame.

constant speed. The car is then an inertial frame. If the car is gaining or losing speed or is rounding a corner, the bob moves from its mark, and the car is then a noninertial reference frame.

5-3 FORCE

A force causes the acceleration of a body. We now wish to define the unit of force carefully, in terms of the acceleration that it gives to a standard reference body. As the standard body, we use (or rather we imagine that we use) the standard kilogram of Fig. 1-6. This body has been assigned, exactly and by definition, a mass of 1 kg.

We put the standard body on a horizontal frictionless table and pull the body to the right (Fig. 5-2) so that by trial and error, it eventually experiences a measured acceleration of 1 m/s². We then declare, as a matter of definition, that the force we are exerting on the standard body has a magnitude of 1 newton (abbreviated N).

We can exert a 2 N force on our standard body by pulling it so that its measured acceleration is 2 m/s², and so on. Thus in general, if our standard body of 1 kg mass has an acceleration a, we know that force F must be acting on it and that the magnitude of the force (in newtons) is equal to the magnitude of the acceleration (in meters per second per second).

Thus, a force is measured by the acceleration it produces. But acceleration is a vector quantity, with both magnitude and direction. Is force also a vector quantity? We can easily assign a direction to a force (just assign the direction of the acceleration), but that is not sufficient. We must prove by experiment that forces are vector quantities. Actually, that has been done: forces are indeed vector quantities; they have magnitudes and directions and they combine according to the vector rules of Chapter 3.

Henceforth, we shall use boldface letters, most often **F**, to represent forces. And we shall use the symbol $\Sigma\mathbf{F}$ for the vector sum of several forces, which we call the **resultant force** or the **net force**. As with other vectors, a force or a net force can have components along coordinate axes. Finally, we note that Newton's first law holds not only when there is no force on a body, but also when the net force is equal to zero.

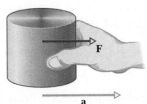

FIGURE 5-2 A force **F** on the standard kilogram gives that body an acceleration **a**.

CHECKPOINT **1:** In the figure, two perpendicular forces **F**₁ and **F**₂ are combined in six different ways. Which ways may be used to correctly determine the net force $\Sigma\mathbf{F}$?

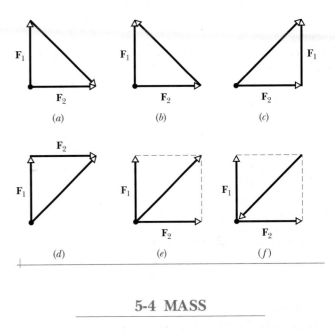

5-4 MASS

Everyday experience tells us that a given force produces different magnitudes of acceleration for different bodies. Put a baseball and a bowling ball on the floor and give both the same sharp kick. Even if you don't actually do this, you know the result: the baseball receives a noticeably larger acceleration than the bowling ball. The two accelerations differ because the mass of the baseball differs from the mass of the bowling ball. But what, exactly, is mass?

We can explain how to measure mass by imagining a series of experiments. In the first experiment we exert a force on our standard body, whose mass m_0 is defined to be 1.0 kg. Suppose that the standard body accelerates at 1.0 m/s². We can then say the force on that body is 1.0 N.

We next apply that same force (we would need some way of being certain it is the same force) to a second body, body X, whose mass is not known. Suppose we find that this body X accelerates at 0.25 m/s². We know that a *less massive* baseball receives a *greater acceleration* than a more massive bowling ball when the same force (kick) is applied to both. Let us then make the following conjecture: the ratio of the masses of two bodies is equal to the inverse of the ratio of their accelerations when the same force is applied to both. For body X and the standard body, this tells us that

$$\frac{m_X}{m_0} = \frac{a_0}{a_X}.$$

Solving for m_X yields

$$m_X = m_0 \frac{a_0}{a_X} = (1.0 \text{ kg}) \frac{1.0 \text{ m/s}^2}{0.25 \text{ m/s}^2} = 4.0 \text{ kg}.$$

Our conjecture will be useful, of course, only if it continues to hold when we change the applied force to other values. For example, if we apply an 8.0 N force to the standard body, we obtain an acceleration of 8.0 m/s². And when the 8.0 N force is applied to body X, we obtain an acceleration of 2.0 m/s². Our conjecture then gives us

$$m_X = m_0 \frac{a_0}{a_X} = (1.0 \text{ kg}) \frac{8.0 \text{ m/s}^2}{2.0 \text{ m/s}^2} = 4.0 \text{ kg},$$

consistent with our first experiment. Many experiments yielding similar results indicate that our conjecture provides a consistent and reliable means of assigning a mass to any given body.

Our measurement experiments indicate that mass is an *intrinsic* characteristic of a body—that is, a characteristic that automatically comes with the existence of the body. They also indicate that mass is a scalar quantity. However, the nagging question remains: What, exactly, is mass?

Since the word *mass* is used in everyday English, we should have some intuitive understanding of it, maybe something that we can physically sense. Is it a body's size, weight, or density? The answer is no, although those characteristics are sometimes confused with mass. We can say only that *the mass of a body is the characteristic that relates a force on the body to the resulting acceleration.* Mass has no more familiar definition than that; you can have a physical sensation of mass only when you attempt to accelerate a body, as in the kicking of a baseball or a bowling ball.

5-5 NEWTON'S SECOND LAW

All the definitions, experiments, and observations that we have described so far can be summarized in a simple vector equation, which is called Newton's second law of motion:

$$\sum \mathbf{F} = m\mathbf{a} \qquad \text{(Newton's second law).} \qquad (5\text{-}1)$$

In using Eq. 5-1, we must first be quite certain what body we are applying it to. Then $\sum\mathbf{F}$ in Eq. 5-1 is the vector sum, or net force, of *all* the forces that act *on* that body. Only forces that act *on* the body are to be included, not forces acting on other bodies that might be involved in a given problem. Finally, $\sum\mathbf{F}$ includes only *external* forces, that is, forces exerted on the body by other bodies. We do not include internal forces, in which one part of the body exerts a force on another part.

Like other vector equations, Eq. 5-1 is equivalent to three scalar equations:

$$\sum F_x = ma_x, \qquad \sum F_y = ma_y, \qquad \sum F_z = ma_z. \qquad (5\text{-}2)$$

These equations relate the three components of the net force acting on a body to the three components of the acceleration of that body.

You should note that Newton's second law includes the formal statement of Newton's first law as a special case. That is, if no force acts on a body, Eq. 5-1 tells us that the body will not be accelerated.

For SI units, Eqs. 5-2 tell us that

$$1 \text{ N} = (1 \text{ kg})(1 \text{ m/s}^2) = 1 \text{ kg} \cdot \text{m/s}^2, \qquad (5\text{-}3)$$

consistent with our discussion in Section 5-3. Although we shall use SI units almost exclusively from now on, other systems of units are still in use. Chief among these are the British system and the CGS (centimeter–gram–second) system. Table 5-1 shows the units in which Eqs. 5-1 and 5-2 are expressed in these systems. (See also Appendix D.)

To solve problems with Newton's second law, we often draw a **free-body diagram,** representing the body by a dot and each external force (or the net force $\sum\mathbf{F}$) that acts on the body by a vector with its tail on the dot. (Instead of the dot, we can sketch the body.) A set of coordinate axes is included, and sometimes a vector representing the acceleration of the body is also included.

We then should start the solution with the vector relation of Eq. 5-1. However, we shall usually quickly switch our attention to one or more of the scalar relations of Eqs. 5-2 and work along one axis at a time. The first relation of Eqs. 5-2 tells us that the sum of all the force components along the x axis causes the x component a_x of the body's acceleration but causes no acceleration in the y and z directions. Or, turned around, it says that the acceleration component a_x is caused only by the sum of the force components along the x axis. Similarly, the acceleration component a_y along the y axis is caused only by the sum of the force components along the y axis; and the acceleration component a_z along the z axis is caused only by the sum of the force components along the z axis. In general we have:

The acceleration component along a given axis is caused only by the sum of the force components along that *same* axis and not by force components along some other axis.

In Sample Problem 5-1 we shall be concerned with a single force that acts along an x axis. So, we shall not need to find several force components. However, in Sample Problem 5-2, three forces act on a body and two make

TABLE 5-1 UNITS IN NEWTON'S SECOND LAW (Eqs. 5-1 and 5-2)

SYSTEM	FORCE	MASS	ACCELERATION
SI	newton (N)	kilogram (kg)	m/s^2
CGS[a]	dyne	gram (g)	cm/s^2
British[b]	pound (lb)	slug	ft/s^2

[a]$1 \text{ dyne} = 1 \text{ g} \cdot cm/s^2$. [b]$1 \text{ lb} = 1 \text{ slug} \cdot ft/s^2$.

nonzero angles with an x axis and a y axis. In this two-dimensional situation, we must find the force components along the x axis and the y axis, and then use, separately, the first two relations of Eqs. 5-2.

Sample Problem 5-2 is also an example of a general type of problem: the body does not accelerate ($\mathbf{a} = 0$) in spite of the external forces on it. In such a situation, Eq. 5-1 tells us that $\Sigma\mathbf{F} = 0$. That is, the net force is zero, the forces acting on the body *balance* each other, and the body is said to be in *equilibrium*.

However, if we switch our attention to Eqs. 5-2, we see something that is more useful to problem solving. The lack of acceleration means that $a_x = 0$ and thus that $\Sigma F_x = 0$. That is, the x components of the forces also balance each other. If we replace ΣF_x with the actual x components of the forces involved, we then have an algebraic relation we can use in the solution. Similarly, $a_y = 0$ and thus $\Sigma F_y = 0$; after replacing ΣF_y with the actual y components of the forces involved, we have another algebraic relation.

In some problems you might find that the force components along one axis balance, but those along a second axis do not. So, there must be acceleration along that second axis only.

How will you know what to do in any given situation? Experience helps, and that is why this chapter contains many Sample Problems.

CHECKPOINT **2:** The figure shows two horizontal forces moving a block along a frictionless floor. Assume that a third horizontal force $\mathbf{F}_3$ also acts on the block. What are the magnitude and direction of $\mathbf{F}_3$ when the block is (a) stationary and (b) moving to the left with a constant speed of 5 m/s?

3 N ◄━━━ ▭ ━━━► 5 N

SAMPLE PROBLEM 5-1

A student (with cleated boots) pushes a loaded sled whose mass m is 240 kg for a distance d of 2.3 m over the frictionless surface of a frozen lake. He exerts a constant horizontal force

F, with magnitude $F = 130$ N, as he does so (see Fig. 5-3*a*).

(a) If the sled starts from rest, what is its final velocity?

SOLUTION: Figure 5-3*b* is a free-body diagram for the situation. We lay out a horizontal x axis, we take the direction of increasing x to be to the right, and we treat the sled as a particle, represented by a dot. We assume that the x component F_x of the force **F** exerted by the student is the only horizontal force acting on the sled. We can then find the magnitude of the acceleration a_x of the sled from Newton's second law:

$$a_x = \frac{F_x}{m} = \frac{130 \text{ N}}{240 \text{ kg}} = 0.542 \text{ m/s}^2.$$

Because the acceleration is constant, we can use Eq. 2-16, $v^2 = v_0^2 + 2a(x - x_0)$, to find the final velocity. Putting $v_0 = 0$ and $x - x_0 = d$, and identifying a_x as a, we solve for v:

$$v = \sqrt{2ad}$$
$$= \sqrt{(2)(0.542 \text{ m/s}^2)(2.3 \text{ m})} = 1.6 \text{ m/s}. \quad \text{(Answer)}$$

The force, the acceleration, the displacement, and thus the final velocity of the sled are all positive, which means that they all point to the right in Fig. 5-3*b*.

(b) The student now wants to reverse the direction of the velocity of the sled in 4.5 s. With what constant force must he push on the sled to do so?

SOLUTION: Let us first find the constant acceleration required to reverse the sled's velocity in 4.5 s, using Eq. 2-11 ($v = v_0 + at$). Solving for a gives

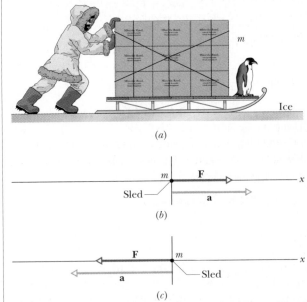

FIGURE 5-3 Sample Problem 5-1. (*a*) A student pushes a loaded sled over a frictionless surface. (*b*) A free-body diagram for part (a) in the problem, showing the net force acting on the sled and the acceleration the net force produces. (*c*) A free-body diagram for part (b). The student now pushes in the opposite direction on the sled, reversing its acceleration.

$$a = \frac{v - v_0}{t} = \frac{(-1.6 \text{ m/s}) - (1.6 \text{ m/s})}{4.5 \text{ s}}$$

$$= -0.711 \text{ m/s}^2.$$

This is larger in magnitude than the acceleration in (a), namely 0.542 m/s², so it stands to reason that the student must push with a greater force this time. We find this greater force from Eqs. 5-2, with a_x being a:

$$F_x = ma_x = (240 \text{ kg})(-0.711 \text{ m/s}^2)$$

$$= -171 \text{ N.} \qquad \text{(Answer)}$$

The minus sign shows that the student must push the sled in the direction of decreasing x, that is, to the left in Fig. 5-3c, the free-body diagram for this situation.

SAMPLE PROBLEM 5-2

In a two-dimensional tug-of-war, Alex, Betty, and Charles pull on an automobile tire, at angles as shown in Fig. 5-4a, which is an overhead view. The tire remains stationary in spite of the three pulls. Alex pulls with force $\mathbf{F}_A$ of magnitude 220 N, and Charles pulls with force $\mathbf{F}_C$ of magnitude 170 N. The direction of $\mathbf{F}_C$ is not given. What is the magnitude of Betty's force $\mathbf{F}_B$?

SOLUTION: Figure 5-4b is a free-body diagram for the tire. Because the acceleration of the tire is zero, Eq. 5-1 tells us that the net force on the tire must also be zero:

$$\Sigma \mathbf{F} = \mathbf{F}_A + \mathbf{F}_B + \mathbf{F}_C = m\mathbf{a} = 0.$$

This vector relation is equivalent to the first two scalar relations of Eqs. 5-2. Along the x axis we have

$$\Sigma F_x = F_{Ax} + F_{Bx} + F_{Cx} = 0, \qquad (5\text{-}4)$$

and along the y axis we have

$$\Sigma F_y = F_{Ay} + F_{By} + F_{Cy} = 0. \qquad (5\text{-}5)$$

Using the given data and the angles in Fig. 5-4b, we now substitute into Eqs. 5-4 and 5-5, using signs to indicate the directions of the vector components. Equation 5-4 becomes

$$\Sigma F_x = -F_A \cos 47.0° + 0 + F_C \cos \phi = 0.$$

Substituting known values yields

$$-(220 \text{ N})(\cos 47.0°) + 0 + (170 \text{ N})(\cos \phi) = 0,$$

which gives us

$$\phi = \cos^{-1} \frac{(220 \text{ N})(\cos 47.0°)}{170 \text{ N}} = 28.0°.$$

Similarly, Eq. 5-5 becomes

$$\Sigma F_y = F_A \sin 47.0° - F_B + F_C \sin \phi = 0,$$

in which we have substituted $-F_B$ for F_{By} because Betty's pull is entirely along the negative direction of y. Substituting known values here yields

$$(220 \text{ N})(\sin 47.0°) - F_B + (170 \text{ N})(\sin \phi) = 0.$$

We next substitute 28.0° for ϕ and solve the equation for F_B, finding that

$$F_B = (220 \text{ N})(\sin 47.0°) + (170 \text{ N})(\sin 28.0°)$$

$$= 241 \text{ N.} \qquad \text{(Answer)}$$

Note that we first had to solve the equations along the x axis to use $\phi = 28.0°$ in the equations along the y axis. If we had started with the equations along the y axis, we would have just gotten "stuck." So, if you get stuck on one axis, try another.

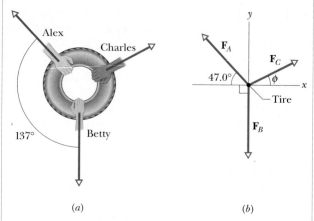

(a) $\qquad\qquad$ (b)

FIGURE 5-4 Sample Problem 5-2. (a) An overhead view of three people pulling on a tire. (b) A free-body diagram for the tire.

SAMPLE PROBLEM 5-3

Figure 5-5a shows an overhead view of a 2 kg cookie tin being accelerated at 8 m/s² across a frictionless surface by three horizontal forces. Forces $\mathbf{F}_1$ and $\mathbf{F}_2$ have magnitudes of 10 N and 12 N, respectively. Figure 5-5b is an incomplete free-body diagram for the situation, with the acceleration $\mathbf{a}$ included. What is the third force $\mathbf{F}_3$ in unit vector notation?

SOLUTION: The net force of the three horizontal forces causes the horizontal acceleration, and Eq. 5-1 gives us

$$\Sigma \mathbf{F} = \mathbf{F}_1 + \mathbf{F}_2 + \mathbf{F}_3 = m\mathbf{a}.$$

From Eqs. 5-2, we have, along the x axis,

$$\Sigma F_x = F_{1x} + F_{2x} + F_{3x} = ma_x, \qquad (5\text{-}6)$$

and along the y axis

$$\Sigma F_y = F_{1y} + F_{2y} + F_{3y} = ma_y. \qquad (5\text{-}7)$$

By rewriting Eq. 5-6 in terms of magnitudes and angles, and including signs to indicate directions, we obtain

$$-F_1 \cos 60° + 0 + F_{3x} = ma \sin 30°.$$

Substitution of the given data yields

$$-(10 \text{ N}) \cos 60° + 0 + F_{3x} = (2 \text{ kg})(8 \text{ m/s}^2) \sin 30°,$$

which tells us that

$$F_{3x} = (10 \text{ N}) \cos 60° + (2 \text{ kg})(8 \text{ m/s}^2) \sin 30°$$
$$= 13 \text{ N}.$$

Similarly, Eq. 5-7 becomes

$$-F_1 \sin 60° + F_2 + F_{3y} = -ma \cos 30°$$

and then

$$-(10 \text{ N}) \sin 60° + 12 \text{ N} + F_{3y} = -(2 \text{ kg})(8 \text{ m/s}^2) \cos 30°,$$

which tells us that

$$F_{3y} = -17.2 \text{ N} \approx 17 \text{ N}.$$

Thus, the third force is

$$\mathbf{F}_3 = (13 \text{ N})\mathbf{i} - (17 \text{ N})\mathbf{j}. \qquad \text{(Answer)}$$

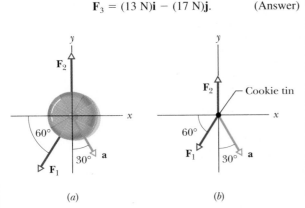

(a)   (b)

FIGURE 5-5 Sample Problem 5-3. (a) An overhead view of a cookie tin that is being accelerated by three horizontal forces, two of which are shown. (b) A free-body diagram for the cookie tin.

CHECKPOINT 3: The figure shows overhead views of four situations in which two forces accelerate the same block across a frictionless floor. Rank the situations according to the magnitudes of (a) the net force on the block and (b) the acceleration of the block, greatest first.

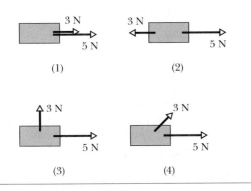

PROBLEM SOLVING TACTICS

TACTIC 1: *Reading Force Problems*

Read the problem statement several times until you have a good mental picture of what the situation is, what data are given, and what is requested. In Sample Problem 5-1, for example, you should tell yourself: "Someone is pushing a sled. Its speed changes, so acceleration is involved. The motion is along a straight line. A force is given in one part and asked for in the other, and so the situation looks like Newton's second law applied to one-dimensional motion."

If you know what the problem is about but don't know what to do next, put the problem aside and reread the text. If you are hazy about Newton's second law, reread that section. Study the sample problems. The one-dimensional-motion parts of Sample Problem 5-1 and the constant acceleration should send you back to Chapter 2 and especially to Table 2-1, which displays all the equations you are likely to need.

TACTIC 2: *Draw Two Types of Figures*

You may need two figures. One is a rough sketch of the actual real-world situation. When you draw the forces on it, place the tail of each force vector either on the boundary of or within the body feeling that force. The other figure is a free-body diagram in which the forces on a *single* body are drawn, with the body represented with a dot or a sketch. Place the tail of each force vector on the dot or sketch.

TACTIC 3: *What Is Your System?*

If you are using Newton's second law, you must know what body or system you are applying it to. In Sample Problem 5-1 it is the sled (not the student or the ice). In Sample Problem 5-2, it is the tire (not the ropes or the people). In Sample Problem 5-3, it is the cookie tin.

TACTIC 4: *Choose Your Axes Wisely*

In Sample Problem 5-2, we saved a lot of work by choosing one of our coordinate axes to coincide with one of the forces (the y axis with $\mathbf{F}_B$). Then if you get stuck in the equations along one axis, go to another axis.

5-6 SOME PARTICULAR FORCES

Weight

The **weight W** of a body is a force that pulls the body directly toward a nearby astronomical body; in everyday circumstances that astronomical body is the Earth. The force is primarily due to an attraction—called a **gravitational attraction**—between the two bodies; we discuss it in detail in Chapter 14. For now we consider only situations in which a body with mass m is located at a point where the free-fall acceleration has magnitude g. Then the magnitude W of the weight (force) vector acting on the body is

$$W = mg. \qquad (5\text{-}8)$$

The *weight vector* itself can be written either as

$$\mathbf{W} = -mg\mathbf{j} = -W\mathbf{j} \qquad (5\text{-}9)$$

(where $+\mathbf{j}$ points upward, away from the Earth), or as

$$\mathbf{W} = m\mathbf{g}, \qquad (5\text{-}10)$$

where $\mathbf{g}$ represents the free-fall acceleration vector. In many cases, the choice of notation is up to you, but you need to understand what you mean by it and not get tripped up by, say, writing Eq. 5-8 when you mean Eq. 5-10.

Since weight is a force, its SI unit is the newton. *It is not mass,* and its magnitude at any given location depends on the value of g there. A bowling ball might weigh 71 N on the Earth, but only 12 N on the Moon, where the free-fall acceleration is different. The ball's mass, 7.2 kg, is the same in either place, because mass is an intrinsic property of the ball alone. (If you want to lose weight, climb a mountain. Not only will the exercise reduce your mass, but the increased elevation means you are farther from the center of the Earth, and that means the value of g is less. So your weight will be less.)

Normally we assume that weight is measured from an inertial frame. If it is, instead, measured from a noninertial frame (an example comes up in Sample Problems 5-11*b* and *c*), the measurement gives an **apparent weight** instead of the actual weight.

We can *weigh* a body by placing it on one of the pans of an equal-arm balance (Fig. 5-6) and then adding reference bodies (whose masses are known) on the other pan until we strike a balance. The masses on the pans then match, and we know the mass m of the body. If we know the value of g for the location of the balance, we can find the weight of the body with Eq. 5-8.

We can also weigh a body with a spring scale (Fig. 5-7). The body stretches a spring, moving a pointer along a scale that has been calibrated and marked in either mass or weight units. (Most bathroom scales in the United States work this way and read in pounds.) If the scale is marked in mass units, it is accurate only where the value of g is the same as where the scale was calibrated.

The Normal Force

When a body is pressed against a surface, the body experiences a force that is perpendicular to the surface. The force is called the **normal force N,** the name coming from the mathematical term *normal,* meaning "perpendicular."

If a body rests on a horizontal surface as in Fig. 5-8*a*, $\mathbf{N}$ is directed upward and the body's weight $\mathbf{W} = m\mathbf{g}$ is

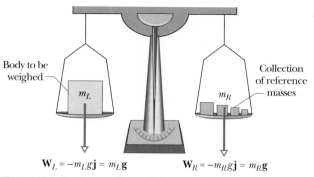

$$\mathbf{W}_L = -m_L g\mathbf{j} = m_L\mathbf{g} \qquad\qquad \mathbf{W}_R = -m_R g\mathbf{j} = m_R\mathbf{g}$$

FIGURE 5-6 An equal-arm balance. When the device is in balance, the masses on the left (*L*) and right (*R*) pans are equal.

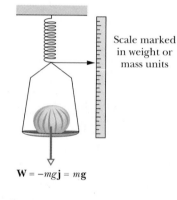

FIGURE 5-7 A spring scale. The reading is proportional to the *weight* of the object placed on the pan, and the scale gives that weight if marked in weight units. If, instead, it is marked in mass units, the reading is accurate only if the free-fall acceleration g is the same as where the scale was calibrated.

$$\mathbf{W} = -mg\mathbf{j} = m\mathbf{g}$$

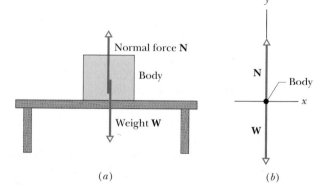

FIGURE 5-8 (*a*) A body resting on a tabletop experiences a normal force $\mathbf{N}$ perpendicular to the tabletop. (*b*) The corresponding free-body diagram.

directed downward. For this *particular* arrangement, we find the magnitude of $\mathbf{N}$ from the second of Eqs. 5-2:

$$\sum F_y = N - mg = ma_y, \qquad (5\text{-}11)$$

and so, with $a_y = 0$,

$$N = mg. \qquad (5\text{-}12)$$

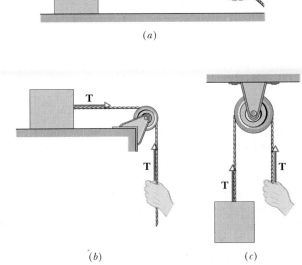

CHECKPOINT **4:** In Fig. 5-8, is the magnitude of the normal force **N** greater than, less than, or equal to the weight magnitude mg if the body and table are in an elevator that is moving upward (a) at a constant speed and (b) at an increasing speed?

FIGURE 5-10 (*a*) The cord, pulled taut, is under tension. It pulls on the body and the hand with force **T**, even if it runs around a massless, frictionless pulley as in (*b*) and (*c*).

Friction

If we slide or attempt to slide a body over a surface, the motion is resisted by a bonding between the body and surface. (We discuss this more in the next chapter.) The resistance is considered to be a single force **f**, called the **frictional force,** or simply **friction.** This force is directed along the surface, opposite the direction of the intended motion (Fig. 5-9). Sometimes, to simplify a situation, friction is assumed to be negligible, and the surface is said to be *frictionless*.

Tension

When a cord (or a rope, cable, or other such object) is attached to a body and pulled taut, the cord is said to be under **tension.** It pulls on the body with a force **T**, whose direction is away from the body and along the cord at the point of attachment (Fig. 5-10a).

A cord is often said to be *massless* (meaning its mass is negligible compared to the body's mass) and unstretchable. The cord exists only as a connection between two bodies. It pulls on both bodies with the same magnitude T, even if the bodies and the cord are accelerating and even if the cord runs around a *massless, frictionless pulley* (Figs. 5-10b and c). Such a pulley has negligible mass compared to the bodies and has negligible friction on its axle opposing its rotation.

CHECKPOINT **5:** The body that is suspended by a rope in Fig. 5-10c has a weight of 75 N. Is T equal to, greater than, or less than 75 N when the body is moving upward (a) at constant speed, (b) at increasing speed, and (c) at decreasing speed?

SAMPLE PROBLEM 5-4

Let us return to John Massis and the railroad cars, and assume that Massis pulled (with his teeth) on his end of the rope with a constant force that was 2.5 times his body weight, at an angle θ of 30° from the horizontal. His mass m was 80 kg. The weight W of the passenger cars was 7.0×10^5 N (about 80 tons), and he moved them 1.0 m along the rails. Assume that the rolling wheels encountered no retarding force from the rails. What was the train's speed at the end of the pull?

SOLUTION: Figure 5-11 is a free-body diagram for the cars, which are represented by a dot. The x axis runs along the rails. From Eqs. 5-2, we have

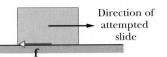

FIGURE 5-9 A frictional force **f** opposes the attempted slide of a body over a surface.

Direction of attempted slide

$$\sum F_x = T \cos \theta = Ma_x, \qquad (5\text{-}13)$$

in which M is the mass of the cars. We can find T and M from the given data.

With our assumptions, the pull from Massis is

$$T = 2.5mg = (2.5)(80 \text{ kg})(9.8 \text{ m/s}^2) = 1960 \text{ N}$$

(or 440 lb), which is about what a good middle-weight weight lifter can lift—and far from a superhuman force.

The weight W of the cars is, by Eq. 5-8,

$$W = Mg,$$

so their mass M must be

$$M = \frac{W}{g} = \frac{7.0 \times 10^5 \text{ N}}{9.8 \text{ m/s}^2} = 7.143 \times 10^4 \text{ kg}.$$

Now, from Eq. 5-13, we find their acceleration to be

$$a_x = \frac{T \cos \theta}{M} = \frac{(1960 \text{ N})(\cos 30°)}{7.143 \times 10^4 \text{ kg}}$$
$$= 2.376 \times 10^{-2} \text{ m/s}^2.$$

To solve for the speed of the cars at the end of the pull, we use Eq. 2-16, with subscripts for the x axis and with $v_0 = 0$ and $x - x_0 = 1.0$ m:

$$v_x^2 = v_{0x}^2 + 2a_x(x - x_0),$$

or

$$v_x = \sqrt{0 + (2)(2.376 \times 10^{-2} \text{ m/s}^2)(1.0 \text{ m})}$$
$$= 0.22 \text{ m/s}. \qquad \text{(Answer)}$$

Massis would have done better if the rope had been attached higher on the car, so that it was horizontal. Can you see why?

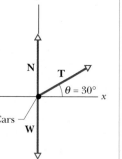

FIGURE 5-11 Sample Problem 5-4. Free-body diagram for the passenger cars pulled by Massis. The vectors are not drawn to scale; the tension in the rope is *much* smaller than the weight and normal force.

5-7 NEWTON'S THIRD LAW

Forces come in pairs. If a hammer exerts a force on a nail, the nail exerts a force of equal magnitude but opposite direction on the hammer. If you lean against a brick wall, the wall pushes back on you.

Let body A in Fig. 5-12 exert a force $\mathbf{F}_{BA}$ on body B; experiment shows that body B then exerts a force $\mathbf{F}_{AB}$ on body A. These two forces are equal in magnitude and op-

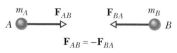

FIGURE 5-12 Newton's third law. Body A exerts a force $\mathbf{F}_{BA}$ on body B, while body B exerts a force $\mathbf{F}_{AB}$ on body A, where $\mathbf{F}_{AB} = -\mathbf{F}_{BA}$.

positely directed. That is,

$$\mathbf{F}_{AB} = -\mathbf{F}_{BA} \qquad \text{(Newton's third law)}. \qquad (5\text{-}14)$$

Note the order of the subscripts. $\mathbf{F}_{AB}$, for example, is the force exerted *on* body A *by* body B. Equation 5-14 holds regardless of whether the bodies move or remain stationary.

Equation 5-14 sums up Newton's third law of motion. Commonly, one of these forces (it does not matter which) is called the **action force.** The other member of the pair is then called the **reaction force.** Every time you find a force, a good question is: Where is its reaction force?

The words "To every action there is always an equal and opposite reaction" have become enshrined in the popular language and mean various things to various speakers. In physics, however, these words mean Eq. 5-14 and nothing else. In particular, cause and effect are not involved; either force can be the action force.

You may think: "If every force has an associated force that is equal in magnitude and opposite in direction, why don't they cancel each other? How can anything ever get moving?" The answer is simple. As Fig. 5-12 shows:

> The forces of an action–reaction pair *always* act on *different* bodies; thus they do not combine to give a net force and cannot cancel each other.

Two forces that act on the *same* body are *not* an action–reaction pair, even though they may be equal in magnitude

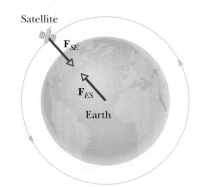

FIGURE 5-13 A satellite in Earth orbit. The forces shown are an action–reaction pair. Note that they act on different bodies.

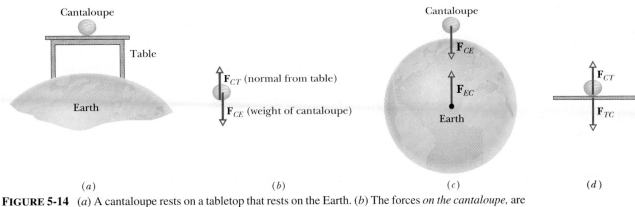

FIGURE 5-14 (*a*) A cantaloupe rests on a tabletop that rests on the Earth. (*b*) The forces *on the cantaloupe*, are $\mathbf{F}_{CT}$ and $\mathbf{F}_{CE}$. The cantaloupe is stationary because these two forces balance. (*c*) The action–reaction pair for the cantaloupe–Earth forces. (*d*) The action–reaction pair for the cantaloupe–table forces.

and opposite in direction. Let's identify the action–reaction pairs in two examples.

An Orbiting Satellite

Figure 5-13 shows an orbiting satellite. The only force that acts on it is $\mathbf{F}_{SE}$, the force exerted *on* the satellite *by* the gravitational pull of the Earth. Where is the corresponding reaction force? It is $\mathbf{F}_{ES}$, the force acting on the Earth due to the gravitational pull of the satellite; the pull is taken to be at the center of the Earth.

You may think that the tiny satellite cannot exert much of a gravitational pull on the Earth but it does, exactly as Newton's third law requires. That is, considering magnitudes only, $F_{ES} = F_{SE}$. The force $\mathbf{F}_{ES}$ causes the Earth to accelerate, but, because of the Earth's large mass, its acceleration is too small to be detected.

A Cantaloupe Resting on a Table

Figure 5-14*a* shows a cantaloupe at rest on a table.* The Earth pulls downward on the cantaloupe with a force $\mathbf{F}_{CE}$, the cantaloupe's weight. The cantaloupe does not accelerate because this force is canceled by an equal and opposite normal force $\mathbf{F}_{CT}$ exerted on the cantaloupe by the table. (See Fig. 5-14*b*.) However, $\mathbf{F}_{CE}$ and $\mathbf{F}_{CT}$ do *not* form an action–reaction pair *because they act on the same body, the cantaloupe.*

The reaction force to $\mathbf{F}_{CE}$ is $\mathbf{F}_{EC}$, the (gravitational) force with which the cantaloupe attracts the Earth. This action–reaction pair is shown in Fig. 5-14*c*.

The reaction force to $\mathbf{F}_{CT}$ is $\mathbf{F}_{TC}$, the force exerted on the table by the cantaloupe. This action–reaction pair is shown in Fig. 5-14*d*. The action–reaction pairs in this

problem, and the bodies on which they act, are then

first pair: $\mathbf{F}_{CE} = -\mathbf{F}_{EC}$ (cantaloupe and Earth)

and

second pair: $\mathbf{F}_{CT} = -\mathbf{F}_{TC}$ (cantaloupe and table).

CHECKPOINT 6: Suppose that the cantaloupe and table of Fig. 5-14 are in an elevator cab that begins to accelerate upward. (a) Do the magnitudes of forces $\mathbf{F}_{TC}$ and $\mathbf{F}_{CT}$ increase, decrease, or stay the same? (b) Are those two forces still equal in magnitude but opposite in direction? (c) Do the magnitudes of forces $\mathbf{F}_{CE}$ and $\mathbf{F}_{EC}$ increase, decrease, or stay the same? (d) Are those two forces still equal in magnitude but opposite in direction?

5-8 APPLYING NEWTON'S LAWS

The rest of this chapter consists of sample problems. You should pore over them, learning not just their particular answers but, instead, the procedures for attacking a problem. Especially important is knowing how to translate a sketch of a situation into a free-body diagram with appropriate axes, so that Newton's laws can be applied. We begin with Sample Problem 5-5, which is worked out in exhaustive detail, using a question-and-answer format.

SAMPLE PROBLEM 5-5

Figure 5-15 shows a block (the *sliding block*) whose mass M is 3.3 kg. It is free to move along a horizontal frictionless surface such as an air table. The sliding block is connected by a cord that extends around a massless, frictionless pulley to a second block (the *hanging block*), whose mass m is 2.1 kg. The hang-

---

*We ignore small complications caused by the rotation of the Earth.

ing block falls and the sliding block accelerates to the right. Find (a) the acceleration of the sliding block, (b) the acceleration of the hanging block, and (c) the tension in the cord.

Q *What is this problem all about?*

You are given two massive objects, the sliding block and the hanging block. It might not occur to you, but you are also given the Earth, which pulls on each of these objects; without the Earth, nothing would happen. A total of five forces act on the blocks, as shown in Fig. 5-16:

1. The cord pulls to the right on the sliding block with a force of magnitude *T*.

2. The cord pulls upward on the hanging block with a force of the same magnitude *T*. This upward force keeps the hanging block from falling freely, which it would otherwise do. We assume that the cord has the same tension throughout its length; the pulley just serves to change the direction of this force, without changing its magnitude.

3. The Earth pulls down on the sliding block with a force *M***g**, the weight of the sliding block.

4. The Earth pulls down on the hanging block with a force *m***g**, the weight of the hanging block.

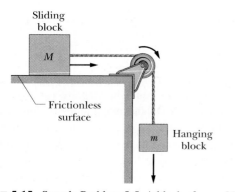

FIGURE 5-15 Sample Problem 5-5. A block of mass *M* on a horizontal frictionless surface is connected to a block of mass *m* by a cord that wraps over a pulley. Both cord and pulley are massless. The pulley is frictionless. The arrows indicate the motion when the system is released from rest.

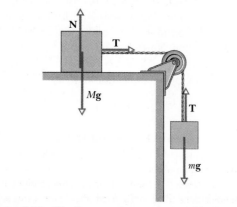

FIGURE 5-16 The forces acting on the two blocks.

5. The table pushes up on the sliding block with a normal force **N**.

There is another thing that you should note. We assume that the cord does not stretch, so that if the hanging block falls 1 mm in a certain time, the sliding block moves 1 mm to the right in that same interval. The blocks move together and their accelerations have the same magnitude *a*.

Q *How do I classify this problem? Should it suggest a particular law of physics to me?*

Yes, it should. Forces, masses, and accelerations are involved, and that should suggest Newton's second law of motion, $\Sigma\mathbf{F} = m\mathbf{a}$.

Q *If I apply that law to this problem, to what body should I apply it?*

We focus on two bodies in this problem, the sliding block and the hanging block. Although they are extended objects, we can treat each block as a particle because every small part of it (every atom, say) moves in exactly the same way. Apply Newton's second law separately to each block.

Q *What about the pulley?*

We cannot represent the pulley as a particle because different parts of it move in different ways. When we discuss rotation, we shall deal with pulleys in detail. Meanwhile, we get around the problem by using a pulley whose mass is negligible compared with the masses of the two blocks.

Q *OK. Now how do I apply $\Sigma\mathbf{F} = m\mathbf{a}$ to the sliding block?*

Represent the sliding block as a particle of mass *M* and draw *all* the forces that act *on* it, as in Fig. 5-17. This is the block's *free-body diagram*. There are three forces. Next, locate a set of axes. It makes sense to draw the *x* axis parallel to the table, in the direction in which the block moves.

Q *Thanks, but you still haven't told me how to apply $\Sigma\mathbf{F} = m\mathbf{a}$ to the sliding block. All you have done is explain how to draw a free-body diagram.*

Right you are. The expression $\Sigma\mathbf{F} = m\mathbf{a}$ is a vector equation and you can write it as three scalar equations:

$$\Sigma F_x = Ma_x, \quad \Sigma F_y = Ma_y, \quad \Sigma F_z = Ma_z, \quad (5\text{-}15)$$

in which ΣF_x, ΣF_y, and ΣF_z are the components of the net force. Because the sliding block does not move vertically, we know there is no net force in the *y* direction: the weight $\mathbf{W} = M\mathbf{g}$ of the sliding block is balanced by the upward-acting normal force **N** on the block. No force acts in the *z* direction, which is perpendicular to the page. We can, however, apply the first of Eqs. 5-15.

In the *x* direction, there is only one force component, so $\Sigma F_x = Ma_x$ becomes

$$T = Ma. \quad (5\text{-}16)$$

This equation contains two unknowns, *T* and *a*, so we cannot yet solve it. Recall, however, that we have not said anything about the hanging block.

Q *I agree. How do I apply $\Sigma\mathbf{F} = m\mathbf{a}$ to the hanging block?*

Draw a free-body diagram for the block, as in Fig. 5-18. This time, we use the second of Eqs. 5-15, finding

$$\sum F_y = T - mg = -ma, \qquad (5\text{-}17)$$

where the minus sign on the right side of the equation indicates that the hanging block accelerates downward, in the negative direction of the y axis. Equation 5-17 yields

$$mg - T = ma. \qquad (5\text{-}18)$$

This contains the same two unknowns as Eq. 5-16 does. If you add these equations, T will cancel out. Solving for a then yields

$$a = \frac{m}{M + m} g. \qquad (5\text{-}19)$$

Substituting this result into Eq. 5-16 yields

$$T = \frac{Mm}{M + m} g. \qquad (5\text{-}20)$$

Putting in the numbers gives, for these two quantities,

$$a = \frac{m}{M + m} g = \frac{2.1 \text{ kg}}{3.3 \text{ kg} + 2.1 \text{ kg}} (9.8 \text{ m/s}^2)$$

$$= 3.8 \text{ m/s}^2 \qquad \text{(Answer)}$$

and

$$T = \frac{Mm}{M + m} g = \frac{(3.3 \text{ kg})(2.1 \text{ kg})}{3.3 \text{ kg} + 2.1 \text{ kg}} (9.8 \text{ m/s}^2)$$

$$= 13 \text{ N}. \qquad \text{(Answer)}$$

Q *The problem is now solved, right?*

That's a fair question, but we are here not only to solve problems but also to learn physics. This problem is not really finished until we have studied the results to see if they make sense. This is often a much more confidence-building experience than simply getting the right answer.

Look first at Eq. 5-19. Note that it is dimensionally correct and also that the acceleration a will always be less than g. This is as it must be, because the hanging block is not in free fall. The cord pulls upward on it.

Look now at Eq. 5-20, which we rewrite in the form

$$T = \frac{M}{M + m} mg. \qquad (5\text{-}21)$$

In this form, it is easier to see that this equation is also dimensionally correct, because both T and mg are forces. Equation 5-21 also lets us see that the tension in the cord is always less than mg, the weight of the hanging block. That is a comforting thought because, if T were *greater* than mg, the hanging block would accelerate upward.

We can also check the results by studying special cases, in which we can guess what the answers must be. A simple example is to put $g = 0$, as if the experiment were carried out in interstellar space. We know that in that case, the blocks would not move from rest and there would be no tension in the cord. Do the formulas predict this? Yes, they do. If you put $g = 0$ in Eqs. 5-19 and 5-20, you find $a = 0$ and $T = 0$. Two more special cases that you might try are $M = 0$ and $m \to \infty$.

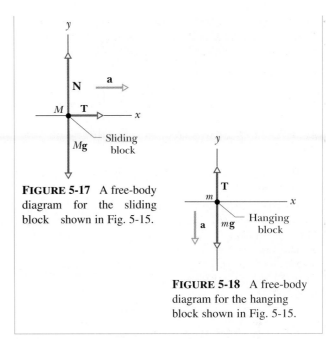

FIGURE 5-17 A free-body diagram for the sliding block shown in Fig. 5-15.

FIGURE 5-18 A free-body diagram for the hanging block shown in Fig. 5-15.

SAMPLE PROBLEM 5-5—ANOTHER WAY

The acceleration a of the blocks of Fig. 5-15 can be found in two lines of algebra if we (a) employ an unconventional axis, call it u, that runs through *both* blocks and along the cord as shown in Fig. 5-19*a*, and then (b) mentally straighten out the u axis as in Fig. 5-19*b* and treat the blocks as being portions of a single composite body with mass $M + m$. A free-body diagram for the two-block system is shown in Fig. 5-19*c*.

SOLUTION: Note that there is only one force acting on the composite body along the u axis, and that is the force mg in the positive direction of the axis. The tension T of Fig. 5-16 is now internal to the composite body and so does not enter into Newton's second law. The force exerted by the pulley on the cord is perpendicular to the u axis; so it too does not enter.

Using Eqs. 5-2 as a guide, we write a component equation for the acceleration along the u axis:

$$\sum F_u = (M + m)a_u,$$

where the mass of the body is $M + m$. The acceleration of the composite body along the u axis (and of the individual blocks, since they are connected) has magnitude a. The only force on the composite body along the u axis has a magnitude of mg. So our equation becomes

$$mg = (M + m)a,$$

or $\qquad\qquad a = \dfrac{m}{M + m} g, \qquad (5\text{-}22)$

which matches Eq. 5-19.

To find T, we apply Newton's second law to either block, obtaining either Eq. 5-16 or Eq. 5-18. We then substitute for a from Eq. 5-22 and solve for T, getting Eq. 5-20.

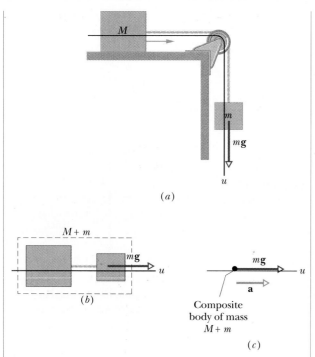

(a)

(b)

Composite
body of mass
$M + m$

(c)

FIGURE 5-19 (a) An "axis" u runs through the blocks-and-cord system of Fig. 5-15. (b) The blocks are rearranged to straighten u and then are treated as a single body with mass $M + m$. (c) The associated free-body diagram, considering only forces along u. There is one such force.

SAMPLE PROBLEM 5-6

A block whose mass M is 33 kg is pushed across a frictionless surface by a stick whose mass m is 3.2 kg, as in Fig. 5-20a. The block is moved (from rest) a distance $d = 77$ cm in 1.7 s at constant acceleration.

(a) Identify all horizontal action–reaction force pairs in this problem.

SOLUTION: As the exploded view of Fig. 5-20b shows, there are two action–reaction pairs:

first pair: $\mathbf{F}_{HS} = -\mathbf{F}_{SH}$ (hand and stick)

second pair: $\mathbf{F}_{SB} = -\mathbf{F}_{BS}$ (stick and block).

The force $\mathbf{F}_{HS}$ on the hand from the stick is the force that you would feel if the hand in Fig. 5-20 were yours.

(b) What force must the hand apply to the stick?

SOLUTION: This is the force that accelerates the block and stick. To find it, we must first find the constant acceleration a, using Eq. 2-15:

$$x - x_0 = v_0 t + \tfrac{1}{2} a t^2.$$

Setting $v_0 = 0$ and $x - x_0 = d$, and solving for a, give

$$a = \frac{2d}{t^2} = \frac{(2)(0.77 \text{ m})}{(1.7 \text{ s})^2} = 0.533 \text{ m/s}^2.$$

To find the force that the hand exerts, we apply Newton's second law to a system consisting of the stick and the block taken together. Thus,

$$F_{SH} = (M + m)a = (33 \text{ kg} + 3.2 \text{ kg})(0.533 \text{ m/s}^2)$$
$$= 19.3 \text{ N} \approx 19 \text{ N}. \qquad \text{(Answer)}$$

(c) With what force does the stick push on the block?

SOLUTION: To find this force, we apply Newton's second law to the block alone:

$$F_{BS} = Ma = (33 \text{ kg})(0.533 \text{ m/s}^2)$$
$$= 17.6 \text{ N} \approx 18 \text{ N}. \qquad \text{(Answer)}$$

(d) What is the net force on the stick?

SOLUTION: We can find the magnitude F of this force in two ways. First, using results from (b) and (c), we have

$$F = F_{SH} - F_{SB} = 19.3 \text{ N} - 17.6 \text{ N}$$
$$= 1.7 \text{ N}. \qquad \text{(Answer)}$$

Note that we have used Newton's third law here, in assuming that $\mathbf{F}_{SB}$, the force on the stick from the block, has the same magnitude (17.6 N before rounding) as $\mathbf{F}_{BS}$.

The second way to arrive at an answer is to apply Newton's second law to the stick directly. We have

$$F = ma = (3.2 \text{ kg})(0.533 \text{ m/s}^2) = 1.7 \text{ N}, \qquad \text{(Answer)}$$

in agreement with our first result. This is as it must be because the two methods are algebraically identical; check it out.

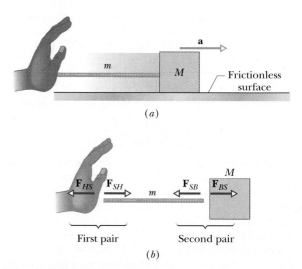

(a)

(b)

FIGURE 5-20 Sample Problem 5-6. (a) A block of mass M is pushed over a frictionless surface by a stick of mass m. (b) An exploded view, showing the action–reaction pairs between the hand and the stick (first pair) and between the stick and the block (second pair).

SAMPLE PROBLEM 5-7

Figure 5-21a shows a block of mass $m = 15$ kg hanging from three cords. What are the tensions in the cords?

SOLUTION: In the free-body diagram for the block (Fig. 5-21b), tension $\mathbf{T}_C$ from cord C pulls upward while the block's weight $m\mathbf{g}$ is directed downward. Since the system is at rest, Newton's second law for the block yields

$$\sum \mathbf{F} = \mathbf{T}_C + m\mathbf{g} = 0.$$

Because forces $\mathbf{T}_C$ and $m\mathbf{g}$ are only vertical, this equation gives us one scalar equation:

$$\sum F_y = T_C - mg = 0.$$

Substituting known values, we find

$$T_C = mg = (15 \text{ kg})(9.8 \text{ m/s}^2)$$
$$= 147 \text{ N} \approx 150 \text{ N}. \qquad \text{(Answer)}$$

The clue to our next step is to realize that the knot where the three cords join is the only point at which all three forces act, and it is this knot to which we should apply Newton's second law. Figure 5-21c is the free-body diagram for the knot. Since the knot is not accelerated, the net force acting on it must be zero. Thus

$$\sum \mathbf{F} = \mathbf{T}_A + \mathbf{T}_B + \mathbf{T}_C = 0.$$

This vector equation is equivalent to the two scalar equations

$$\sum F_y = T_A \sin 28° + T_B \sin 47° - T_C = 0 \quad (5\text{-}23)$$

and

$$\sum F_x = -T_A \cos 28° + T_B \cos 47° = 0. \quad (5\text{-}24)$$

Note carefully that, when we write the x component of $\mathbf{T}_A$ as $T_A \cos 28°$, we must include a minus sign to show that it extends in the negative direction of the x axis.

Substituting numerical values into Eqs. 5-23 and 5-24 leads to

$$T_A(0.469) + T_B(0.731) = 147 \text{ N} \quad (5\text{-}25)$$

and

$$T_B(0.682) = T_A(0.883). \quad (5\text{-}26)$$

From Eq. 5-26, we have

$$T_B = \frac{0.883}{0.682} T_A = 1.29 T_A.$$

Substituting this into Eq. 5-25 and solving for T_A, we obtain

$$T_A = \frac{147 \text{ N}}{0.469 + (1.29)(0.731)}$$
$$= 104 \text{ N} \approx 100 \text{ N}. \qquad \text{(Answer)}$$

Finally, T_B is found from

$$T_B = 1.29 T_A = (1.29)(104 \text{ N})$$
$$= 134 \text{ N} \approx 130 \text{ N}. \qquad \text{(Answer)}$$

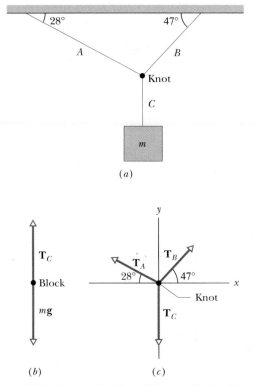

FIGURE 5-21 Sample Problem 5-7. (a) A block of mass m hangs from three cords. (b) A free-body diagram for the block. (c) A free-body diagram for the knot at the intersection of the three cords.

SAMPLE PROBLEM 5-8

Figure 5-22a shows a block of mass $m = 15$ kg held by a cord on a frictionless inclined plane. What is the tension in the cord if $\theta = 27°$? What force does the plane exert on the block?

SOLUTION: Figure 5-22b is the free-body diagram for the block. The following forces act on it: (1) a normal force $\mathbf{N}$, exerted outward on the block by the plane on which it rests, (2) the tension $\mathbf{T}$ in the cord, and (3) the weight $\mathbf{W}$ $(= m\mathbf{g})$ of the block. Because the acceleration of the block is zero, the net force acting on the block must also be zero by Newton's second law:

$$\sum \mathbf{F} = \mathbf{T} + \mathbf{N} + m\mathbf{g} = 0. \quad (5\text{-}27)$$

We choose a coordinate system with the x axis parallel to the plane. With this choice, not one but two forces ($\mathbf{N}$ and $\mathbf{T}$) line up with coordinate axes (a bonus). Note that the angle between the weight vector and the negative direction of the y axis equals the slant angle of the plane. The x and y components of that vector are found with the triangle of Fig. 5-22c. The component versions of Eq. 5-27 are

$$\sum F_x = T - mg \sin \theta = 0$$

and

$$\sum F_y = N - mg \cos \theta = 0.$$

Thus

$$T = mg \sin \theta$$
$$= (15 \text{ kg})(9.8 \text{ m/s}^2)(\sin 27°)$$
$$= 67 \text{ N} \qquad \text{(Answer)}$$

and

$$N = mg \cos \theta$$
$$= (15 \text{ kg})(9.8 \text{ m/s}^2)(\cos 27°)$$
$$= 131 \text{ N} \approx 130 \text{ N}. \qquad \text{(Answer)}$$

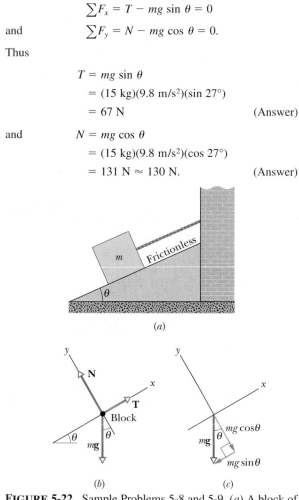

(a)

(b)   (c)

FIGURE 5-22 Sample Problems 5-8 and 5-9. (a) A block of mass m rests on a smooth plane, held there by a cord. (b) A free-body diagram for the block. Note how the coordinate axes are placed. (c) Finding the x and y components of $m\mathbf{g}$.

CHECKPOINT 7: In the figure, horizontal force **F** is applied to the block. (a) Is the component of **F** that is perpendicular to the ramp $F \cos \theta$ or $F \sin \theta$? (b) Does the presence of **F** increase or decrease the magnitude of the normal force on the block from the ramp?

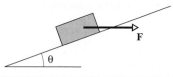

SAMPLE PROBLEM 5-9

Suppose you cut the cord holding the block on the plane in Fig. 5-22a. With what acceleration will the block move?

SOLUTION: Cutting the cord removes the tension **T** in Fig. 5-22b. The two remaining forces do not cancel and, indeed, they cannot, because they do not act along the same line. Applying Newton's second law to the x components of the forces **N** and $m\mathbf{g}$ in Fig. 5-22b now yields

$$\sum F_x = 0 - mg \sin \theta = ma,$$

so

$$a = -g \sin \theta. \qquad (5\text{-}28)$$

Note that the normal force **N** plays no role in producing the acceleration because its x component is zero.

Equation 5-28 yields

$$a = -(9.8 \text{ m/s}^2)(\sin 27°) = -4.4 \text{ m/s}^2. \qquad \text{(Answer)}$$

The minus sign indicates that the acceleration is in the direction of decreasing x, that is, down the plane.

Equation 5-28 reveals that the acceleration of the block is independent of its mass, just as the acceleration of a freely falling body is independent of the mass of the falling body. Indeed, Eq. 5-28 shows that an inclined plane can be used to "dilute" gravitation—to "slow down" free fall. For $\theta = 90°$, Eq. 5-28 yields $a = -g$; for $\theta = 0°$, it yields $a = 0$. Both are expected results.

SAMPLE PROBLEM 5-10

Figure 5-23a shows two blocks connected by a cord that passes over a massless, frictionless pulley (the arrangement is known as *Atwood's machine*). Let $m = 1.3$ kg and $M = 2.8$ kg. Find the tension in the cord and the (common) magnitude of the acceleration of the two blocks.

SOLUTION: Figures 5-23b and 5-23c are free-body diagrams for the blocks. We are given $M > m$, so we expect M to fall and m to rise. That information allows us to assign the proper algebraic signs to the accelerations of the blocks.

Before we begin the calculations, we note that the tension in the cord must be less than the weight of block M (otherwise, that block would not fall from rest) and greater than the weight of block m (otherwise, that block would not rise). The vectors in the two free-body diagrams of Fig. 5-23 are drawn to represent these facts.

Applying Newton's second law to the block with mass m, which has an acceleration of magnitude a in the positive direction of the y axis, we find

$$T - mg = ma. \qquad (5\text{-}29)$$

For the block with mass M, which has an acceleration of magnitude a in the negative direction of the y axis, we have

$$T - Mg = -Ma, \qquad (5\text{-}30)$$

or

$$-T + Mg = Ma. \qquad (5\text{-}31)$$

By adding Eqs. 5-29 and 5-31 (or eliminating T through substitution), we obtain

$$a = \frac{M - m}{M + m} g. \qquad (5\text{-}32)$$

Substituting this result in either Eq. 5-29 or Eq. 5-31 and solving for T, we obtain

$$T = \frac{2mM}{M + m} g. \qquad (5\text{-}33)$$

Inserting the given data, we obtain

$$a = \frac{M - m}{M + m} g = \frac{2.8 \text{ kg} - 1.3 \text{ kg}}{2.8 \text{ kg} + 1.3 \text{ kg}} (9.8 \text{ m/s}^2)$$

$$= 3.6 \text{ m/s}^2 \qquad \text{(Answer)}$$

and

$$T = \frac{2Mm}{M + m} g = \frac{(2)(2.8 \text{ kg})(1.3 \text{ kg})}{2.8 \text{ kg} + 1.3 \text{ kg}} (9.8 \text{ m/s}^2)$$

$$= 17 \text{ N.} \qquad \text{(Answer)}$$

You can show that the weights of the two blocks are 13 N ($= mg$) and 27 N ($= Mg$). So the tension ($= 17$ N) does indeed lie between these two values.

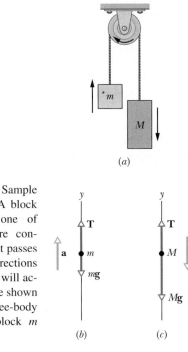

(a)

FIGURE 5-23 Sample Problem 5-10. (a) A block of mass M and one of smaller mass m are connected by a cord that passes over a pulley. The directions in which the system will accelerate from rest are shown by the arrows. Free-body diagrams for (b) block m and (c) block M.

(b)

(c)

SAMPLE PROBLEM 5-10—ANOTHER WAY

Just as we redid Sample Problem 5-5 with an unconventional axis u, we can redo Sample Problem 5-10 with u.

SOLUTION: Run the u axis through the system as shown in Fig. 5-24a. Straighten out the axis as in Fig. 5-24b, and treat the blocks as a single body with a mass of $M + m$. Then draw a free-body diagram as in Fig. 5-24c. Note that along the u

axis there are two forces on the composite of the blocks: mg in the negative direction of the u axis and Mg in the positive direction. (The force exerted on the cord by the pulley is perpendicular to the u axis and so doesn't enter our calculation.) The two forces along the u axis give the composite body (and each block) an acceleration **a**. Newton's second law in component form for motion along u is

$$\sum F_u = Mg - mg = (M + m)a, \qquad (5\text{-}34)$$

which yields

$$a = \frac{M - m}{M + m} g,$$

as previously. To get T we apply Newton's second law to either block, using a conventional axis y. For the block with mass m, we obtain Eq. 5-29. With the above result for a substituted into Eq. 5-29, we obtain Eq. 5-33.

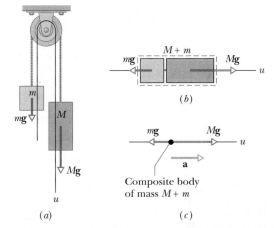

(a)

(b)

(c)

FIGURE 5-24 (a) An "axis" u runs through the system of Fig. 5-23. (b) The blocks are rearranged to straighten u and then treated as a single body with mass $M + m$. (c) The associated free-body diagram, considering only forces along u. There are two such forces.

SAMPLE PROBLEM 5-11

A passenger of mass $m = 72.2$ kg stands on a platform scale in an elevator cab (Fig. 5-25). What is the scale reading for the acceleration values given in the figure?

SOLUTION: We consider this problem from the point of view of an observer in an (inertial) reference frame fixed with respect to the Earth. Let this observer apply Newton's second law to the accelerating passenger. Figure 5-25a–c presents free-body diagrams for the passenger, treated as a particle (some particle!), for various accelerations of the cab.

Regardless of the acceleration of the cab, the Earth pulls downward on the passenger with a force having magnitude mg, in which $g = 9.80$ m/s^2 is the free-fall acceleration in the reference frame of the Earth. The elevator pushes upward on the scale platform. The scale platform, in turn, pushes upward on the passenger with a normal force whose magnitude N equals the reading of the scale. The weight that the accelerating passenger would judge himself to have is what he reads on the scale. This quantity is often called the *apparent weight;* the term *weight* (or *true weight*) is reserved for the quantity mg.

Newton's second law yields

$$N - mg = ma,$$

or

$$N = m(g + a). \qquad (5\text{-}35)$$

(a) What does the scale read if the cab is at rest or moving with constant speed? (See Fig. 5-25a.)

SOLUTION: Here $a = 0$ and we have

$$N = m(g + a) = (72.2 \text{ kg})(9.80 \text{ m/s}^2 + 0)$$
$$= 708 \text{ N}. \qquad \text{(Answer)}$$

This is the weight of the passenger.

(b) What does the scale read if the cab has an upward acceleration of magnitude 3.20 m/s²? (See Fig. 5-25b.)

SOLUTION: An upward acceleration means that the cab is either moving up with increasing speed or down with decreasing speed. In either case, the cab is a noninertial frame and Eq. 5-35 yields

$$N = m(g + a) = (72.2 \text{ kg})(9.80 \text{ m/s}^2 + 3.20 \text{ m/s}^2)$$
$$= 939 \text{ N}. \qquad \text{(Answer)}$$

The passenger presses down on the scale with a greater force than if he were at rest. The passenger—reading the scale—might conclude that he has gained 231 N!

(c) What does the scale read if the cab has a downward acceleration of magnitude 3.20 m/s²? (See Fig. 5-25c.)

SOLUTION: A downward acceleration means that the cab is either moving up with decreasing speed or down with increas-

ing speed. Again the cab is a noninertial frame. Equation 5-35 now yields

$$N = m(g + a) = (72.2 \text{ kg})(9.80 \text{ m/s}^2 - 3.20 \text{ m/s}^2)$$
$$= 477 \text{ N}. \qquad \text{(Answer)}$$

The passenger presses down on the scale with less force than if the cab were at rest. He seems to have lost 231 N.

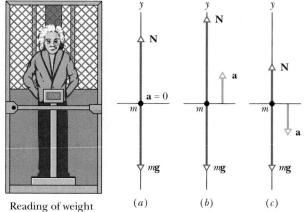

Reading of weight or apparent weight

FIGURE 5-25   Sample Problem 5-11. A passenger with mass m in an elevator, standing on a spring scale which indicates his weight or apparent weight. The free-body diagrams are for the cases in which (*a*) the acceleration of the elevator cab is zero, (*b*) $a = +3.20$ m/s², and (*c*) $a = -3.20$ m/s².

$\mathbf{C}$HECKPOINT **8:** What does the scale read if the cable breaks, so that the cab falls freely? That is, what is the apparent weight of the passenger in free fall?

REVIEW & SUMMARY

Newtonian Mechanics

The velocity of a particle or a particle-like body can change—that is, the particle can accelerate—when the particle is acted on by one or more **forces**—pushes or pulls—exerted by other objects. *Newtonian mechanics* relates accelerations and forces.

Force

The magnitudes of forces are defined in terms of the acceleration they give the standard kilogram. A force that accelerates that standard body by exactly 1 m/s² is defined to have a magnitude of 1 N. The direction of the force is the direction of the acceleration. Forces are experimentally found to be vector quantities, so they are combined according to the rules of vector algebra. The **net force** on a body is the vector sum of all the forces acting on it.

Mass

The **mass** of a body is the characteristic of that body that relates the body's acceleration to the force (or net force) causing the acceleration. Masses are scalar quantities.

Newton's First Law

If there is no net force on a body, the body must remain at rest if it is initially at rest, or move in a straight line at constant speed if it is in motion. For such a body, there are reference frames, called *inertial frames,* from which the body's acceleration **a** will be measured as being zero.

Newton's Second Law

The net force $\Sigma \mathbf{F}$ on a body with mass m is related to the body's acceleration **a** by

$$\sum \mathbf{F} = m\mathbf{a}, \tag{5-1}$$

which may be written in its scalar component version:

$$\sum F_x = ma_x, \quad \sum F_y = ma_y, \quad \text{and} \quad \sum F_z = ma_z. \tag{5-2}$$

The second law indicates that in SI units

$$1 \text{ N} = 1 \text{ kg} \cdot \text{m/s}^2. \tag{5-3}$$

A **free-body diagram** is helpful in solving problems with the second law: it is a stripped-down diagram in which only *one* body is considered. That body is represented by a dot. The external forces on the body are drawn as vectors, and a coordinate system is superimposed, oriented so as to simplify the solution.

Some Particular Forces

A body's **weight W** is the force on the body from a nearby astronomical body:

$$\mathbf{W} = m\mathbf{g}, \tag{5-10}$$

where **g** is the free-fall acceleration. Usually the astronomical body is the Earth.

A **normal force N** is the force exerted on a body by a surface against which the body is pressed. The normal force is always perpendicular to the surface.

A **frictional force f** is the force on a body when the body slides or attempts to slide along a surface. The force is parallel to the surface and directed so as to oppose the motion of the body. A **frictionless surface** is one where the frictional force is negligible.

A **tension T** is the force on a body from a taut cord at its point of attachment. The force points along the cord, away from the body. For **massless** cords (their mass is negligible), the pulls at both ends of the cord have the same magnitude T, even if the cord runs around a **massless, frictionless pulley** (the pulley's mass is negligible and the pulley has negligible friction on its axle opposing its rotation).

Newton's Third Law

If body A exerts a force $\mathbf{F}_{BA}$ on body B, then B must exert a force $\mathbf{F}_{AB}$ on body A. The forces are equal in magnitude and opposite in direction:

$$\mathbf{F}_{AB} = -\mathbf{F}_{BA}. \tag{5-14}$$

These forces act on *different* bodies.

QUESTIONS

1. If two forces act on a body, could the body possibly move (a) at constant speed or (b) at constant velocity? Could the body's velocity be zero (c) for an instant only or (d) continuously?

2. Figure 5-26 shows four forces of equal magnitude. Do any three of them look as if they will leave a body's velocity unchanged when all three act on the body? If so, which three?

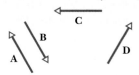

FIGURE 5-26 Question 2.

3. Figure 5-27 shows overhead views of four situations in which forces act on a block that lies on a frictionless floor. If the force magnitudes are chosen properly, in which situations is it possible that the block is (a) stationary and (b) moving with a constant velocity?

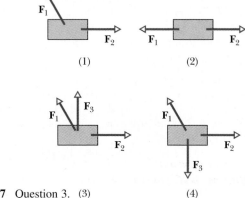

FIGURE 5-27 Question 3. (3)   (4)

4. A vertical force **F** is applied to a block of mass m that lies on a floor. What happens to the magnitude of the normal force **N** on the block from the floor as magnitude F is increased from zero if force **F** is (a) downward and (b) upward?

5. If the body in Fig. 5-8 has a weight of 100 N, and if a body with a weight of 50 N lies on top of it, what is the normal force on (a) the upper body from the lower body and (b) the lower body from the table?

6. (a) In Fig. 5-28, does the vertical component of force **F** help support the box, or does it press the box against the floor? (b) Suppose the mass of the box is m. Then is the magnitude of the normal force on the box equal to, greater than, or less than mg? (c) Is the vertical component of the force $F \sin \theta$ or $F \cos \theta$?

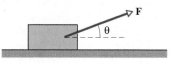

FIGURE 5-28 Question 6.

7. The body that is suspended by a rope in Fig. 5-10c has a weight of 75 N. Is T equal to, greater than, or less than 75 N when the body is moving downward at (a) increasing speed and (b) decreasing speed?

8. An elevator is supported by a single cable and there is no counterweight. The elevator receives passengers at the ground floor and takes them to the top floor, where they disembark. New passengers enter and are taken down to the ground floor. During this round trip, (a) when is the tension in the cable equal to the weight of the elevator plus passengers, (b) when is it greater, and (c) when is it less?

9. In Fig. 5-29, a massless rope is strung over a frictionless pulley. A monkey holds onto the rope, and a mirror, having the same weight as the monkey, is attached to the other side of the rope at the monkey's level. Can the monkey get away from the image it sees in the mirror by (a) climbing up the rope, (b) climbing down the rope, or (c) releasing the rope? Explain.

FIGURE 5-29 Question 9.

10. July 17, 1981, Kansas City: the newly opened Hyatt Regency is packed with people listening and dancing to a band playing favorites from the 1940s. Many of the people are crowded onto the walkways that hang like bridges across the wide atrium. Suddenly two of the walkways collapse, falling onto the merrymakers on the main floor.

The walkways were suspended one above another on vertical rods and held in place by nuts threaded onto the rods. In the original design, only two long rods were to be used, each extending through all three walkways (Fig. 5-30a). If each walkway and the merrymakers on it have a combined weight of W, what is the total weight supported by the threads and two nuts on (a) the lowest walkway and (b) the highest walkway?

Threading nuts on a rod is impossible except at the ends, so the design was changed: instead, six rods were used, each connecting two walkways (Fig. 5-30b). What now is the total weight supported by the threads and two nuts on (c) the lowest walkway and (d) the highest walkway? It was this design that failed.

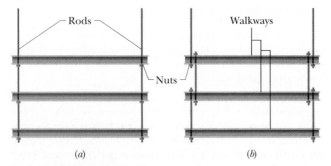

(a)   (b)

FIGURE 5-30 Question 10.

11. In Fig. 5-31, a block is attached by a rope to a bar that is itself rigidly attached to a ramp. Tell whether the magnitudes of the following increase, decrease, or remain the same as the angle θ of the ramp is increased from zero: (a) the block's weight component along the ramp, (b) the tension in the cord, (c) the block's weight component perpendicular to the ramp, and (d) the normal force on the block from the ramp.

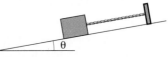

FIGURE 5-31 Question 11.

12. The box of donuts in Fig. 5-32 has a weight component of 5 N along the frictionless ramp. The force on the box from the cord has magnitude T. When the box is (a) stationary, (b) moving up the ramp at constant speed, (c) moving down the ramp at constant speed, (d) moving up the ramp at decreasing speed, and (e) moving down the ramp at decreasing speed, is T equal to, greater than, or less than 5 N?

FIGURE 5-32 Question 12.

13. Figure 5-33 shows four blocks being pulled across a frictionless floor by force $\mathbf{F}$. What total mass is accelerated to the right by (a) force $\mathbf{F}$, (b) cord 3, and (c) cord 1? (d) Rank the blocks according to their accelerations, greatest first. (e) Rank the cords according to their tension, greatest first. (Warmup for Problems 48 and 49)

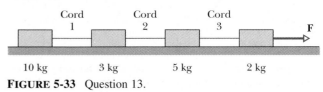

FIGURE 5-33 Question 13.

14. Figure 5-34 shows three blocks being pushed across a frictionless floor by horizontal force $\mathbf{F}$. What total mass is accelerated to the right by (a) force $\mathbf{F}$, (b) force $\mathbf{F}_{21}$ of block 1 on block 2, and (c) force $\mathbf{F}_{32}$ of block 2 on block 3? (d) Rank the blocks according to their accelerations, greatest first. (e) Rank forces $\mathbf{F}$, $\mathbf{F}_{21}$, and $\mathbf{F}_{32}$ according to their magnitude, greatest first. (Warmup for Problem 40)

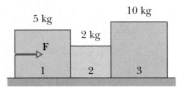

FIGURE 5-34 Question 14.

15. Figure 5-35 shows four choices for the direction of a force of magnitude F to be applied to a block on an inclined plane. The directions are either horizontal or vertical. (For choices a and b, the force is not enough to lift the block off the plane.) Rank the choices according to the magnitude of the normal force on the block from the plane, greatest first.

FIGURE 5-35
Question 15.

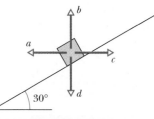

EXERCISES & PROBLEMS

SECTION 5-3   Force

1E. If the 1 kg standard body has an acceleration of 2.00 m/s² at 20° to the positive direction of the x axis, then (a) what are the x and y components of the net force on it, and (b) what is the net force in unit-vector notation?

2E. If the 1 kg standard body is accelerated by $\mathbf{F}_1 = (3.0 \text{ N})\mathbf{i} + (4.0 \text{ N})\mathbf{j}$ and $\mathbf{F}_2 = (-2.0 \text{ N})\mathbf{i} + (-6.0 \text{ N})\mathbf{j}$, then (a) what is the net force in unit-vector notation, and what are the magnitude and direction of (b) the net force and (c) the acceleration?

3P. Suppose that the 1 kg standard body accelerates at 4.00 m/s² at 160° from the positive direction of the x axis, owing to two forces, one of which is $\mathbf{F}_1 = (2.50 \text{ N})\mathbf{i} + (4.60 \text{ N})\mathbf{j}$. What is the other force (a) in unit-vector notation and (b) as a magnitude and direction?

SECTION 5-5   Newton's Second Law

4E. Two horizontal forces act on a 2.0 kg chopping block that can slide over a frictionless kitchen counter, which lies in an xy plane. One force is $\mathbf{F}_1 = (3.0 \text{ N})\mathbf{i} + (4.0 \text{ N})\mathbf{j}$. Find the acceleration of the chopping block in unit-vector notation when the other force is (a) $\mathbf{F}_2 = (-3.0 \text{ N})\mathbf{i} + (-4.0 \text{ N})\mathbf{j}$, (b) $\mathbf{F}_2 = (-3.0 \text{ N})\mathbf{i} + (4.0 \text{ N})\mathbf{j}$, and (c) $\mathbf{F}_2 = (3.0 \text{ N})\mathbf{i} + (-4.0 \text{ N})\mathbf{j}$.

5E. While two forces act on it, a particle is to move continuously with $\mathbf{v} = (3 \text{ m/s})\mathbf{i} - (4 \text{ m/s})\mathbf{j}$. One of the forces is $\mathbf{F}_1 = (2 \text{ N})\mathbf{i} + (-6 \text{ N})\mathbf{j}$. What is the other force?

6E. Three forces act on a particle that moves with an unchanging velocity of $\mathbf{v} = (2 \text{ m/s})\mathbf{i} - (7 \text{ m/s})\mathbf{j}$. Two of the forces are $\mathbf{F}_1 = (2 \text{ N})\mathbf{i} + (3 \text{ N})\mathbf{j} + (-2 \text{ N})\mathbf{k}$ and $\mathbf{F}_2 = (-5 \text{ N})\mathbf{i} + (8 \text{ N})\mathbf{j} + (-2 \text{ N})\mathbf{k}$. What is the third force?

7E. There are two horizontal forces on the 2.0 kg box in the overhead view of Fig. 5-36 but only one is shown. The box moves strictly along the x axis. For each of the following values for the acceleration a_x of the box, find the second force: (a) 10 m/s², (b) 20 m/s², (c) 0, (d) −10 m/s², and (e) −20 m/s².

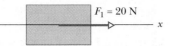
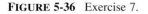

FIGURE 5-36   Exercise 7.

8E. There are two forces on the 2.0 kg box in the overhead view of Fig. 5-37 but only one is shown. The figure also shows the acceleration of the box. Find the second force (a) in unit-vector notation and (b) as a magnitude and direction.

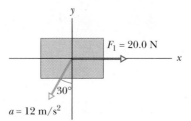

FIGURE 5-37   Exercise 8.

9E. Five forces pull on the 4.0 kg box in Fig. 5-38. Find the box's acceleration (a) in unit-vector notation and (b) as a magnitude and direction.

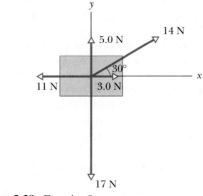

FIGURE 5-38   Exercise 9.

10P. Three astronauts, propelled by jet backpacks, push and guide a 120 kg asteroid toward a processing dock, exerting the forces shown in Fig. 5-39. What is the asteroid's acceleration (a) in unit-vector notation and (b) as a magnitude and direction?

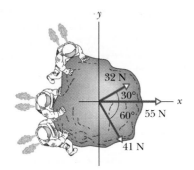

FIGURE 5-39 Problem 10.

11P. Figure 5-40 is an overhead view of a 12 kg tire that is to be pulled by three ropes. One force (F_1, with magnitude 50 N) is indicated. Orient the other two forces F_2 and F_3 so that the magnitude of the resulting acceleration of the tire is least, and find that magnitude if (a) $F_2 = 30$ N, $F_3 = 20$ N; (b) $F_2 = 30$ N, $F_3 = 10$ N; and (c) $F_2 = F_3 = 30$ N.

FIGURE 5-40 Problem 11.

SECTION 5-6 Some Particular Forces

12E. What are the mass and weight of (a) a 1400 lb snowmobile and (b) a 421 kg heat pump?

13E. What are the weight in newtons and the mass in kilograms of (a) a 5.0 lb bag of sugar, (b) a 240 lb fullback, and (c) a 1.8 ton automobile? (1 ton = 2000 lb.)

14E. A space traveler whose mass is 75 kg leaves the Earth. Compute his weight (a) on Earth, (b) on Mars, where $g = 3.8$ m/s², and (c) in interplanetary space, where $g = 0$. (d) What is his mass at each of these locations?

15E. A certain particle has a weight of 22 N at a point where the free-fall acceleration is 9.8 m/s². (a) What are the weight and mass of the particle at a point where the free-fall acceleration is 4.9 m/s²? (b) What are the weight and mass of the particle if it is moved to a point in space where the free-fall acceleration is zero?

16E. A weight-conscious penguin with a mass of 15.0 kg rests on a bathroom scale (Fig. 5-41). What are (a) the penguin's weight **W** and (b) the normal force **N** on the penguin? (c) What is the reading on the scale, assuming it is calibrated in weight units?

FIGURE 5-41 Exercise 16.

17E. Figure 5-42*a* shows a crude mobile hanging from a ceiling, with two metal pieces strung by cords of negligible mass. The masses of the pieces are given. What is the tension in (a) the bottom cord and (b) the top cord? Figure 5-42*b* shows a similar mobile consisting of three metal pieces. The masses of the highest and lowest pieces are given. The tension in the top cord is 199 N. What is the tension in (c) the lowest cord and (d) the middle cord?

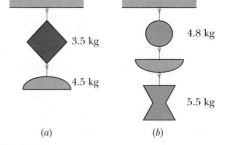

FIGURE 5-42 Exercise 17.

18E. (a) An 11.0 kg salami is supported by a cord that runs to a spring scale, which is supported by another cord from the ceiling (Fig. 5-43*a*). What is the reading on the scale? (b) In Fig. 5-43*b* the salami is supported by a cord that runs around a pulley and to a scale. The opposite end of the scale is attached by a cord to a wall. What is the reading on the scale? (c) In Fig. 5-43*c* the wall has been replaced with a second 11.0 kg salami on the left, and the assembly is stationary. What is the reading on the scale now?

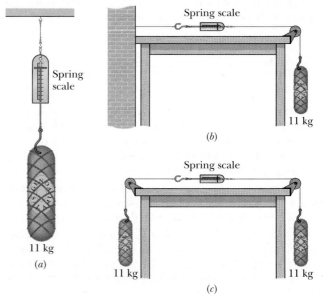

FIGURE 5-43 Exercise 18.

SECTION 5-8 Applying Newton's Laws

19E. When an airplane is in level flight, its weight is balanced by a vertical "lift," which is a force exerted by the air. How large is the lift on an airplane in such flight if the mass of the airplane is 1.20×10^3 kg?

20E. What is the magnitude of the net force acting on a 3800 lb automobile accelerating at 12 ft/s²?

21E. An experimental rocket sled can be accelerated at a constant rate from rest to 1600 km/h in 1.8 s. What is the magnitude of the required average force if the sled has a mass of 500 kg?

22E. A car traveling at 53 km/h hits a bridge abutment. A passenger in the car moves forward a distance of 65 cm (with respect to the road) while being brought to rest by an inflated air bag. What magnitude of force (assumed constant) acts on the passenger's upper torso, which has a mass of 41 kg?

23E. If a nucleus captures a stray neutron, it must bring the neutron to a stop within the diameter of the nucleus by means of the *strong force*. That force, which "glues" the nucleus together, is essentially zero outside the nucleus. Suppose that a stray neutron with an initial speed of 1.4×10^7 m/s is just barely captured by a nucleus with diameter $d = 1.0 \times 10^{-14}$ m. Assuming that the force on the neutron is constant, find the magnitude of that force. The neutron's mass is 1.67×10^{-27} kg.

24E. In a modified tug-of-war game, two people pull in opposite directions, not on a rope, but on a 25 kg sled resting on (frictionless) ice. If the participants exert forces of 90 N and 92 N, what will be the magnitude of the sled's acceleration?

25E. A 450 lb motorcycle accelerates from 0 to 55 mi/h in 6.0 s. (a) What is the magnitude of the motorcycle's constant acceleration? (b) What is the magnitude of the net force causing the acceleration?

26E. Refer to Fig. 5-15 and suppose the two masses are $m = 2.0$ kg and $M = 4.0$ kg. (a) Decide without any calculations which of them should be hanging if the magnitude of the acceleration is to be largest. What then are (b) the magnitude of the acceleration and (c) the associated tension in the cord?

27E. Refer to Fig. 5-22. Let the mass of the block be 8.5 kg and the angle θ equal 30°. Find (a) the tension in the cord and (b) the normal force acting on the block. (c) If the cord is cut, find the magnitude of the block's acceleration.

28E. A jet airplane starts from rest on the runway and accelerates for takeoff at 2.3 m/s². It has two jet engines, each of which exerts a force (thrust) on the airplane of 1.4×10^5 N. What is the weight of the airplane?

29E. *Sunjamming.* A "sun yacht" is a spacecraft with a large sail that is pushed by sunlight. Although such a push is tiny in everyday circumstances, it can be large enough to send the spacecraft outward from the Sun on a cost-free but slow trip. Suppose that the spacecraft has a mass of 900 kg and receives a push of 20 N. (a) What is the magnitude of the resulting acceleration? If the craft starts from rest, (b) how far will it travel in 1 day and (c) how fast will it then be moving?

30E. The tension at which a fishing line snaps is commonly called the line's "strength." What minimum strength is needed for a line that is to stop a 19 lb salmon in 4.4 in. if the fish is initially drifting at 9.2 ft/s? Assume a constant deceleration.

31E. In a laboratory experiment, an initially stationary electron (mass = 9.11×10^{-31} kg) undergoes a constant acceleration through 1.5 cm, reaching a speed of 6.0×10^6 m/s at the end of that distance. (a) What is the magnitude of the force accelerating the electron? (b) What is the weight of the electron?

32E. An electron is projected horizontally at a speed of 1.2×10^7 m/s into an electric field that exerts a constant vertical force of 4.5×10^{-16} N on it. The mass of the electron is 9.11×10^{-31} kg. Determine the vertical distance the electron is deflected during the time it has moved 30 mm horizontally.

33E. A car that weighs 1.30×10^4 N is initially moving at a speed of 40 km/h when the brakes are applied and the car is brought to a stop in 15 m. Assuming that the force that stops the car is constant, find (a) the magnitude of that force and (b) the time required for the change in speed. If the initial speed is, instead, twice as great, and the car experiences the same force during the braking, how are (c) the stopping distance and (d) the stopping time changed? (There could be a lesson here about the danger of driving at high speeds.)

34E. Compute the initial upward acceleration of a rocket of mass 1.3×10^4 kg if the initial upward force produced by its engine (the thrust) is 2.6×10^5 N. Do not neglect the weight of the rocket.

35E. A rocket and its payload have a total mass of 5.0×10^4 kg. How large is the force produced by the engine (the thrust) when (a) the rocket is "hovering" over the launchpad just after ignition, and (b) when the rocket is accelerating upward at 20 m/s²?

36P. A 40 kg girl and an 8.4 kg sled are on the surface of a frozen lake, 15 m apart. By means of a rope, the girl exerts a horizontal 5.2 N force on the sled, pulling it toward her. (a) What is the acceleration of the sled? (b) What is the acceleration of the girl? (c) How far from the girl's initial position do they meet, assuming that no frictional forces act?

37P. A 160 lb firefighter slides down a vertical pole with an acceleration of 10 ft/s², directed downward. What are the magnitudes and directions of the vertical forces (a) exerted by the pole on the firefighter and (b) exerted by the firefighter on the pole?

38P. A sphere of mass 3.0×10^{-4} kg is suspended from a cord. A steady horizontal breeze pushes the sphere so that the cord makes an angle of 37° with the vertical when at rest. Find (a) the magnitude of that push and (b) the tension in the cord.

39P. A worker drags a crate across a factory floor by pulling on a rope tied to the crate (Fig. 5-44). The worker exerts a force of 450 N on the rope, which is inclined at 38° to the horizontal, and the floor exerts a horizontal force of 125 N that opposes the motion. Calculate the acceleration of the crate (a) if its mass is 310 kg and (b) if its weight is 310 N.

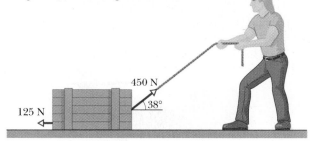

FIGURE 5-44 Problem 39.

40P. Two blocks are in contact on a frictionless table. A horizontal force is applied to one block, as shown in Fig. 5-45. (a) If

$m_1 = 2.3$ kg, $m_2 = 1.2$ kg, and $F = 3.2$ N, find the force between the two blocks. (b) Show that if a force of the same magnitude F is applied to m_2 but in the opposite direction, the force between the blocks is 2.1 N, which is not the same value calculated in (a). Explain the difference.

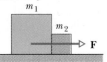

FIGURE 5-45   Problem 40.

41P. You pull a short refrigerator with a constant force **F** across a greased (frictionless) floor, either with **F** horizontal (case 1) or with **F** tilted upward at an angle θ (case 2). (a) What is the ratio of the refrigerator's speed in case 2 to its speed in case 1 if you pull for a certain time t? (b) What is this ratio if you pull for a certain distance d?

42P. For sport, a 12-kg armadillo runs onto a large pond of level, frictionless ice with an initial velocity of 5.0 m/s along the positive direction of the x axis. Take its initial position on the ice as being the origin. It slips over the ice while being pushed by a wind with a force of 17 N in the positive direction of the y axis. In unit-vector notation, what are the animal's (a) velocity and (b) position vectors when it has slid for 3.0 s?

43P. An elevator and its load have a combined mass of 1600 kg. Find the tension in the supporting cable when the elevator, originally moving downward at 12 m/s, is brought to rest with constant acceleration in a distance of 42 m.

44P. An object is hung from a spring balance attached to the ceiling of an elevator. The balance reads 65 N when the elevator is standing still. What is the reading when the elevator is moving upward (a) with a constant speed of 7.6 m/s and (b) with a speed of 7.6 m/s while decelerating at a rate of 2.4 m/s²?

45P. A 1400 kg jet engine is fastened to the fuselage of a passenger jet by just three bolts (this is the usual practice). Assume that each bolt supports one-third of the load. (a) Calculate the force on each bolt as the plane waits in line for clearance to take off. (b) During flight, the plane encounters turbulence, which suddenly imparts an upward vertical acceleration of 2.6 m/s² to the plane. Calculate the force on each bolt now.

46P. In Fig. 5-46 a 15,000 kg helicopter is lifting a 4500 kg truck with an upward acceleration of 1.4 m/s². Calculate (a) the force the air exerts on the helicopter blades and (b) the tension in the upper supporting cable.

FIGURE 5-46   Problem 46.

47P. An 80 kg man jumps down to a concrete patio from a window ledge only 0.50 m above the ground. He neglects to bend his knees on landing, so that his motion is arrested in a distance of 2.0 cm. (a) What is the average acceleration of the man from the time his feet first touch the patio to the time he is brought fully to rest? (b) With what force does this jump jar his bone structure?

48P. Three blocks are connected, as shown in Fig. 5-47, on a horizontal frictionless table and pulled to the right with a force $T_3 = 65.0$ N. If $m_1 = 12.0$ kg, $m_2 = 24.0$ kg, and $m_3 = 31.0$ kg, calculate (a) the acceleration of the system and (b) the tensions T_1 and T_2 in the interconnecting cords.

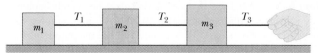

FIGURE 5-47   Problem 48.

49P. Figure 5-48 shows four penguins that are being playfully pulled along very slippery (frictionless) ice by a curator. The masses of three penguins and the tension in two of the cords are given. Find the penguin mass that is not given.

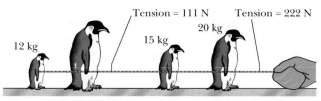

FIGURE 5-48   Problem 49.

50P. An elevator weighing 6240 lb is pulled upward by a cable with an acceleration of 4.00 ft/s². (a) Calculate the tension in the cable. (b) What is the tension when the elevator is decelerating at the rate of 4.00 ft/s² but is still moving upward?

51P. An 80-kg person is parachuting and experiencing a downward acceleration of 2.5 m/s². The mass of the parachute is 5.0 kg. (a) What upward force is exerted on the open parachute by the air? (b) What downward force is exerted by the person on the parachute?

52P. An 85-kg man lowers himself to the ground from a height of 10.0 m by holding onto a rope that runs over a frictionless pulley to a 65-kg sandbag. (a) With what speed does the man hit the ground if he started from rest? (b) Is there anything he could do to reduce the speed with which he hits the ground?

53P. A new 26-ton Navy jet (Fig. 5-49) requires an airspeed of 280 ft/s for liftoff (1 ton = 2000 lb). The engine develops a maximum force of 24,000 lb, but that is insufficient for reaching takeoff speed in the 300 ft available on an aircraft carrier. What minimum force (assumed constant) is needed from the catapult that is used to help launch the jet? Assume that the catapult and the jet's engine each exert a constant force over the 300 ft distance used for takeoff.

FIGURE 5-49 Problem 53.

54P. Imagine a landing craft approaching the surface of Callisto, one of Jupiter's moons. If the engine provides an upward force (thrust) of 3260 N, the craft descends at constant speed; if the engine provides only 2200 N, the craft accelerates downward at 0.39 m/s². (a) What is the weight of the landing craft in the vicinity of Callisto's surface? (b) What is the mass of the craft? (c) What is the free-fall acceleration near the surface of Callisto?

55P. A 52 kg circus performer is to slide down a rope that will snap if the tension exceeds 425 N. (a) What happens if the performer hangs stationary on the rope? (b) At what magnitude of acceleration does the performer just avoid breaking the rope?

56P. A chain consisting of five links, each of mass 0.100 kg, is lifted vertically with a constant acceleration of 2.50 m/s², as shown in Fig. 5-50. Find (a) the forces acting between adjacent links, (b) the force **F** exerted on the top link by the person lifting the chain, and (c) the *net* force accelerating each link.

FIGURE 5-50 Problem 56.

57P. A 1.0 kg mass on a 37° incline is connected to a 3.0 kg mass on a horizontal surface (Fig. 5-51). The surfaces and the pulley are frictionless. If $F = 12$ N, what is the tension in the connecting cord?

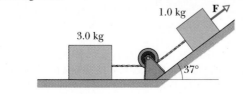

FIGURE 5-51 Problem 57.

58P. A block of mass $m_1 = 3.70$ kg on a frictionless inclined plane of angle 30.0° is connected by a cord over a massless, frictionless pulley to a second block of mass $m_2 = 2.30$ kg hanging vertically (Fig. 5-52). What are (a) the magnitude of the acceleration of each block and (b) the direction of the acceleration of m_2? (c) What is the tension in the cord?

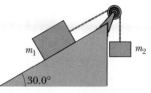

FIGURE 5-52 Problem 58.

59P. You need to lower a bundle of old roofing material weighing 100 lb to the ground with a rope that will snap if the tension in it exceeds 87 lb. (a) How can you avoid snapping the rope during the descent? (b) If the descent is 20 ft and you just barely avoid snapping the rope, with what speed will the bundle hit the ground?

60P. A block is projected up a frictionless inclined plane with initial speed v_0. The angle of incline is θ. (a) How far up the plane does it go? (b) How long does it take to get there? (c) What is its speed when it gets back to the bottom? Find numerical answers for $\theta = 32.0°$ and $v_0 = 3.50$ m/s.

61P. A spaceship lifts off vertically from the moon, where the free-fall acceleration is 1.6 m/s². If the spaceship has an upward acceleration of 1.0 m/s² as it lifts off, what is the force of the spaceship on an astronaut who weighs 735 N on Earth?

62P. A lamp hangs vertically from a cord in a descending elevator that decelerates at 2.4 m/s². (a) If the tension in the cord is 89 N, what is the lamp's mass? (b) What is the cord's tension when the elevator ascends with an upward acceleration of 2.4 m/s²?

63P. A 100 kg crate is pushed at constant speed up the frictionless 30.0° ramp shown in Fig. 5-53. (a) What horizontal force **F** is required? (b) What force is exerted by the ramp on the crate?

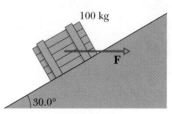

FIGURE 5-53 Problem 63.

64P. A 10 kg monkey climbs up a massless rope that runs over a frictionless tree limb and back down to a 15 kg package on the ground (Fig. 5-54). (a) What is the magnitude of the least acceleration the monkey must have if it is to lift the package off the ground? If, after the package has been lifted, the monkey stops its climb and holds onto the rope, what are (b) the monkey's acceleration and (c) the tension in the rope?

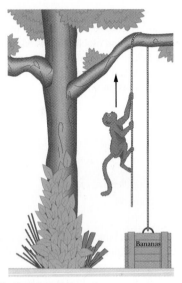

FIGURE 5-54   Problem 64.

65P. Figure 5-55 shows a section of an alpine cable-car system. The maximum permissible mass of each car with occupants is 2800 kg. The cars, riding on a support cable, are pulled by a second cable attached to each pylon. What is the difference in tension between adjacent sections of pull cable if the cars are at the maximum permissible mass and are being accelerated up the 35° incline at 0.81 m/s²?

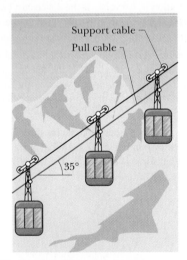

FIGURE 5-55   Problem 65.

66P. An interstellar ship has a mass of 1.20×10^6 kg and is initially at rest relative to a star system. (a) What constant acceleration is needed to bring the ship up to a speed of $0.10c$ (where c is the speed of light) relative to the star system in 3.0 days? (You need not consider Einstein's special relativity.) (b) What is that acceleration in g units? (c) What force is required for the acceleration? (d) If the engines are shut down when $0.10c$ is reached, how long does the ship take (start to finish) to journey 5.0 light-months, the distance that light travels in 5.0 months?

67P. The elevator in Fig. 5-56 consists of a 1150 kg cage (A), a 1400 kg counterweight (B), a driving mechanism (C), a cable, and two pulleys. In operation, mechanism C grabs the cable, either forcing it along or retarding its motion. This process means that tension T_1 in the cable on one side of C differs from tension T_2 on the other side. Suppose that the upward acceleration of A and the downward acceleration of B have the magnitude $a = 2.0$ m/s². Neglecting the pulleys and the mass of the cable, find (a) T_1, (b) T_2, and (c) the force on the cable produced by C.

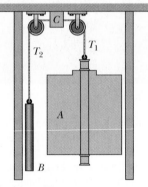

FIGURE 5-56   Problem 67.

68P. A 5.00 kg block is pulled along a horizontal frictionless floor by a cord that exerts a force $F = 12.0$ N at an angle $\theta = 25.0°$ above the horizontal, as shown in Fig. 5-57. (a) What is the acceleration of the block? (b) The force F is slowly increased. What is its value just before the block is lifted (completely) off the floor? (c) What is the acceleration of the block just before it is lifted (completely) off the floor?

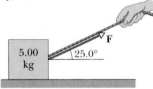

FIGURE 5-57   Problem 68.

69P. In earlier days, horses pulled barges down canals in the manner shown in Fig. 5-58. Suppose that the horse pulls on the rope with a force of 7900 N at an angle of 18° to the direction of motion of the barge, which is headed straight along the canal. The mass of the barge is 9500 kg, and its acceleration is 0.12 m/s². Calculate the force exerted by the water on the barge.

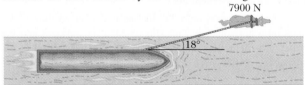

FIGURE 5-58   Problem 69.

70P. A hot-air balloon of mass M is descending vertically with downward acceleration a (Fig. 5-59). How much mass must be thrown out to give the balloon an upward acceleration a (same magnitude but opposite direction)? Assume that the upward force from the air (the lift) does not change because of the mass (the ballast) that is lost.

FIGURE 5-59   Problem 70.

71P.   A certain force gives mass m_1 an acceleration of 12.0 m/s² and mass m_2 an acceleration of 3.30 m/s². What acceleration would the force give to an object with a mass of (a) $m_2 - m_1$ and (b) $m_2 + m_1$?

72P.   A rocket with mass 3000 kg is fired from the ground at an angle of elevation of 60°. The motor creates a force (thrust) on the rocket of 6.0×10^4 N at a constant angle of 60° to the horizontal for 50 s and then cuts out. As a rough approximation, ignore the mass of fuel consumed and neglect forces from the air. Calculate (a) the altitude of the rocket at motor cutout and (b) the total horizontal distance from firing point to eventual impact with the ground (assumed to be level).

73P.   A block of mass M is pulled along a horizontal frictionless surface by a rope of mass m, as shown in Fig. 5-60. A horizontal force **F** is applied to one end of the rope. (a) Show that the rope *must* sag, even if only by an imperceptible amount. Then, assuming that the sag is negligible, find (b) the acceleration of rope and block, (c) the force that the rope exerts on the block, and (d) the tension in the rope at its midpoint.

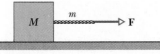

FIGURE 5-60   Problem 73.

74P.   Figure 5-61 shows a man sitting in a bosun's chair that dangles from a massless rope, which runs over a massless, frictionless pulley and back down to the man's hand. The combined mass of the man and chair is 95.0 kg. (a) With what force must the man pull on the rope for him to rise at constant speed? (b) What force is needed for an upward acceleration of 1.30 m/s²? (c) Suppose, instead, that the rope on the right is held by a person on the ground. Repeat (a) and (b) for this new situation. (d) In each of the four cases, what is the force exerted on the ceiling by the pulley system?

Electronic Computation

75.   Fig. 5-62 shows two blocks on a frictionless plane, with inclination angle θ, and a third block. Blocks m_1 and m_2 are con-

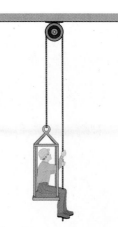

FIGURE 5-61   Problem 74.

nected by a cord with tension T_1; blocks m_2 and m_3 are connected by a cord (around a frictionless, massless pulley) with tension T_2. If $\theta = 20°$, $m_1 = 2.00$ kg, $m_2 = 1.00$ kg, and $m_3 = 3.00$ kg, find T_1, T_2, and the acceleration of the blocks. (*Hint:* Use Newton's second law for each block to write three simultaneous equations, which can then be solved electronically.)

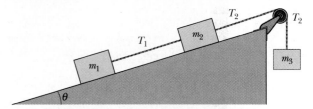

FIGURE 5-62   Problem 75.

76.   A 2.00 kg object is subjected to three forces that give it an acceleration $\mathbf{a} = -(8.00$ m/s²$)\mathbf{i} + (6.00$ m/s²$)\mathbf{j}$. If two of the three forces are $\mathbf{F}_1 = (30.0$ N$)\mathbf{i} + (16.0$ N$)\mathbf{j}$ and $\mathbf{F}_2 = -(12.0$ N$)\mathbf{i} + (8.00$ N$)\mathbf{j}$, find the third force.

Cats, who enjoy sleeping on window sills, are often kept in apartment buildings. When a cat accidentally falls out of a window and onto a sidewalk, the extent of injury (such as the number of fractured bones or the certainty of death) <u>*decreases*</u> *with height if the fall is more than seven or eight floors. (There is even a record of a cat who fell 32 floors and suffered only slight damage to its thorax and one tooth.) How can that be?*

6-1 FRICTION

Frictional forces are unavoidable in our daily lives. Left to act alone, they would stop every moving object and bring to a halt every rotating shaft. In an automobile, about 20% of the gasoline is used to counteract friction in the engine and in the drive train. On the other hand, if friction were totally absent, we could not get an automobile to go anywhere, and we could not walk or ride a bicycle. We could not hold a pencil and, if we could, it would not write. Nails and screws would be useless, woven cloth would fall apart, and knots would come undone.

Here we deal with the frictional forces that exist between dry solid surfaces, moving across each other at slow speeds. Consider two simple experiments:

1. *First experiment.* Send a book sliding across a tabletop. A frictional force, exerted by the tabletop on the bottom of the sliding book, slows the book and eventually stops it. If you want to make the book move across the table with constant velocity, you must push or pull it with a steady force of equal magnitude and opposite direction to that of the frictional force, which opposes the motion.

2. *Second experiment.* A heavy crate is resting on the floor of a warehouse. You push on it horizontally with a steady force but it does not move. That is because the force that you apply is balanced by a frictional force, exerted horizontally on the bottom of the crate by the floor and opposite the direction of your push. Remarkably, this frictional force automatically adjusts itself, in both magnitude and direction, to cancel exactly whatever force you decide to apply. Of course, if you can push hard enough, you will be able to move the crate (see the first experiment).

Figure 6-1 shows a similar situation in detail. In Fig. 6.1a, a block rests on a tabletop, its weight **W** balanced by an equal but opposite normal force **N**. In Fig. 6-1b, you exert a force **F** on the block, attempting to pull it to the left. In response, a frictional force **f**$_s$ arises, pointing to the right, exactly matching the force that you have applied. The force **f**$_s$ is called the **static frictional force.**

Figures 6-1c and 6-1d show that as you increase your applied force, the static frictional force **f**$_s$ increases also and the block remains at rest. When the applied force reaches a certain value, however, the block ''breaks away'' from its intimate contact with the tabletop and accelerates leftward (Fig. 6-1e). The frictional force that then opposes the motion is called the **kinetic frictional force f**$_k$.

Usually, the kinetic frictional force, which acts when there is motion, is less than the maximum value of the static frictional force, which acts when there is no motion. So, if you wish the block to move across the surface with a

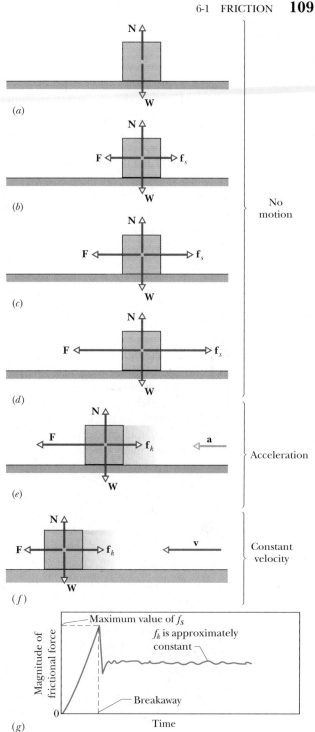

FIGURE 6-1 (*a*) The forces on a stationary block. (*b–d*) An external force **F**, applied to the block, is counterbalanced by an equal but opposite static frictional force **f**$_s$. As **F** is increased, **f**$_s$ also increases, until **f**$_s$ reaches a certain maximum value. (*e*) The block then ''breaks away,'' accelerating suddenly to the left. (*f*) If the block is now to move with constant velocity, the applied force **F** must be reduced from the maximum value it had just before the block broke away. (*g*) Some experimental results for the sequence (*a*) through (*f*).

constant speed, you must usually decrease the applied force once the block begins to move, as in Fig. 6-1*f*. For example, Fig. 6-1*g* shows the results of an experiment in which the force on a block was slowly increased until breakaway occurred. Note the reduced force needed to keep the block moving at constant speed after breakaway.

Basically, the frictional force is a force acting between the surface atoms of one body and those of another. If two highly polished and carefully cleaned metal surfaces are brought together in a very good vacuum, they cannot be made to slide over each other. Instead, they *cold-weld* together instantly, forming a single piece of metal. If a machinist's specially polished gage blocks are brought together in air, they stick almost as firmly to each other and can be separated only by means of a wrenching motion. Usually, however, such intimate atom-to-atom contact is not possible. Even a highly polished metal surface is far from being flat on the atomic scale. Moreover, the surfaces of everyday objects have layers of oxides and other contaminants that reduce cold-welding.

When two such surfaces are placed together, only the high points touch each other. (It is like having the Alps of Switzerland turned over and placed down on the Alps of Austria.) The actual *micro*scopic area of contact is much less than the apparent *macro*scopic contact area, perhaps by a factor of 10^4. Nonetheless, many contact points cold-weld together. These welds produce static friction when an applied force attempts to slide the surfaces.

When the surfaces are pulled across each other, there is first a tearing of welds (breakaway) and then a continuous tearing apart and re-forming of welds as additional chance contacts are made (Fig. 6-2). The kinetic friction $\mathbf{f}_k$ is the vector sum of forces at those many chance contacts. Often, the motion is "jerky," because the two surfaces

briefly stick together and then slip. Such repetitive *stick-and-slip* can produce squeaking or squealing, as when tires skid on dry pavement, fingernails scratch along a chalkboard, a rusty hinge is opened, and a bow is drawn across a violin string.

6-2 PROPERTIES OF FRICTION

Experiment shows that when a dry and unlubricated body is pressed against a surface in the same condition, and an applied force **F** attempts to slide the body along the surface, the resulting frictional force has three properties:

PROPERTY 1. If the body does not move, then the static frictional force $\mathbf{f}_s$ and the component of **F** that is parallel to the surface are equal in magnitude, and $\mathbf{f}_s$ is directed opposite that component of **F**.

PROPERTY 2. The magnitude of $\mathbf{f}_s$ has a maximum value $f_{s,\text{max}}$ that is given by

$$f_{s,\text{max}} = \mu_s N, \tag{6-1}$$

where μ_s is the **coefficient of static friction** and N is the magnitude of the normal force. If the magnitude of the component of **F** that is parallel to the surface exceeds $f_{s,\text{max}}$, then the body begins to slide along the surface.

PROPERTY 3. If the body begins to slide along the surface, the magnitude of the frictional force rapidly decreases to a value f_k given by

$$f_k = \mu_k N, \tag{6-2}$$

where μ_k is the **coefficient of kinetic friction.** Thereafter during the sliding, the kinetic frictional force $\mathbf{f}_k$ has a magnitude given by Eq. 6-2.

Properties 1 and 2 are worded in terms of a single applied force **F**, but they also hold for the resultant of several applied forces acting on the body. Equations 6-1 and 6-2 are *not* vector equations; the direction of $\mathbf{f}_s$ or $\mathbf{f}_k$ is always parallel to the surface and opposite the intended motion, and **N** is perpendicular to the surface.

The coefficients μ_s and μ_k are dimensionless and must be determined experimentally. Their values depend on certain properties of both the body and the surface; hence, they are usually referred to with the preposition "between," as in "the value of μ_s *between* a sled and asphalt is 0.5." We assume that the value of μ_k does not depend on the speed at which the body slides along the surface.

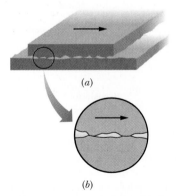

FIGURE 6-2 The mechanism of sliding friction. (*a*) The upper surface is sliding to the right over the lower surface in this enlarged view. (*b*) A detail, showing two spots where cold-welding has occurred. Force is required to break these welds and maintain the motion.

CHECKPOINT **1:** A block lies on a floor. (a) What is the magnitude of the frictional force on it from the floor? (b) If a horizontal force of 5 N is now applied to the block, but the block does not move, what is the magnitude of the frictional force on it? (c) If the maximum value $f_{s,\text{max}}$ of the static frictional force on the block is 10 N, will the block move if the horizontally applied force is 8 N? (d) If the force is 12 N?

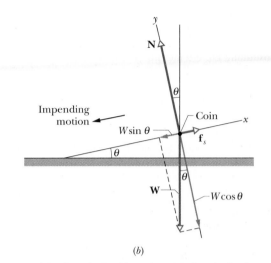

(b)

FIGURE 6-3 Sample Problem 6-1. (*a*) A coin is about to slide from rest down the cover of a book. (*b*) A free-body diagram for the coin, showing the three forces (drawn to scale) that act on it. The weight **W** is shown resolved into its components along the *x* and the *y* axes, whose orientations are chosen to simplify the problem.

SAMPLE PROBLEM 6-1

Figure 6-3*a* shows a coin resting on a book that has been tilted at an angle θ with the horizontal. By trial and error you find that when θ is increased to 13°, the coin begins to slide down the book. What is the coefficient of static friction μ_s between the coin and the book?

SOLUTION: Figure 6-3*b* is a free-body diagram for the coin when it is on the verge of sliding. The forces on the coin are the normal force **N**, pushing outward from the plane of the book, the weight **W** of the coin, and the frictional force $\mathbf{f}_s$, which points up the plane because the impending motion is down the plane. Since the coin is in equilibrium, the net force on it must be zero. From Newton's second law, we have

$$\sum \mathbf{F} = \mathbf{f}_s + \mathbf{W} + \mathbf{N} = 0. \quad (6\text{-}3)$$

For the *x* components, this vector equation gives us

$$\sum F_x = f_s - W \sin \theta = 0, \quad \text{or} \quad f_s = W \sin \theta. \quad (6\text{-}4)$$

For the *y* components, we have

$$\sum F_y = N - W \cos \theta = 0, \quad \text{or} \quad N = W \cos \theta. \quad (6\text{-}5)$$

When the coin is on the verge of sliding (*and only then*) the magnitude of the static frictional force acting on it has the maximum value $\mu_s N$. Substituting $\mu_s N$ into Eq. 6-4 and dividing by Eq. 6-5, we obtain

$$\frac{f_s}{N} = \frac{\mu_s N}{N} = \frac{W \sin \theta}{W \cos \theta} = \tan \theta,$$

or

$$\mu_s = \tan \theta = \tan 13° = 0.23. \quad \text{(Answer)} \quad (6\text{-}6)$$

You can easily measure μ_s for a coin and this text; you do not need a protractor. You can measure the two lengths shown in Fig. 6-3*a* with a ruler, and their ratio h/d is $\tan \theta$.

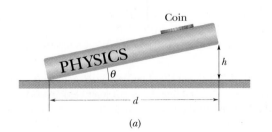

(a)

SAMPLE PROBLEM 6-2

If a car's wheels are "locked" (kept from rolling) during emergency braking, the car slides along the road. Ripped-off bits of tire and small melted sections of road form the "skid marks" that reveal the cold-welding during the slide. The record for the longest skid marks on a public road was reportedly set in 1960 by a Jaguar on the M1 highway in England— the marks were 290 m (about 950 ft) long! Assuming that $\mu_k = 0.60$, how fast was the car going when the wheels became locked?

SOLUTION: Figure 6-4*a* depicts the car's travel; Fig. 6-4*b* is the car's free-body diagram during deceleration, showing the car's weight **W**, the normal force **N**, and the kinetic frictional force $\mathbf{f}_k$ acting on the car. We can use Eq. 2-16,

$$v^2 = v_0^2 + 2a_x(x - x_0),$$

with $v = 0$ and $x - x_0 = d$, to find the car's initial speed v_0. Substituting these values and rearranging yield

$$v_0 = \sqrt{-2a_x d}. \quad (6\text{-}7)$$

To find a_x, we apply Newton's second law along the *x* axis. If we ignore the effects of the air on the car, the only force with a component along the *x* axis is $\mathbf{f}_k$, and that component is $-f_k$. So we have

$$-f_k = ma_x, \quad \text{or} \quad a_x = -\frac{f_k}{m} = -\frac{\mu_k N}{m}, \quad (6\text{-}8)$$

where *m* is the car's mass, and Eq. 6-2 tells us that $f_k = \mu_k N$.

The normal force **N** has magnitude $N = W = mg$. Substituting this result into Eq. 6-8 yields

$$a_x = -\frac{\mu_k mg}{m} = -\mu_k g. \tag{6-9}$$

Substituting Eq. 6-9 into Eq. 6-7, we find

$$v_0 = \sqrt{2\mu_k g d} = \sqrt{(2)(0.60)(9.8 \text{ m/s}^2)(290 \text{ m})}$$
$$= 58 \text{ m/s} = 210 \text{ km/h} \qquad \text{(Answer)}$$

(about 130 mi/h). In obtaining the answer, we implicitly assumed that $v = 0$ at the far end of the skid marks. Actually, the marks ended only because the Jaguar left the road after 290 m. So v_0 was at least 210 km/h, and possibly much more.

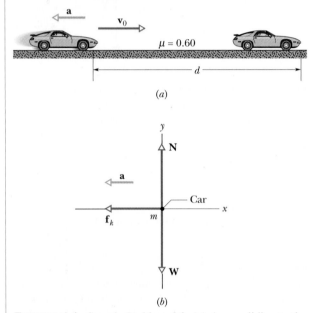

(a)

(b)

FIGURE 6-4 Sample Problem 6-2. (*a*) A car, sliding to the right and finally stopping after a displacement *d*. (*b*) A free-body diagram for the decelerating car. The acceleration vector points to the left, in the direction of the frictional force $\mathbf{f}_k$.

SAMPLE PROBLEM 6-3

A woman pulls a loaded sled of mass $m = 75$ kg along a horizontal surface at constant velocity. The coefficient of kinetic friction μ_k between the runners and the snow is 0.10, and the angle ϕ in Fig. 6-5 is 42°.

(a) What is the tension T in the rope?

SOLUTION: Figure 6-5*b* is the free-body diagram for the sled. Applying Newton's second law in the horizontal direction yields

$$T \cos \phi - f_k = ma_x = 0, \tag{6-10}$$

where a_x is zero because the velocity is constant. In the vertical direction, we have

$$T \sin \phi + N - mg = ma_y = 0, \tag{6-11}$$

in which mg is the weight of the sled. From Eq. 6-2,

$$f_k = \mu_k N. \tag{6-12}$$

These three equations contain T, N, and f_k as unknowns. Eliminating N and f_k will allow us to find the remaining variable T. We start by adding Eqs. 6-10 and 6-12, obtaining

$$T \cos \phi = \mu_k N,$$

or

$$N = \frac{T \cos \phi}{\mu_k}. \tag{6-13}$$

Substituting this into Eq. 6-11 and solving for T, we obtain

$$T = \frac{\mu_k mg}{\cos \phi + \mu_k \sin \phi} \tag{6-14}$$
$$= \frac{(0.10)(75 \text{ kg})(9.8 \text{ m/s}^2)}{\cos 42° + (0.10)(\sin 42°)}$$
$$= 91 \text{ N}, \qquad \text{(Answer)}$$

which is considerably less than the weight of the sled.

(b) What is the normal force with which the snow pushes vertically upward on the sled?

SOLUTION: We can find N by substituting $T = 91$ N and given data into either Eq. 6-11 or 6-13. With Eq. 6-11 we find

$$N = mg - T \sin \phi$$
$$= (75 \text{ kg}) (9.8 \text{ m/s}^2) - (91 \text{ N}) \sin 42°$$
$$= 670 \text{ N}. \qquad \text{(Answer)}$$

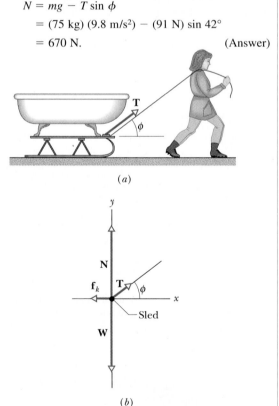

(a)

(b)

FIGURE 6-5 Sample Problem 6-3. (*a*) A woman exerts a force **T** on a sled, pulling it at a constant velocity. (*b*) A free-body diagram for the sled and its load.

CHECKPOINT **2:** In the figure, horizontal force $\mathbf{F}_1$ of magnitude 10 N is applied to a box on a floor, but the box does not slide. Then, as the magnitude of vertically applied force $\mathbf{F}_2$ is increased from zero, but before the box begins to slide, do the following quantities increase, decrease, or stay the same: (a) the magnitude of the frictional force on the box; (b) the magnitude of the normal force on the box from the floor; (c) the maximum value $f_{s,\text{max}}$ of the static frictional force on the box?

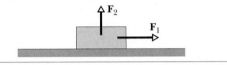

SAMPLE PROBLEM 6-4

In Fig. 6-6a, a crate of dilled pickles with mass $m_1 = 14$ kg moves along a plane that makes an angle of $\theta = 30°$ with the horizontal. That crate is connected to a crate of pickled dills with mass $m_2 = 14$ kg by a taut, massless cord that runs around a frictionless, massless pulley. The hanging crate of dills descends with constant velocity.

(a) What are the magnitude and direction of the frictional force exerted on m_1 by the plane?

SOLUTION: The fact that m_2 descends indicates that m_1 moves up the plane, and so a *kinetic* frictional force $\mathbf{f}_k$ must point down the plane.

We cannot use Eq. 6-2 to find the magnitude of $\mathbf{f}_k$, because we do not know the coefficient of kinetic friction μ_k between m_1 and the plane. However, we can use the techniques of Chapter 5. To start, we draw the free-body diagrams for m_1 and m_2 in Figs. 6-6b and 6-6c, where $\mathbf{T}$ is the pull from the tension in the cord and the weight vectors are $\mathbf{W}_1 = m_1\mathbf{g}$ and $\mathbf{W}_2 = m_2\mathbf{g}$.

With $\mathbf{W}_1$ resolved into x and y components we have, from Newton's second law applied to the x axis in Fig. 6-6b,

$$\sum F_x = T - f_k - m_1 g \sin\theta = m_1 a_x = 0, \quad (6\text{-}15)$$

where $a_x = 0$ because m_1 must move at constant velocity. Next, for m_2 we apply Newton's second law along the y axis in Fig. 6-6c and use the fact that m_2 moves at constant velocity. We find

$$\sum F_y = T - m_2 g = m_2 a_y = 0,$$

or $$T = m_2 g. \quad (6\text{-}16)$$

We then substitute T from Eq. 6-16 into Eq. 6-15 and solve for f_k, obtaining

$$f_k = m_2 g - m_1 g \sin\theta$$
$$= (14 \text{ kg})(9.8 \text{ m/s}^2) - (14 \text{ kg})(9.8 \text{ m/s}^2)(\sin 30°)$$
$$= 68.6 \text{ N} \approx 69 \text{ N}. \quad \text{(Answer)}$$

(b) What is μ_k?

SOLUTION: We can use Eq. 6-2 to find μ_k, but first we need the magnitude of the normal force $\mathbf{N}$ acting on m_1. To find N, we apply Newton's second law for m_1 along the y axis in Fig. 6-6b:

$$\sum F_y = N - m_1 g \cos\theta = m_1 a_y = 0,$$

or $$N = m_1 g \cos\theta.$$

From Eq. 6-2 we now have

$$\mu_k = \frac{f_k}{N} = \frac{f_k}{m_1 g \cos\theta}$$

$$= \frac{68.6 \text{ N}}{(14 \text{ kg})(9.8 \text{ m/s}^2)(\cos 30°)} = 0.58. \quad \text{(Answer)}$$

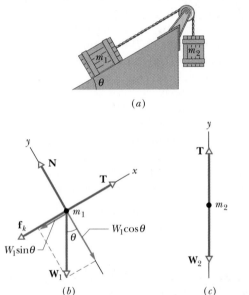

FIGURE 6-6 Sample Problem 6-4. (a) Mass m_1 moves up the plane while mass m_2 descends at constant velocity. (b) A free-body diagram for mass m_1. (c) A free-body diagram for mass m_2.

6-3 THE DRAG FORCE AND TERMINAL SPEED

A **fluid** is anything that can flow—generally either a gas or a liquid. When there is a relative velocity between a fluid and a body (either because the body moves through the fluid or because the fluid moves past the body), the body experiences a **drag force D** that opposes the relative motion and points in the direction in which the fluid flows relative to the body.

Here we examine only cases in which air is the fluid, the body is blunt (like a baseball) rather than slender (like a javelin), and the relative motion is fast enough so that the air becomes turbulent (breaks up into swirls) behind the body. In such cases, the magnitude of the drag force **D** is related to the relative speed v by an experimentally determined **drag coefficient** C according to

$$D = \tfrac{1}{2} C \rho A v^2, \tag{6-17}$$

where ρ is the air density (mass per volume) and A is the **effective cross-sectional area** of the body (the area of a cross section taken perpendicular to the velocity **v**). The drag coefficient C (typical values range from 0.4 to 1.0) is not truly a constant for a given body, because if v varies significantly, the value of C can vary as well. Here, we ignore such complications.

Downhill speed skiers know well that drag depends on A and v^2. To reach high speeds a skier must reduce D as much as possible by, for example, riding the skis in the "egg position" (Fig. 6-7) to minimize A.

Equation 6-17 indicates that when a blunt object falls from rest through air, D gradually increases from zero as the speed of the body increases. As suggested in Fig. 6-8, if the body falls far enough, D eventually equals the body's weight $W (= mg)$, and the net vertical force on the body is then zero. By Newton's second law, the acceleration must also be zero then, and so the body's speed no longer increases. The body then falls at a constant **terminal speed** v_t, which we find by setting $D = mg$ in Eq. 6-17, obtaining

$$\tfrac{1}{2} C \rho A v_t^2 = mg,$$

which gives

$$v_t = \sqrt{\frac{2mg}{C\rho A}}. \tag{6-18}$$

FIGURE 6-7   This skier crouches in an "egg position" so as to minimize her effective cross-sectional area and thus the air drag acting on her.

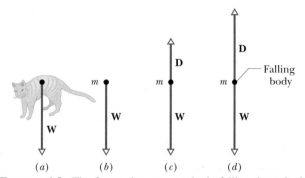

FIGURE 6-8   The forces that act on a body falling through air: (*a*) the body when it has just begun to fall, (*b*) the free-body diagram just then, and (*c*) the free-body diagram a little later, after a drag force has developed. (*d*) The drag force has increased until it balances the weight of the body. The body now falls at its constant terminal speed.

Table 6-1 gives values of v_t for some common objects.

According to calculations* based on Eq. 6-17, a cat must fall about six floors to reach terminal speed. Until it does so, $W > D$ and the cat accelerates downward because of the net downward force. Recall from Chapter 2 that your body is an accelerometer, not a speedometer. Because the cat too senses the acceleration, it is frightened and keeps its feet underneath its body, its head tucked in, and its spine bent upward, making A small, v_t large, and injury on landing likely.

However, if the cat does reach v_t, the acceleration vanishes and the cat relaxes somewhat, stretching its legs

TABLE 6-1   SOME TERMINAL SPEEDS IN AIR

OBJECT	TERMINAL SPEED (m/s)	95% DISTANCE[a] (m)
16 lb Shot	145	2500
Sky diver (typical)	60	430
Baseball	42	210
Tennis ball	31	115
Basketball	20	47
Ping-Pong ball	9	10
Raindrop (radius = 1.5 mm)	7	6
Parachutist (typical)	5	3

[a]This is the distance through which the body must fall from rest to reach 95% of its terminal speed.

Source: Adapted from Peter J. Brancazio, *Sport Science,* Simon & Schuster, New York, 1984.

*W. O. Whitney and C. J. Mehlhaff, "High-rise syndrome in cats," *The Journal of the American Veterinary Medical Association,* Vol. 191, pages 1399–1403 (1987).

and neck horizontally outward and straightening its spine (it then resembles a flying squirrel). These actions increase A and D, and the cat begins to slow because now $D > W$ —the net force is upward—until a new, smaller v_t is reached. The decrease in v_t reduces the possibility of serious injury on landing. Just before the end of the fall, when it sees it is nearing the ground, the cat pulls its legs back beneath its body to prepare for the landing.

SAMPLE PROBLEM 6-5

If a falling cat reaches a first terminal speed of 60 mi/h while it is tucked in and then stretches out, doubling A, how fast is it falling when it reaches a new terminal speed?

SOLUTION: We let v_{to} and v_{tn} represent the original and new terminal speeds, and A_o and A_n the original and new areas. We then use Eq. 6-18 to set up a ratio of speeds:

$$\frac{v_{tn}}{v_{to}} = \frac{\sqrt{2mg/C\rho A_n}}{\sqrt{2mg/C\rho A_o}} = \sqrt{\frac{A_o}{A_n}} = \sqrt{\frac{A_o}{2A_o}} = \sqrt{0.5} \approx 0.7,$$

which means that $v_{tn} \approx 0.7 v_{to}$, or about 40 mi/h.

In April 1987, during a jump, parachutist Gregory Robertson noticed that fellow parachutist Debbie Williams had been knocked unconscious in a collision with a third sky diver and was unable to open her parachute. Robertson, who was well above Williams at the time and who had not yet opened his parachute for the 13,500-ft plunge, reoriented his body head-down so as to minimize A and maximize his downward speed. Reaching an estimated v_t of 200 mi/h, he caught up with Williams and then went into a horizontal "spread eagle" (as in Fig. 6-9) to increase D so that he could grab her. He opened her parachute and then, after releasing her, his own, with a scant 10 s before impact. Williams received extensive internal injuries due to her lack of control on landing but survived.

FIGURE 6-9 A parachutist in a horizontal "spread eagle" maximizes the air drag.

SAMPLE PROBLEM 6-6

A raindrop with radius $R = 1.5$ mm falls from a cloud that is at height $h = 1200$ m above the ground. The drag coefficient C for the drop is 0.60. Assume that the drop is spherical throughout its fall. The density of water ρ_w is 1000 kg/m³, and the density of air ρ_a is 1.2 kg/m³.

(a) What is the terminal speed of the drop?

SOLUTION: The volume of a sphere is $\frac{4}{3}\pi R^3$, and its effective area A is that of a circle with radius R. So, for the drop:

$$m = \tfrac{4}{3}\pi R^3 \rho_w \quad \text{and} \quad A = \pi R^2.$$

Then, from Eq. 6-18, we find

$$v_t = \sqrt{\frac{2mg}{C\rho_a A}} = \sqrt{\frac{8\pi R^3 \rho_w g}{3C\rho_a \pi R^2}} = \sqrt{\frac{8R\rho_w g}{3C\rho_a}}$$

$$= \sqrt{\frac{(8)(1.5 \times 10^{-3}\ \text{m})(1000\ \text{kg/m}^3)(9.8\ \text{m/s}^2)}{(3)(0.60)(1.2\ \text{kg/m}^3)}}$$

$$= 7.4\ \text{m/s} \ (= 17\ \text{mi/h}). \qquad \text{(Answer)}$$

Note that the height of the cloud does not enter into the calculation. The raindrop (see Table 6-1) reaches terminal speed after falling just a few meters.

(b) What would have been the speed just before impact if there had been no drag force?

SOLUTION: From Eq. 2-23, with $h = -(y - y_0)$ and $v_0 = 0$, we find

$$v = \sqrt{2gh} = \sqrt{(2)(9.8\ \text{m/s}^2)(1200\ \text{m})}$$

$$= 150\ \text{m/s} \ (= 340\ \text{mi/h}). \qquad \text{(Answer)}$$

Under these conditions, Shakespeare would scarcely have written, "it droppeth as the gentle rain from heaven, upon the place beneath."

CHECKPOINT 3: Near the ground, is the speed of large raindrops greater than, less then, or the same as that of small raindrops, assuming that both large and small raindrops are spherical?

6-4 UNIFORM CIRCULAR MOTION

Recall that when a body moves in a circle (or a circular arc) at constant speed v, it is said to be in uniform circular motion. Also recall that the body has a centripetal acceleration (directed toward the center of the circle), of constant magnitude given by Eq. 4-22:

$$a = \frac{v^2}{r} \qquad \text{(centripetal acceleration)}, \qquad (6\text{-}19)$$

where r is the radius of the circle.

This centripetal acceleration is caused by a **centripetal force** that acts on the body and that is directed toward the center of the circle. The magnitude F of this force is constant and is given by Newton's second law as

$$F = ma = \frac{mv^2}{r} \qquad \text{(centripetal force).} \qquad (6\text{-}20)$$

If this force is not present, the body cannot undergo uniform circular motion. Both the centripetal acceleration and the centripetal force are vector quantities whose magnitudes are constant but whose directions are always changing so as to point toward the center of the circle.

If the body in uniform circular motion is, say, a hockey puck whirled around on the end of a string as in Fig. 6-10, the centripetal force is provided by the tension in the string. For the Moon in its (nearly) uniform circular motion around the Earth, the centripetal force is the gravitational attraction of the Earth. Thus a centripetal force is not a new kind of force; it can be a tension force, a gravitational force, or any other force.

Let us compare two familiar examples of uniform circular motion:

1. *Rounding a curve in a car.* You are sitting in the center of the rear seat of a car moving at high speed over flat road. When the driver suddenly turns left, rounding a corner in a circular arc, you are slid across the seat toward the right and jammed against the interior wall of the car. What is going on?

While the car is moving in the circular arc, it is in uniform circular motion. The centripetal force that causes the motion is a frictional force exerted by the roadway on the tires. This force points radially inward and—spread over the four tires—has a magnitude given by Eq. 6-20.

You would have been in uniform circular motion in the center of the seat if the frictional force exerted on you

by the seat had been great enough. However, it was not, and so you slid across the seat. Viewed from the reference frame of the ground, you actually continued moving in a straight line while the seat slid beneath you, until you met the car's wall. Its push on you was a centripetal force, and you then joined the car's uniform circular motion.

2. *Orbiting the Earth.* This time you are a passenger in the space shuttle *Atlantis,* orbiting the Earth and experiencing "weightlessness." What is going on in this case?

The centripetal force that keeps you and the shuttle in uniform circular motion is the gravitational attraction of the Earth for you and the shuttle. This force is directed radially inward, toward the center of the Earth, and has a magnitude given by Eq. 6-20.

In both car and shuttle you are in uniform circular motion, acted on by a centripetal force. Yet your experiences in the two situations are quite different. In the car, jammed up against the wall, you are aware of being compressed by the wall. In the orbiting shuttle, on the other hand, you are floating around with no sensation of any force acting on you. Why this great difference?

The difference is due to the nature of the two centripetal forces. In the car, the centripetal force is a *contact force,* exerted by the wall externally on the part of your body touching the wall. In the shuttle, the centripetal force is a *volume force,* due to the Earth's gravitational pull on every atom of your body and on every atom of the shuttle, in proportion to the mass of that atom. Thus there is no compression on any one part of your body, nor even any sensation of a force acting on you.

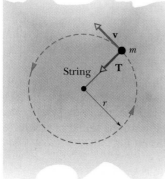

FIGURE 6-10 A hockey puck of mass m moves with constant speed v in a circular path on a horizontal frictionless surface. The centripetal force acting on the puck is **T**, the pull from the string.

SAMPLE PROBLEM 6-7

Igor is a cosmonaut-engineer in the spacecraft *Vostok II,* orbiting the Earth at an altitude h of 520 km with a speed v of 7.6 km/s. Igor's mass m is 79 kg.

(a) What is his acceleration?

SOLUTION: Igor is in uniform circular motion in a circle of radius $R_E + h$, where R_E is the radius of the Earth. His centripetal acceleration is given by Eq. 6-19:

$$a = \frac{v^2}{r} = \frac{v^2}{R_E + h}$$

$$= \frac{(7.6 \times 10^3 \text{ m/s})^2}{6.37 \times 10^6 \text{ m} + 0.52 \times 10^6 \text{ m}}$$

$$= 8.38 \text{ m/s}^2 \approx 8.4 \text{ m/s}^2, \qquad \text{(Answer)}$$

which is the value of the free-fall acceleration at Igor's altitude. If he were lifted to that altitude and released, instead of being put into orbit there, he would fall toward the Earth's center, starting out with that value for the acceleration. The difference in the two situations is that when he orbits the Earth, he always has a "sideways" motion as well: as he falls, he also moves to the side, so that he ends up moving along a curved path around the Earth.

(b) What (centripetal) gravitational force does the Earth exert on Igor?

SOLUTION: The centripetal force is

$$F = ma = (79 \text{ kg})(8.38 \text{ m/s}^2)$$
$$= 660 \text{ N} \approx 150 \text{ lb}. \qquad \text{(Answer)}$$

If Igor were to stand on a scale placed on the top of a tower with height $h = 520$ km, the scale would read 660 N or 150 lb. In orbit, the scale (if Igor could "stand" on it) would read zero because he and the scale are in free fall together, and therefore his feet do not actually press against it.

PROBLEM SOLVING TACTICS

TACTIC 1: *Looking Things Up*

In Sample Problem 6-7, we had to know the radius of the Earth, which was not given in the problem statement. You need to become familiar with sources of this kind of information, starting with this book. Many useful data are given in the inside covers, in the various appendixes, and in tables. The *Handbook of Chemistry and Physics,* updated every year by the publisher, CRC Press, is an invaluable resource.

For practice, see if you can track down the density of iron, the series expansion of e^x, the number of centimeters in a mile, the mean distance of Saturn from the Sun, the mass of the proton, the speed of light, and the atomic number of samarium. You can find them all in this book.

CHECKPOINT **4:** When you ride in a Ferris wheel at constant speed, what are the directions of your acceleration **a** and the normal force **N** on you (from the seat) as you pass through (a) the highest point and (b) the lowest point of the ride?

SAMPLE PROBLEM 6-8

In a 1901 circus performance, Allo "Dare Devil" Diavolo introduced the stunt of riding a bicycle in a loop-the-loop (Fig. 6-11*a*). Assuming that the loop is a circle with radius $R = 2.7$ m, what is the least speed v Diavolo could have at the top of the loop if he was to remain in contact with it there?

SOLUTION: Figure 6-11*b* is a free-body diagram for Diavolo and the bicycle (taken together as being a single particle) at the top of the loop, showing the normal force **N** exerted on them by the loop, and their weight **W** = m**g**. Their accelera-

tion **a** points downward, toward the center of the loop and, according to Eq. 6-19, has magnitude $a = v^2/R$. Applying Newton's second law along the y axis, we have

$$\sum F = -N - mg = -ma = -m \frac{v^2}{R}.$$

If he is on the verge of losing contact with the loop, $N = 0$, and so we then have

$$mg = m \frac{v^2}{R},$$

or
$$v = \sqrt{gR} = \sqrt{(9.8 \text{ m/s}^2)(2.7 \text{ m})}$$
$$= 5.1 \text{ m/s}. \qquad \text{(Answer)}$$

To avoid losing contact, Diavolo made certain that his speed at the top of the loop was greater than 5.1 m/s. Then $N > 0$ and, more important to him, he and the bicycle exerted a force of magnitude N on the track.

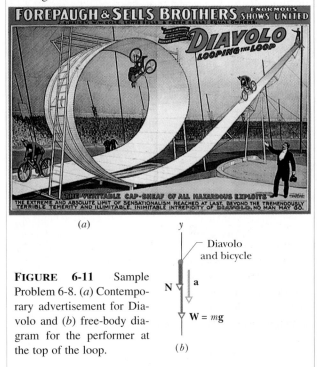

FIGURE 6-11 Sample Problem 6-8. (*a*) Contemporary advertisement for Diavolo and (*b*) free-body diagram for the performer at the top of the loop.

SAMPLE PROBLEM 6-9

Figure 6-12*a* shows a *conical pendulum*. Its bob, whose mass m is 1.5 kg, whirls around in a horizontal circle at constant speed v at the end of a cord whose length L, measured to the center of the bob, is 1.7 m. The cord makes an angle θ of 37° with the vertical. As the bob swings around in a circle, the cord sweeps out the surface of a cone. Find the period τ of the pendulum bob.

SOLUTION: Figure 6-12*b*, the free-body diagram for the bob, shows the forces on the bob: the pull **T** from the cord due to the cord's tension, and the bob's weight **W** (= m**g**). We place the origin of axes at the center of the bob, as shown. Instead of the usual x axis (which is stationary), we use a radial axis r that always points from the bob toward the center of the circle.

The y and r components of $\mathbf{T}$ are $T \cos \theta$ and $T \sin \theta$, respectively. Since $a_y = 0$, Newton's second law gives

$$T \cos \theta - mg = ma_y = 0, \quad \text{or} \quad T \cos \theta = mg. \quad (6\text{-}21)$$

There must be a net force along the r axis to provide the centripetal acceleration for the bob. The only force component in that direction is $T \sin \theta$. So along the r axis, Eq. 6-20 gives

$$T \sin \theta = ma_r = \frac{mv^2}{R}, \quad (6\text{-}22)$$

where R is the radius of the bob's circular path. Dividing Eq. 6-22 by Eq. 6-21 and solving for v, we obtain

$$v = \sqrt{\frac{gR \sin \theta}{\cos \theta}}.$$

We can substitute $2\pi R/\tau$ (the distance around the circle divided by the period) for the speed v of the bob. Doing so and solving for τ, we obtain

$$\tau = 2\pi \sqrt{\frac{R \cos \theta}{g \sin \theta}}. \quad (6\text{-}23)$$

Now, from Fig. 6-12a we see that $R = L \sin \theta$. Making this substitution in Eq. 6-23 yields

$$\tau = 2\pi \sqrt{\frac{L \cos \theta}{g}} \quad (6\text{-}24)$$

$$= 2\pi \sqrt{\frac{(1.7 \text{ m})(\cos 37°)}{9.8 \text{ m/s}^2}} = 2.3 \text{ s.} \quad \text{(Answer)}$$

From Eq. 6-24 we see that the period τ does not depend on the mass of the bob, but only on $L \cos \theta$, the vertical distance of the bob from its point of support. Thus, as shown in Fig. 6-12c, if several conical pendulums of different lengths are swung from the same support *with the same period,* their bobs will all lie in the same horizontal plane.

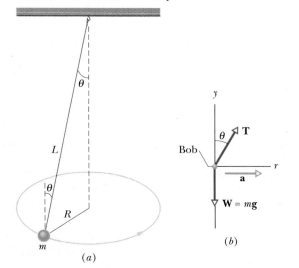

(a)

(b)

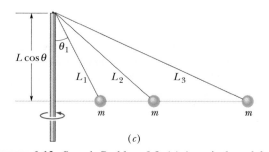

(c)

FIGURE 6-12 Sample Problem 6-9. (*a*) A conical pendulum, its cord making an angle θ with the vertical. (*b*) A free-body diagram for the pendulum bob. The axes point in the vertical and radial directions. The resultant force (and thus the acceleration) points radially inward toward the center of the circle. (*c*) Three pendulums, of different lengths, are whirled around by a rotating shaft; their bobs circulate in the same horizontal plane, as Eq. 6-24 predicts.

SAMPLE PROBLEM 6-10

Figure 6-13a represents a car of mass $m = 1600$ kg traveling at a constant speed $v = 20$ m/s along a flat, circular road of radius $R = 190$ m. What is the minimum value of μ_s between the tires of the car and the road that will prevent the car from slipping?

SOLUTION: The centripetal force that causes the car to travel in a circle is the radial frictional force $\mathbf{f}_s$ exerted on the tires by the road. (Although the car is moving, it is not sliding radially, and so the frictional force is $\mathbf{f}_s$ and not $\mathbf{f}_k$.)

Figure 6-13b, the free-body diagram for the car, shows the forces on the car: $\mathbf{f}_s$, $\mathbf{N}$, and $\mathbf{W} = m\mathbf{g}$. Since the car is not accelerating vertically, $a_y = 0$, and Newton's second law leads to the now familiar result $N = W = mg$.

In the radial direction, however, there must be a net force $\Sigma \mathbf{F}_r$ to give the car a centripetal acceleration $\mathbf{a}_r$. (Otherwise the car would move off the road along a straight line.) According to Eq. 6-20, $\Sigma F_r = mv^2/R$. Because the only radial force we have is f_s, it must be true that

$$f_s = \frac{mv^2}{R}. \quad (6\text{-}25)$$

Next, recall that a body is on the verge of slipping when f_s reaches its maximum value, $\mu_s N$. Since the problem concerns just such a critical situation, we set $f_s = \mu_s N$ in Eq. 6-25 and then substitute mg for N. We get

$$\mu_s mg = \frac{mv^2}{R},$$

or

$$\mu_s = \frac{v^2}{gR} \quad (6\text{-}26)$$

$$= \frac{(20 \text{ m/s})^2}{(9.8 \text{ m/s}^2)(190 \text{ m})} = 0.21. \quad \text{(Answer)}$$

If $\mu_s \geq 0.21$, the car will be held in a circle by $\mathbf{f}_s$. But if $\mu_s < 0.21$, the car will slide out of the circle.

Note two additional features of Eq. 6-26. First, the value

of μ_s depends on the square of v. That means that much more friction is required as the turning speed increases. You may have noted this effect if you have ever taken a flat turn too fast and suddenly felt the tires slip. Second, the mass m dropped out in our derivation of Eq. 6-26. That means Eq. 6-26 holds for a vehicle of any mass, from a kiddy car to a bicycle to a heavy truck.

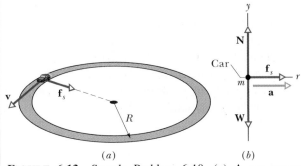

(a)   (b)

FIGURE 6-13 Sample Problem 6-10. (*a*) A car moves around a flat curved road at constant speed. The frictional force $\mathbf{f}_s$ provides the necessary centripetal force. (*b*) A free-body diagram (not to scale) for the car, in the vertical plane.

CHECKPOINT 5: In Fig. 6-13, suppose the car is on the verge of sliding when the radius of the circle is R_1. (a) If we double the car's speed, what is the least radius that would now keep the car from sliding? (b) If we also double the weight of the car (say by adding sandbags), what is the least radius that would now keep the car from sliding?

SAMPLE PROBLEM 6-11

You cannot always count on friction to get your car around a curve, especially if the road is icy or wet. That is why highway curves are banked. As in Sample Problem 6-10, suppose that a car of mass m moves at a constant speed v of 20 m/s around a curve, now banked, whose radius R is 190 m (Fig. 6-14*a*). What bank angle θ makes reliance on friction unnecessary?

SOLUTION: The centripetal acceleration and the required centripetal force ΣF_r are the same as in the preceding sample problem. The effect of banking is to tilt the normal force $\mathbf{N}$ toward the center of curvature of the road, so that now it has an inward radial component N_r that can supply the required centripetal force.

Since there is no acceleration in the vertical direction,

$$N_y = N \cos \theta = W = mg. \qquad (6-27)$$

In the radial direction the only force component is N_r (we assume a frictional force is unnecessary). So, by Eq. 6-20,

$$N_r = N \sin \theta = \frac{mv^2}{R}. \qquad (6-28)$$

Dividing Eq. 6-28 by Eq. 6-27 gives

$$\tan \theta = \frac{v^2}{gR}.$$

Thus we have

$$\theta = \tan^{-1} \frac{v^2}{gR} \qquad (6-29)$$

$$= \tan^{-1} \frac{(20 \text{ m/s})^2}{(9.8 \text{ m/s}^2)(190 \text{ m})} = 12°. \qquad \text{(Answer)}$$

Equations 6-26 and 6-29 tell us that the critical coefficient of friction for an unbanked road is the same as the tangent of the bank angle for a banked road. The road must produce a certain centripetal force one way or the other—either with friction or by being banked.

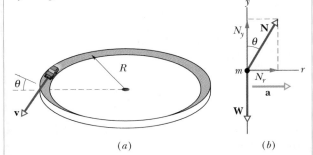

(a)   (b)

FIGURE 6-14 Sample Problem 6-11. (*a*) A car moves around a curved banked road at constant speed. The bank angle is exaggerated for clarity. (*b*) A free-body diagram for the car, assuming that friction between tires and road is zero. The radially inward component of the normal force provides the necessary centripetal force. The resulting acceleration also points radially inward.

SAMPLE PROBLEM 6-12

Even some seasoned roller coaster riders blanch at the thought of riding the Rotor, which is essentially a large hollow cylinder that is rotated rapidly around its central axis (Fig. 6-15). Before the ride begins, a rider enters the cylinder through a door on the side and stands on a floor, up against a canvas-covered wall. The door is closed, and as the cylinder begins to turn, the rider, wall, and floor move in unison. When the rider's speed reaches some predetermined value, the floor abruptly and alarmingly falls away. The rider does not fall with it but instead is pinned to the wall while the cylinder rotates, as if an unseen (and somewhat unfriendly) agent is pressing the body to the wall. Later, the floor is eased back to the rider's feet, the cylinder slows, and the rider sinks a few centimeters to regain footing on the floor. (Some riders consider all this to be fun.)

Suppose that the coefficient of static friction μ_s between the rider's clothing and the canvas is 0.40 and that the cylinder's radius R is 2.1 m.

(a) What minimum speed v must the cylinder and rider have if the rider is not to fall when the floor drops?

SOLUTION: The rider will not be pulled down by her weight **W** provided the magnitudes of **W** and the frictional force $\mathbf{f}_s$, which is exerted upward on her by the wall, are equal. At the minimum required speed, she is on the verge of slipping, which means that f_s must be at its maximum value of $\mu_s N$. So in this critical situation

$$\mu_s N = mg, \tag{6-30}$$

where m is her mass.

The normal force **N** is, as usual, perpendicular to the surface against which the body (here, the woman) is pressed, but note that now it points horizontally toward the central axis. That force accounts for the centripetal force that provides the woman's centripetal acceleration a_r and keeps her moving in a circle. Thus, by Eq. 6-20,

$$N = \frac{mv^2}{R}. \tag{6-31}$$

Substituting this expression for N in Eq. 6-30 and solving for v, we have

$$v = \sqrt{\frac{gR}{\mu_s}} = \sqrt{\frac{(9.8 \text{ m/s}^2)(2.1 \text{ m})}{0.40}}$$
$$= 7.17 \text{ m/s} \approx 7.2 \text{ m/s}. \tag{Answer}$$

Note that the result is independent of the rider's mass; it holds for anyone riding the Rotor, from a child to a sumo wrestler.

(b) If the rider's mass is 49 kg, what is the magnitude of the centripetal force on her?

SOLUTION: According to Eq. 6-31,

$$N = \frac{mv^2}{R} = \frac{(49 \text{ kg})(7.17 \text{ m/s})^2}{2.1 \text{ m}}$$
$$\approx 1200 \text{ N}. \tag{Answer}$$

FIGURE 6-15 Sample Problem 6-12. A Rotor in an amusement park, showing the forces on a rider. The centripetal force is the normal force with which the wall pushes inward on the rider.

Although this force is directed toward the central axis, the rider has an overwhelming sensation that the force pinning her against the wall is directed radially outward. Her sensation stems from the fact that she is in a noninertial frame (she and it are accelerating). As measured from such frames, forces can be illusionary. The illusion is part of the Rotor's attraction.

CHECKPOINT **6:** If the Rotor of Sample Problem 6-12 initially moves at the minimum required speed for the rider not to fall and then its speed is increased in steps, do the following increase, decrease, or remain the same: (a) the magnitude of $\mathbf{f}_s$; (b) the magnitude of **N**; (c) the value of $f_{s,\text{max}}$?

6-5 THE FORCES OF NATURE

We have used **F** as a generic symbol for force. We have also used other symbols: **W** for the weight of a body, **T** for the pull from a cord under tension, **f** for a frictional force, **N** for a normal force, and **D** for the drag force exerted, for example, by air on a sky diver. At the fundamental level, all these forces fall into two types: (1) the **gravitational force,** of which weight is our only example, and (2) the **electromagnetic force,** which includes—without exception—all the others. The electromagnetic force is the combination of electrical forces and magnetic forces. The force that makes an electrically charged balloon stick to a wall and the force with which a magnet picks up an iron nail are other examples of it. In fact, aside from the gravitational force, *all* forces that we can experience directly as a push or pull are electromagnetic in nature. That is, all such forces, including frictional forces, normal forces, contact forces, and tension forces, involve, fundamentally, electromagnetic forces exerted by one atom on another. The tension in a taut rope, for example, exists only because the atoms of the rope attract one another.

Only two other fundamental forces are known, and they both act over such short distances that we cannot ex-

Banked tracks are needed for turns that are taken so quickly that friction alone cannot provide enough centripetal force.

TABLE 6-2 THE QUEST FOR THE SUPERFORCE—A PROGRESS REPORT

DATE	RESEARCHER	ACHIEVEMENT
1687	Newton	Showed that the same laws apply to astronomical bodies and to objects on Earth. Unified celestial and terrestrial mechanics.
1820	Oersted	Showed, by brilliant experiments, that the then separate sciences of electricity and magnetism are intimately linked.
1830s	Faraday	
1873	Maxwell	Unified the sciences of electricity, magnetism, and optics into the single subject of electromagnetism.
1979	Glashow, Salam, Weinberg	Received the Nobel prize for showing that the weak force and the electromagnetic force could be viewed as different aspects of a single *electroweak force.* This combination of forces reduced the number of fundamental forces from four to three.
1984	Rubbia, van der Meer	Received the Nobel prize for verifying experimentally the predictions of the theory of the electroweak force.

Work in Progress

Grand unification theories (GUTs), seek to unify the electroweak force and the strong force.

Supersymmetry theories: seek to unify all forces, including the gravitational force, within a single framework.

Superstring theories: interpret pointlike particles, such as electrons, as being unimaginably tiny, closed loops. Strangely, extra dimensions beyond the familiar four dimensions of spacetime appear to be required.

perience them directly through our senses. They are the **weak force,** which is involved in certain kinds of radioactive decay, and the **strong force,** which binds together the quarks that make up protons and neutrons and is the "glue" that holds together an atomic nucleus.

Physicists have long believed that nature has an underlying simplicity and that the number of fundamental forces can be reduced. Einstein spent most of his working life trying to interpret these forces as different aspects of a single *superforce.* He failed, but in the 1960s and 1970s, other physicists showed that the weak force and the electromagnetic force are different aspects of a single **electroweak force.** The quest for further reduction continues today, at the very forefront of physics. Table 6-2 lists the progress that has been made toward **unification** (as the goal is called) and gives some hints about the future.

REVIEW & SUMMARY

Friction

When a force **F** attempts to slide a body along a surface, a **frictional force** is exerted on the body by the surface. The frictional force is parallel to the surface and directed so as to oppose the sliding. It is due to bonding between the body and the surface.

If the body does not slide, the frictional force is a **static frictional force** f_s. If there is sliding, the frictional force is a **kinetic frictional force** f_k.

Properties of Friction

Property 1. If the body does not move, then the static frictional force f_s and the component of **F** that is parallel to the surface are equal in magnitude, and f_s is directed opposite that component. If that parallel component increases, f_s also increases.

Property 2. The magnitude of f_s has a maximum value $f_{s,max}$ that is given by

$$f_{s,max} = \mu_s N, \qquad (6\text{-}1)$$

where μ_s is the **coefficient of static friction** and N is the magnitude of the normal force. If the component of **F** that is parallel to the surface exceeds $f_{s,max}$, then the body slides on the surface.

Property 3. If the body begins to slide along the surface, the

magnitude of the frictional force rapidly decreases to a constant value f_k given by

$$f_k = \mu_k N, \qquad (6\text{-}2)$$

where μ_k is the **coefficient of kinetic friction.**

Drag Force

When there is a relative velocity between air (or some other fluid) and a body, the body experiences a **drag force D** that opposes the relative motion and points in the direction in which the fluid flows relative to the body. The magnitude of **D** is related to the relative speed v by an experimentally determined **drag coefficient** C according to

$$D = \tfrac{1}{2} C \rho A v^2, \qquad (6\text{-}17)$$

where ρ is the fluid density (mass per volume) and A is the **effective cross-sectional area** of the body (the area of a cross section taken perpendicular to the relative velocity **v**).

Terminal Speed

When a blunt object falls far enough through air, the magnitudes of the drag force and the object's weight are equal. The body then falls at a constant **terminal speed** v_t given by

$$v_t = \sqrt{\frac{2mg}{C\rho A}}, \qquad (6\text{-}18)$$

where m is the body's mass.

Uniform Circular Motion

If a particle moves in a circle or a circular arc with radius r at constant speed v, it is said to be in **uniform circular motion.** It then has a **centripetal acceleration** with a magnitude given by

$$a = \frac{v^2}{r}, \qquad (6\text{-}19)$$

which is due to a **centripetal force** with a magnitude given by

$$F = \frac{mv^2}{r}, \qquad (6\text{-}20)$$

where m is the particle's mass. The vectors $\mathbf{a}$ and $\mathbf{F}$ point toward the center of curvature of the particle's path.

Fundamental Forces

The myriad examples of forces can be reduced to three fundamental types: **gravitational, electroweak** (a combination of the historic grouping of **electric** and **magnetic** forces with the **weak** force), and **strong.** Only the gravitational, electric, and magnetic forces are readily apparent in the everyday world. Physicists hope to reduce the list of three fundamental forces to a single force, the elusive *superforce* that would include all others.

QUESTIONS

1. Figure 6-16 shows four blocks arranged on a board. The board will be lifted by its right end (like the book in Fig. 6-3*a*) until the blocks begin to slide down it. The blocks are made of the same material and their masses are:

block 1, 5 kg	block 3, 10 kg
block 2, 10 kg	block 4, 5 kg

In which order, left to right, should the blocks be placed for them to begin sliding at the smallest possible angle between the board and the horizontal?

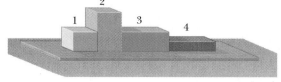

FIGURE 6-16 Question 1.

2. In Fig. 6-17, horizontal force $\mathbf{F}_1$ of magnitude 10 N is applied to a box on a floor, but the box does not slide. Then, as the magnitude of vertical force $\mathbf{F}_2$ is increased from zero, do the following quantities increase, decrease, or stay the same: (a) the magnitude of the frictional force $\mathbf{f}_s$ on the box; (b) the magnitude of the normal force $\mathbf{N}$ on the box from the floor; (c) the maximum value $f_{s,\max}$ of the static frictional force on the box? (d) Does the box eventually slide?

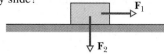

FIGURE 6-17 Question 2.

3. If you press an apple crate against a wall so hard that the crate cannot slide down the wall, what is the direction of (a) the static frictional force $\mathbf{f}_s$ on the crate from the wall and (b) the normal force $\mathbf{N}$ on the crate from the wall? If you increase your push, what happens to (c) f_s, (d) N, and (e) $f_{s,\max}$?

4. A box is on a ramp that is at angle θ to the horizontal. As θ is increased from zero, and before the box slips, do the following increase, decrease, or remain the same: (a) the weight component of the box along the ramp, (b) the magnitude of the static frictional force on the box from the ramp, (c) the weight component of the box perpendicular to the ramp, (d) the normal force on the box from the ramp, and (e) the maximum value $f_{s,\max}$ of the static frictional force?

5. In Fig. 6-18, a block is held stationary on a ramp by the frictional force on it from the ramp. A force $\mathbf{F}$, directed up the ramp, is then applied to the block and gradually increased in magnitude from zero. During the increase, what happens to the direction and magnitude of the frictional force on the block?

FIGURE 6-18 Question 5.

6. Reconsider Question 5 but with the force $\mathbf{F}$ now directed down the ramp. As the magnitude of $\mathbf{F}$ is increased from zero, what happens to the direction and magnitude of the frictional force on the block?

7. In Fig. 6-19, if the angle θ of force $\mathbf{F}$ on the stationary box is increased, do the following quantities increase, decrease, or remain the same: (a) F_x; (b) $\mathbf{f}_s$; (c) $\mathbf{N}$; (d) $f_{s,\max}$?

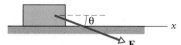

FIGURE 6-19 Question 7.

8. Repeat Question 7 for force $\mathbf{F}$ angled upward instead of downward.

9. In the conical pendulum of Sample Problem 6-9, what (a) period and (b) speed are associated with $\theta = 90°$?

10. A particle is made to move around three circular arcs, with the following speeds and radii of curvature:

ARC	SPEED	RADIUS
1	$2v_0$	r_0
2	$3v_0$	$3r_0$
3	$2v_0$	$4r_0$

Rank the arcs according to the magnitude of the centripetal force acting on the particle, greatest first.

11. Figure 6-20 shows an overhead view of a amusement park ride that travels at constant speed through five circular arcs of radii R_0, $2R_0$, and $3R_0$. Rank the arcs according to the magnitude of the centripetal force acting on a rider while traveling in the arcs, greatest first.

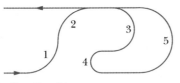

FIGURE 6-20   Question 11.

12. Figure 6-21 shows a section of a circular space station that rotates about its center so as to give an apparent weight to the crew. One of the crew is shown at the outer wall of the station, which has velocity $\mathbf{v}_s$. (a) If the astronaut moves to a point closer to the center of the station (say by taking an elevator), does his apparent weight increase, decrease, or remain the same? (b) If,

instead, the astronaut runs along the outer wall in the direction opposite $\mathbf{v}_s$ (with a speed less than the magnitude of $\mathbf{v}_s$), does his apparent weight increase, decrease, or remain the same?

FIGURE 6-21   Question 12.

13. Figure 6-22 shows overhead views of two stones that travel in circles over a frictionless surface. Each stone is tied to a cord whose opposite end is anchored at the center of the circle. Is the tension in the longer cord greater than, less than, or the same as that in the shorter cord if the stones travel (a) at the same speed and (b) with the same period of motion?

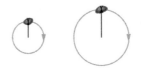

FIGURE 6-22   Question 13.

14. A coin lies on a turntable whose speed can be gradually increased from zero in steps. What happens to the magnitude of the frictional force on the coin from the turntable as the speed is increased to a large value?

EXERCISES & PROBLEMS

SECTION 6-2 Properties of Friction

1E. A bedroom bureau with a mass of 45 kg, including drawers and clothing, rests on the floor. (a) If the coefficient of static friction between the bureau and the floor is 0.45, what is the minimum horizontal force a person must apply to start the bureau moving? (b) If the drawers and clothing, with 17 kg mass, are removed before the bureau is pushed, what is the new minimum magnitude?

2E. A baseball player with mass $m = 79$ kg, sliding into second base, is retarded by a force of friction $f = 470$ N. What is the coefficient of kinetic friction μ_k between the player and the ground?

3E. The coefficient of static friction between Teflon and scrambled eggs is about 0.04. What is the smallest angle from the horizontal that will cause the eggs to slide across the bottom of a Teflon-coated skillet?

4E. A 100 N force, directed at an angle θ above the horizontal, is applied to a 25.0 kg chair sitting on the floor. (a) For each of the following angles θ, calculate the magnitude of the normal force of the floor on the chair and the horizontal component of the applied force: (i) 0°, (ii) 30.0°, (iii) 60.0°. (b) Take the coefficient of static friction between the chair and the floor to be 0.420 and, for each of the values of θ, decide if the chair remains at rest or slides.

5E. In Nevada and southern California, stones leave trails in the hard-baked desert floor, as if they had been migrating (Fig. 6-23). For years curiosity mounted about the unseen motion that caused the trails. The answer finally came in the 1970s: when an occasional storm hits the desert, a thin layer of mud may form over a still-firm base, greatly reducing the coefficient of friction between the stones and ground. If a strong wind accompanies the storm, it pushes the stones, leaving trails that are later baked hard by the Sun. Suppose a stone's mass is 300 kg (about the greatest stone mass that has left a trail) and the coefficient of static friction is reduced to 0.15. Of what magnitude is the force from a horizontal gust that is needed to move the stone?

FIGURE 6-23   Exercise 5.

6E. What is the greatest acceleration that can be generated by a runner if the coefficient of static friction between shoes and track is 0.95? (Only one foot is on the track during the acceleration.)

7E. A worker pushes horizontally on a 35 kg crate with a 110 N force. The coefficient of static friction between the crate and the floor is 0.37. (a) What is the frictional force exerted on the crate by the floor? (b) What is the maximum magnitude $f_{s,\text{max}}$ of the static frictional force under the circumstances? (c) Does the crate move? (d) Suppose, next, that a second worker pulls directly upward on the crate to help out. What is the least pull she can exert that will allow the first worker's 110 N push to move the crate? (e) If, instead, the second worker pulls horizontally to help out, what is the least pull she can exert to get the crate moving?

8E. A person pushes horizontally with a force of 220 N on a 55 kg crate to move it across a level floor. The coefficient of kinetic friction is 0.35. (a) What is the magnitude of the frictional force? (b) What is the acceleration of the crate?

9E. A trunk with a weight of 220 N rests on the floor. The coefficient of static friction between the trunk and the floor is 0.41, while the coefficient of kinetic friction is 0.32. (a) What is the minimum magnitude for a horizontal force with which a person must push on the trunk to start it moving? (b) Once the trunk is moving, what magnitude of horizontal force must the person apply to keep it moving with constant velocity? (c) If the person continued to push with the force used to start the motion, what would be the acceleration of the trunk?

10E. A filing cabinet with a weight of 556 N rests on the floor. The coefficient of static friction between it and the floor is 0.68, and the coefficient of kinetic friction is 0.56. In four different attempts to move it, it is pushed with horizontal forces of (a) 222 N, (b) 334 N, (c) 445 N, and (d) 556 N. For each attempt, determine whether the cabinet moves, and calculate the magnitude of the frictional force the floor exerts on it. The cabinet is initially at rest for each attempt.

11E. A horizontal force F of 12 N pushes a block weighing 5.0 N against a vertical wall (Fig. 6-24). The coefficient of static

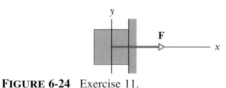

FIGURE 6-24   Exercise 11.

friction between the wall and the block is 0.60, and the coefficient of kinetic friction is 0.40. Assume that the block is not moving initially. (a) Will the block start moving? (b) In unit-vector notation, what is the force exerted on the block by the wall?

12E. A 49 kg rock climber is climbing a "chimney" between two rock slabs as shown in Fig. 6-25. The static coefficient of friction between her shoes and the rock is 1.2; between her back and the rock it is 0.80. She has reduced her push against the rock until her back and her shoes are on the verge of slipping. (a) What is her push against the rock? (b) What fraction of her weight is supported by the frictional force on her shoes?

FIGURE 6-25   Exercise 12.

13E. A house is built on the top of a hill with a nearby 45° slope (Fig. 6-26). An engineering study indicates that the slope angle should be reduced because the top layers of soil along the slope might slip past the lower layers. If the static coefficient of friction between two such layers is 0.5, what is the least angle ϕ through which the present slope should be reduced to prevent slippage?

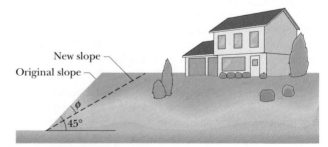

FIGURE 6-26   Exercise 13.

14E. The coefficient of kinetic friction in Fig. 6-27 is 0.20. What is the acceleration of the block if (a) it is sliding down the slope and (b) it has been given an upward shove and is still sliding up the slope?

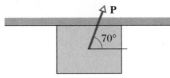

FIGURE 6-27 Exercise 14.

15E. A 110 g hockey puck slides on the ice for 15 m before it stops. (a) If its initial speed was 6.0 m/s, what was the magnitude of the frictional force on the puck during the sliding? (b) What was the coefficient of friction between the puck and the ice?

16P. A student, crazed with final exams, uses a force **P** of magnitude 80 N to push a 5.0 kg block across the ceiling of his room, as shown in Fig. 6-28. If the coefficient of kinetic friction between the block and surface is 0.40, what is the magnitude of the acceleration of the block?

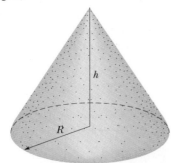

FIGURE 6-28 Problem 16.

17P. A student wants to determine the coefficients of static friction and kinetic friction between a box and a plank. She places the box on the plank and gradually raises one end of the plank. When the angle of inclination with the horizontal reaches 30°, the box starts to slip, and it slides 2.5 m down the plank in 4.0 s. What are the coefficients of friction?

18P. A worker wishes to pile a cone of sand onto a circular area in his yard. The radius of the circle is R, and no sand is to spill onto the surrounding area (Fig. 6-29). If μ_s is the static coefficient of friction between each layer of sand along the slope and the sand beneath it (along which it might slip), show that the greatest volume of sand that can be stored in this manner is $\pi\mu_s R^3/3$. (The volume of a cone is $Ah/3$, where A is the base area and h is the cone's height.)

FIGURE 6-29 Problem 18.

19P. A ski that is placed on snow will stick to the snow. However, when the ski is moved along the snow, the rubbing warms and partially melts the snow, reducing the coefficient of friction and promoting sliding. Waxing the ski makes it water repellent

and reduces friction with the resulting layer of water. A magazine reports that a new type of plastic ski is especially water repellent and that, on a gentle 200 m slope in the Alps, a skier reduced his top-to-bottom time from 61 s with standard skis to 42 s with the new skis. (a) Determine the magnitudes of his average acceleration with each pair of skis. (b) Assuming a 3.0° slope, compute the coefficient of kinetic friction for each case.

20P. An 11 kg block of steel is at rest on a horizontal table. The coefficient of static friction between block and table is 0.52. (a) What is the magnitude of the horizontal force that will just start the block moving? (b) What is the magnitude of a force acting upward 60° from the horizontal that will just start the block moving? (c) If the force acts down at 60° from the horizontal, how large can its magnitude be without causing the block to move?

21P. A railroad flatcar is loaded with crates having a coefficient of static friction of 0.25 with the floor. If the train is moving at 48 km/h, in how short a distance can the train be stopped at constant deceleration without causing the crates to slide?

22P. A block slides down an inclined plane of slope angle θ with constant velocity. It is then projected up the same plane with an initial speed v_0. (a) How far up the incline will it move before coming to rest? (b) Will it slide down again? Give an argument to back your answer.

23P. A 68 kg crate is dragged across a floor by pulling on a rope inclined 15° above the horizontal. (a) If the coefficient of static friction is 0.50, what minimum tension in the rope is required to start the crate moving? (b) If $\mu_k = 0.35$, what is the magnitude of the initial acceleration of the crate?

24P. A pig slides down a 35° incline (Fig. 6-30) in twice the time it would take to slide down a frictionless 35° incline. What is the coefficient of kinetic friction between the pig and the incline?

FIGURE 6-30 Problem 24.

25P. In Fig. 6-31, A and B are blocks with weights of 44 N and 22 N, respectively. (a) Determine the minimum weight (block C) that must be placed on A to keep it from sliding, if μ_s between A

and the table is 0.20. (b) Block C suddenly is lifted off A. What is the acceleration of block A, if μ_k between A and the table is 0.15?

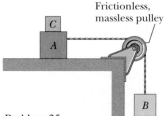

FIGURE 6-31 Problem 25.

26P. A 3.5 kg block is pushed along a horizontal floor by a force $F = 15$ N that makes an angle $\theta = 40°$ with the horizontal (Fig. 6-32). The coefficient of kinetic friction between the block and floor is 0.25. Calculate (a) the magnitude of the frictional force exerted on the block and (b) the acceleration of the block.

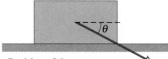

FIGURE 6-32 Problem 26.

27P. In Fig. 6-33 a fastidious worker pushes directly along the handle of a mop with a force $\mathbf{F}$. The handle is at an angle θ with the vertical, and μ_s and μ_k are the coefficients of static and kinetic friction between the head of the mop and the floor. Ignore the mass of the handle and assume that all the mop's mass m is in its head. (a) If the mop head moves along the floor with a constant velocity, then what is F? (b) Show that if θ is less than a certain value θ_0, then $\mathbf{F}$ (still directed along the handle) is unable to move the mop head. Find θ_0.

FIGURE 6-33 Problem 27.

28P. A 5.0 kg block on an inclined plane is acted on by a horizontal force $\mathbf{F}$ with magnitude 50 N (Fig. 6-34). The coefficient of kinetic friction between block and plane is 0.30. The coefficient of static friction is not given (but you might still know something about it). (a) What is the acceleration of the block if it is moving up the plane? (b) With the horizontal force still acting, how far up the plane will the block go if it has an initial upward speed of 4.0 m/s? (c) What happens to the block after it reaches the highest point? Give an argument to back your answer.

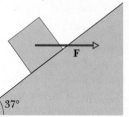

FIGURE 6-34 Problem 28.

29P. Figure 6-35 shows the cross section of a road cut into the side of a mountain. The solid line AA' represents a weak bedding plane along which sliding is possible. Block B directly above the highway is separated from uphill rock by a large crack (called a *joint*), so that only the force of friction between the block and the bedding plane prevents sliding. The mass of the block is 1.8×10^7 kg, the *dip angle* θ of the failure plane is 24°, and the coefficient of static friction between block and plane is 0.63. (a) Show that the block will not slide. (b) Water seeps into the joint and expands upon freezing, exerting on the block a force $\mathbf{F}$ parallel to AA'. What minimum value of F will trigger a slide?

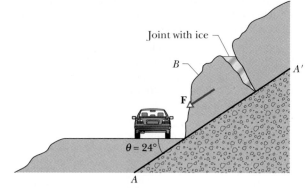

FIGURE 6-35 Problem 29.

30P. A block weighing 80 N rests on a plane inclined at 20° to the horizontal (Fig. 6-36). The coefficient of static friction is 0.25, and the coefficient of kinetic friction is 0.15. (a) What is the minimum magnitude of the force $\mathbf{F}$, parallel to the plane, that will prevent the block from slipping down the plane? (b) What is the minimum magnitude F that will start the block moving up the plane? (c) What value of F is required to move the block up the plane at constant velocity?

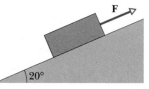

FIGURE 6-36 Problem 30.

31P. Block B in Fig. 6-37 weighs 711 N. The coefficient of static friction between block and horizontal surface is 0.25. Find the maximum weight of block A for which the system will be stationary.

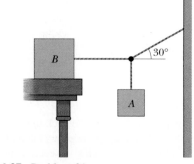

FIGURE 6-37 Problem 31.

32P. Body *A* in Fig. 6-38 weighs 102 N, and body *B* weighs 32 N. The coefficients of friction between *A* and the incline are $\mu_s = 0.56$ and $\mu_k = 0.25$. Angle θ is 40°. Find the acceleration of the system if (a) *A* is initially at rest, (b) *A* is moving up the incline, and (c) *A* is moving down the incline.

33P. Two blocks are connected over a pulley as shown in Fig. 6-38. The mass of block *A* is 10 kg and the coefficient of kinetic friction is 0.20. Angle θ is 30°. Block *A* slides down the incline at constant speed. What is the mass of block *B*?

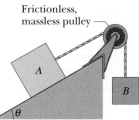

FIGURE 6-38 Problems 32 and 33.

34P. Block m_1 in Fig. 6-39 has a mass of 4.0 kg and m_2 has a mass of 2.0 kg. The coefficient of kinetic friction between m_2 and the horizontal plane is 0.50. The inclined plane is frictionless. Find (a) the tension in the cord and (b) the acceleration of the blocks.

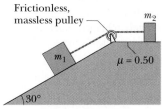

FIGURE 6-39 Problem 34.

35P. Two blocks, of weights 8.0 lb and 16 lb, are connected by a massless string and slide down a 30° inclined plane. The coefficient of kinetic friction between the 8.0 lb block and the plane is 0.10; that between the 16 lb block and the plane is 0.20. Assuming that the 8.0 lb block leads, find (a) the acceleration of the blocks and (b) the tension in the string. (c) Describe the motion if the blocks are reversed.

36P. Two masses, $m_1 = 1.65$ kg and $m_2 = 3.30$ kg, attached by a massless rod parallel to the inclined plane on which both slide (Fig. 6-40), travel down along the plane with m_1 trailing m_2. The angle of incline is $\theta = 30°$. The coefficient of kinetic friction between m_1 and the incline is $\mu_1 = 0.226$; that between m_2 and the incline is $\mu_2 = 0.113$. Compute (a) the tension in the rod and (b) the common acceleration of the two masses. (c) How would the answers to (a) and (b) change if m_2 trailed m_1?

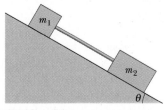

FIGURE 6-40 Problem 36.

37P. A 4.0 kg block is put on top of a 5.0 kg block. To cause the top block to slip on the bottom one, while the bottom one is held fixed, a horizontal force of at least 12 N must be applied to the top block. The assembly of blocks is now placed on a horizontal, frictionless table (Fig. 6-41). Find (a) the magnitude of the maximum horizontal force **F** that can be applied to the lower block so that the blocks will move together and (b) the resulting acceleration of the blocks.

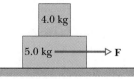

FIGURE 6-41 Problem 37.

38P. The two blocks (with $m = 16$ kg and $M = 88$ kg) shown in Fig. 6-42 are not attached. The coefficient of static friction between the blocks is $\mu_s = 0.38$, but the surface beneath *M* is frictionless. What is the minimum magnitude of the horizontal force **F** required to hold *m* against *M*?

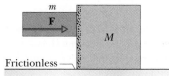

FIGURE 6-42 Problem 38.

39P. A 40 kg slab rests on a frictionless floor. A 10 kg block rests on top of the slab (Fig. 6-43). The coefficient of static friction μ_s between the block and the slab is 0.60, whereas the kinetic coefficient μ_k is 0.40. The 10 kg block is pulled by a horizontal force with a magnitude of 100 N. What are the resulting acceleration magnitudes of (a) the block and (b) the slab?

FIGURE 6-43 Problem 39.

40P. A crate slides down an inclined right-angled trough as in Fig. 6-44. The coefficient of kinetic friction between the crate and the trough is μ_k. What is the acceleration of the crate in terms of μ_k, θ, and g?

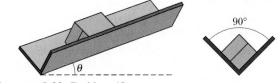

FIGURE 6-44 Problem 40.

41P. A locomotive accelerates a 25-car train along a level track. Every car has a mass of 50 metric tons and is subject to a frictional force $f = 250v$, where the speed *v* is in meters per second and the force *f* is in newtons. At the instant when the speed of the train is 30 km/h, the acceleration is 0.20 m/s². (a) What is the tension in the coupling between the first car and the locomotive? (b) If this tension is the maximum force the locomotive can exert on the train, what is the steepest grade up which the locomotive can pull the train at 30 km/h?

42P. An initially stationary box of sand is to be pulled across a floor by means of a cord in which the tension should not exceed 1100 N. The coefficient of static friction between the box and floor is 0.35. (a) What should be the angle between the cord and the horizontal in order to pull the greatest possible amount of sand, and (b) what is the weight of the sand and box in that situation?

43P*. A 1000 kg boat is traveling at 90 km/h when its engine is shut off. The magnitude of the frictional force $\mathbf{f}_k$ between boat and water is proportional to the speed v of the boat: $f_k = 70v$, where v is in meters per second and f_k is in newtons. Find the time required for the boat to slow down to 45 km/h.

SECTION 6-3 The Drag Force and Terminal Speed

44E. Calculate the drag force on a missile 53 cm in diameter cruising with a speed of 250 m/s at low altitude, where the density of air is 1.2 kg/m³. Assume $C = 0.75$.

45E. The terminal speed of a sky diver in the spread-eagle position is 160 km/h. In the nosedive position, the terminal speed is 310 km/h. Assuming that C does not change from one position to the other, find the ratio of the effective cross-sectional area A in the slower position to that in the faster position.

46E. Calculate the ratio of the drag force on a passenger jet flying with a speed of 1000 km/h at an altitude of 10 km to the drag force on a prop-driven transport flying at half the speed and half the altitude of the jet. At 10 km the density of air is 0.38 kg/m³ and at 5.0 km it is 0.67 kg/m³. Assume that the airplanes have the same effective cross-sectional area and the same drag coefficient C.

47P. From the data in Table 6-1, deduce the diameter of the 16 lb shot. Assume that $C = 0.49$ and the density of air is 1.2 kg/m³.

SECTION 6-4 Uniform Circular Motion

48E. If the coefficient of static friction for tires on a road is 0.25, at what maximum speed can a car round a level curve of 47.5 m radius without slipping?

49E. What is the smallest radius of an unbanked curve around which a bicyclist can travel if her speed is 18 mi/h and the coefficient of static friction between the tires and the road is 0.32?

50E. During an Olympic bobsled run, a European team takes a turn of radius 25 ft at a speed of 60 mi/h. How many g's do the riders experience during the turn?

51E. A car weighing 10.7 kN and traveling at 13.4 m/s attempts to round an unbanked curve with a radius of 61.0 m. (a) What force of friction is required to keep the car on its circular path? (b) If the coefficient of static friction between the tires and road is 0.35, is the attempt at taking the curve successful?

52E. A circular curve of highway is designed for traffic moving at 60 km/h. (a) If the radius of the curve is 150 m, what is the correct angle of banking of the road? (b) If the curve were not banked, what would be the minimum coefficient of friction between tires and road that would keep traffic from skidding at this speed?

53E. A banked circular highway curve is designed for traffic moving at 60 km/h. The radius of the curve is 200 m. Traffic is moving along the highway at 40 km/h on a rainy day. What is the minimum coefficient of friction between tires and road that will allow cars to negotiate the turn without sliding off the road?

54E. A child places a picnic basket on the outer rim of a merry-go-round that has a radius of 4.6 m and revolves once every 30 s. (a) What is the speed of a point on that rim? (b) How large must the coefficient of static friction between the basket and the merry-go-round be for the basket to stay on the ride?

55E. A conical pendulum is formed by attaching a 50 g mass to a 1.2 m string. The mass swings around a horizontal circle of radius 25 cm. (a) What is the speed of the mass? (b) What is the acceleration of the mass? (c) What is the tension in the string?

56E. In the Bohr model of the hydrogen atom, the electron revolves in a circular orbit around the nucleus. If the radius is 5.3×10^{-11} m and the electron circles 6.6×10^{15} times per second, find (a) the speed of the electron, (b) the acceleration (magnitude and direction) of the electron, and (c) the centripetal force acting on the electron. (This force is the result of the attraction between the positively charged nucleus and the negatively charged electron.) The electron's mass is 9.11×10^{-31} kg.

57E. A mass m on a frictionless table is attached to a hanging mass M by a cord through a hole in the table (Fig. 6-45). Find the speed with which m must move in order for M to stay at rest.

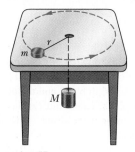

FIGURE 6-45 Exercise 57.

58E. A stuntman drives a car over the top of a hill, the cross section of which can be approximated by a circle of radius 250 m, as in Fig. 6-46. What is the greatest speed at which he can drive without the car leaving the road at the top of the hill?

FIGURE 6-46 Exercise 58.

59P. A small coin is placed on a flat, horizontal turntable. The turntable is observed to make three revolutions in 3.14 s. (a) What is the speed of the coin when it rides without slipping at a distance of 5.0 cm from the center of the turntable? (b) What is the acceleration (magnitude and direction) of the coin? (c) What is the magnitude of the frictional force acting on the coin if the coin has a mass of 2.0 g? (d) What is the coefficient of static friction between the coin and the turntable if the coin is observed to slide off the turntable when it is more than 10 cm from the center of the turntable?

60P. A small object is placed 10 cm from the center of a phonograph turntable. It remains in place when the table rotates at $33\frac{1}{3}$ rev/min but slides off when the table rotates at 45 rev/min. Between what limits must the coefficient of static friction between the object and the surface of the turntable lie?

61P. A bicyclist travels in a circle of radius 25.0 m at a constant speed of 9.00 m/s. The combined mass of the bicycle and rider is 85.0 kg. Calculate the magnitudes of (a) the force of friction exerted by the road on the bicycle and (b) the total force exerted by the road on the bicycle.

62P. A 150 lb student on a steadily rotating Ferris wheel has an apparent weight of 125 lb at the highest point. (a) What is the student's apparent weight at the lowest point? (b) What is the student's apparent weight at the highest point if the wheel's speed is doubled?

63P. A car is rounding a flat curve of radius $R = 220$ m at the curve's maximum design speed $v = 94.0$ km/h. What *total* force does a passenger with mass $m = 85.0$ kg exert on the seat cushion?

64P. A stone tied to the end of a string is whirled around in a vertical circle of radius R. Find the critical speed below which the string would become slack at the highest point.

65P. A certain string can withstand a maximum tension of 9.0 lb without breaking. A child ties a 0.82 lb stone to one end and, holding the other end, whirls the stone in a vertical circle of radius 3.0 ft, slowly increasing the speed until the string breaks. (a) Where is the stone on its path when the string breaks? (b) What is the speed of the stone as the string breaks?

66P. An airplane is flying in a horizontal circle at a speed of 480 km/h. If the wings of the plane are tilted 40° to the horizontal, what is the radius of the circle in which the plane is flying? (See Fig. 6-47.) Assume that the required force is provided entirely by an "aerodynamic lift" that is perpendicular to the wing surface.

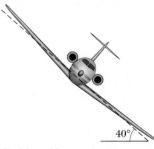

40°

FIGURE 6-47 Problem 66.

67P. A frigate bird is soaring in a horizontal circular path. Its bank angle (relative to the horizontal) is estimated to be 25° and the bird takes 13 s to complete one circle. (a) How fast is the bird flying? (b) What is the radius of the circle?

68P. A model airplane of mass 0.75 kg is flying at constant speed in a horizontal circle at one end of a 30 m cord and at a height of 18 m. The other end of the cord is tethered to the ground. The airplane circles 4.4 times per minute and has its wings horizontal so that the air is pushing vertically upward. (a) What is the acceleration of the plane? (b) What is the tension in the cord? (c) What is the total upward force (lift) on the plane's wings?

69P. An old streetcar rounds a corner on unbanked tracks. If the radius of the tracks is 30 ft and the car's speed is 10 mi/h, what angle with the vertical will be made by the loosely hanging hand straps?

70P. As shown in Fig. 6-48, a 1.34 kg ball is connected by means of two massless strings to a vertical, rotating rod. The strings are tied to the rod, are taut, and form two sides of an equilateral triangle. The tension in the upper string is 35 N. (a) Draw the free-body diagram for the ball. (b) What is the tension in the lower string? (c) What is the net force on the ball at the instant shown in Fig. 6-48? (d) What is the speed of the ball?

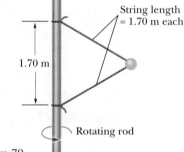

String length = 1.70 m each

1.70 m

Rotating rod

FIGURE 6-48 Problem 70.

71P. Assume that the standard kilogram mass would weigh exactly 9.80 N at sea level on the Earth's equator if the Earth did not rotate. Then take into account the fact that the Earth does rotate, so that this mass moves in a circle of radius 6.40×10^6 m (the Earth's radius) at a constant speed of 465 m/s. (a) Determine the centripetal force needed to keep the standard mass moving in its circular path. (b) Determine the force exerted by the standard mass on a spring balance from which it is suspended at the equator (that force is its "apparent weight").

72P. Suppose that the space station in Question 12 has a radius of 500 m. (a) If a crew member weighs 600 N on Earth, what must be the speed v_s of the outer wall of the station if that crew member is to have an apparent weight of 300 N when standing near the outer wall (as in Fig. 6-21)? (b) What is the apparent weight if the crew member sprints along the outer wall at 10 m/s (relative to the outer wall) in the same direction as $\mathbf{v}_s$?

In the weight-lifting competition of the 1976 Olympics, Vasili Alexeev astounded the world by lifting a record-breaking 562 lb (2500 N) from the floor to over his head (about 2 m). In 1957 Paul Anderson stooped beneath a reinforced wood platform, placed his hands on a short stool to brace himself, and then pushed upward on the platform with his back, lifting the platform and its load about a centimeter. On the platform were auto parts and a safe filled with lead; the composite weight of the load was 6270 lb (27,900 N)! Who, Alexeev or Anderson, did more work on the objects they lifted?

7-1 KINETIC ENERGY

We begin this chapter with a definition: **energy** is a scalar quantity that is associated with a state of one or more objects. The term *state* here has its common meaning: it is the condition of an object. In this chapter we shall focus on one form of energy, **kinetic energy** K, which is associated with the *state of motion* of an object. The faster the object moves, the greater is its kinetic energy. And when the object is stationary, its kinetic energy is zero.

For an object of mass m and whose speed v is well below the speed of light, we define kinetic energy as

$$K = \tfrac{1}{2}mv^2 \qquad \text{(kinetic energy).} \qquad (7\text{-}1)$$

Kinetic energy can never be negative because m and v^2 can never be negative.

The SI unit of kinetic energy (and every other type of energy) is the **joule** (J), named for James Prescott Joule, an English scientist of the 1800s. It is derived directly from the units for mass and velocity:

$$1 \text{ joule} = 1 \text{ J} = 1 \text{ kg} \cdot \text{m}^2/\text{s}^2. \qquad (7\text{-}2)$$

A convenient unit of energy for dealing with atoms or with subatomic particles is the **electron-volt** (eV):

$$1 \text{ electron-volt} = 1 \text{ eV} = 1.60 \times 10^{-19} \text{ J}. \qquad (7\text{-}3)$$

Three common multiples of this unit are the kiloelectron-volt (1 keV = 10^3 eV), the megaelectron-volt (1 MeV = 10^6 eV), and the gigaelectron-volt (1 GeV = 10^9 eV).

SAMPLE PROBLEM 7-1

In 1896 in Waco, Texas, William Crush of the "Katy" railroad parked two locomotives at opposite ends of a 6.4 km track, fired them up, tied their throttles open, and then allowed them to crash head-on at full speed (Fig. 7-1), in front of 30,000 spectators. Hundreds of people were hurt by flying debris; several were killed. Assuming the weight of each locomotive was 1.2×10^6 N and its acceleration prior to the collision was a constant 0.26 m/s², what was the total kinetic energy of the two locomotives just before the collision?

SOLUTION: To find the kinetic energy of each locomotive, we need its mass and its speed just before the collision. To find the speed v, we use Eq. 2-16,

$$v^2 = v_0^2 + 2a(x - x_0).$$

With $v_0 = 0$ and $x - x_0 = 3.2 \times 10^3$ m (half the initial separation), this yields

$$v^2 = 0 + 2(0.26 \text{ m/s}^2)(3.2 \times 10^3 \text{ m}),$$

or

$$v = 40.8 \text{ m/s}$$

(about 90 mi/h). To find the mass m of each locomotive, we divide its weight by g:

$$m = \frac{1.2 \times 10^6 \text{ N}}{9.8 \text{ m/s}^2} = 1.22 \times 10^5 \text{ kg}.$$

Now, using Eq. 7-1, we find the total kinetic energy of both locomotives just before the collision as:

$$K = 2(\tfrac{1}{2}mv^2) = (1.22 \times 10^5 \text{ kg})(40.8 \text{ m/s})^2$$
$$= 2.0 \times 10^8 \text{ J}. \qquad \text{(Answer)}$$

This is equivalent to a detonation of about 100 lb of TNT.

FIGURE 7-1 Sample Problem 7-1. The aftermath of an 1896 crash of two locomotives.

7-2 WORK

The energy of an object changes if an exchange of energy occurs between the object and its environment. Such a transfer can occur due to a force or due to an exchange of heat. We discuss the exchange of heat in Chapter 19. Here, we discuss the transfer of energy via a force, a process known as doing **work.**

If you accelerate an object to a greater speed by applying a force to the object, you increase the kinetic energy K ($= \tfrac{1}{2}mv^2$) of the object. Similarly, if you decelerate the object to a lesser speed by applying a force, you decrease the kinetic energy of the object. We account for these changes in kinetic energy by saying that your force has transferred energy *to* the object from yourself or *from* the object to yourself.

In such a transfer of energy via a force, *work W* is said to be *done on the object by the force*. More formally, we define work as follows:

> Work W is energy transferred to or from an object by means of a force acting on the object. Energy transferred to the object is positive work, and energy transferred from the object is negative work.

"Work," then, is transferred energy; "doing work" is the act of transferring the energy. Work has the same units as energy and is a scalar quantity.

The term *transfer* can be misleading; it does not mean that anything material flows into or out of the object. That is, the transfer is not like a flow of water. Rather it is like the electronic transfer of money between two bank accounts: the number in one account goes up while the number in the other account goes down, with nothing material passing between the two accounts.

Note that we are not concerned here with the common meaning of the word "work," which implies that *any* physical or mental labor is work. For example, if you push hard against a wall, you tire owing to the continuously repeated contractions of your muscles that are required and you are, in the common sense, working. But such effort does not cause an energy transfer to or from the wall and thus is not work done on the wall as defined here.

To avoid confusion in this chapter, we shall use the symbol W only for work and shall represent weight with $m\mathbf{g}$ or mg.

7-3 WORK AND KINETIC ENERGY

Let us now relate the work done on an object by a force and the corresponding change in the kinetic energy of the object. If the force changes the speed of the object, it also changes the kinetic energy of the object. If the kinetic energy is the only type of energy of the object being changed by the force, then the change ΔK in kinetic energy is equal to the work W done by the force:

$$\Delta K = K_f - K_i = W \qquad \text{(work–kinetic energy theorem).} \qquad (7\text{-}4)$$

Here K_i is the initial kinetic energy ($= \frac{1}{2}mv_0^2$) and K_f is the kinetic energy ($+\frac{1}{2}mv^2$) after the work—the energy transfer—is done.

The right-hand equality of Eq. 7-4 can also be written

$$K_f = K_i + W. \qquad (7\text{-}5)$$

Equations 7-4 and 7-5 are equivalent statements of the **work–kinetic energy theorem.**

If the object's energy other than kinetic energy is being changed by the force, then Eqs. 7-4 and 7-5 do not apply. A kinetic frictional force, for example, can change

both the kinetic energy and the *thermal energy* of an object. (Thermal energy is associated with the random motions of atoms and molecules within an object.) Equations 7-4 and 7-5 then do not apply.

Some forces cause a transfer of energy within the object itself, and again Eqs. 7-4 and 7-5 do not apply. For example, suppose you push yourself away from a wall while ice skating; then the push (a force) transfers energy internally from a biological type of energy in your muscles to kinetic energy of your body as a whole. We shall consider kinetic frictional forces in Chapter 8 and internal energy transfers in Chapter 9.

CHECKPOINT **1:** A particle moves along an x axis. Does the kinetic energy of the particle increase, decrease, or remain the same if the particle's velocity changes (a) from -3 m/s to -2 m/s and (b) from -2 m/s to 2 m/s? (c) In each situation, is the work done on the particle positive, negative, or zero?

Work Done by a Single Force

We would now like to relate the change ΔK in kinetic energy of an object to the magnitude F of the force causing the change. We start with a particle; the only type of energy this simplest type of object can have is kinetic energy. In Fig. 7-2, a particle moves along an x axis on a horizontal frictionless floor while a constant force $\mathbf{F}$ acts on it at a constant angle ϕ to the particle's path. Because the horizontal force component $F \cos \phi$ gives the particle an acceleration a_x along the path, the force changes the particle's velocity from its initial value $\mathbf{v}_0$. Thus the force also changes the particle's kinetic energy.

Suppose the force acts on the particle through a displacement $\mathbf{d}$, giving it a velocity $\mathbf{v}$ whose magnitude is, by Eq. 2-16,

$$v^2 = v_0^2 + 2a_x d. \qquad (7\text{-}6)$$

Multiplying both sides of Eq. 7-6 by the mass m of the particle and rearranging yield

$$\tfrac{1}{2}mv^2 - \tfrac{1}{2}mv_0^2 = ma_x d. \qquad (7\text{-}7)$$

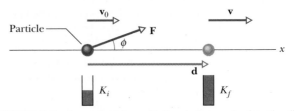

FIGURE 7-2 A constant force $\mathbf{F}$ directed at angle ϕ to the displacement $\mathbf{d}$ of a particle accelerates the particle along that path, changing the velocity of the particle from $\mathbf{v}_0$ to $\mathbf{v}$. A "kinetic energy gauge" indicates the resulting change in the kinetic energy of the particle, from the value K_i to K_f.

The left side of Eq. 7-7 is the difference between the initial kinetic energy K_i ($= \frac{1}{2}mv_0^2$) of the particle and the final kinetic energy K_f ($= \frac{1}{2}mv^2$). This difference is the change ΔK in the kinetic energy of the particle due to the force **F**. Making that substitution, and substituting $F \cos \phi$ for the product ma_x according to Newton's second law, we get

$$\Delta K = Fd \cos \phi. \qquad (7-8)$$

Now by comparing Eq. 7-8 with Eq. 7-4, we see that the right side of Eq. 7-8 gives the work W done on the particle by force **F**. Thus, we may write

$$W = Fd \cos \phi \quad \substack{\text{(work done by a} \\ \text{constant force).}} \qquad (7-9)$$

If the angle ϕ is less than 90°, then the work W is positive, which means that energy is transferred *to* the particle and the kinetic energy of the particle *increases*. If ϕ is greater than 90° (up to 180°), then the work W is negative, which means that energy is transferred *from* the particle and the kinetic energy of the particle *decreases*.

From Eq. 7-9 we see that in addition to the joule, another SI unit of work is the newton-meter (N·m). The corresponding unit in the British system is the foot-pound (ft·lb). Thus, we can extend Eq. 7-2 to

$$1 \text{ J} = 1 \text{ kg·m}^2/\text{s}^2 = 1 \text{ N·m} = 0.738 \text{ ft·lb}. \quad (7-10)$$

The right side of Eq. 7-9 is equivalent to the scalar (or dot) product **F·d**. So in vector form Eq. 7-9 is

$$W = \mathbf{F} \cdot \mathbf{d} \quad \substack{\text{(work by a} \\ \text{constant force).}} \qquad (7-11)$$

(This is our first application of a scalar product; you may wish to review its discussion in Section 3-7.) Equation 7-11 is especially useful when **F** and **d** are given in unit-vector notation.

As derived, Eqs. 7-9 and 7-11 give the work done by a constant force that changes the kinetic energy of a particle. However, in certain cases we can extend them to objects that are clearly not particles:

If a force acting alone on an object changes only the kinetic energy of the object (and no other energy of the object), then the work done by the force is given by Eqs. 7-9 and 7-11.

Such objects are said to be *particle-like.*

For example, Fig. 7-3 shows a student propelling a bed in an intramural bed race. If his force on the bed is constant, do Eqs. 7-9 and 7-11 give the work done by the force? We can easily see that most but not all of the trans-

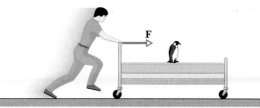

FIGURE 7-3 A bed race. We can approximate the bed as being a particle for the purpose of calculating the work done on the bed by the force applied by the student.

ferred energy goes into the kinetic energy of the bed and penguin-rider moving along the street. Some small amount goes into the rotation of the wheels. However, if we choose to neglect that small amount, then Eqs. 7-9 and Eq. 7-11 give the work done by the force applied by the student, and the bed and rider are particle-like.

CHECKPOINT **2:** The figure shows four situations in which a force acts on a box while the box slides rightward a distance d across a frictionless floor. The magnitudes of the forces are identical; their orientations are as shown. Rank the situations according to the work done on the box during the displacement, from most positive to most negative.

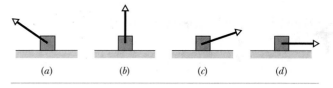

(a)   (b)   (c)   (d)

Work Done by Multiple Forces

If several forces act on a particle, we can replace **F** in Eq. 7-11 with the net force $\Sigma \mathbf{F}$, where

$$\sum \mathbf{F} = \mathbf{F}_1 + \mathbf{F}_2 + \mathbf{F}_3 + \cdots, \qquad (7-12)$$

where $\mathbf{F}_j$ are the individual forces. Then

$$W = \left(\sum \mathbf{F}\right) \cdot \mathbf{d} \qquad (7-13)$$

is the work done by the net force during a displacement **d** of the particle. With Eq. 7-12, we can rewrite Eq. 7-13 as

$$W = \mathbf{F}_1 \cdot \mathbf{d} + \mathbf{F}_2 \cdot \mathbf{d} + \mathbf{F}_3 \cdot \mathbf{d} + \cdots$$
$$= W_1 + W_2 + W_3 + \cdots \quad \text{(total work).} \quad (7-14)$$

This equation tells us that the total work done on the particle is the sum of the work done by all the forces acting on the particle.

We can now rewrite Eq. 7-4 (the work–kinetic energy theorem) as

$$\Delta K = K_f - K_i = W_1 + W_2 + W_3 + \cdots . \quad (7-15)$$

This tells us that the change ΔK in the kinetic energy of a particle is equal to the total work done by all the forces acting on the particle.

SAMPLE PROBLEM 7-2

Figure 7-4a shows two industrial spies sliding an initially stationary 225 kg floor safe a distance of 8.50 m along a straight line toward their truck. The push $\mathbf{F}_1$ of Spy 001 is 12.0 N, directed at an angle of 30° downward from the horizontal; the pull $\mathbf{F}_2$ of Spy 002 is 10.0 N, directed at 40° above the horizontal. The floor and safe make frictionless contact.

(a) What is the total work done on the safe by forces $\mathbf{F}_1$ and $\mathbf{F}_2$ during the 8.50 m displacement $\mathbf{d}$?

SOLUTION: Figure 7-4b is a free-body diagram for the safe, considered to be a particle. We can find the total work done on the safe by finding the work done by each force and then adding the results. From Eq. 7-9, the work done by $\mathbf{F}_1$ is

$$W_1 = F_1 d \cos \phi_1 = (12.0 \text{ N})(8.50 \text{ m})(\cos 30°)$$
$$= 88.33 \text{ J},$$

and the work done by $\mathbf{F}_2$ is

$$W_2 = F_2 d \cos \phi_2 = (10.0 \text{ N})(8.50 \text{ m})(\cos 40°)$$
$$= 65.11 \text{ J}.$$

So, from Eq. 7-14, the total work W is

$$W = W_1 + W_2 = 88.33 \text{ J} + 65.11 \text{ J}$$
$$= 153.4 \text{ J} \approx 153 \text{ J}. \qquad \text{(Answer)}$$

Thus during the 8.50 m displacement, the spies transfer 153 J of energy to the kinetic energy of the safe.

(b) During the displacement, what is the work W_g done on the safe by its weight $m\mathbf{g}$ and what is the work W_N done on the safe by the normal force $\mathbf{N}$ due to the floor?

SOLUTION: Both these forces are perpendicular to the displacement. Thus Eq. 7-9 tells us that

$$W_g = mgd \cos 90° = mgd(0) = 0 \quad \text{(Answer)}$$

and $\qquad W_N = Nd \cos 90° = Nd(0) = 0.$ (Answer)

These forces do not transfer any energy to or from the safe.

(c) The safe is initially stationary. What is its speed v at the end of the 8.50 m displacement?

SOLUTION: The speed of the safe changes because its kinetic energy is changed when energy is transferred to it by the forces. We relate the speed to the work done by combining Eqs. 7-4 and 7-1:

$$W = K_f - K_i = \tfrac{1}{2}mv^2 - \tfrac{1}{2}mv_0^2.$$

The initial speed v_0 is zero, and we now know that the work done is 153.4 J. Solving for v and then substituting the known data, we find that

$$v = \sqrt{\frac{2W}{m}} = \sqrt{\frac{2(153.4 \text{ J})}{225 \text{ kg}}}$$
$$= 1.17 \text{ m/s}. \qquad \text{(Answer)}$$

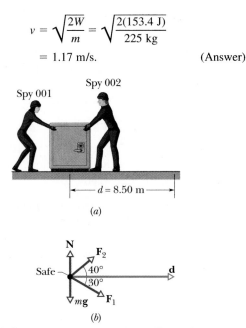

FIGURE 7-4 Sample Problem 7-2. (a) Two spies move a floor safe. (b) A free-body diagram for the safe, with the displacement $\mathbf{d}$ of the safe included.

SAMPLE PROBLEM 7-3

A runaway crate of prunes slides over a floor toward you. To slow the crate you push against it with a force $\mathbf{F} = (2.0 \text{ N})\mathbf{i} + (-6.0 \text{ N})\mathbf{j}$ while running backward (Fig. 7-5). During your pushing, the crate goes through a displacement $\mathbf{d} = (-3.0 \text{ m})\mathbf{i}$.

(a) How much work has your force done on the crate during the displacement?

SOLUTION: From Eq. 7-11, the work is

$$W = \mathbf{F} \cdot \mathbf{d} = [(2.0 \text{ N})\mathbf{i} + (-6.0 \text{ N})\mathbf{j}] \cdot [(-3.0 \text{ m})\mathbf{i}].$$

Of the possible unit-vector dot products, only $\mathbf{i} \cdot \mathbf{i}$, $\mathbf{j} \cdot \mathbf{j}$, and $\mathbf{k} \cdot \mathbf{k}$ are nonzero (see Section 3-7). Here we have

$$W = (2.0)(-3.0 \text{ m})\mathbf{i} \cdot \mathbf{i} + (-6.0 \text{ N})(-3.0 \text{ m})\mathbf{j} \cdot \mathbf{i}$$
$$= (-6.0 \text{ J})(1) + 0 = -6.0 \text{ J}. \qquad \text{(Answer)}$$

Thus the force transfers 6.0 J of energy from the kinetic energy of the crate.

(b) If the crate has a kinetic energy of 10 J at the beginning of the displacement $\mathbf{d}$, what is its kinetic energy at the end of the displacement?

SOLUTION: Using Eq. 7-5 with $K_i = 10$ J and $W = -6$ J, we find

$$K_f = K_i + W = 10 \text{ J} + (-6.0 \text{ J}) = 4.0 \text{ J}. \quad \text{(Answer)}$$

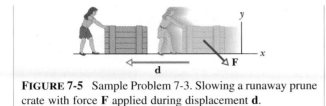

FIGURE 7-5 Sample Problem 7-3. Slowing a runaway prune crate with force **F** applied during displacement **d**.

7-4 WORK DONE BY WEIGHT

We next examine the work done on an object by a particular type of force, namely, its weight. Figure 7-6 shows a particle-like tomato of mass m that is thrown upward with initial speed v_0 and thus with initial kinetic energy $K_i = \frac{1}{2}mv_0^2$. As the tomato rises, it slows because a constant downward force, its weight mg, acts on it.

Because the tomato slows, its kinetic energy decreases. We know from experimental observation that if the weight mg is the only force acting on the tomato (air drag is somehow eliminated), then kinetic energy is the only energy of the tomato that changes. Thus, to find the work W_g done on the tomato by its weight, we substitute mg for F in Eq. 7-9, finding

$$W_g = mgd \cos \phi \quad \begin{matrix}\text{(work done}\\\text{by weight).}\end{matrix} \quad (7\text{-}16)$$

For such a rising object, the weight mg is directed opposite the displacement **d**, as indicated in Fig. 7-6. Then $\phi = 180°$ and

$$W_g = mgd \cos (180°) = mgd(-1) = -mgd. \quad (7\text{-}17)$$

The minus sign tells us that during the object's rise, the weight of the object transfers energy in the amount mgd from the kinetic energy of the object. This is consistent with the slowing of the object as it rises.

After the object has reached its maximum height and

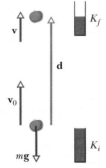

FIGURE 7-6 A particle-like tomato of mass m thrown upward slows from velocity $\mathbf{v}_0$ to velocity $\mathbf{v}$ during displacement **d** because its weight mg acts on it. A kinetic energy gauge indicates the resulting change in the kinetic energy of the object, from K_i ($= \frac{1}{2}mv_0^2$) to K_f ($= \frac{1}{2}mv^2$).

is falling back down, the angle ϕ between the weight mg and the displacement **d** is zero. Thus,

$$W_g = mgd \cos (0°) = mgd(+1) = +mgd. \quad (7\text{-}18)$$

The plus sign tells us that the weight now transfers energy in the amount mgd to the kinetic energy of the object. This is consistent with the speeding up of the object as it falls. (Actually, as we shall see in Chapter 8, energy transfers associated with lifting and lowering an object involve not just the object, but the full object-Earth system. Without the Earth, of course, "lifting" would be meaningless.)

Work Done in Lifting and Lowering an Object

Now suppose we lift a particle-like object by applying a force **F** to it. During the upward displacement, our applied force does positive work W_a on the object while the object's weight also does negative work W_g on it. That is, our force tends to transfer energy to the object while its weight tends to transfer energy from it. By Eq. 7-15, the change ΔK in the kinetic energy of the object due to these two energy transfers is

$$\Delta K = K_f - K_i = W_a + W_g, \quad (7\text{-}19)$$

in which K_f is the kinetic energy at the end of the displacement and K_i is that at the start of the displacement. This equation also applies if we lower the object; but then the weight tends to transfer energy *to* the object while our force tends to transfer energy *from* it.

In one common situation the object is stationary before and after the lift—for example, when you lift a book from the floor to a shelf. Then K_f and K_i are both zero, and Eq. 7-19 reduces to

$$W_a + W_g = 0$$

or $$W_a = -W_g. \quad (7\text{-}20)$$

Note that we get the same result if K_f and K_i are not zero but are still equal. Either way, the result means that the work done by the applied force is the negative of the work done by the weight. That is, the applied force transfers the same amount of energy to the object as its weight transfers from the object. Using Eq. 7-16, we can rewrite Eq. 7-20 as

$$W_a = -mgd \cos \phi \quad \begin{matrix}\text{(work in lifting and}\\\text{lowering; } K_f = K_i),\end{matrix} \quad (7\text{-}21)$$

with ϕ being the angle between mg and **d**. If the displacement is vertically upward (Fig. 7-7a), then $\phi = 180°$ and the work done by our force equals mgd. If the displacement is vertically downward (Fig. 7-7b), then $\phi = 0°$ and the work done by the applied force equals $-mgd$.

Equations 7-20 and 7-21 apply to any situation in

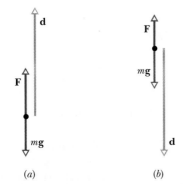

(a)   (b)

FIGURE 7-7 (a) An applied force **F** lifts an object. The displacement **d** of the object makes an angle $\phi = 180°$ with the weight vector $m\mathbf{g}$ of the object. The applied force does positive work on the object. (b) An applied force **F** lowers an object. The displacement **d** of the object makes an angle $\phi = 0°$ with weight vector $m\mathbf{g}$. The applied force does negative work on the object.

which an object is lifted or lowered, with the object stationary before and after the lift. They are independent of the magnitude of the force used. For example, when Alexeev lifted the record-breaking weight of 2500 N, his force on the weights varied considerably during the lift. Still, because the weights were stationary before and after the lift, the work he did is given by Eqs. 7-20 and 7-21, where, in Eq. 7-21, mg is the weight he lifted and d is the distance he lifted that weight.

SAMPLE PROBLEM 7-4

Let us return to the weight-lifting feats of Vasili Alexeev and Paul Anderson.

(a) When Alexeev lifted a weight of 2500 N a distance of 2.0 m, how much work was done on the weights by their weight $m\mathbf{g}$?

SOLUTION: The magnitude of the weight vector $m\mathbf{g}$ is mg. The angle ϕ between that vector and the displacement vector **d** is 180°. From Eq. 7-16, the work done by $m\mathbf{g}$ is

$$W_g = mgd \cos \phi = (2500 \text{ N})(2.0 \text{ m})(\cos 180°)$$
$$= -5000 \text{ J}. \qquad \text{(Answer)}$$

(b) How much work was done by Alexeev's force during the lift?

SOLUTION: Because the weights were stationary at the start and end of the lift, we can use Eq. 7-20, finding

$$W_{VA} = -W_g = +5000 \text{ J}. \qquad \text{(Answer)}$$

(c) While Alexeev held the weights stationary above his head, how much work was done by his force on the weights?

SOLUTION: When he supported the weights, they were stationary. Thus their displacement $d = 0$ and, by Eq. 7-9, the

Using a harness, Paul Anderson lifts 30 people having a combined weight of about 2400 lb.

work done on the weights was zero (even though supporting the weights was a very tiring task).

(d) How much work was done by the force Paul Anderson applied to lift a weight of 27,900 N a distance of 1.0 cm?

SOLUTION: From Eq. 7-21, with $mg = 27,900$ N and $d = 1.0$ cm, we find

$$W_{PA} = -mgd \cos \phi = -mgd \cos 180°$$
$$= -(27,900 \text{ N})(0.010 \text{ m})(-1) = 280 \text{ J}. \quad \text{(Answer)}$$

Anderson's lift required a tremendous upward force but only a small energy transfer of 280 J, owing to the short displacement involved.

SAMPLE PROBLEM 7-5

An initially stationary 15.0 kg crate of cheese is pulled, via a cable, a distance $L = 5.70$ m up a frictionless ramp, to a height h of 2.50 m, where it stops (Fig. 7-8a).

(a) How much work is done on the crate by its weight $m\mathbf{g}$ during the lift?

SOLUTION: We calculate this work with Eq. 7-16, using L for the magnitude of the displacement. The angle between $m\mathbf{g}$ and the displacement is $\theta + 90°$ (see the free-body diagram of Fig. 7-8b). We have

$$W_g = mg \, L \cos (\theta + 90°) = -mgL \sin \theta.$$

From Fig. 7-8a, we see that $L \sin \theta$ is the height h moved by the crate. So we have

$$W_g = -mgh. \qquad (7-22)$$

This means that the work done by the weight depends on the

vertical displacement of the crate and not on the horizontal displacement. (We shall return to this point in Chapter 8.) Inserting the given data in Eq. 7-22 yields

$$W_g = -(15.0 \text{ kg})(9.8 \text{ m/s}^2)(2.50 \text{ m})$$
$$= -368 \text{ J.} \qquad \text{(Answer)}$$

(b) How much work is done on the crate by the force **T** applied by the cable, which pulls the crate up the ramp?

SOLUTION: Because the crate is stationary before and after the lift, the change ΔK in its kinetic energy must be zero. Then from Eq. 7-15,

$$\Delta K = W_1 + W_2 + W_3 + \cdots, \qquad (7\text{-}23)$$

we know that the total work done by all the forces acting on it must be zero. Besides the weight of the crate, there are only two other forces acting on the crate: the normal force **N** due to the ramp and the force **T** from the cable. Because **N** is perpendicular to the displacement of the crate along the ramp, the normal force does zero work on the crate. So, with W_T representing the work done by the force **T**, Eq. 7-23 becomes

$$0 = W_g + W_T.$$

Substituting -368 J for W_g, we find

$$W_T = 368 \text{ J.} \qquad \text{(Answer)}$$

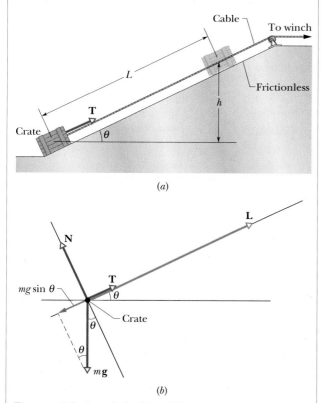

(a)

(b)

FIGURE 7-8 Sample Problem 7-5. (*a*) A crate is pulled up a frictionless ramp by a force parallel to the ramp. (*b*) A free-body diagram for the crate, showing all the forces that act on it. Its displacement **L** is also shown.

CHECKPOINT 3: Suppose we raise the crate in Sample Problem 7-5 by the same height h but with a longer ramp. (a) Is the work done by force **T** now greater, smaller, or the same as before? (b) Is the magnitude of **T** needed to move the crate now greater, smaller, or the same as before?

SAMPLE PROBLEM 7-6

A 500 kg elevator cab is descending with speed $v_i = 4.0$ m/s when the cable that controls it begins to slip, allowing it to fall with constant acceleration $\mathbf{a} = \mathbf{g}/5$ (Fig. 7-9*a*).

(a) During its fall through a distance $d = 12$ m, what is the work W_1 done on the cab by its weight $m\mathbf{g}$?

SOLUTION: The cab's free-body diagram during the 12-m fall is shown in Fig. 7-9*b*. Note that the angle between the cab's displacement **d** and its weight $m\mathbf{g}$ is 0°. With Eq. 7-16 we find

$$W_1 = mgd \cos 0° = (500 \text{ kg})(9.8 \text{ m/s}^2)(12 \text{ m})(1)$$
$$= 5.88 \times 10^4 \text{ J} \approx 5.9 \times 10^4 \text{ J.} \qquad \text{(Answer)}$$

(b) During the 12 m fall, what is the work W_2 done on the cab by the upward pull **T** exerted by the elevator cable?

SOLUTION: The situation here differs from those in Sample Problems 7-4 and 7-5, because the kinetic energy of the cab at the beginning of the displacement does not equal that at the end of the displacement. So Eqs. 7-20 and 7-21 do not apply; the work done by force **T** on the cab *is not* the negative of the

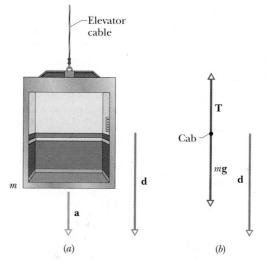

(a)

(b)

FIGURE 7-9 Sample Problem 7-6. An elevator cab, descending with speed v_i, suddenly begins to accelerate downward. (*a*) It moves through a displacement **d** with constant acceleration $\mathbf{a} = \mathbf{g}/5$. (*b*) A free-body diagram for the cab, with the displacement included.

work done by the weight of the cab. We must calculate the work W_2 done by **T** with Eq. 7-9 ($W = Fd \cos \phi$). To do so, we first must find the magnitude of **T**. Applying Newton's second law to the cab yields

$$\sum F = T - mg = ma.$$

Since **a** has magnitude $g/5$ and is directed downward, we get

$$T = m(g + a) = m(g - g/5)$$
$$= (500 \text{ kg})(\tfrac{4}{5})(9.8 \text{ m/s}^2) = 3920 \text{ N}.$$

We can now use Eq. 7-9 to find the work done by **T**. The angle between **T** and the cab's displacement **d** is 180°. The magnitude of **T** is 3920 N. And the magnitude of the displacement is 12 m. So

$$W_2 = Td \cos 180° = (3920 \text{ N})(12 \text{ m})(-1)$$
$$= -4.70 \times 10^4 \text{ J}. \qquad \text{(Answer)}$$

(c) What is the total work W done on the cab in the 12 m fall?

SOLUTION: The total work is the algebraic sum of the work done by the two forces (by Eq. 7-14):

$$W = W_1 + W_2 = 5.88 \times 10^4 \text{ J} - 4.70 \times 10^4 \text{ J}$$
$$= 1.18 \times 10^4 \text{ J} \approx 1.2 \times 10^4 \text{ J}. \qquad \text{(Answer)}$$

Thus, during the 12 m fall, a net energy of 1.2×10^4 J is transferred to the cab.

We can also find W a different way. We first find the net force on the cab, using Newton's second law:

$$\sum F = ma = (500 \text{ kg}) \left(-\frac{9.8 \text{ m/s}^2}{5} \right) = -980 \text{ N}.$$

We next find the work done on the cab by that net force, which acts downward and thus at an angle of 0° with **d**:

$$W = (980 \text{ N})(12 \text{ m}) \cos 0°$$
$$= 1.18 \times 10^4 \text{ J} \approx 1.2 \times 10^4 \text{ J}. \qquad \text{(Answer)}$$

(d) What is the cab's kinetic energy at the end of the 12 m fall?

SOLUTION: The kinetic energy K_i at the start of the fall, when the speed is $v_i = 4.0$ m/s, is

$$K_i = \tfrac{1}{2}mv_i^2 = \tfrac{1}{2}(500 \text{ kg})(4.0 \text{ m/s})^2 = 4000 \text{ J}.$$

The kinetic energy K_f at the end of the fall is given by Eq. 7-5,

$$K_f = K_i + W = 4000 \text{ J} + 1.18 \times 10^4 \text{ J}$$
$$= 1.58 \times 10^4 \text{ J} \approx 1.6 \times 10^4 \text{ J}. \qquad \text{(Answer)}$$

(e) What is the speed v_f of the cab at the end of the 12 m fall?

SOLUTION: From Eq. 7-1, we have

$$K_f = \tfrac{1}{2}mv_f^2,$$

which we solve for v_f:

$$v_f = \sqrt{\frac{2K_f}{m}} = \sqrt{\frac{(2)(1.58 \times 10^4 \text{ J})}{500 \text{ kg}}}$$
$$= 7.9 \text{ m/s}. \qquad \text{(Answer)}$$

7-5 WORK DONE BY A VARIABLE FORCE

One-Dimensional Analysis

Let us return to the situation of Fig. 7-2 but now consider the force to be directed along the x axis and the force magnitude to vary with position x. Thus, as the particle moves, the magnitude of the force doing work on it changes. Only the magnitude of this **variable force** changes, not its direction. Moreover, its magnitude changes with the position of the particle, but not over time.

Figure 7-10a shows a plot of such a one-dimensional variable force. How do we find the work done on the particle by this force as the particle moves from an initial point x_i to a final point x_f? We cannot use Eq. 7-9, because it applies only for a constant force **F**. To develop a new approach, let us divide the total displacement of the particle into a number of intervals of width Δx. We choose Δx small enough to permit us to take the force $F(x)$ as being reasonably constant over that interval. We let $\overline{F_j(x)}$ be the average value of $F(x)$ within the jth interval.

The increment (small amount) of work ΔW_j done by the force in the jth interval is now given by Eq. 7-9 and is

$$\Delta W_j = \overline{F_j(x)}\, \Delta x. \qquad (7\text{-}24)$$

On the graph of Fig. 7-10b, $\overline{F_j(x)}$ is the height of the jth strip, and Δx is its width; ΔW_j is then equal in magnitude to the area of the strip.

To approximate the total work W done by the force as the particle moves from x_i to x_f, we add the areas of all the strips between x_i and x_f in Fig. 7-10b. That is,

$$W = \sum \Delta W_j = \sum \overline{F_j(x)}\, \Delta x. \qquad (7\text{-}25)$$

Equation 7-25 is an approximation because the broken "skyline" formed by the tops of the rectangular strips in Fig. 7-10b only approximates the actual curve.

We can make the approximation better by reducing the strip width Δx and using more strips, as in Fig. 7-10c. In the limit, we let the strip width approach zero; the number of strips then becomes infinitely large and we have, as an exact result,

$$W = \lim_{\Delta x \to 0} \sum \overline{F_j(x)}\, \Delta x. \qquad (7\text{-}26)$$

This limit is exactly what we mean by the integral of the function $F(x)$ between the limits x_i and x_f. Thus Eq. 7-26 becomes

$$W = \int_{x_i}^{x_f} F(x)\, dx \qquad \text{(work: variable force).} \quad (7\text{-}27)$$

If we know the function $F(x)$, we can substitute it into

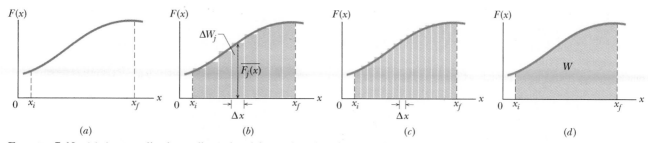

FIGURE 7-10 (a) A generalized one-dimensional force plotted against the displacement x of a particle on which it acts. The particle moves from x_i to x_f. (b) Same as (a) but with the area under the curve divided into narrow strips. (c) Same as (b) but with the area divided into narrower strips. (d) The limiting case. The work done by the force is given by Eq. 7-27 and is represented geometrically by the shaded area between the curve and the x axis and between x_i and x_f.

Eq. 7-27, introduce the proper limits of integration, carry out the integration, and thus find the work. (Appendix E contains a list of common integrals.) Geometrically, the work is equal to the area between the $F(x)$ curve and the x axis, between the limits x_i and x_f (shaded in Fig. 7-10d).

Three-Dimensional Analysis

Consider now a particle that is acted on by a three-dimensional force

$$\mathbf{F} = F_x\mathbf{i} + F_y\mathbf{j} + F_z\mathbf{k}, \tag{7-28}$$

and let some or all of the components F_x, F_y, and F_z depend on the position of the particle, which means that they are functions of that position. Furthermore, let the particle move through an incremental displacement

$$d\mathbf{r} = dx\mathbf{i} + dy\mathbf{j} + dz\mathbf{k}. \tag{7-29}$$

The increment of work dW done on the particle by $\mathbf{F}$ during the displacement $d\mathbf{r}$ is, by Eq. 7-11,

$$dW = \mathbf{F} \cdot d\mathbf{r} = F_x\,dx + F_y\,dy + F_z\,dz. \tag{7-30}$$

The work W done by $\mathbf{F}$ while the particle moves from an initial position r_i with coordinates (x_i, y_i, z_i) to a final position r_f with coordinates (x_f, y_f, z_f) is then

$$W = \int_{r_i}^{r_f} dW = \int_{x_i}^{x_f} F_x\,dx + \int_{y_i}^{y_f} F_y\,dy + \int_{z_i}^{z_f} F_z\,dz. \tag{7-31}$$

If $\mathbf{F}$ has only an x component, then the y and z terms in Eq. 7-31 are zero and the equation reduces to Eq. 7-27.

Work–Kinetic Energy Theorem with a Variable Force

Equation 7-27 gives the work done by a variable force on a particle in a one-dimensional situation. Let us now make certain that the work calculated with Eq. 7-27 is indeed

equal to the change in kinetic energy of the particle, as the work–kinetic energy theorem states.

Consider a particle of mass m, moving along the x axis and acted on by a net force $F(x)$ that points along that axis. The work done on the particle by this force as the particle moves from an initial position x_i to a final position x_f is given by Eq. 7-27 as

$$W = \int_{x_i}^{x_f} F(x)\,dx = \int_{x_i}^{x_f} ma\,dx, \tag{7-32}$$

in which we use Newton's second law to replace $F(x)$ with ma. We can write the quantity $ma\,dx$ in Eq. 7-32 as

$$ma\,dx = m\frac{dv}{dt}\,dx. \tag{7-33}$$

From the "chain rule" of calculus, we have

$$\frac{dv}{dt} = \frac{dv}{dx}\frac{dx}{dt} = \frac{dv}{dx}v, \tag{7-34}$$

and Eq. 7-33 becomes

$$ma\,dx = m\frac{dv}{dx}v\,dx = mv\,dv. \tag{7-35}$$

Substituting Eq. 7-35 into Eq. 7-32 yields

$$W = \int_{v_i}^{v_f} mv\,dv = m\int_{v_i}^{v_f} v\,dv$$
$$= \tfrac{1}{2}mv_f^2 - \tfrac{1}{2}mv_i^2. \tag{7-36}$$

Note that when we change the variable from x to v we are required to express the limits on the integral in terms of the new variable. Note also that because the mass m is a constant, we are able to move it outside the integral.

Recognizing the terms on the right of Eq. 7-36 as kinetic energies allows us to write this equation as

$$W = K_f - K_i = \Delta K,$$

which is the work–kinetic energy theorem.

SAMPLE PROBLEM 7-7

Force $\mathbf{F} = (3x \text{ N})\mathbf{i} + (4 \text{ N})\mathbf{j}$, with x in meters, acts on a particle, changing only the kinetic energy of the particle. How much work is done on the particle as it moves from coordinates (2 m, 3 m) to (3 m, 0 m)? Does the speed of the particle increase, decrease, or remain the same?

SOLUTION: From Eq. 7-31 we have

$$W = \int_2^3 3x \, dx + \int_3^0 4 \, dy = 3\int_2^3 x \, dx + 4\int_3^0 dy.$$

Using the list of integrals in Appendix E, we obtain

$$W = 3\left[\tfrac{1}{2}x^2\right]_2^3 + 4\left[y\right]_3^0$$
$$= \tfrac{3}{2}[3^2 - 2^2] + 4[0 - 3]$$
$$= -4.5 \text{ J} \approx -5 \text{ J.} \qquad \text{(Answer)}$$

The negative result tells us that energy is transferred from the particle by force $\mathbf{F}$. Because the kinetic energy of the particle decreases, its speed must decrease.

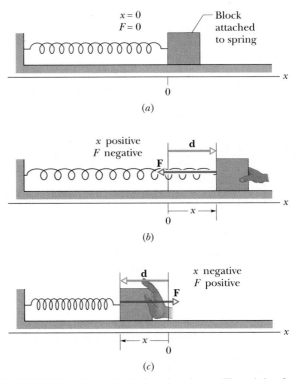

FIGURE 7-11 (a) A spring in its relaxed state. The origin of an x axis has been placed at the end of the spring that is attached to a block. (b) The block is displaced by $\mathbf{d}$, and the spring is stretched by an amount x. Note the restoring force $\mathbf{F}$ exerted by the spring. (c) The spring is compressed by an amount x. Again, note the restoring force.

7-6 WORK DONE BY A SPRING FORCE

We next want to examine the work done on a particle-like object by a particular type of variable force, namely, a **spring force** — the force exerted by a spring. Many forces in nature have the same mathematical form as the spring force. So, by examining this one force, you can gain an understanding of many others.

The Force Exerted by a Spring

Figure 7-11a shows a spring in its **relaxed state,** that is, neither compressed nor extended. One end is fixed, and a particle-like object, say, a block, is attached to the other, free end. In Fig. 7-11b, we stretch the spring by pulling the block to the right. In reaction, the spring pulls on the block toward the left, in the direction that will restore the relaxed state. (A spring's force is sometimes said to be a *restoring force.*) In Fig. 7-11c, we compress the spring by pushing the block to the left. The spring now pushes on the block toward the right, again so as to restore the relaxed state.

To a good approximation for many springs, the force $\mathbf{F}$ exerted by the spring is proportional to the displacement $\mathbf{d}$ of the free end from its position when the spring is in the relaxed state. That is, the *spring force* is given by

$$\mathbf{F} = -k\mathbf{d} \qquad \text{(Hooke's law),} \qquad (7\text{-}37)$$

which is known as **Hooke's law** after Robert Hooke, an English scientist of the late 1600s. The minus sign in Eq. 7-37 indicates that the spring force is always opposite in direction from the displacement of its free end. The con-

stant k is called the **spring constant** (or **force constant**) and is a measure of the stiffness of the spring. The larger k is, the stiffer the spring; that is, the stronger the spring will pull or push for a given displacement. The SI unit for k is the newton per meter.

In Fig. 7-11 an x axis has been placed parallel to the length of the spring, with the origin ($x = 0$) at the position of the free end when the spring is in its relaxed state. For this common arrangement, Eq. 7-37 becomes

$$F = -kx \qquad \text{(Hooke's law).} \qquad (7\text{-}38)$$

Note that a spring force is a variable force because it depends on the position of the free end: F can be symbolized as $F(x)$, as in Section 7-5. Also note that Hooke's law is a linear relationship between the force magnitude F and the position x of the free end.

Work Done by a Spring Force

Suppose we give the block in Fig. 7-11a an abrupt rightward jerk to provide the block with kinetic energy. And let us assume that the contact between block and floor is frictionless, that the spring has negligible mass compared to the block (the spring is *massless*), and that the spring obeys Hooke's law exactly (it is *ideal*).

As the block moves rightward, the spring force **F** slows it and thus decreases its kinetic energy. Experimentally we find that when our assumptions about the spring and floor hold, the only energy of the block that is changed by **F** is the kinetic energy. Thus the work–kinetic energy theorem applies to this situation. To find the work, we must use Eq. 7-27 because the spring force is a variable force, with F given by Eq. 7-38. So, as the block moves from a position x_i to x_f, the work done on the block by the spring force is

$$W_s = \int_{x_i}^{x_f} F \, dx = \int_{x_i}^{x_f} (-kx) \, dx = -k \int_{x_i}^{x_f} x \, dx$$

$$= (-\tfrac{1}{2}k) \left[x^2\right]_{x_i}^{x_f} = (-\tfrac{1}{2}k)(x_f^2 - x_i^2). \qquad (7\text{-}39)$$

Multiplied out this yields

$$W_s = \tfrac{1}{2}kx_i^2 - \tfrac{1}{2}kx_f^2 \qquad \begin{matrix}\text{(work by a}\\ \text{spring force).}\end{matrix} \qquad (7\text{-}40)$$

This work W_s done by the spring force can have a positive or negative value, depending on whether the *net* transfer of energy is to or from the block as the block moves from x_i to x_f. If $x_i = 0$ and if we call the final position x, then Eq. 7-40 becomes

$$W_s = -\tfrac{1}{2}kx^2 \qquad \begin{matrix}\text{(work by a}\\ \text{spring force).}\end{matrix} \qquad (7\text{-}41)$$

Now suppose that we displace the block along x while continuing to apply a force $\mathbf{F}_a$ to it. During the displacement, our applied force does work W_a on the block while the spring force does work W_s. By Eq. 7-15, the change ΔK in the kinetic energy of the block due to these two energy transfers is

$$\Delta K = K_f - K_i = W_a + W_s, \qquad (7\text{-}42)$$

in which K_f is the kinetic energy at the end of the displacement and K_i is that at the start of the displacement. If the block is stationary before and after the displacement, then K_f and K_i are both zero and Eq. 7-42 reduces to

$$W_a + W_s = 0,$$

or $$W_a = -W_s. \qquad (7\text{-}43)$$

The result means that the work done on the block by the applied force is the negative of the work done on the block by the spring force.

Note that the length of the spring does not appear explicitly in the expressions for the spring force (Eqs. 7-37 and 7-38) and for the work done by the spring force (Eqs. 7-40 and 7-41). The length of the spring is one of several factors that determine the spring constant k; thus, the length is in those equations implicitly.

SAMPLE PROBLEM 7-8

You apply a 4.9 N force F_a to a block attached to the free end of a spring to keep the spring stretched from its relaxed length by 12 mm, as in Fig. 7-11*b*.

(a) What is the spring constant of the spring?

SOLUTION: The stretched spring pulls with a force of -4.9 N. From Eq. 7-38, with $x = 12$ mm, we have

$$k = -\frac{F}{x} = -\frac{-4.9 \text{ N}}{12 \times 10^{-3} \text{ m}}$$

$$= 408 \text{ N/m} \approx 410 \text{ N/m}. \qquad \text{(Answer)}$$

Note that we do not need to know the length of the spring to find k. The plot of Eq. 7-38 in Fig. 7-12 refers to this spring. The slope of the line is -410 N/m.

(b) What force does the spring exert on the block if you stretch the spring by 17 mm?

SOLUTION: From Eq. 7-38 we have

$$F = -kx = -(408 \text{ N/m})(17 \times 10^{-3} \text{ m})$$

$$= -6.9 \text{ N}. \qquad \text{(Answer)}$$

The dot on the curve of Fig. 7-12 represents this force and the corresponding displacement. Note that x is positive and F is negative, as required by Eq. 7-38.

(c) How much work does the spring force do on the block as the spring is stretched 17 mm as in (b)?

SOLUTION: Because the spring is initially in its relaxed state, we can use Eq. 7-41:

$$W_s = -\tfrac{1}{2}kx^2 = -(\tfrac{1}{2})(408 \text{ N/m})(17 \times 10^{-3} \text{ m})^2$$

$$= -5.9 \times 10^{-2} \text{ J} = -59 \text{ mJ}. \qquad \text{(Answer)}$$

The shaded area in Fig. 7-12 represents this work. The work is negative because the spring force and the displacement of the block are in opposite directions. Note that the amount of work done by the spring force would be the same if the spring had been compressed (rather than stretched) by 17 mm.

(d) With the spring initially stretched by 17 mm, you allow the block to return to $x = 0$ (the spring returns to its relaxed state); you then compress the spring by 12 mm. How much work does the spring force do on the block during this total displacement of the block?

SOLUTION: For this situation, we have $x_i = +17$ mm (the spring is initially stretched) and $x_f = -12$ mm (the spring is finally compressed). Equation 7-40 becomes

$$W_s = \tfrac{1}{2}kx_i^2 - \tfrac{1}{2}kx_f^2 = \tfrac{1}{2}k(x_i^2 - x_f^2)$$

$$= \tfrac{1}{2}(408 \text{ N/m})[(17 \times 10^{-3} \text{ m})^2 - (-12 \times 10^{-3} \text{ m})^2]$$

$$= 0.030 \text{ J} = 30 \text{ mJ}. \qquad \text{(Answer)}$$

This work done on the block by the spring force is positive because the spring force does more positive work as the block moves from $x_i = +17$ mm to $x = 0$ than it does negative work as the block moves from $x = 0$ to $x_f = -12$ mm.

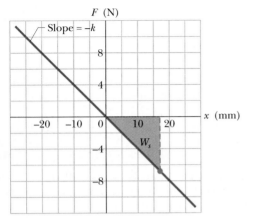

FIGURE 7-12   The force–distance plot for the spring of Sample Problem 7-8. The spring obeys Hooke's law (Eqs. 7-37 and 7-38) and has a spring constant $k = 410$ N/m. For the significance of the dot and the shaded area marked W_s, see Sample Problem 7-8(b) and 7-8(c), respectively.

CHECKPOINT **4:** For three situations, the initial and final positions, respectively, along the x axis for the block in Fig. 7-11 are (a) -3 cm, 2 cm; (b) 2 cm, 3 cm; and (c) -2 cm, 2 cm. In each situation is the work done by the spring force on the block positive, negative, or zero?

SAMPLE PROBLEM 7-9

A block whose mass m is 5.7 kg slides on a horizontal frictionless tabletop with a constant speed v of 1.2 m/s. It is brought momentarily to rest as it compresses a spring in its path (Fig. 7-13). By what distance d is the spring compressed? The spring constant k is 1500 N/m.

SOLUTION: From Eq. 7-41, the work done *by* the spring force *on* the block as the spring is compressed a distance d from its rest state is given by

$$W_s = -\tfrac{1}{2}kd^2.$$

The change in the kinetic energy of the block as it is stopped is

$$\Delta K = K_f - K_i = 0 - \tfrac{1}{2}mv^2.$$

The work–kinetic energy theorem (Eq. 7-4) requires that these two quantities be equal. Setting them so and solving for d, we obtain

$$d = v\sqrt{\frac{m}{k}} = (1.2 \text{ m/s})\sqrt{\frac{5.7 \text{ kg}}{1500 \text{ N/m}}}$$

$$= 7.4 \times 10^{-2} \text{ m} = 7.4 \text{ cm.} \qquad \text{(Answer)}$$

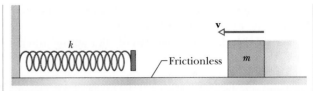

FIGURE 7-13   Sample Problem 7-9. A block moves toward a spring with velocity **v**. When it is momentarily stopped by the spring, it will have compressed the spring by a distance d.

PROBLEM SOLVING TACTICS

TACTIC 1: *Derivatives and Integrals; Slopes and Areas*

If you know a function $y = F(x)$, you can find the value of its derivative (for any value of x) or of its integral (between any two values of x) from the rules of calculus. If you do not know the function analytically but have a plot of it, you can find the values of both its derivative and its integral graphically. Finding a derivative graphically is shown in Tactic 5 of Chapter 2. Here we evaluate an integral graphically.

Figure 7-14 is a plot of a particular force function $F(x)$. Let us calculate graphically the work W done by this force as the particle on which it acts moves from $x_i = 2.0$ cm to $x_f = 5.0$ cm. According to Eq. 7-27, the work is

$$W = \int_{x_i}^{x_f} F(x)\, dx,$$

which is equal to the (total) shaded area shown under the curve between the two points.

You can approximate this area with a rectangle formed by drawing a horizontal line across Fig. 7-14. Draw it at a level such that the areas marked "1" and "2" appear to be equal. A line at $F = 44$ N is about right, and the area of the equivalent rectangle ($= W$) is then

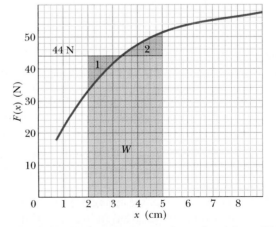

FIGURE 7-14   The graph of a one-dimensional force $F(x)$. The shaded area under the curve (which represents the work done by F) is approximated by a rectangle formed by excluding area 2 and including (approximately equal) area 1.

$$W = \text{height} \times \text{base} = (44 \text{ N})(5.0 \text{ cm} - 2.0 \text{ cm})$$
$$= 132 \text{ N} \cdot \text{cm} \approx 1.3 \text{ N} \cdot \text{m} = 1.3 \text{ J}.$$

You can also find the area by counting the small squares underneath the curve. The shaded area contains about 260 squares, and each square represents $(2 \text{ N})(0.25 \text{ cm}) = 0.5$ N·cm. The work is then

$$W = (260 \text{ squares}) \left(\frac{0.5 \text{ N} \cdot \text{cm}}{1 \text{ square}} \right) = 130 \text{ N} \cdot \text{cm}$$
$$= 1.3 \text{ J},$$

just as above. *Remember:* On a two-dimensional graph, a derivative is represented by a slope, and an integral by an area.

7-7 POWER

A contractor wishes to lift a load of bricks from the sidewalk to the top of a building by means of a winch. We can now calculate how much work the force applied by the winch must do on the load to make the lift. The contractor, however, is much more interested in the *rate* at which that work is done. Will the job take 5 minutes (acceptable) or a week (unacceptable)?

The rate at which work is done by a force is said to be the **power** due to the force. If an amount of work W is done in an amount of time Δt by a force, the **average power** due to the force during that time interval is

$$\overline{P} = \frac{W}{\Delta t} \qquad \text{(average power).} \qquad (7\text{-}44)$$

The **instantaneous power** P is the instantaneous rate of doing work, which we can write as

$$P = \frac{dW}{dt} \qquad \text{(instantaneous power).} \qquad (7\text{-}45)$$

The SI unit of power is the joule per second. This unit is used so often that it has a special name, the **watt** (W), after James Watt, who greatly improved the rate at which steam engines could do work. In the British system, the unit of power is the foot-pound per second. Often the horsepower is used. Some relations among these units are

$$1 \text{ watt} = 1 \text{ W} = 1 \text{ J/s} = 0.738 \text{ ft} \cdot \text{lb/s} \qquad (7\text{-}46)$$

and

$$1 \text{ horsepower} = 1 \text{ hp} = 550 \text{ ft} \cdot \text{lb/s}$$
$$= 746 \text{ W}. \qquad (7\text{-}47)$$

Inspection of Eq. 7-44 shows that work can be expressed as power multiplied by time, as in the common unit, the kilowatt-hour. Thus

$$1 \text{ kilowatt-hour} = 1 \text{ kW} \cdot \text{h} = (10^3 \text{ W})(3600 \text{ s})$$
$$= 3.60 \times 10^6 \text{ J} = 3.60 \text{ MJ.} \qquad (7\text{-}48)$$

Perhaps because of our utility bills, the watt and the kilowatt-hour have become identified as electrical units. They can be used equally well as units for other examples of power and energy. Thus if you pick up this book from the floor and put it on a tabletop, you are free to report the work that you have done as $4 \times 10^{-6} \text{ kW} \cdot \text{h}$ (or more conveniently as 4 mW·h).

We can also express the rate at which a force does work on a particle (or particle-like object) in terms of that force and the body's velocity. For a particle moving along a straight line (say, the x axis) and acted on by a constant force $\mathbf{F}$ directed at some angle ϕ to that line, Eq. 7-45 becomes

$$P = \frac{dW}{dt} = \frac{F \cos \phi \, dx}{dt} = F \cos \phi \left(\frac{dx}{dt} \right),$$

or

$$P = Fv \cos \phi. \qquad (7\text{-}49)$$

Reorganizing the right side of Eq. 7-49 as the dot product $\mathbf{F} \cdot \mathbf{v}$, we may also write Eq. 7-49 as

$$P = \mathbf{F} \cdot \mathbf{v} \qquad \text{(instantaneous power).} \qquad (7\text{-}50)$$

For example, the truck in Fig. 7-15 exerts a force $\mathbf{F}$ on the trailing load, which has velocity $\mathbf{v}$ at some instant. The instantaneous power due to $\mathbf{F}$ is the rate at which $\mathbf{F}$ transfers energy to the load at that instant and is given by Eqs. 7-49 and 7-50. Saying that this power is "the power of the truck" is often acceptable, but we should keep in mind what is meant: power is the rate at which the applied *force* transfers energy.

FIGURE 7-15 The power due to the truck's applied force on the trailing load is the rate at which that force does work on the load.

SAMPLE PROBLEM 7-10

Figure 7-16 shows forces $\mathbf{F}_1$ and $\mathbf{F}_2$ acting on a box as the box slides rightward across a frictionless floor. Force $\mathbf{F}_1$ is horizontal, with magnitude 2.0 N; force $\mathbf{F}_2$ is angled upward by 60° to the floor and has magnitude 4.0 N. The speed v of the box at a certain instant is 3.0 m/s.

(a) What is the power due to each force acting on the box at that instant, and what is the net power? Is the net power changing at that instant?

SOLUTION: We use Eq. 7-49 to find the power due to each force. For force $\mathbf{F}_1$, at angle $\phi_1 = 180°$ to velocity $\mathbf{v}$, we have

$$P_1 = F_1 v \cos \phi_1 = (2.0 \text{ N})(3.0 \text{ m/s}) \cos 180°$$
$$= -6.0 \text{ W.} \qquad \text{(Answer)}$$

This result tells us that force $\mathbf{F}_1$ is transferring energy from the box at the rate of 6.0 J/s.

For force $\mathbf{F}_2$, at angle $\phi_2 = 60°$ to velocity $\mathbf{v}$, we have

$$P_2 = F_2 v \cos \phi_2 = (4.0 \text{ N})(3.0 \text{ m/s}) \cos 60°$$
$$= 6.0 \text{ W.} \qquad \text{(Answer)}$$

This result tells us that force $\mathbf{F}_2$ is transferring energy to the box at the rate of 6.0 J/s.

The net power is the sum of the individual powers:

$$P_{\text{net}} = P_1 + P_2$$
$$= -6.0 \text{ W} + 6.0 \text{ W} = 0, \qquad \text{(Answer)}$$

which tells us that the net rate of transfer of energy to or from the box is zero. Thus, the kinetic energy ($K = \frac{1}{2}mv^2$) of the box is not changing, and so the speed of the box will remain at 3.0 m/s. With neither the forces $\mathbf{F}_1$ and $\mathbf{F}_2$ nor the velocity $\mathbf{v}$ changing, we see from Eq. 7-50 that P_1 and P_2 are not changing and thus neither is P_{net}.

(b) If the magnitude of $\mathbf{F}_2$ is, instead, 6.0 N, what now is the power due to each force acting on the box at the given instant, and what is the net power? Is the net power changing?

SOLUTION: For force $\mathbf{F}_2$, we now have

$$P_2 = F_2 v \cos \theta_2 = (6.0 \text{ N})(3.0 \text{ m/s}) \cos 60°$$
$$= 9.0 \text{ W.} \qquad \text{(Answer)}$$

The power of force $\mathbf{F}_1$ is still

$$P_1 = -6.0 \text{ W.} \qquad \text{(Answer)}$$

So, the net power is now

$$P_{\text{net}} = P_1 + P_2 = -6.0 \text{ W} + 9.0 \text{ W}$$
$$= 3.0 \text{ W,} \qquad \text{(Answer)}$$

which tells us that the net rate of transfer of energy to the box has a positive value. Thus, the kinetic energy of the box is increasing, and so also is the speed of the box. With the speed increasing, we see from Eq. 7-50 that the values of P_1 and P_2, and thus also of P_{net}, will be changing. Hence, this net power of 3.0 W is the net power only at the instant the speed is the given 3.0 m/s.

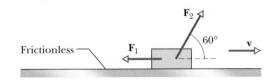

FIGURE 7-16 Sample Problem 7-10. Two forces $\mathbf{F}_1$ and $\mathbf{F}_2$ act on a box that slides rightward across a frictionless floor. The velocity of the box is $\mathbf{v}$.

CHECKPOINT **5**: A block moves with uniform circular motion because a cord tied to the block is anchored at the center of a circle. Is the power of the force exerted on the block by the cord positive, negative, or zero?

7-8 KINETIC ENERGY AT HIGH SPEEDS (OPTIONAL)

In Section 4-10 we saw that for particles moving at speeds near the speed of light, Newtonian mechanics fails and must be replaced by Einstein's theory of special relativity. One consequence is that we can no longer use the expression $K = \frac{1}{2}mv^2$ for the kinetic energy of a particle. Instead, we must use

$$K = mc^2\left(\frac{1}{\sqrt{1 - (v/c)^2}} - 1\right), \qquad (7\text{-}51)$$

in which c is the speed of light.

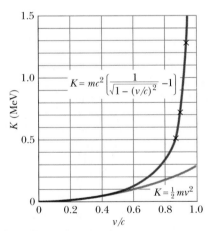

FIGURE 7-17 The relativistic (Eq. 7-51) and classical (Eq. 7-1) equations for the kinetic energy of an electron, plotted as a function of v/c, where v is the speed of the electron and c is the speed of light. Note that the two curves blend together at low speeds and diverge widely at high speeds. Experimental data (at the × marks) show that at high speeds the relativistic curve agrees with experiment but the classical curve does not.

Figure 7-17 shows that these two formulas, which look so different, do indeed give widely different results at high speeds. Experiment shows that—beyond any doubt—the relativistic expression (Eq. 7-51) is correct and the classical expression (Eq. 7-1) is not. At low speeds, however, the two formulas merge, yielding the same result. In particular, both formulas yield $K = 0$ for $v = 0$.

All relativistic formulas must reduce to their classical counterparts at low speeds. To see how this comes about for Eq. 7-51, let us write that equation in the form

$$K = mc^2[(1 - \beta^2)^{-1/2} - 1]. \qquad (7\text{-}52)$$

Here, for convenience, we have substituted the dimensionless *speed parameter* β for the speed ratio v/c.

At very low speeds, $v \ll c$ and therefore $\beta \ll 1$. At low speeds, we can then expand the quantity $(1 - \beta^2)^{-1/2}$ by the binomial theorem, obtaining (see Tactic 2, below)

$$(1 - \beta^2)^{-1/2} = 1 + \tfrac{1}{2}\beta^2 + \cdots. \qquad (7\text{-}53)$$

Substituting Eq. 7-53 into Eq. 7-52 leads to

$$K = mc^2[(1 + \tfrac{1}{2}\beta^2 + \cdots) - 1]. \qquad (7\text{-}54)$$

For very small β, the terms represented by the dots in Eq. 7-54 decrease rapidly in size. We can then, with little error, replace the entire sum in parentheses with its first two terms, obtaining

$$K \approx (mc^2)[(1 + \tfrac{1}{2}\beta^2) - 1]$$

or, with $\beta = v/c$,

$$K \approx (mc^2)(\tfrac{1}{2}\beta^2) = \tfrac{1}{2}mv^2,$$

which is exactly what we set out to show.

PROBLEM SOLVING TACTICS

TACTIC 2: *Approximations with the Binomial Theorem*

Often we have a quantity written in the form $(a + b)^n$ and we want to find its approximate value for the case in which $b \ll a$. To do so, it is simplest to recast the quantity in the form $(1 + x)^n$, where x is dimensionless and is much less than unity. Thus we can put

$$(a + b)^n = a^n(1 + b/a)^n = (a^n)(1 + x)^n.$$

This is of the desired form, with $x \,(= b/a)$ being dimensionless. We can then evaluate $(1 + x)^n$ with the binomial theorem, keeping only as many terms as are appropriate to the problem. (That decision takes some experience.)

The binomial theorem (given in Appendix E) is

$$(1 + x)^n = 1 + \frac{n}{1!}x + \frac{n(n-1)}{2!}x^2 + \cdots. \qquad (7\text{-}55)$$

The exclamation marks in Eq. 7-55 identify factorials, or products of all the whole numbers from the given number

down to 1. As an example, $4! = 4 \times 3 \times 2 \times 1 = 24$; you probably have a key for factorials on your calculator.

We applied the binomial theorem in Eq. 7-53, letting $x = -\beta^2$ and $n = -\tfrac{1}{2}$.

As an exercise, evaluate $(1 + 0.045)^{-2.3}$ both on your calculator and by expansion, using Eq. 7-55 with $x = 0.045$ and $n = -2.3$. Check the various terms in the binomial sum to see how rapidly they decrease.

7-9 REFERENCE FRAMES

We apply Newton's laws of mechanics only in inertial reference frames. Recall that these frames move at constant velocity.

For some physical quantities, observers in different inertial reference frames will measure the exact same values. In Newtonian mechanics, these *invariant* quantities (as they are called) are force, mass, acceleration, and time. As an example, if an observer in one inertial frame finds that a certain particle has a mass of 3.15 kg, observers in all other inertial frames will measure the same mass for the particle. For other physical quantities, such as the displacement and velocity of a particle, observers in different inertial frames will measure different values; these quantities are *not* invariant.

If the displacement of a particle depends on the reference frame of the observer, so must the work done on the particle because work $(W = \mathbf{F} \cdot \mathbf{d})$ is defined in terms of displacement. If the displacement of a particle during a given interval is measured to be $+2.47$ m in one reference frame, it might be measured to be zero in another and -3.64 m in a third. Because the force $\mathbf{F}$ does not change (it is invariant), work that is positive in one reference frame can be zero in another and negative in a third.

What about the kinetic energy of the particle? If its velocity depends on the reference frame of the observer, then so must its kinetic energy because kinetic energy $(K = \tfrac{1}{2}mv^2)$ is defined in terms of velocity. Does that mean that the work–kinetic energy theorem is in trouble?

From Galileo to Einstein, physicists have come to believe in the **principle of invariance:**

The laws of physics must have the same form in all inertial reference frames.

That is, even though some *physical quantities* have different values in different reference frames, the *laws of physics* must hold true in all frames. Behind this formal statement of invariance is a feeling that—in some deep sense—if different observers look at the same event, they must see nature working in the same way.

Among the laws to which this invariance principle applies is the work–kinetic energy theorem. Thus, even though different observers watching the same moving particle might measure different values for work and for kinetic energy, they would all find that the work–kinetic energy theorem holds in their respective frames. Let us look at a simple example.

In Fig. 7-18, Sally rides up in an elevator cab at constant speed v, holding a book. Steve, standing on a facing balcony, observes her as the cab rises through a height h. From these two references frames, what are the work–kinetic energy relations that apply to the book?

1. *Sally's report.* "My reference frame is the elevator cab. I am exerting an upward force on the book, but this force does no work because the book is not moving in my reference frame. The weight of the book, acting downward, does no work, for the same reason. Thus the total work done on the book by all the forces that act on it is zero. According to the work–kinetic energy theorem, the kinetic energy of the book should not change. That is what I observe; the kinetic energy of the book is zero in my reference frame and remains so. Everything fits together."

2. *Steve's report.* "My reference frame is the balcony. I see that Sally exerts a force **F** on her book. In my frame, the point of application of **F** is moving and the work done by **F** as the book moves upward through a height h is $+mgh$. I also know that the book's weight does work on the book, in the amount $-mgh$. Thus the total work done on the book during its rise is zero. According to the work–kinetic energy theorem, the kinetic energy of the book

FIGURE 7-18 Sally rides up in an elevator, holding a book. Steve watches her. Both check the work–kinetic energy theorem in their respective reference frames, as it applies to the motion of the book.

should not change. That is what I observe. In my frame, the kinetic energy is a constant $\frac{1}{2}mv^2$. Everything fits."

Although Steve and Sally do not agree about the displacement of the book and its kinetic energy, they do agree that the work–kinetic energy theorem holds in their respective reference frames.

It does not matter what (inertial) reference frame you pick in which to solve a problem as long as you (1) make sure you know what that frame is and (2) use that frame consistently throughout the problem.

REVIEW & SUMMARY

Kinetic Energy

The **kinetic energy** K associated with the motion of a particle of mass m and speed v, where v is well below the speed of light, is

$$K = \tfrac{1}{2}mv^2 \quad \text{(kinetic energy).} \quad (7\text{-}1)$$

Work

Work W is energy transferred to or from an object via a force acting on the object. Energy transferred to the object is positive work, and that transferred from the object is negative work.

Work and Kinetic Energy

If the only type of energy of an object being changed by work W is kinetic energy, then we can write the change ΔK in kinetic energy of the object as

$$\Delta K = K_f - K_i = W \quad \begin{matrix}\text{(work–kinetic}\\ \text{energy theorem),}\end{matrix} \quad (7\text{-}4)$$

in which K_i is the initial kinetic energy of the object and K_f is the kinetic energy after the work is done. Equation 7-4 rearranged gives us

$$K_f = K_i + W. \quad (7\text{-}5)$$

Work Done by a Constant Force

The work done on a particle by a constant force **F** during displacement **d** of the particle is

$$W = Fd \cos \phi = \mathbf{F} \cdot \mathbf{d} \quad \begin{matrix}\text{(work,}\\ \text{constant force),}\end{matrix} \quad (7\text{-}9,\ 7\text{-}11)$$

in which ϕ is the constant angle between **F** and **d**. When several forces $\mathbf{F}_1, \mathbf{F}_2, \mathbf{F}_3, \ldots$ act on a particle, the work done by the net force $\Sigma\mathbf{F}$ is

$$W = \left(\Sigma \mathbf{F} \right) \cdot \mathbf{d} \quad \begin{matrix}\text{(work by constant}\\ \text{net force),}\end{matrix} \quad (7\text{-}13)$$

This work is equal to the total work done by all the forces:

$$W = \mathbf{F}_1 \cdot \mathbf{d} + \mathbf{F}_2 \cdot \mathbf{d} + \mathbf{F}_3 \cdot \mathbf{d} + \cdots$$
$$= W_1 + W_2 + W_3 + \cdots \quad \text{(total work).} \quad (7\text{-}14)$$

Substituting this total work into the work–kinetic energy theorem, we have

$$\Delta K = K_f - K_i = W_1 + W_2 + W_3 + \cdots, \quad (7\text{-}15)$$

which gives us the change in the kinetic energy of the particle due to the total work.

Work Done by Weight

The work W_g done by the weight $m\mathbf{g}$ of a particle-like object during a displacement $\mathbf{d}$ of the object is given by

$$W_g = mgd \cos \phi, \quad (7\text{-}16)$$

in which ϕ is the angle between $m\mathbf{g}$ and $\mathbf{d}$.

Work Done in Lifting and Lowering an Object

The work W_a done by an applied force during a lifting or lowering of a particle-like object is related to the work W_g done by the object's weight and the change ΔK in the object's kinetic energy by

$$\Delta K = K_f - K_i = W_a + W_g. \quad (7\text{-}19)$$

If the kinetic energy at the beginning of a lift equals that at the end of the lift, then Eq. 7-19 reduces to

$$W_a = -W_g, \quad (7\text{-}20)$$

which tells us that the applied force transfers as much energy to the object as the weight of the object transfers from the object.

Work Done by a Variable Force

When the force $\mathbf{F}$ on a particle-like object depends on the position of the object, the work done by $\mathbf{F}$ on the object while the object moves from an initial position r_i with coordinates (x_i, y_i, z_i) to a final position r_f with coordinates (x_f, y_f, z_f) is

$$W = \int_{x_i}^{x_f} F_x \, dx + \int_{y_i}^{y_f} F_y \, dy + \int_{z_i}^{z_f} F_z \, dz. \quad (7\text{-}31)$$

If $\mathbf{F}$ has only an x component, then Eq. 7-31 reduces to

$$W = \int_{x_i}^{x_f} F(x) \, dx. \quad (7\text{-}27)$$

Spring Force

The force $\mathbf{F}$ exerted by a spring is

$$\mathbf{F} = -k\mathbf{d} \quad \text{(Hooke's law),} \quad (7\text{-}37)$$

where $\mathbf{d}$ is the displacement of its free end from the position when the spring is in its **relaxed state** (neither compressed or extended)

and k is the **spring constant** (a measure of the spring's stiffness). If an x axis lies along the spring, with the origin at the location of the spring's free end when the spring is in its relaxed state, Eq. 7-37 can be written as

$$F = -kx \quad \text{(Hooke's law).} \quad (7\text{-}38)$$

Work Done by a Spring Force

If an object is attached to the spring's free end, the work W_s done on the object by the spring force when the object is moved from an initial position x_i to a final position x_f is

$$W_s = \tfrac{1}{2} k x_i^2 - \tfrac{1}{2} k x_f^2. \quad (7\text{-}40)$$

If $x_i = 0$ and $x_f = x$, then Eq. 7-40 becomes

$$W_s = -\tfrac{1}{2} k x^2. \quad (7\text{-}41)$$

Power

The **power** due to a force is the *rate* at which that force does work on an object. If the force does work W during a time interval Δt, the *average power* due to the force over that time interval is

$$\overline{P} = \frac{W}{\Delta t}. \quad (7\text{-}44)$$

Instantaneous power is the instantaneous rate of doing work:

$$P = \frac{dW}{dt}. \quad (7\text{-}45)$$

If a force $\mathbf{F}$ is at an angle ϕ to the direction of travel of the object, the instantaneous power is

$$P = Fv \cos \phi = \mathbf{F} \cdot \mathbf{v}, \quad (7\text{-}49, 7\text{-}50)$$

in which $\mathbf{v}$ is the instantaneous velocity of the object.

Relativistic Kinetic Energy

The kinetic energy of objects moving at speeds near the speed of light c must be calculated using the relativistic expression

$$K = mc^2 \left(\frac{1}{\sqrt{1 - (v/c)^2}} - 1 \right). \quad (7\text{-}51)$$

This equation reduces to Eq. 7-1 when v is much less than c.

The Principle of Invariance

Some quantities (such as mass, force, acceleration, and time in Newtonian mechanics) are *invariant;* that is, they have the same numerical values when measured in different inertial reference frames. Others (for example, velocity, kinetic energy, and work) have different values in different frames. However, the *laws* of physics have the same *form* in all inertial reference frames. This is called the **principle of invariance.**

QUESTIONS

1. Rank the following velocities according to the kinetic energy a particle will have with each velocity, greatest first: (a) $\mathbf{v} = 4\mathbf{i} + 3\mathbf{j}$, (b) $\mathbf{v} = -4\mathbf{i} + 3\mathbf{j}$, (c) $\mathbf{v} = -3\mathbf{i} + 4\mathbf{j}$, (d) $\mathbf{v} = 3\mathbf{i} - 4\mathbf{j}$, (e) $\mathbf{v} = 5\mathbf{i}$, and (f) $v = 5$ m/s at $30°$ to the horizontal.

2. For $t > 0$, is the kinetic energy of a particle increasing, decreasing, or remaining the same if the particle's velocity is given by (a) $\mathbf{v} = 3\mathbf{i}$ and (b) $v = -4t$?

3. For $t > 0$, is the kinetic energy of a particle increasing, decreasing, or remaining the same if the particle's position is given by (a) $x = 4t^2 - 2$, (b) $x = -3t + 14$, (c) $\mathbf{r} = 2\mathbf{i} - 3t\mathbf{j}$, and (d) $\mathbf{r} = (2t^2 - 3)\mathbf{i} + (4t - 2)\mathbf{j}$?

4. A force $\mathbf{F}$ that is directed along an x axis moves a particle 5 m along that axis. Can we use Eqs. 7-9 and 7-11 to find the work done if the magnitude (in newtons) of $\mathbf{F}$ is given by (a) $F = 3$; (b) $F = 2x$; (c) $F = 2t$?

5. Is positive or negative work done by a constant force $\mathbf{F}$ on a particle during a straight-line displacement $\mathbf{d}$ if (a) the angle between $\mathbf{F}$ and $\mathbf{d}$ is $30°$; (b) the angle is $100°$; (c) $\mathbf{F} = 2\mathbf{i} - 3\mathbf{j}$ and $\mathbf{d} = -4\mathbf{i}$?

6. If a particle-like object of mass 5 kg travels in uniform circular motion at speed 4 m/s, how much work is done on the particle by the centripetal force acting on it (a) at any instant and (b) during one complete revolution?

7. Figure 7-19 shows six situations in which two forces act simultaneously on a box after the box has been sent sliding over a frictionless surface, either to the left or to the right. The forces are either 1 N or 2 N in magnitude, as indicated by the vector lengths. For each situation, is the work done on the box by the net force during the indicated displacement $\mathbf{d}$ positive, negative, or zero?

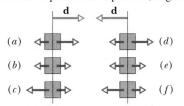

FIGURE 7-19 Question 7.

8. In overhead views, Fig. 7-20 shows three situations in which a box is acted on by two forces ($\mathbf{F}_1$ and $\mathbf{F}_2$) of equal magnitude. As the box moves, the forces maintain their orientation relative to the velocity $\mathbf{v}$. For each situation, is the total work done on the box positive, negative, or zero?

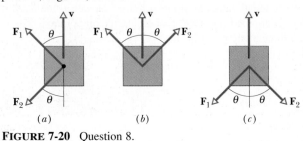

FIGURE 7-20 Question 8.

9. In Fig. 7-21, a greased pig has a choice of three frictionless slides along which to slide to the ground. Rank the slides according to how much work the pig's weight does on the pig during the descent, greatest first.

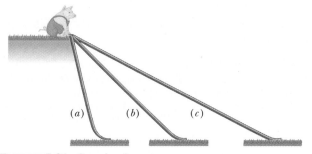

FIGURE 7-21 Question 9.

10. You lift an armadillo to a shelf. Does the work done by your force on the armadillo depend on (a) the mass of the armadillo, (b) the weight of the armadillo, (c) the height of the shelf, (d) the time you take, or (e) whether you move the armadillo sideways or directly upward?

11. Figure 7-22 shows a bundle of magazines that is lifted by a cord through distance d. The table shows six pairs of values of the initial speed v_0 and final speed v (in meters per second) of the bundle at the beginning and end of distance d. Rank the pairs according to the work done by the cord's force on the bundle over distance d, most positive first, most negative last.

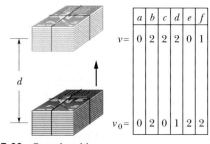

FIGURE 7-22 Question 11.

12. Figure 7-23 shows four graphs (drawn to the same scale) of the values of a variable force $\mathbf{F}$ (directed along an x axis) versus the position x of a particle on which the force acts. Rank the graphs according to the work done by $\mathbf{F}$ on the particle from $x = 0$ to x_1, from most positive work to most negative work.

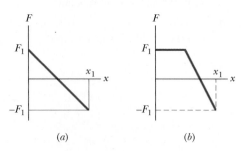

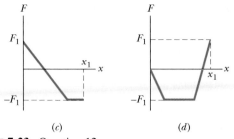

FIGURE 7-23 Question 12.

13. Figure 7-24 shows the values of a force **F**, directed along an x axis, that will act on a particle at the corresponding values of x. If the particle begins at rest at $x = 0$, what is the particle's coordinate when it has (a) its greatest kinetic energy, (b) its greatest speed, and (c) zero speed? (d) What is its direction of travel after it reaches $x = 6$ m?

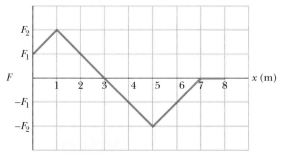

FIGURE 7-24 Question 13.

14. A block is attached to a relaxed spring as in Fig. 7-25*a*. The spring constant k of the spring is such that for a certain rightward displacement **d** of the block, the spring force acting on the block has magnitude F_1 and has done work W_1 on the block. A second, identical spring is then attached on the opposite side of the block,

as shown in Fig. 7-25*b*; both springs are in their relaxed state in the figure. If the block is again displaced by **d**, (a) what is the magnitude of the net force on it from both springs and (b) how much work has been done on it by the spring forces?

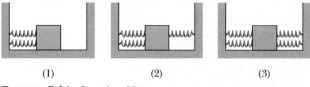

FIGURE 7-25 Question 14.

15. Spring A is stiffer than spring B, that is, $k_A > k_B$. The spring force of which spring does more work if the springs are compressed (a) the same distance and (b) by the same applied force?

16. Figure 7-26 shows three arrangements of a block attached to identical springs that are in their relaxed state when the block is centered as shown. Rank the arrangements according to the magnitude of the net force on the block, largest first, when the block is displaced by distance d (a) to the right and (b) to the left. Rank the arrangements according to the work done on the block by the spring forces, greatest first, when the block is displaced by d (c) to the right and (d) to the left.

FIGURE 7-26 Question 16.

17. If you cut a spring in half, the spring constant of either half is what multiple of the spring constant of the original spring? (*Hint:* Consider the amount by which each half stretches for a given value of force.)

EXERCISES & PROBLEMS

SECTION 7-1 Kinetic Energy

1E. If a Saturn V rocket with an Apollo spacecraft attached has a combined mass of 2.9×10^5 kg and is to reach a speed of 11.2 km/s, how much kinetic energy will it then have?

2E. If an electron (mass $m = 9.11 \times 10^{-31}$ kg) in copper near the lowest possible temperature has a kinetic energy of 6.7×10^{-19} J, what is the speed of the electron?

3E. Calculate the kinetic energies of the following objects moving at the given speeds: (a) a 110 kg football linebacker running at 8.1 m/s; (b) a 4.2 g bullet at 950 m/s; (c) the aircraft carrier *Nimitz*, 91,400 tons at 32 knots.

4E. On August 10, 1972, a large meteorite skipped across the atmosphere above western United States and Canada, much like a stone skipped across water. The accompanying fireball was so bright that it could be seen in the daytime sky (Fig. 7-27). The meteorite's mass was about 4×10^6 kg; its speed was about 15

FIGURE 7-27 Exercise 4 A large meteorite skips across the atmosphere in the sky above the mountains (upper right).

km/s. Had it entered the atmosphere vertically, it would have hit the Earth's surface with about the same speed. (a) Calculate the meteorite's loss of kinetic energy (in joules) that would have been associated with the vertical impact. (b) Express the energy as a multiple of the explosive energy of 1 megaton of TNT, which is 4.2×10^{15} J. (c) The energy associated with the atomic bomb explosion over Hiroshima was equivalent to 13 kilotons of TNT. To how many "Hiroshima bombs" would the meteorite impact have been equivalent?

5E. An explosion at ground level leaves a crater with a diameter that is proportional to the energy of the explosion raised to the $\frac{1}{3}$ power; an explosion of 1 megaton of TNT leaves a crater with a 1 km diameter. Below Lake Huron in Michigan there appears to be an ancient impact crater with a 50 km diameter. What was the kinetic energy associated with that impact, in terms of (a) megatons of TNT and (b) Hiroshima bomb equivalents (see Exercise 4)? (Such meteorite or comet impacts may have significantly altered the Earth's climate and contributed to the extinction of the dinosaurs and other life-forms.)

6P. A proton (mass $m = 1.67 \times 10^{-27}$ kg) is being accelerated along a straight line at 3.6×10^{15} m/s² in a machine called a linear accelerator. If the proton has an initial speed of 2.4×10^{7} m/s and travels 3.5 cm, what then is (a) its speed and (b) the increase in its kinetic energy in electron-volts?

7P. A man racing his son has half the kinetic energy of the son, who has half the mass of the father. The man speeds up by 1.0 m/s and then has the same kinetic energy as the son. What were the original speeds of man and son?

8P. What is the kinetic energy associated with the Earth's orbiting of the Sun? (See Appendix C for numerical data.)

SECTION 7-3 Work and Kinetic Energy

9E. A 102 kg object moves in a horizontal straight line with an initial speed of 53 m/s. If it is stopped along that line with a deceleration of 2.0 m/s², (a) what magnitude of force is required, (b) what distance does it travel while decelerating, and (c) what work is done by the decelerating force on the object? (d) Answer questions (a) to (c) for a deceleration of 4.0 m/s².

10E. To pull a 50 kg crate across a frictionless floor, a worker applies a force of 210 N, directed 20° above the horizontal. As the crate moves 3.0 m, what work is done on the crate by (a) the worker's force, (b) the weight of the crate, and (c) the normal force exerted by the floor on the crate? (d) What is the total work done on the crate?

11E. A floating ice block is pushed through a displacement $\mathbf{d} = (15 \text{ m})\mathbf{i} - (12 \text{ m})\mathbf{j}$ along a straight embankment by rushing water, which exerts a force $\mathbf{F} = (210 \text{ N})\mathbf{i} - (150 \text{ N})\mathbf{j}$ on the block. How much work does the force do on the block during the displacement?

12E. A particle moves along a straight path through a displacement $\mathbf{d} = (8 \text{ m})\mathbf{i} + c\mathbf{j}$ while a force $\mathbf{F} = (2 \text{ N})\mathbf{i} - (4 \text{ N})\mathbf{j}$ acts on it. (Other forces also act on the particle.) What is the value of c if the work done by $\mathbf{F}$ on the particle is (a) zero, (b) positive, and (c) negative?

13E. An initially stationary proton is accelerated in a cyclotron to a final speed of 3.0×10^{6} m/s. How much work, in electron-volts, is done on the proton by the (electric) force accelerating it? (Although the speed is 1% the speed of light, relativity theory is not required here.)

14E. A single force acts on a body that moves along a straight line. Figure 7-28 shows the velocity versus time for the body. Find the sign (plus or minus) of the work done by the force on the body in each of the intervals *AB, BC, CD,* and *DE.*

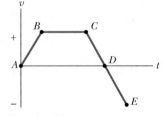

FIGURE 7-28 Exercise 14.

15E. A firehose 12 m long (Fig. 7-29) is uncoiled by pulling the nozzle end horizontally along a frictionless surface at the steady speed of 2.3 m/s. The mass of 1.0 m of the hose is 0.25 kg. How much work has been done on the hose by the applied force when the entire hose is moving?

FIGURE 7-29 Exercise 15.

16P. Figure 7-30 shows three forces applied to a greased trunk that moves leftward by 3.00 m over a frictionless floor. The force magnitudes are $F_1 = 5.00$ N, $F_2 = 9.00$ N, and $F_3 = 3.00$ N. During the displacement, (a) what is the net work done on the trunk by the three forces and (b) does the kinetic energy of the trunk increase or decrease?

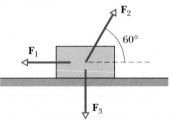

FIGURE 7-30 Problem 16.

17P. A force acts on a 3.0 kg particle-like object in such a way that the position of the object as a function of time is given by $x = 3.0t - 4.0t^2 + 1.0t^3$, with x in meters and t in seconds. Find the work done on the object by the force from $t = 0$ to $t = 4.0$ s. (*Hint:* What are the speeds at those times?)

18P. Figure 7-31 shows an overhead view of three horizontal forces acting on a cargo canister that is initially stationary but which now moves across a frictionless floor. The force magnitudes are $F_1 = 3.00$ N, $F_2 = 4.00$ N, and $F_3 = 10.0$ N. What is the net work done on the canister by the three forces during the first 4.00 m of displacement?

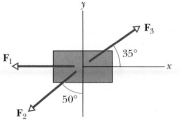

FIGURE 7-31 Problem 18.

19P. A constant force of magnitude 10 N makes an angle of 150° (measured counterclockwise) with the direction of increasing x as the force acts on a 2.0 kg object. How much work is done on the object by the force as the object moves from the origin to the point with position vector $(2.0 \text{ m})\mathbf{i} - (4.0 \text{ m})\mathbf{j}$?

SECTION 7-4 Work Done by Weight

20E. (a) In 1975 the roof of Montreal's Velodrome, with a weight of 41,000 tons, was lifted by 4.0 in. so that it could be centered. How much work was done on the roof by the forces making the lift? (b) In 1960 Mrs. Maxwell Rogers of Tampa, Florida, reportedly raised one end of a car that had fallen onto her son when a jack failed. If her panic lift effectively raised 900 lb of the 3600 lb car by 2 in., how much work was done by her force acting on the car?

21E. To push a 25.0 kg crate up a frictionless incline, angled at 25.0° to the horizontal, a worker exerts a force of 209 N, parallel to the incline. As the crate slides 1.50 m, how much work is done on the crate by (a) the worker's applied force, (b) the weight of the crate, and (c) the normal force exerted by the incline on the crate? (d) What is the total work done on the crate?

22E. A 45 kg block of ice slides down a frictionless incline 1.5 m long and 0.91 m high. A worker pushes up against the ice, parallel to the incline, so that the block slides down at constant speed. (a) Find the force exerted by the worker. How much work is done on the block by (b) the worker's force, (c) the weight of the block, (d) the normal force exerted by the surface of the incline on the block, and (e) the resultant force on the block?

23E. In Fig. 7-32, a cord runs around two massless, frictionless pulleys; a canister with mass $m = 20$ kg hangs from one pulley; and you exert a force **F** on the free end of the cord. (a) What must

FIGURE 7-32 Exercise 23.

be the magnitude of **F** if you are to lift the canister at a constant speed? (b) To lift the canister by 2.0 cm, how far must you pull the free end of the cord? During that lift, what is the work done on the canister by (c) your force (via the cord) and (d) the weight $m\mathbf{g}$ of the canister? (*Hint:* When a cord loops around a pulley as shown, it pulls on the pulley with a net force that is twice the tension in the cord.)

24P. A helicopter lifts a 72 kg astronaut 15 m vertically from the ocean by means of a cable. The acceleration of the astronaut is $g/10$. How much work is done on the astronaut by (a) the force from the helicopter and (b) her weight? What are (c) the kinetic energy and (d) the speed of the astronaut just before she reaches the helicopter?

25P. A cord is used to vertically lower an initially stationary block of mass M at a constant downward acceleration of $g/4$. When the block has fallen a distance d, find (a) the work done by the cord's force on the block, (b) the work done by the weight of the block, (c) the kinetic energy of the block, and (d) the speed of the block.

26P. A cave rescue team lifts an injured spelunker directly upward and out of a sinkhole by means of a motor-driven cable. The lift is performed in three stages, each requiring a vertical distance of 10.0 m: (1) the initially stationary spelunker is accelerated to a speed of 5.00 m/s; (2) he is then lifted at the constant speed of 5.00 m/s; (3) finally he is decelerated to zero speed. How much work is done on the 80.0 kg rescuee by the force lifting him during each stage?

SECTION 7-5 Work Done by a Variable Force

27E. A 5.0 kg block moves in a straight line on a horizontal frictionless surface under the influence of a force that varies with position as shown in Fig. 7-33. How much work is done by the force as the block moves from the origin to $x = 8.0$ m?

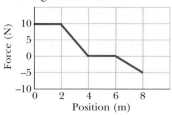

FIGURE 7-33 Exercise 27.

28E. A 10 kg mass moves along an x axis. Its acceleration as a function of its position is shown in Fig. 7-34. What is the net work performed on the mass by the force causing the acceleration as the mass moves from $x = 0$ to $x = 8.0$ m?

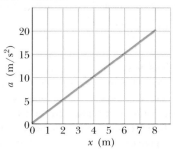

FIGURE 7-34 Exercise 28.

29P. (a) Estimate the work done by the force represented by the graph of Fig. 7-35 in displacing a particle from $x = 1$ m to $x = 3$ m. Refine your method to see how close you can come to the exact answer of 6 J. (b) The curve is given by $F = a/x^2$, with $a = 9$ N·m². Calculate the work using integration.

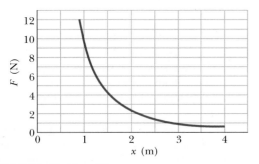

FIGURE 7-35 Problem 29.

30P. The force on a particle is $F = F_0(x/x_0 - 1)$. Find the work done by the force in moving the particle from $x = 0$ to $x = 2x_0$ by (a) plotting $F(x)$ and measuring the work from the graph and (b) integrating $F(x)$.

31P. What work is done by a force $\mathbf{F} = (2x \text{ N})\mathbf{i} + (3 \text{ N})\mathbf{j}$, with x in meters, that moves a particle from a position $\mathbf{r}_i = (2 \text{ m})\mathbf{i} + (3 \text{ m})\mathbf{j}$ to a position $\mathbf{r}_f = -(4 \text{ m})\mathbf{i} - (3 \text{ m})\mathbf{j}$?

32P. A force $\mathbf{F}$ in the direction of increasing x acts on an object moving along the x axis. If the magnitude of the force is $F = 10e^{-x/2.0}$ N, where x is in meters, find the work done by $\mathbf{F}$ as the object moves from $x = 0$ to $x = 2.0$ m (a) by plotting $F(x)$ and finding the area under the curve and (b) by integrating to find the work analytically.

33P. The only force acting on a 2.0 kg body as it moves along the x axis varies as shown in Fig. 7-36. The velocity of the body at $x = 0$ is 4.0 m/s. (a) What is the kinetic energy of the body at $x = 3.0$ m? (b) At what value of x will the body have a kinetic energy of 8.0 J? (c) What is the maximum kinetic energy attained by the body between $x = 0$ and $x = 5.0$ m?

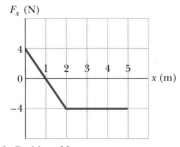

FIGURE 7-36 Problem 33.

34P. A 230 kg crate hangs from the end of a 12.0 m rope. You push horizontally on the crate with a varying force $\mathbf{F}$ to move it 4.00 m to the side (Fig. 7-37). (a) What is the magnitude of $\mathbf{F}$ when the crate is in this final position? During the crate's displacement, what are (b) the total work done on it, (c) the work done by the weight of the crate, and (d) the work done by the pull on the crate from the rope? (e) Knowing that the crate is motion-

less before and after its displacement, use the answers to (b), (c), and (d) to find the work your force $\mathbf{F}$ does on the crate. (f) Why is the work of your force not equal to the product of the horizontal displacement and the answer to (a)?

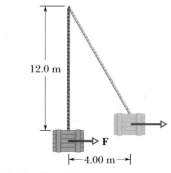

FIGURE 7-37 Problem 34.

SECTION 7-6 Work Done by a Spring Force

35E. A spring with a spring constant of 15 N/cm has a cage attached to one end (Fig. 7-38). (a) How much work does the spring force do on the cage if the spring is stretched from its relaxed length by 7.6 mm? (b) How much additional work is done by the spring force if it is stretched by an additional 7.6 mm?

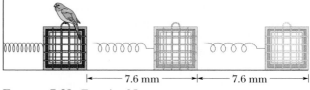

FIGURE 7-38 Exercise 35.

36E. During spring semester at MIT, residents of the parallel buildings of the East Campus dorms battle one another with large catapults that are made with surgical hose mounted on a window frame. A balloon filled with dyed water is placed in a pouch attached to the hose, which is then stretched through the width of the room. Assume that the stretching of the hose obeys Hooke's law and has a spring constant of 100 N/m. If the hose is stretched by 5.00 m and then released, how much work does the force from the hose do on the balloon in the pouch by the time the hose reaches its relaxed length?

37P. The only force acting on a 2.0 kg body as it moves along the positive x axis has an x component $F_x = -6x$ N, where x is in meters. The velocity of the body at $x = 3.0$ m is 8.0 m/s. (a) What is the velocity of the body at $x = 4.0$ m? (b) At what positive value of x will the body have a velocity of 5.0 m/s?

38P. A spring with a pointer attached is hanging next to a scale marked in millimeters. Three different weights are hung from the spring, in turn, as shown in Fig. 7-39. (a) If all weight is removed from the spring, which mark on the scale will the pointer indicate? (b) What is the weight W?

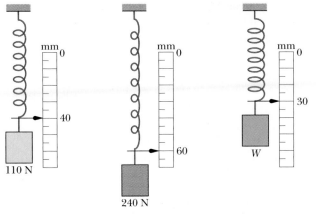

FIGURE 7-39   Problem 38.

39P. In Fig. 7-40, two identical springs, each with a relaxed length of 50 cm and a spring constant of 500 N/m, are connected by a short cord of length 10 cm. The upper spring is attached to the ceiling; a box that weighs 100 N hangs from the lower spring. Two additional cords, each 85 cm long, are also tied to the assembly; they are limp. (a) If the short cord is cut, so that the box then hangs from the springs and the two longer cords, does the box move up or down? (b) How far does the box move? (c) How much total work do the two spring forces (one directly, the other via a cord) do on the box during that move?

FIGURE 7-40   Problem 39.

40P. A 250 g block is dropped onto a vertical spring with spring constant $k = 2.5$ N/cm (Fig. 7-41). The block becomes attached to the spring, and the spring compresses 12 cm before momentarily stopping. While the spring is being compressed, what work is

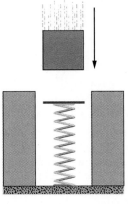

FIGURE 7-41   Problem 40.

done on the block (a) by its weight and (b) by the spring force? (c) What is the speed of the block just before it hits the spring? (Assume that friction is negligible.) (d) If the speed at impact is doubled, what is the maximum compression of the spring?

SECTION 7-7   Power

41E. The loaded cab of an elevator has a mass of 3.0×10^3 kg and moves 210 m up the shaft in 23 s at constant speed. At what average rate does the force from the cable do work on the cab?

42E. If a ski lift raises 100 passengers averaging 150 lb a height of 500 ft in 60 s, at constant speed, what average power is required of the force making the lift?

43E. An elevator cab in the New York Marriott Marquis has a mass of 4500 kg and can carry a maximum load of 1800 kg. If the cab is moving upward at full load at 3.80 m/s, what power is required of the force moving the cab to maintain that speed?

44E. (a) At a certain instant, a particle experiences a force $\mathbf{F} = (4.0 \text{ N})\mathbf{i} - (2.0 \text{ N})\mathbf{j} + (9.0 \text{ N})\mathbf{k}$ while having a velocity $\mathbf{v} = -(2.0 \text{ m/s})\mathbf{i} + (4.0 \text{ m/s})\mathbf{k}$. What is the instantaneous rate at which the force does work on the particle? (b) At some other time, the velocity consists of only a $\mathbf{j}$ component. If the force is unchanged, and the instantaneous power is -12 W, what is the velocity of the particle just then?

45P. A 100 kg block is pulled at a constant speed of 5.0 m/s across a horizontal floor by an applied force of 122 N directed 37° above the horizontal. What is the rate at which the force does work on the block?

46P. A horse pulls a cart with a force of 40 lb at an angle of 30° above the horizontal and moves along at a speed of 6.0 mi/h. (a) How much work does the force do in 10 min? (b) What is the average power (in horsepower) of the force?

47P. An initially stationary 2.0 kg object accelerates horizontally and uniformly to a speed of 10 m/s in 3.0 s. (a) In that 3.0 s interval, how much work is done on the object by the force accelerating it? What is the instantaneous power due to that force (b) at the end of the interval and (c) at the end of the first half of the interval?

48P. A force of 5.0 N acts on a 15 kg body initially at rest. Compute (a) the work done by the force in the first, second, and third seconds and (b) the instantaneous power due to the force at the end of the third second.

49P. A fully loaded, slow-moving freight elevator has a cab with a total mass of 1200 kg, which is required to travel upward 54 m in 3.0 min. The elevator's counterweight has a mass of only 950 kg. So, the elevator motor must help pull the cab upward. What average power (in horsepower) is required of the force that motor exerts on the cab via the cable?

50P. The force (but not the power) required to tow a boat at constant velocity is proportional to the speed. If a speed of 2.5 mi/h requires 10 hp, how much horsepower does a speed of 7.5 mi/h require?

51P. A 0.30 kg mass sliding on a horizontal frictionless surface is attached to one end of a horizontal spring (with $k = 500$ N/m) whose other end is fixed. The mass has a kinetic energy of 10 J as it passes through its equilibrium position (the point at which the spring force is zero). (a) At what rate is the spring doing work on the mass as the mass passes through its equilibrium position? (b) At what rate is the spring doing work on the mass when the spring is compressed 0.10 m and the mass is moving away from the equilibrium position?

SECTION 7-8 Kinetic Energy at High Speeds

52E. An electron moves through 5.1 cm in 0.25 ns. (a) What is its speed in terms of the speed of light? (b) What is its kinetic energy in electron-volts? (c) What percentage error do you make if you use the classical formula to compute the kinetic energy?

53E. The work–kinetic energy theorem applies to particles at all speeds. How much work, in electron-volts, must be done to increase the speed of an electron from rest to (a) $0.500c$, (b) $0.990c$, and (c) $0.999c$?

54P. An electron has a speed of $0.999c$. (a) What is its kinetic energy? (b) If its speed is increased by 0.05%, by what percentage does its kinetic energy increase?

Electronic Computation

55P. A force $\mathbf{F} = (3.00 \text{ N})\mathbf{i} + (7.00 \text{ N})\mathbf{j} + (7.00 \text{ N})\mathbf{k}$ acts on a 2.00 kg object which moves from an initial position of $\mathbf{d}_i = (3.00 \text{ m})\mathbf{i} - (2.00 \text{ m})\mathbf{j} + (5.00 \text{ m})\mathbf{k}$ to a final position of $\mathbf{d}_f = -(5.00 \text{ m})\mathbf{i} + (4.00 \text{ m})\mathbf{j} + (7.00 \text{ m})\mathbf{k}$ in 4.00 s. Find (a) the work done on the object by the force in the 4.00 s interval, (b) the average power due to the force during the 4.00 s interval, and (c) the angle between vectors $\mathbf{d}_i$ and $\mathbf{d}_f$.

8
Potential Energy and Conservation of Energy

In the dangerous "sport" of bungee-jumping, a participant attaches a specially designed elastic cord to the ankles and leaps from a tall structure. How can we tell just how far down a jumper will go? The answer is, of course, important to the jumper, who is keenly aware of what lies below. ❓

8-1 POTENTIAL ENERGY

In this chapter we continue the discussion of energy that we began in Chapter 7. To do so, we define a second form of energy: **potential energy** U is energy that can be associated with the configuration (or arrangement) of a system of objects that exert a force on one another. If the configuration of the system changes, then the potential energy of the system also changes.

One type of potential energy is the **gravitational potential energy** that is associated with the state of separation between objects, which attract one another via the gravitational force. For example, when Vasili Alexeev lifted 562 lb above his head in the 1976 Olympics, he increased the separation between the weights and the Earth. His effort changed the gravitational potential energy of the weights–Earth system by changing the configuration of the system, that is, by shifting the relative locations of the Earth and the weights (Fig. 8-1).

Another type of potential energy is **elastic potential energy,** which is associated with the state of compression or extension of an elastic (springlike) object. The force involved here is the spring force. If you compress or extend a spring, you change the relative locations of the coils within the spring. A spring force resists that change, and the result of your effort is an increase in the elastic potential energy of the spring.

Work and Potential Energy

In Chapter 7 we discussed the relation between work and a change in kinetic energy. Here we discuss the relation between work and a change in potential energy.

Let us throw a tomato upward (Fig. 8-2). We already know that as the tomato rises, the work W_g done on the tomato by its weight (the gravitational force acting on it) is negative because the weight transfers energy *from* the ki-

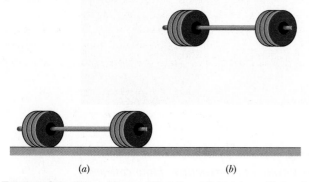

(a)   (b)

FIGURE 8-1 When Alexeev lifted the weights above his head, he increased the separation between the weights and the Earth and thus changed the configuration of the weights–Earth system from that in (a) to that in (b).

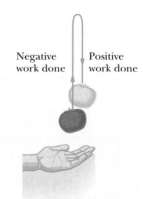

Negative work done · Positive work done

FIGURE 8-2 A tomato is thrown upward. As it rises, its weight does negative work on it, decreasing its kinetic energy by transferring energy to the gravitational potential energy of the tomato–Earth system. As the tomato descends, its weight does positive work on it, increasing its kinetic energy by transferring energy from the gravitational potential energy of the system.

netic energy of the tomato. We can now finish the story by saying that this energy is transferred by the weight *to* the gravitational potential energy of the tomato–Earth system.

The tomato slows, stops, and then begins to fall back down because of its weight. During the fall, the transfer is reversed: the work W_g done on the tomato by its weight is now positive—the weight transfers energy *from* the gravitational potential energy of the tomato–Earth system *to* the kinetic energy of the tomato.

For either rise or fall, the change ΔU in the gravitational potential energy is defined to equal the negative of the work done on the tomato by its weight. Using the general symbol W for work, we write this as

$$\Delta U = -W. \tag{8-1}$$

The same relationship applies to a block–spring system, as in Fig. 8-3. If we abruptly shove the block to send it rightward, the spring force does negative work on the block during the rightward motion, transferring energy from the kinetic energy of the block to the elastic potential energy of the spring. The block slows, stops, and then begins to move in the opposite direction because of the spring force. The transfer of energy is then reversed.

Conservative and Nonconservative Forces

Let us list the key elements of the two situations we just discussed:

1. The *system* consists of two or more objects.

2. A *force* acts between a particle-like object (tomato or block) in the system and the rest of the system.

3. When the system configuration changes, the force does *work* (call it W_1) on the particle-like object, transferring

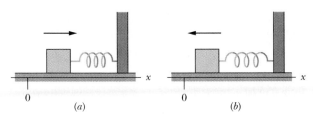

FIGURE 8-3 A block, attached to a spring and initially at rest at $x = 0$, is set in motion toward the right. (a) As the block moves rightward (as indicated by the arrow), the spring force does negative work on it, decreasing its kinetic energy by transferring energy to the elastic potential energy of the spring. The block momentarily stops when it has zero kinetic energy. (b) Then, as the block moves back toward $x = 0$, the spring force does positive work on it, increasing its kinetic energy by transferring energy from the elastic potential energy.

energy between the kinetic energy K of the object and some other form of energy of the system.

4. When the configuration change is reversed, the force reverses the energy transfer, doing work W_2 in the process.

If $W_1 = -W_2$ is always true, then the other form of energy is a potential energy, and the force is said to be a **conservative force.** As you might suspect, the gravitational force and the spring force are both conservative (since otherwise we could not have spoken of gravitational potential energy and elastic potential energy, as above.)

Not all forces are conservative. For an example, let us send a block sliding across a floor that is not frictionless. During the sliding, a kinetic frictional force from the floor does negative work on the block, slowing the block by transferring energy from its kinetic energy to thermal energy of the block–floor system. We know from experiment that this energy transfer cannot be reversed (thermal energy cannot be transferred back to kinetic energy of the block by the kinetic frictional force). So, although we have a system, a force that acts between parts of the system, and a transfer of energy by the force, the force is not conservative (it is **nonconservative**). Thus the thermal energy of the block–floor system is not a potential energy.

When only conservative forces act on a particle-like object, we can greatly simplify otherwise difficult problems involving motion of the object. The next section, in which we develop a test for identifying conservative forces, provides one means for simplifying such problems.

8-2 PATH INDEPENDENCE OF CONSERVATIVE FORCES

The primary test for determining whether a force is conservative or nonconservative is this: let the force act (alone) on a particle that moves along a *closed path* begin-

ning at some initial position and eventually returning to that position (so that the particle makes a *round trip* beginning and ending at the initial position). The force is conservative only if the total energy it transfers to and from the particle during the round trip is zero. In other words:

> The net work done by a conservative force on a particle moving around any closed path is zero.

We know from experiment that the gravitational force passes this test. An example is the tossed tomato of Fig. 8-2. The tomato leaves the launch point with a speed v_0 and kinetic energy $\frac{1}{2}mv_0^2$. The gravitational force acting on the tomato slows it, stops it, and then makes it fall back down. When the tomato returns to the launch point, it again has speed v_0 and kinetic energy $\frac{1}{2}mv_0^2$. Thus the gravitational force transfers as much energy *from* the tomato during the ascent as it transfers *to* the tomato during the descent back to the launch point. The net work done on the tomato by the gravitational force during the round trip is zero.

Figure 8-4a shows an arbitrary round trip for a particle acted upon by a single force. The particle moves from an initial point a to point b along path 1 and then back to point a along path 2. The force does work on the particle as the particle moves along each path. Without worrying about where positive work is done and where negative work is done, let us just represent the work done from a to b along path 1 as $W_{ab,1}$ and the work done from b back to a along path 2 as $W_{ba,2}$. If the force is conservative, then the net work done during the round trip must be zero:

$$W_{ab,1} + W_{ba,2} = 0,$$

and thus

$$W_{ab,1} = -W_{ba,2}. \tag{8-2}$$

In words, the work done along the outward path must be the negative of the work done along the path back.

Let us now consider the work $W_{ab,2}$ done on the particle by the force as the particle moves from a to b along path 2, as indicated in Fig. 8-4b. If the force is conservative, that

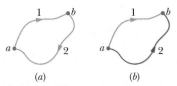

FIGURE 8-4 (a) A particle, acted on by a conservative force, moves in a round trip from point a to point b along path 1 and then back to point a along path 2. (b) The particle can move from point a to point b by following either path 1 or path 2.

work is the negative of $W_{ba,2}$:

$$W_{ab,2} = -W_{ba,2}. \qquad (8\text{-}3)$$

Substituting $W_{ab,2}$ for $-W_{ba,2}$ in Eq. 8-2, we obtain

$$W_{ab,1} = W_{ab,2}. \qquad (8\text{-}4)$$

This simple equation is a powerful one, because it allows us to simplify difficult problems when only a conservative force is involved. It tells us that the work done on a particle by a conservative force as the particle moves from some initial point a to some other point b does not depend on *which* path is taken.

> The work done by a conservative force on a particle moving between two points does not depend on the path taken by the particle.

Suppose you need to calculate the work done by a conservative force along a given path between two points, and the calculation is difficult. You can find the work by using another path between those two points for which the calculation is easier. Sample Problem 8-1 gives an example.

SAMPLE PROBLEM 8-1

Figure 8-5a shows a 2.0 kg block of slippery cheese that slides along a frictionless track from point a to point b. The cheese travels through a total distance of 2.0 m along the track, and a net vertical distance of 0.80 m. How much work is done on the cheese by its weight during the slide?

SOLUTION: From Eq. 7-16, we know that the work done by a weight force $m\mathbf{g}$, over a displacement $\mathbf{d}$ at angle ϕ to $m\mathbf{g}$, is given by $W = mgd \cos \phi$. But using that equation here appears to be impossible because we do not know the actual shape of the track. And even if we did, the calculation would be difficult because the angle ϕ is not constant over the path taken by the cheese.

However, because weight is a conservative force, we can find the work by choosing another path between a and b—one that makes the calculation easy. Let us choose the dashed path in Fig. 8-5b; it consists of two straight segments. Along the horizontal segment, the angle ϕ is a constant 90°. Even though we do not know the displacement along that horizontal segment, Eq. 7-16 tells us that the work W_h done there is

$$W_h = mgd \cos 90° = 0.$$

Along the vertical segment, the displacement d is 0.80 m and, with $m\mathbf{g}$ and $\mathbf{d}$ both downward, the angle ϕ is a constant 0°. So Eq. 7-16 gives us, for the work W_v done along the vertical part of the dashed path,

$$W_v = mgd \cos 0°$$
$$= (2.0 \text{ kg})(9.8 \text{ m/s}^2)(0.80 \text{ m})(1) = 15.7 \text{ J}.$$

The total work done on the cheese by its weight as the cheese moves from point a to point b along the dashed path is then

$$W = W_h + W_v = 0 + 15.7 \text{ J} \approx 16 \text{ J}. \quad \text{(Answer)}$$

This is also the work done as the cheese moves along the track from a to b.

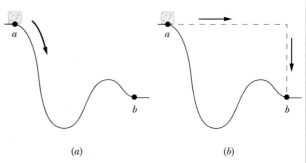

(a) (b)

FIGURE 8-5   Sample Problem 8-1. (a) A block of cheese slides along a frictionless track from point a to point b. (b) Finding the work done on the cheese by its weight is easier along the dashed path than along the actual path taken by the cheese; the result is the same for both paths.

CHECKPOINT 1: The figure shows three paths connecting points a and b. A single force $\mathbf{F}$ does the indicated work on a particle moving along each path in the indicated direction. On the basis of this information, is force $\mathbf{F}$ conservative?

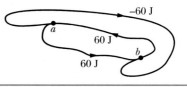

8-3 DETERMINING POTENTIAL ENERGY VALUES

Consider now a particle-like object (say, a melon) that is part of a system (say, a melon–Earth system) in which a conservative force $\mathbf{F}$ acts. When that force does work W on the object, the change ΔU in the potential energy associated with the system is the negative of the work done. We wrote this fact as Eq. 8-1 ($\Delta U = -W$). For the most general case, in which the force may vary with position, we may write the work W as in Eq. 7-27:

$$W = \int_{x_i}^{x_f} F(x) \, dx. \qquad (8\text{-}5)$$

This equation gives the work done by the force when the object moves from point x_i to point x_f, changing the configuration of the system. (Because F is a conservative force, the work is the same for all paths between those two points.)

Substituting Eq. 8-5 into Eq. 8-1, we find that the change in potential energy due to the change in configuration is

$$\Delta U = -\int_{x_i}^{x_f} F(x)\, dx. \qquad (8\text{-}6)$$

Gravitational Potential Energy

Now consider a particle with mass m moving vertically along a y axis. As the particle moves from point y_i to point y_f, its weight $m\mathbf{g}$ does work on it. To find the corresponding change in the gravitational potential energy of the particle–Earth system, we change the integration in Eq. 8-6 to be along the y axis and we substitute $-mg$ for force F. We then have

$$\Delta U = -\int_{y_i}^{y_f} (-mg)\, dy = mg \int_{y_i}^{y_f} dy = mg \left[\, y\, \right]_{y_i}^{y_f},$$

which yields

$$\Delta U = mg(y_f - y_i) = mg\, \Delta y. \qquad (8\text{-}7)$$

Only a change ΔU in gravitational potential energy (or any other type of potential energy) is physically important. However, to simplify a calculation or a discussion, we can say that a certain gravitational potential energy U is associated with any given configuration of the system, with the particle at a given position y. To do so, we rewrite Eq. 8-7 as

$$U - U_i = mg(y - y_i). \qquad (8\text{-}8)$$

Then we take U_i to be the gravitational potential energy of the system when it is in a **reference configuration,** with the particle at a **reference point** y_i. Usually, we set $U_i = 0$ and $y_i = 0$. Doing this changes Eq. 8-8 to

$$U - 0 = mg(y - 0),$$

which gives us

$$U = mgy \qquad \begin{array}{l}\text{(gravitational}\\ \text{potential energy).}\end{array} \qquad (8\text{-}9)$$

Equation 8-9 tells us that the gravitational potential energy associated with a particle–Earth system depends on the vertical position y (or height) of the particle relative to the reference position of $y = 0$, not on the horizontal position.

Elastic Potential Energy

We next consider the block–spring system shown in Fig. 8-3, with the block moving on the end of a spring of spring constant k. As the block moves from point x_i to point x_f, the spring force $F = -kx$ does work on the block. To find the

corresponding change in the elastic potential energy of the block–spring system, we substitute $-kx$ for $F(x)$ in Eq. 8-6. We then have

$$\Delta U = -\int_{x_i}^{x_f} (-kx)\, dx = k \int_{x_i}^{x_f} x\, dx = \tfrac{1}{2}k \left[\, x^2\, \right]_{x_i}^{x_f},$$

and

$$\Delta U = \tfrac{1}{2}kx_f^2 - \tfrac{1}{2}kx_i^2. \qquad (8\text{-}10)$$

To associate a potential energy U with any given configuration of the system, with the block at position x, we set the reference point for the block as $x_i = 0$, which is always at the equilibrium position of the block. And we set the corresponding elastic potential energy of the system as $U_i = 0$. Thus Eq. 8-10 becomes

$$U - 0 = \tfrac{1}{2}kx^2 - 0,$$

which gives us

$$U(x) = \tfrac{1}{2}kx^2 \qquad \begin{array}{l}\text{(elastic potential}\\ \text{energy).}\end{array} \qquad (8\text{-}11)$$

PROBLEM SOLVING TACTICS

TACTIC 1: *Using the Term "Potential Energy"*

A potential energy is associated with a system as a whole. However, you might see statements that associate it with only part of the system. For example, you might read, "An apple hanging in a tree has a gravitational potential energy of 30 J." Such statements are often acceptable, but you should always keep in mind that the potential energy is actually associated with a system—here the apple–Earth system. Also keep in mind that assigning a particular potential energy value, such as 30 J here, to an object or even a system makes sense *only* if the reference potential energy value is known.

SAMPLE PROBLEM 8-2

A 2.0 kg sloth clings to a limb that is 5.0 m above the ground (Fig. 8-6).

(a) What is the gravitational potential energy U of the sloth–Earth system if we take the reference point $y = 0$ to be (1) at the ground, (2) at the bottom of a balcony that is 3.0 m above the ground, (3) at the limb, and (4) 1.0 m above the limb? Take the gravitational potential energy to be zero at $y = 0$.

SOLUTION: With Eq. 8-9, we can calculate U for each choice of $y = 0$. For example, for (1) the sloth is initially at $y = 5.0$ m, and

$$U = mgy = (2.0\text{ kg})(9.8\text{ m/s}^2)(5.0\text{ m})$$

= 98 J   (Answer)

For the other choices, the values of U are

(2) $U = mgy = mg(2.0 \text{ m}) = 39$ J,

(3) $U = mgy = mg(0) = 0$ J,

(4) $U = mgy = mg(-1.0 \text{ m}) = -19.6$ J

   ≈ -20 J.   (Answer)

(b) The sloth drops to the ground. For each choice of reference point, what is the change in the potential energy of the sloth–Earth system due to the fall?

SOLUTION: For all four situations, we have $\Delta y = -5.0$ m. So for (1) to (4), Eq. 8-7 tells us that

$\Delta U = mg \, \Delta y = (2.0 \text{ kg})(9.8 \text{ m/s}^2)(-5.0 \text{ m})$

   $= -98$ J.   (Answer)

Thus, although the value of U depends on the choice of where we let $y = 0$, the *change* in potential energy does not. Remember, only the change ΔU in potential energy is physically important—not the value of U, which depends on arbitrary choices about the reference configuration.

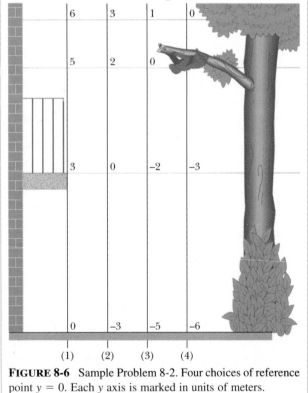

FIGURE 8-6   Sample Problem 8-2. Four choices of reference point $y = 0$. Each y axis is marked in units of meters.

CHECKPOINT 2: A particle is to move along the x axis from $x = 0$ to x_1 while a conservative force, directed along the x axis, acts on the particle. The figure shows three situations in which the magnitude of that force varies with x. The force has the same maximum magnitude F_1 in all three situations. Rank the situations

according to the change in the associated potential energy during the particle's motion, most positive first, most negative last.

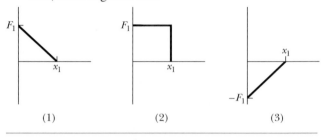

8-4 CONSERVATION OF MECHANICAL ENERGY

The **mechanical energy** E of a system is the sum of its potential energy U and the kinetic energy K of the objects within it:

$$E = K + U \quad \text{(mechanical energy).} \quad (8\text{-}12)$$

In this section, we examine what happens to this mechanical energy when a conservative force acts within the system. (Later we shall consider the effects of nonconservative forces.)

When a conservative force does work W on an object within the system, it transfers energy between kinetic en-

In olden days, a native Alaskan would be tossed via a blanket to be able to see farther over the flat terrain. Nowadays, it is done just for fun. During the ascent of the child in the photograph, energy is transferred from kinetic energy to gravitational potential energy. The maximum height is reached when that transfer is complete. Then the transfer is reversed during the fall.

ergy K of the object and potential energy U of the system. From Eq. 7-4, the change ΔK in kinetic energy is

$$\Delta K = W \qquad (8\text{-}13)$$

and from Eq. 8-1, the change ΔU in potential energy is

$$\Delta U = -W. \qquad (8\text{-}14)$$

Combining Eqs. 8-13 and Eq. 8-14, we find that

$$\Delta K = -\Delta U. \qquad (8\text{-}15)$$

In other words, one of these forms of energy increases exactly as much as the other decreases.

We can rewrite Eq. 8-15 as

$$K_2 - K_1 = -(U_2 - U_1), \qquad (8\text{-}16)$$

where the subscripts 1 and 2 refer to two different instants during the particle's motion, and thus to two different configurations of the system. Rearranging Eq. 8-16 yields

$$K_2 + U_2 = K_1 + U_1 \qquad \begin{array}{l}\text{(conservation of}\\ \text{mechanical energy).}\end{array} \qquad (8\text{-}17)$$

The left side of Eq. 8-17 is the mechanical energy of the system at one instant; the right side is the mechanical energy at another instant and in a different configuration. The equation tells us that these two energies are equal:

> When only a conservative force acts within a system, the kinetic energy and potential energy can change. However, their sum, the mechanical energy E of the system, does not change.

This result is called the **principle of conservation of mechanical energy**. (Now you can see where *conservative* forces got their name.) With the aid of Eq. 8-15, we can write this principle in one more form, as

$$\Delta E = \Delta K + \Delta U = 0. \qquad (8\text{-}18)$$

The principle of conservation of mechanical energy allows us to solve problems that would be quite difficult to solve using only Newton's laws:

> When the mechanical energy of a system is conserved, we can relate the total of kinetic energy and potential energy at one instant to that at another instant *without considering the intermediate motion* and *without finding the work done by the forces involved.*

Figure 8-7 shows an example in which the principle of conservation of mechanical energy can be applied: as a pendulum swings, the energy of the pendulum–Earth system is transferred back and forth between kinetic energy K and gravitational potential energy U, with the sum $K + U$ being constant. If we are given the gravitational potential energy when the pendulum bob is at its highest point (Fig. 8-7c), we can find the kinetic energy of the bob at the lowest point (Fig. 8-7e) by using Eq. 8-17.

For example, let us choose the lowest point as the reference point and set the corresponding gravitational potential energy $U_2 = 0$. Suppose then that the potential energy at the highest point is $U_1 = 20$ J relative to the reference point. Because the bob momentarily stops at its highest point, the kinetic energy there is $K_1 = 0$. Substituting these values into Eq. 8-17 gives us the kinetic energy K_2 at the lowest point:

$$K_2 + 0 = 0 + 20 \text{ J} \quad \text{or} \quad K_2 = 20 \text{ J}.$$

Note that we get this result without considering the motion between the highest and lowest points (such as in Fig. 8-7d) and without finding the work done by any forces involved in the motion.

CHECKPOINT **3:** The figure shows four situations—one in which an initially stationary block is dropped and three in which the block is allowed to slide down frictionless ramps. (a) Rank the situations according to the kinetic energy of the block at point B, greatest first. (b) Rank them according to the speed of the block at point B, greatest first.

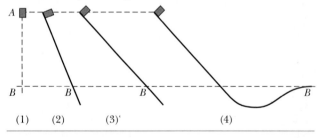

(1) (2) (3)' (4)

SAMPLE PROBLEM 8-3

In Fig. 8-8, a child of mass m is released from rest at the top of a water slide, at height $h = 8.5$ m above the bottom of the slide. Assuming that the slide is frictionless because of the water on it, find the child's speed at the bottom of the slide.

SOLUTION: If we use only the physics of Chapters 2 through 6, this problem is impossible to solve because we are given no information about the shape of the slide. However, with the physics of Chapters 7 and 8, it is easy. We first note that, in the absence of friction, the only force exerted on the child by the slide is a normal force, which is always perpendicular to the surface of the slide. Since the child always moves *along* the slide, this force is always perpendicular to

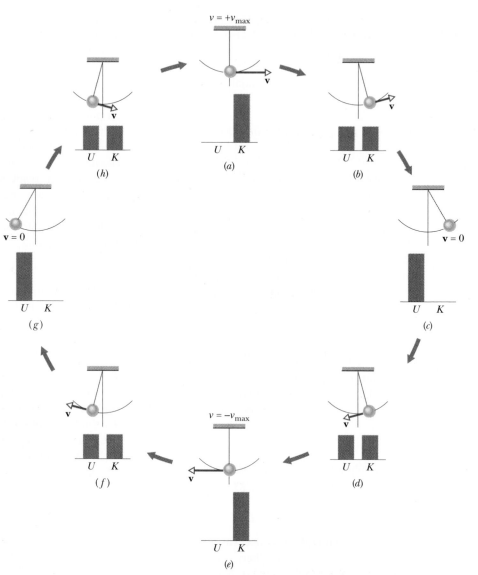

FIGURE 8-7 A pendulum, with its mass concentrated in a bob at the lower end, swings back and forth. One full cycle of the motion is shown. During the cycle the values of the potential and kinetic energies of the pendulum–Earth system vary as the bob rises and falls, but the mechanical energy E of the system remains constant. The energy E can be described as continuously shifting between the kinetic and potential forms. In stages (a) and (e), all the energy is kinetic energy. The bob then has its greatest speed and is at its lowest point. In stages (c) and (g), all the energy is potential energy. The bob then has zero speed and is at its highest point. In stages (b), (d), (f), and (h), half the energy is kinetic energy and half is potential energy. If the swinging involved a frictional force at the point where the pendulum is attached to the ceiling, or a drag force due to the air, then E would not be conserved, and eventually the pendulum would stop.

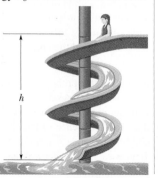

FIGURE 8-8 Sample Problem 8-3. A child slides down a water slide as she descends a height h.

the child's displacement and thus cannot do work on the child. The only force that does work is the weight mg of the child, a conservative force. We therefore have a child–Earth system for which the mechanical energy E is conserved throughout the child's motion.

Thus, the mechanical energy E_b at the bottom of the slide and the mechanical energy E_t at the top are equal: $E_b = E_t$. Writing this conservation of energy equation in the form of Eq. 8-17 gives us

$$K_b + U_b = K_t + U_t,$$

or

$$\tfrac{1}{2}mv_b^2 + mgy_b = \tfrac{1}{2}mv_t^2 + mgy_t.$$

Dividing by m and rearranging yield

$$v_b^2 = v_t^2 + 2g(y_t - y_b).$$

Putting $v_t = 0$ and $y_t - y_b = h$ leads to

$$v_b = \sqrt{2gh} = \sqrt{(2)(9.8 \text{ m/s}^2)(8.5 \text{ m})}$$
$$= 13 \text{ m/s.} \qquad \text{(Answer)}$$

This is the same speed that the child would reach if she fell 8.5 m. On an actual slide, some frictional forces would act and the child would not be moving quite so fast.

This problem is hard to solve directly with Newton's laws. Using conservation of mechanical energy makes the solution much easier. However, if you were asked to find the time taken for the child to reach the bottom of the slide, energy methods would be of no use; you would need to know the shape of the slide and you would have a difficult problem.

SAMPLE PROBLEM 8-4

The spring of a spring gun is compressed a distance d of 3.2 cm from its relaxed state, and a ball of mass $m = 12$ g is put in the barrel. With what speed will the ball leave the barrel when the gun is fired? The spring constant k is 7.5 N/cm. Assume no friction and a horizontal gun barrel. Also assume that the ball leaves the spring and the spring stops when the spring reaches its relaxed length.

SOLUTION: Let E_i be the mechanical energy of the gun–ball system in the initial state (before the gun has been fired), and E_f be the system's mechanical energy in the final state (as the ball leaves the barrel). Initially, the mechanical energy is the spring's potential energy $U_i = \frac{1}{2}kd^2$. In the final state, the mechanical energy is the ball's kinetic energy $K_f = \frac{1}{2}mv^2$. Because mechanical energy is conserved, we have

$$E_i = E_f,$$

$$U_i + K_i = U_f + K_f,$$

and $\frac{1}{2}kd^2 + 0 = 0 + \frac{1}{2}mv^2.$

Solving for v yields

$$v = d\sqrt{\frac{k}{m}} = (0.032 \text{ m})\sqrt{\frac{750 \text{ N/m}}{12 \times 10^{-3} \text{ kg}}}$$

$$= 8.0 \text{ m/s.} \qquad \text{(Answer)}$$

SAMPLE PROBLEM 8-5

A 61.0 kg bungee-cord jumper is on a bridge 45.0 m above a river. In its relaxed state, the elastic bungee cord has length $L = 25.0$ m. Assume that the cord obeys Hooke's law, with a spring constant of 160 N/m.

(a) If the jumper stops before reaching the water, what is the height h of her feet above the water at her lowest point?

SOLUTION: As shown in Fig. 8-9, let d be the amount by which the cord has stretched when she momentarily stops at her lowest point. As a result of her jump, the change ΔU_g in her gravitational potential energy is

$$\Delta U_g = mg \, \Delta y = -mg(L + d),$$

where m is her mass. The change in the elastic potential energy U_e of the cord is

$$\Delta U_e = \frac{1}{2}kd^2.$$

Her kinetic energy K is zero both initially and when she stops.

From Eq. 8-18 and the above, we have for the cord–jumper–Earth system,

$$\Delta K + \Delta U_e + \Delta U_g = 0$$

$$0 + \frac{1}{2}kd^2 - mg(L + d) = 0$$

$$\frac{1}{2}kd^2 - mgL - mgd = 0.$$

Inserting the given data, we obtain

$$\frac{1}{2}(160 \text{ N/m})d^2 - (61.0 \text{ kg})(9.8 \text{ m/s}^2)(25.0 \text{ m})$$

$$- (61.0 \text{ kg})(9.8 \text{ m/s}^2)d = 0,$$

which we solve for d with the quadratic formula, getting

$$d = 17.9 \text{ m.}$$

(The quadratic formula also gives a negative value for d, which has no meaning here.) Her feet are then a distance of $(L + d) = 42.9$ m below their initial height. Thus

$$h = 45.0 \text{ m} - 42.9 \text{ m} = 2.1 \text{ m.} \qquad \text{(Answer)}$$

If she happens to be especially tall, the jumper could be in for a dunking.

(b) What is the net force on her at her lowest point (in particular, is it zero)?

SOLUTION: Her weight mg acts downward and has magnitude $mg = 597.8$ N. The upward force exerted on her by the cord at the moment of stopping is given by Hooke's law, $F = -kx$, where x is the displacement of the free end of the cord. Here, the displacement is downward, so $x = -d$ and

$$F = -kx = -(160 \text{ N/m})(-17.9 \text{ m}) = 2864 \text{ N.}$$

The net force on the jumper is then

$$2864 \text{ N} - 597.8 \text{ N} \approx 2270 \text{ N.} \qquad \text{(Answer)}$$

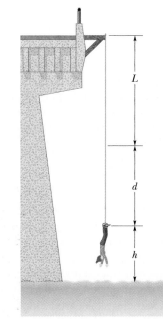

FIGURE 8-9   Sample Problem 8-5. A bungee-cord jumper at the lowest point of the jump.

Thus at the lowest point, where she momentarily stops, there is a net upward force of 2270 N (almost four times her weight) acting on her. This force will yank her back upward.

PROBLEM SOLVING TACTICS

TACTIC 2: *Conservation of Mechanical Energy*

Asking the following questions will help you to solve problems involving the conservation of mechanical energy.

For what system is mechanical energy conserved? You should be able to separate your system from its environment. Imagine drawing a closed surface such that whatever is inside is your system and whatever is outside is the environment of that system. In Sample Problem 8-3 the system is the *child + Earth*. In Sample Problem 8-4, it is the *ball + gun*. In Sample Problem 8-5, it is the *cord + jumper + Earth*.

Is friction or drag present? If friction or drag is present, mechanical energy is not conserved.

Is your system isolated? Conservation of mechanical energy applies only to isolated systems. That means that no *external forces* (forces exerted by objects outside the system) should do work on the objects in the system. In Sample Problem 8-4, if you chose the ball alone as your system, it is not isolated because the spring force does work on it. Thus the mechanical energy of the ball is not conserved.

What are the initial and final states of your system? The system changes from some initial state (or configuration) to some final state. You apply the principle of conservation of mechanical energy by saying that E has the same value in both these states. Be very clear about what these two states are.

8-5 READING A POTENTIAL ENERGY CURVE

Once again consider a particle that is part of a system in which a conservative force acts. This time suppose that the particle is constrained to move along an x axis while the conservative force does work on it. We can learn a lot about the motion of the particle from a plot of the system's potential energy $U(x)$. However, before we discuss such plots, we need one more relationship.

Finding the Force Analytically

Equation 8-6 tells us how to find the potential energy $U(x)$ in a one-dimensional situation if we know the force $F(x)$. But now we want to go the other way. That is, we know the potential energy function $U(x)$ and want to find the force.

For one-dimensional motion, the work W done by a force that acts on a particle as the particle moves through a distance Δx is $F(x)\,\Delta x$. We can then write Eq. 8-1 as

$$\Delta U(x) = -W = -F(x)\,\Delta x.$$

Solving for $F(x)$ and passing to the differential limit yield

$$F(x) = -\frac{dU(x)}{dx} \qquad \text{(one-dimensional motion).} \qquad (8\text{-}19)$$

We can check this result by putting $U(x) = \frac{1}{2}kx^2$, which is the elastic potential energy function for a spring force. Equation 8-19 then yields, as expected, $F(x) = -kx$, which is Hooke's law. Similarly, we can substitute $U(x) = mgx$, which is the gravitational potential energy function for a particle of mass m at height x above the Earth's surface. Equation 8-19 then yields $F = -mg$, which is the weight that acts on the particle.

The Potential Energy Curve

Figure 8-10*a* is a plot of a potential energy function $U(x)$ for a system in which a particle is in one-dimensional motion while a conservative force $F(x)$ does work on it. We can easily find $F(x)$ by (graphically) taking the slope of the $U(x)$ curve at various points, as Eq. 8-19 tells us to do. Figure 8-10*b* is a plot of $F(x)$ found in this way.

Turning Points

In the absence of a nonconservative force, the mechanical energy E of the system has a constant value given by

$$U(x) + K(x) = E. \qquad (8\text{-}20)$$

Here $K(x)$ is the *kinetic energy function* of the particle (this $K(x)$ gives the kinetic energy as a function of the particle's location x). We may rewrite Eq. 8-20 as

$$K(x) = E - U(x). \qquad (8\text{-}21)$$

Suppose that E (which has a constant value, remember) happens to be 5.0 J. It would be represented in Fig. 8-10 by a horizontal line that runs through the value 5.0 J on the energy axis. (It is shown in Fig. 8-10*a*.)

Equation 8-21 tells us how to determine the kinetic energy K for any location x of the particle: on the $U(x)$ curve, find U for that location x and then subtract U from E. For example, if the particle is at any point to the right of x_5, then $K = 1.0$ J. The value of K is greatest (5.0 J) when the particle is at x_2, and least (0 J) when the particle is at x_1.

Since K can never be negative (because v^2 is always positive), the particle can never move to the left of x_1, where $E - U$ is negative. As the particle moves toward x_1 from x_2, K decreases (the particle slows) until $K = 0$ at x_1 (the particle stops).

Note that when the particle reaches x_1, the force on the particle, given by Eq. 8-19, is positive (because the slope dU/dx is negative). This means that the particle does not remain at x_1 but instead begins to move to the right, opposite its earlier motion. Hence x_1 is a **turning point,** a place where $K = 0$ and the particle changes direction. There is

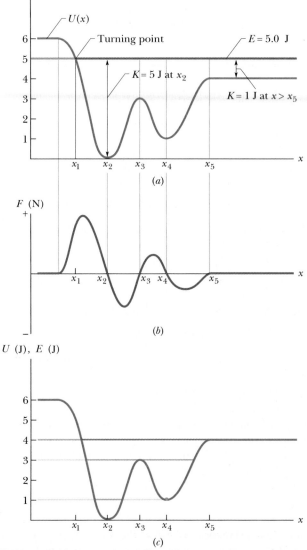

FIGURE 8-10 (*a*) A plot of $U(x)$, the potential energy function of a system containing a particle confined to move along the *x* axis. There is no friction, so mechanical energy is conserved. (*b*) A plot of the force $F(x)$ acting on the particle, derived from the potential energy plot by taking its slope at various points. (*c*) The $U(x)$ plot of (*a*) with three possible values of E shown.

no turning point (where $K = 0$) on the right side of the graph. When the particle heads toward increasing values of *x*, it will continue indefinitely.

Equilibrium Points

Figure 8-10*c* shows three additional values for E superposed on the plot of the potential energy $U(x)$. Let us see how they would change the situation. If $E = 4.0$ J (purple line), the turning point shifts from x_1 to a point between x_1 and x_2. Also, at any point to the right of x_5, the system's

mechanical energy is equal to its potential energy; thus, the particle has no kinetic energy and (by Eq. 8-19) no force acts on it, and so it must be stationary there. A particle at such a position is said to be in **neutral equilibrium.** (A marble placed on a horizontal tabletop is in that state.)

If $E = 3.0$ J (pink line), there are two turning points: one is between x_1 and x_2, and the other is between x_4 and x_5. In addition, x_3 is a point at which $K = 0$. If the particle is located exactly there, the force on it is also zero, and the particle remains stationary. However, if it is displaced even slightly in either direction, a nonzero force pushes it farther in the same direction, and the particle continues to move. A particle at such a position is said to be in **unstable equilibrium.** (A marble balanced on top of a bowling ball is an example.)

Next consider the particle's behavior if $E = 1.0$ J (green line). If we place it at x_4, it is stuck there. It cannot move left or right on its own because to do so would require a negative kinetic energy. If we push it slightly left or right, a restoring force appears that moves it back to x_4. A particle at such a position is said to be in **stable equilibrium.** (A marble placed at the bottom of a hemispherical bowl is an example.) If we place the particle in the cuplike *potential well* centered at x_2, it can still move left and right somewhat, but only partway to x_1 or x_3.

CHECKPOINT **4:** The figure gives the potential energy function $U(x)$ for a system in which a particle is in one-dimensional motion. (a) Rank regions *AB*, *BC*, and *CD* according to the magnitude of the force on the particle, greatest first. (b) What is the direction of the force when the particle is in region *AB*?

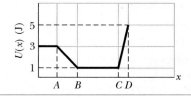

8-6 WORK DONE BY NONCONSERVATIVE FORCES

When only a conservative force does work on an object within a system, the mechanical energy E of the system stays constant. However, when a nonconservative force does work on an object, the mechanical energy of the system changes. In that situation we *cannot* use Eqs. 8-17 and 8-18. What, then, can we say about the energy changes? We answer this question by examining nonconservative forces of two types: an applied force (applied by, say, you) and a kinetic frictional force.

Work Done by an Applied Force

Suppose you push a bowling ball directly upward. During the upward motion and before the ball leaves your hand, your applied force does work W_{app} on the ball while the weight of the ball does work W_g on the ball. From observation we find that the ball can be treated as a particle. Thus, we can relate the work done on the ball to the change ΔK in its kinetic energy by using the work–kinetic energy theorem (Eq. 7-15):

$$W_{app} + W_g = \Delta K. \qquad (8\text{-}22)$$

Because the weight of the ball is a conservative force, we can use Eq. 8-1 to relate W_g to the corresponding change ΔU in the gravitational potential energy of the ball–Earth system:

$$W_g = -\Delta U. \qquad (8\text{-}23)$$

Substituting for W_g in Eq. 8-22, we get

$$W_{app} = \Delta K + \Delta U. \qquad (8\text{-}24)$$

The right side of this equation is the change ΔE in the mechanical energy of the ball–Earth system, so

$$W_{app} = \Delta E. \qquad (8\text{-}25)$$

Equation 8-25 says that the work done by your applied force changes the mechanical energy of the ball–Earth system. Specifically, your force transfers energy to the kinetic energy of the ball (you make it move faster) and to the gravitational potential energy of ball–Earth system (you increase the separation between the ball and the Earth).

Previously we defined work as being energy transferred to or from an object via a force. Now we can expand that definition to include situations in which the object is part of a system:

> Work is energy transferred to or from a system via a force.

In some situations, the system consists of just an object, as in Chapter 7; in other situations, the system consists of two (or more) objects, as in the ball–Earth system here.

Work Done by a Kinetic Frictional Force

Figure 8-11 shows a block of mass m and initial speed v_0 sliding across a floor that is not frictionless. A kinetic frictional force $\mathbf{f}_k$ stops the block during displacement $\mathbf{d}$. From experiment we know that this force reduces the kinetic energy K of the block by transferring energy to thermal energy E_{int} of the block and the portion of the floor along which the block slides. Thermal energy is said to be an

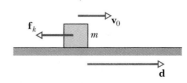

FIGURE 8-11 A block of mass m and initial speed v_0 slides across a floor. A kinetic frictional force $\mathbf{f}_k$ does work on the block, stopping it during displacement $\mathbf{d}$.

internal energy of an object because it involves random motions of the atoms and molecules within the object; hence, we represent it with the symbol E_{int}.

Because we also know from experiment that the transfer to thermal energy is not reversible, we say that the block's kinetic energy is **dissipated** by force $\mathbf{f}_k$. The work W_f done on the block by $\mathbf{f}_k$ is not the entire amount of dissipated energy, but only the part that is transferred *from* the block to the floor. The rest of the dissipated energy remains within the block as thermal energy.

As an example, suppose we find experimentally that the change in kinetic energy of the block is $\Delta K = -100$ J. Then 100 J is dissipated by $\mathbf{f}_k$, and the change in thermal energy of the block and floor is $\Delta E_{int} = +100$ J. Suppose we also find experimentally that 40 J is transferred (internally) to thermal energy of the block and 60 J is transferred (externally) to thermal energy of the floor. Then the work W_f done on the block by $\mathbf{f}_k$ is $W_f = -60$ J.

The situation in Fig. 8-11 is comparable to that in Fig. 7-2 (Section 7-3), so Eq. 7-8 applies here with two changes. The force magnitude is now f_k instead of F, and the angle ϕ between the displacement and the force is $180°$. With those changes, Eq. 7-8 becomes

$$\Delta K = f_k d \cos 180° = -f_k d. \qquad (8\text{-}26)$$

Thus the product $-f_k d$ tells us the amount by which $\mathbf{f}_k$ decreases the kinetic energy K of the block in Fig. 8-11. If a potential energy were also involved (say, the block were sliding along a ramp), the product $-f_k d$ would tell us the amount by which $\mathbf{f}_k$ decreases (by dissipating) the mechanical energy E of the system:

$$\Delta E = -f_k d \qquad \text{(dissipated mechanical energy).} \qquad (8\text{-}27)$$

Only a portion of this dissipated energy is transferred from the block to the floor, and only that portion is equal to the work W_f done by $\mathbf{f}_k$. *Without additional information, we cannot determine W_f.*

SAMPLE PROBLEM 8-6

Figure 8-12 shows a disabled robot of mass $m = 40$ kg being dragged by a cable up the $30°$ inclined wall inside a volcano crater. The applied force **F** exerted on the robot by the cable has a magnitude of 380 N. The kinetic frictional force $\mathbf{f}_k$ acting on the robot has a magnitude of 140 N. The robot moves through a displacement $\mathbf{d}$ of magnitude 0.50 m along the crater wall.

(a) How much of the mechanical energy of the robot–Earth system is dissipated by the kinetic frictional force $\mathbf{f}_k$ during displacement $\mathbf{d}$?

SOLUTION: From Eq. 8-27 and the given data, we have

$$\Delta E = -f_k d = -(140 \text{ N})(0.50 \text{ m}) = -70 \text{ J}. \quad \text{(Answer)}$$

This result means that during the displacement, 70 J is transferred to thermal energy of the robot and the wall along which it slides.

(b) What is the work W_g done on the robot by its weight during the displacement?

SOLUTION: From Eqs. 8-1 and 8-7, we know that

$$W_g = -\Delta U = -mg \, \Delta y. \quad (8\text{-}28)$$

In Fig. 8-12, the change Δy in the height of the robot is $h = d \sin 30°$. Substituting this and given data into Eq. 8-28, we find

$$\begin{aligned} W_g &= -mgh = -mgd \sin 30° \\ &= -(40 \text{ kg})(9.8 \text{ m/s}^2)(0.50 \text{ m})(0.5) \\ &= -98 \text{ J}. \quad \text{(Answer)} \end{aligned}$$

This result means that during the displacement, 98 J is transferred to gravitational potential energy of the robot–Earth system.

(c) What is the work W_{app} done by the applied force $\mathbf{F}$?

SOLUTION: We use Eq. 7-9 ($W = Fd \cos \phi$) to find the work W_{app} done by $\mathbf{F}$ in Fig. 8-12. The angle ϕ between $\mathbf{F}$ and $\mathbf{d}$ is 0°. Substituting this and given data into Eq. 7-9, we find

$$\begin{aligned} W_{\text{app}} &= Fd \cos \phi = (380 \text{ N})(0.50 \text{ m}) \cos 0° \\ &= 190 \text{ J}. \quad \text{(Answer)} \end{aligned}$$

This result means that during the displacement, 190 J is transferred by the applied force $\mathbf{F}$ to the robot–Earth system.

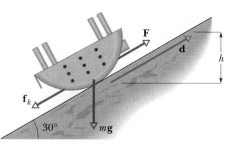

FIGURE 8-12   Sample Problem 8-6. A disabled robot is dragged up the wall inside a volcano crater, through a displacement $\mathbf{d}$ and vertical distance h, by force $\mathbf{F}$. A kinetic frictional force $\mathbf{f}_k$ opposes the motion.

CHECKPOINT 5: The figure shows three choices for the orientation of a plane that is not frictionless and for the direction in which a block slides along the plane. The block begins with the same speed in all three choices and slides until the kinetic frictional force has stopped it. Rank the choices according to the amount of mechanical energy that is dissipated, greatest first.

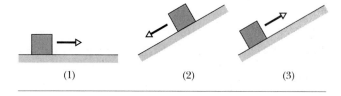

(1)   (2)   (3)

8-7   CONSERVATION OF ENERGY

Picture a block sliding over a frictionless floor. We can consider the block to be an isolated system in the sense that no external force transfers energy into or out of the system. So, a conservation principle applies to our system (albeit a simple one): the energy of the block is conserved.

Now suppose the block slides over a floor that is not frictionless. If we again take the block to be our system, we must say that the system is no longer isolated: Now an external force—the kinetic frictional force—acts on the block, and part of the energy transferred by that force is transferred to the floor. So now the energy of the system (the block) is not conserved.

However, we can expand the system to include both the block and the floor. For this new system, the kinetic frictional force is no longer an external force, and so it no longer transfers energy from the system, only within the system. So again we have an isolated system within which energy is conserved. Expanding what we mean by *the system* allows us to write a new conservation principle, one that helps us account for the energy transfers.

To find this new conservation principle, we use Eq. 8-27 to write the amount of mechanical energy dissipated by $\mathbf{f}_k$. With the block sliding across the floor, the only mechanical energy involved is kinetic energy, so we write Eq. 8-27 as

$$\Delta K = -f_k d, \quad (8\text{-}29)$$

in which ΔK is the change (decrease) in kinetic energy. This entire amount of energy is transferred to thermal energy of the block and floor. Thus, the change ΔE_{int} in thermal energy is

$$\Delta E_{\text{int}} = -\Delta K, \quad (8\text{-}30)$$

which gives us

$$\Delta K + \Delta E_{\text{int}} = 0. \quad (8\text{-}31)$$

So, although the mechanical energy of the block is not conserved, the sum of the mechanical energy of the block and the thermal energy of the block and floor *is* conserved. That sum is called the **total energy** E_{tot} of the block–floor system, and our new conservation principle is called the **law of conservation of energy.** Using Eq. 8-31, we can write this law as

$$\Delta E_{tot} = \Delta K + \Delta E_{int} = 0. \qquad (8\text{-}32)$$

If, in addition, conservative forces act on parts of our system, we can again expand the meaning of *the system* to include any transfers to and from potential energy by those forces. Then the total energy of this new, expanded isolated system does not change and our conservation law is written as

$$\Delta E_{tot} = \Delta K + \Delta U + \Delta E_{int} = 0 \qquad \begin{matrix}\text{(isolated system,}\\ \text{conservation of}\\ \text{total energy).}\end{matrix}$$
$$(8\text{-}33)$$

If we discover other forms of energy, we can always include them in our conservation law by rewriting Eq. 8-33 as

$$\Delta E_{tot} = \Delta K + \Delta U + \Delta E_{int}$$
$$+ \begin{pmatrix}\text{changes in other}\\ \text{forms of energy}\end{pmatrix} = 0. \qquad (8\text{-}34)$$

Note that this law of conservation of energy is *not* something we derived. Instead, it is a law based on countless experiments. Scientists and engineers have never found an exception to it. The law is extremely powerful in analyzing difficult situations because it tells us something that does not change when various types of energies *are* changing. In words, it says:

In an isolated system, energy can be transferred from one type to another, but the total energy of the system remains constant.

In less formal wording, it says: energy cannot magically appear or disappear.

For example, in Fig. 8-13, we can consider the Earth and the climber and her gear to be one system. As the climber rappels down the rock face, changing the configuration of the system, she needs to control the transfer of energy from the gravitational potential energy of the system (it cannot just magically disappear). Some of it is transferred to her kinetic energy, but she obviously does not want much transferred to that form of energy. So she has wrapped the rope around metal rings. The resulting frictional force between the rope and the rings dissipates much of the mechanical energy of the system in a controlled way by slowly transferring energy to thermal energy of the rope and the rings.

If a system is not isolated and external forces transfer energy to or from the system, then Eqs. 8-33 and 8-34 do not apply. Instead we account for those transfers by writing

$$W = \Delta E_{tot} = \Delta K + \Delta U + \Delta E_{int}, \qquad (8\text{-}35)$$

in which W is the work done on the system by the external forces. For example, in Fig. 8-13, if we consider the rope to be external to the system, then the frictional force exerted by the rope on the metal rings of the system does work W on the system, transferring energy from the system to thermal energy of the rope while, within the system, the values of K, U, and E_{int} change.

Power

Now that you have seen how energy can be transferred from one form to another and have been given a hint that additional forms of energy exist, we can expand the definition of power given in Section 7-7. There it is the rate at which work is done by a force. In a more general sense, power P is the rate at which energy is transferred by a force from one form to another. If an amount of energy ΔE is transferred in an amount of time Δt, the **average power** due to the force is

$$\overline{P} = \frac{\Delta E}{\Delta t}. \qquad (8\text{-}36)$$

Similarly, the **instantaneous power** due to the force is

$$P = \frac{dE}{dt}. \qquad (8\text{-}37)$$

FIGURE 8-13 To descend, the rock climber must transfer energy from the gravitational potential energy of a system consisting of her, her gear, and the Earth. She has wrapped the rope around metal rings so that the rope rubs against the rings. This allows most of the transferred energy to go to the thermal energy of the rope and bars rather than to her kinetic energy.

SAMPLE PROBLEM 8-7

In Fig. 8-14, a circus beagle of mass 6.0 kg runs onto the left end of a curved ramp with speed $v_0 = 7.8$ m/s at height $y_0 = 8.5$ m above the floor. It then slides to the right and comes to a momentary stop when it reaches a height $y = 11.1$ m from the floor. What is the increase in thermal energy of the beagle and ramp due to the sliding?

SOLUTION: The isolated system to which Eq. 8-33 applies is the beagle–Earth–ramp system, because the only forces acting on the beagle are its weight $m\mathbf{g}$ (due to the Earth) and frictional and normal forces (due to the ramp). The weight is, of course, a conservative force with which we associate a change ΔU in gravitational potential energy. The frictional force dissipates that potential energy and the beagle's kinetic energy during the slide, increasing the thermal energy of the beagle and ramp by an amount ΔE_{int}. The normal force causes no energy transfers because it is always perpendicular to the path of the beagle.

At the stopping point, the speed of the beagle is zero, and thus so is its kinetic energy. Applying Eq. 8-33 to the beagle–Earth–ramp system, we find that

$$\Delta K + \Delta U + \Delta E_{int} = 0$$

and

$$(0 - \tfrac{1}{2}mv_0^2) + mg(y - y_0) + \Delta E_{int} = 0.$$

Solving for ΔE_{int}, we get

$$\begin{aligned}\Delta E_{int} &= \tfrac{1}{2}mv_0^2 - mg(y - y_0) \\ &= \tfrac{1}{2}(6.0 \text{ kg})(7.8 \text{ m/s})^2 \\ &\quad -(6.0 \text{ kg})(9.8 \text{ m/s}^2)(11.1 \text{ m} - 8.5 \text{ m}) \\ &\approx 30 \text{ J.} \qquad \text{(Answer)}\end{aligned}$$

FIGURE 8-14 Sample Problem 8-7. A beagle slides along a curved ramp, starting with speed v_0 at height y_0, and reaching a height y where it comes to a momentary stop.

SAMPLE PROBLEM 8-8

A steel ball whose mass m is 5.2 g is fired vertically downward from a height h_1 of 18 m with an initial speed v_0 of 14 m/s (Fig. 8-15a). It buries itself in sand to a depth h_2 of 21 cm.

(a) What is the change in the mechanical energy of the ball?

SOLUTION: At the stopping depth h_2, the speed of the ball is zero, and thus so is its kinetic energy. The change in the mechanical energy of the ball is given by

$$\Delta E = \Delta K + \Delta U, \qquad (8\text{-}38)$$

or, with Eq. 8-8 ($\Delta U = mg\,\Delta y$),

$$\Delta E = (0 - \tfrac{1}{2}mv_0^2) - mg(h_1 + h_2),$$

where $-(h_1 + h_2)$ is the total *downward* displacement of the ball. Inserting the given data, we find

$$\begin{aligned}\Delta E &= -\tfrac{1}{2}(5.2 \times 10^{-3} \text{ kg})(14 \text{ m/s})^2 \\ &\quad -(5.2 \times 10^{-3} \text{ kg})(9.8 \text{ m/s}^2)(18 \text{ m} + 0.21 \text{ m}) \\ &= -1.437 \text{ J} \approx -1.4 \text{ J.} \qquad \text{(Answer)}\end{aligned}$$

(Note that here we assigned the potential energy to the ball alone rather than, more properly, to the ball–Earth system— recall Problem Solving Tactic 1.)

(b) What is the change in the internal energy of the ball–Earth–sand system?

SOLUTION: This system is an isolated system because once the ball has been fired, the only forces that act on it are its weight $m\mathbf{g}$ (due to the Earth) and the average upward force $\mathbf{F}$ exerted on it by the sand (Fig. 8-15b). So Eq. 8-33 applies. Substituting Eq. 8-38 into Eq. 8-33, we find that for the ball–Earth–sand system

$$\Delta E + \Delta E_{int} = 0,$$

or

$$\Delta E_{int} = -\Delta E = -(-1.437 \text{ J}) \approx 1.4 \text{ J.} \quad \text{(Answer)}$$

As the ball moves through the sand, $\mathbf{F}$ dissipates all the ball's mechanical energy, transferring that energy to thermal energy of the ball and sand.

(c) What is the magnitude of the average force $\mathbf{F}$ exerted on the ball by the sand?

SOLUTION: The mechanical energy of the ball is conserved until it reaches the sand. Then, as the ball moves through a distance h_2 in the sand, its mechanical energy is changed by ΔE. So Eq. 8-27 ($-f_k d = \Delta E$) can be rewritten as

$$-Fh_2 = \Delta E.$$

Solving this for F, we find

$$F = \frac{\Delta E}{-h_2} = \frac{-1.437 \text{ J}}{-0.21 \text{ m}} = 6.84 \text{ N} \approx 6.8 \text{ N.} \quad \text{(Answer)}$$

We could also find F by first using the techniques of Chapter 2 to find the ball's speed at the surface of the sand and its average deceleration within the sand. Then, using Newton's second law, we would have F. Obviously, more algebraic steps would be required.

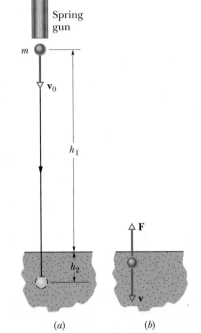

FIGURE 8-15   Sample Problem 8-8. (*a*) A ball is fired downward, coming to rest in sand. Mechanical energy is conserved over path h_1 but not over path h_2, where a nonconservative force **F** acts on the ball. (*b*) Force **F**, due to the sand, transfers energy primarily to thermal energy of the ball and the sand through which the ball travels.

8-8 MASS AND ENERGY (OPTIONAL)

The science of chemistry was built up by assuming that in chemical reactions, energy and mass are each conserved. In 1905, Einstein showed that as a consequence of his theory of special relativity, mass can be considered to be another form of energy. Thus the law of conservation of energy is really the law of conservation of mass–energy.

In a chemical reaction, the amount of mass that is transferred into other forms of energy (or vice versa) is such a tiny fraction of the total mass involved that there is no hope of measuring the mass change with even the best laboratory balances. Mass and energy truly *seem* to be separately conserved. In a nuclear reaction, however, the energy released is often about a million times greater than in a chemical reaction, and the change in mass can easily be measured. Taking mass–energy transfers into account where nuclear reactions are involved becomes a matter of necessary laboratory routine.

Mass and energy are related by what is certainly the best-known equation in physics (see Fig. 8-16), namely,

FIGURE 8-16   In 1979, students of Shenandoah Junior High School, in Miami, Florida, honored Albert Einstein on the 100th anniversary of his birth by spelling out his famous formula with their bodies.

$$E = mc^2, \tag{8-39}$$

in which E is the energy equivalent (called the **mass energy**) of mass m, and c is the speed of light. (If you continue your study of physics beyond this book, you will see more refined discussions of the relation of mass and energy. You might even encounter disagreements about just what that relation is and means.)

Table 8-1 shows the energy equivalents of the masses of a few objects. The amount of energy lying dormant in ordinary objects is enormous. The energy equivalent of the mass of a penny, for example, would cost over a million dollars to purchase from your local utility company. The mass equivalents of some energies are equally striking. The entire annual U.S. electrical energy production, for example, corresponds to a mass of only a few hundred kilograms of matter (stones, potatoes, anything!).

In applying Eq. 8-39 to nuclear or chemical reactions between particles, we can rewrite it as

$$Q = -\Delta m\, c^2, \tag{8-40}$$

TABLE 8-1   THE ENERGY EQUIVALENTS OF A FEW OBJECTS

OBJECT	MASS (kg)	ENERGY EQUIVALENT
Electron	9.11×10^{-31}	8.2×10^{-14} J (= 511 keV)
Proton	1.67×10^{-27}	1.5×10^{-10} J (= 938 MeV)
Uranium atom	4.0×10^{-25}	3.6×10^{-8} J (= 225 GeV)
Dust particle	1×10^{-13}	1×10^4 J (= 2 kcal)
Penny	3.1×10^{-3}	2.8×10^{14} J (= 78 GW·h)

in which Q (called simply the *Q of the reaction*) is the energy released (positive value) or absorbed (negative value) in the reaction, and Δm is the corresponding decrease or increase in the mass of the particles as a result of the reaction. If the reaction is nuclear fission (a nucleus splits into smaller nuclei), less than 0.1% of the mass initially present is transformed into other forms of energy. If the reaction is a chemical one, the percentage is much less, typically a million times less.

In practice, SI units are rarely used with Eq. 8-40, because they are too large to be convenient. Masses are usually measured in atomic mass units (abbreviated u; see Section 1-6), where

$$1 \text{ u} = 1.66 \times 10^{-27} \text{ kg}, \qquad (8-41)$$

and energies are usually measured in electron-volts or multiples thereof. Equation 7-3 tells us that

$$1 \text{ eV} = 1.60 \times 10^{-19} \text{ J.} \qquad (8-42)$$

In the units of Eqs. 8-41 and 8-42, the multiplying constant c^2 has the values

$$c^2 = 9.315 \times 10^8 \text{ eV/u} = 9.315 \times 10^5 \text{ keV/u}$$
$$= 931.5 \text{ MeV/u.} \qquad (8-43)$$

SAMPLE PROBLEM 8-9

In the *nuclear fission* reaction

$$\text{n} + {}^{235}\text{U} \rightarrow {}^{140}\text{Ce} + {}^{94}\text{Zr} + \text{n} + \text{n},$$

a neutron (n) combines with a uranium nucleus (^{235}U), making the nucleus unstable and causing it to split (to fission) into two smaller nuclei (^{140}Ce and ^{94}Zr) and to release two neutrons. The masses involved are

mass(^{235}U) = 235.04 u   mass(^{94}Zr) = 93.91 u

mass(^{140}Ce) = 139.91 u   mass(n) = 1.00867 u

(a) What is the fractional change in the mass of the two interacting particles?

SOLUTION: To find the mass change Δm, we subtract the mass of the reacting particles from the mass of the product particles:

$$\Delta m = (139.91 + 93.91 + 2 \times 1.00867)$$
$$- (235.04 + 1.00867)$$
$$= -0.211 \text{ u.}$$

The mass of the reacting particles is

$$M = 235.04 \text{ u} + 1.00867 \text{ u} = 236.05 \text{ u,}$$

so the fractional decrease in mass is

$$\frac{|\Delta m|}{M} = \frac{0.211 \text{ u}}{236.05 \text{ u}}$$
$$= 0.00089, \text{ or about 0.1\%.} \qquad \text{(Answer)}$$

Although this is small, it is easily measurable.

(b) How much energy is released during each fission reaction?

SOLUTION: From Eq. 8-40 we have

$$Q = -\Delta m\, c^2 = -(-0.211 \text{ u})(931.5 \text{ MeV/u})$$
$$= 197 \text{ MeV,} \qquad \text{(Answer)}$$

where the value for c^2 is taken from Eq. 8-43. The energy release of 197 MeV per reaction is much larger than the energy releases of a few electron-volts per reaction that are typical of chemical reactions.

SAMPLE PROBLEM 8-10

The nucleus of an atom of deuterium (or heavy hydrogen) is called a **deuteron.** It is composed of a proton and a neutron. If a deuteron is torn apart, how much energy is involved in the process? Is energy absorbed or released by the reaction? The masses involved are

deuteron: m_d = 2.01355 u

proton: m_p = 1.00728 u $\Big\}$ 2.01595 u
neutron: m_n = 1.00867 u

SOLUTION: Since the total mass of the separated proton and neutron is greater than the deuteron mass, energy must be added to the deuteron to cause the reaction. The increase in mass due to the reaction is

$$\Delta m = (m_p + m_n) - m_d$$
$$= (1.00728 + 1.00867) - (2.01355) = 0.00240 \text{ u.}$$

The corresponding energy is then, from Eq. 8-40,

$$Q = -\Delta m\, c^2 = -(0.00240 \text{ u})(931.5 \text{ MeV/u})$$
$$= -2.24 \text{ MeV.} \qquad \text{(Answer)}$$

The minus sign means that energy must be added to the deuteron. The quantity 2.24 MeV is called the *binding energy* of the deuteron. The binding energy of any nucleus is the amount of energy that must be supplied to separate the nucleus into its basic units of protons and neutrons. Conversely, it is the amount of energy that would be released if the nucleus could somehow be constructed from those basic units.

CHECKPOINT 6: If we could somehow bring together six protons and six neutrons to form the nucleus of a ^{12}C atom, would the total mass of those 12 particles decrease or increase?

8-9 QUANTIZED ENERGY

So far, we have assumed that the energy within a system can take on *any* value. That seems a reasonable assumption for everyday systems involving springs and weights and such. However, it is not true for such microscopic systems as an atom and a *quantum dot* (a laboratory-produced confinement of electrons that resembles an atom). For these microscopic systems, the energy internal to the system is **quantized;** that is, the energy is restricted to certain values. When the system has a particular energy value, it is said to be in the **quantum state** that is characterized by that value.

Figure 8-17 is a typical diagram of the energy values (or *energy levels*) for an atom: energy is plotted vertically and each of the five lowest possible energy values E_0, E_1, . . . , E_4 is represented by a horizontal line (hence the term *level*). The atom cannot have any intermediate value of energy. The quantum state associated with the lowest energy, labeled E_0 in Fig. 8-17 and assigned the arbitrary value of zero, is called the *ground state* of the atom. The quantum states with greater energies are called *excited states* of the atom: the state with energy E_1 is the first excited state, the state with energy E_2 is the second excited state, and so on.

The atom tends to be in its ground state so as to have the lowest allowed energy (much as a ball tends to roll downhill). The atom can move to an excited state, with greater energy, only if an external source provides energy equal to the energy difference between the ground state and the excited state. The atom can gain energy in a collision with another atom or a free electron. Or it can gain energy by absorbing light. However, in either process the atom must gain enough energy to put it into one of the higher energy levels. Thus, not only the energy values of an atom are quantized, but also the amounts of energy that an atom can gain (or lose) as it makes *quantum jumps* (or transitions) from one energy level to another are quantized.

The upward arrow in Fig. 8-17 represents a quantum jump from the ground state to the second excited state, which (according to the diagram) requires that the atom gain 2.5 eV of energy. If it gains the energy by absorbing light, that light ceases to exist and its energy is fully transferred to the energy of the atom. Thus, that light must have an energy of 2.5 eV if it is to be absorbed by the atom. If the light has slightly more or slightly less energy, it cannot be absorbed.

> If an atom is to absorb light, the energy of that light must equal the energy difference between the initial energy level of the atom and a higher level.

When the atom reaches an excited state, it does not stay there but quickly *de-excites* by decreasing its energy, either in a collision or by emitting light. In the latter process, the emitted light is actually created by the atom (the

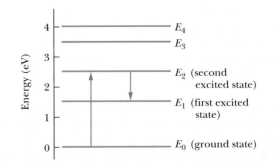

FIGURE 8-17 An energy level diagram for an atom showing the quantized energy values (or levels). Each energy level corresponds to a quantum state of the atom. The lowest energy, marked E_0, is associated with the ground state. The upward arrow represents a quantum jump of the atom from the ground state to the second excited state, which has energy E_2. The downward arrow represents a quantum jump from there to the first excited state, which has energy E_1.

light did not exist prior to de-excitation). In either process, the atom must lose enough energy to reach one of its lower energy levels.

> If an atom emits light, the energy of that light equals the energy difference between the initial energy level of the atom and a lower level.

As an example, the downward arrow in Fig. 8-17 represents a quantum jump from the second excited state to the first excited state. The light that is emitted in the jump has an energy of 1.0 eV. The atom must next jump to the ground state with a second emission of light having an energy of 1.5 eV. Instead of this two-jump process, the atom could have jumped from the second excited state directly to the ground state with a single emission of light having an energy of 2.5 eV.

Checkpoint 7: An atom that is being monitored emits light four times without being re-excited. Three of the emissions are detected at energies of 1.1 eV, 1.4 eV, and 2.8 eV. The sequence of the emissions is not known. Here are the energy levels of the atom:

$E_8 = 7.7$ eV	$E_5 = 4.8$ eV	$E_2 = 2.7$ eV
$E_7 = 6.6$	$E_4 = 4.2$	$E_1 = 1.3$
$E_6 = 5.5$	$E_3 = 3.9$	$E_0 = 0$

(a) What was the initial state of the atom and (b) what was the energy of the fourth emission?

REVIEW & SUMMARY

Conservative Forces

A force is a **conservative force** if the net work it does on a particle moving along a closed path from an initial point and then back to that point is zero. Or, equivalently, it is conservative if its work on a particle moving between two points does not depend on the path taken by the particle. The gravitational force (weight) and the spring force are conservative forces; the kinetic frictional force is a **nonconservative force.**

Potential Energy

A **potential energy** is energy that is associated with the configuration of a system in which a conservative force acts. When the conservative force does work W on a particle within the system, the change ΔU in the potential energy of the system is

$$\Delta U = -W. \tag{8-1}$$

If the particle moves from point x_i to point x_f, the change in the potential energy of the system is

$$\Delta U = -\int_{x_i}^{x_f} F(x) \, dx. \tag{8-6}$$

Gravitational Potential Energy

The potential energy associated with a system consisting of the Earth and a nearby particle is the **gravitational potential energy.** If the particle moves from height y_i to height y_f, the change in the gravitational potential energy of the particle–Earth system is

$$\Delta U = mg(y_f - y_i) = mg \, \Delta y. \tag{8-7}$$

If the **reference position** of the particle is set as $y_i = 0$ and the corresponding gravitational potential energy of the system is set as $U_i = 0$, then the gravitational potential energy U when the particle is at any position y is

$$U = mgy. \tag{8-9}$$

Elastic Potential Energy

Elastic potential energy is the energy associated with the state of compression or extension of an elastic object. For a spring that exerts a spring force $F = -kx$, the elastic potential energy is

$$U(x) = \tfrac{1}{2}kx^2. \tag{8-11}$$

The reference configuration is with the spring at its relaxed length, with $x = 0$ and $U = 0$.

Mechanical Energy

The **mechanical energy** E of a system is the sum of its kinetic energy K and its potential energy U:

$$E = K + U. \tag{8-12}$$

If only a conservative force within the system does work, then the mechanical energy E of the system cannot change. This **conservation of mechanical energy** is written as

$$K_2 + U_2 = K_1 + U_1, \tag{8-17}$$

in which the subscripts refer to different instants during an energy transfer process. This conservation can also be written as

$$\Delta E = \Delta K + \Delta U = 0. \tag{8-18}$$

Potential Energy Curves

If we know the **potential energy function** $U(x)$ for a system in which a force F acts on a particle, we can find the force as

$$F(x) = -\frac{dU(x)}{dx}. \tag{8-19}$$

If $U(x)$ is given on a graph, then at any value of x, the force F is the negative of the slope of the curve there and the kinetic energy of the particle is given by

$$K(x) = E - U(x), \tag{8-21}$$

where E is the mechanical energy of the system. A **turning point** is a point x where the particle reverses its motion (there, $K = 0$). The particle is in **equilibrium** at points where the slope of the $U(x)$ curve is zero (there, $F(x) = 0$).

Work by Nonconservative Forces

If a nonconservative applied force **F** does work on particle that is part of a system having a potential energy, then the work W_{app} done on the system by **F** is equal to the change ΔE in the mechanical energy of the system:

$$W_{app} = \Delta K + \Delta U = \Delta E. \tag{8-24, 8-25}$$

If a kinetic frictional force $\mathbf{f}_k$ does work on an object, the change ΔE in the total mechanical energy of the object and any system containing it is given by

$$\Delta E = -f_k d, \tag{8-27}$$

in which d is the displacement of the object during the work. The mechanical energy lost by this transfer is said to be **dissipated** by $\mathbf{f}_k$. A portion of the dissipated energy is transferred from the object; that amount is equal to the work done by $\mathbf{f}_k$.

Principle of Conservation of Energy

In an isolated system, energy may be transferred from one type to another, but the total energy E_{tot} of the system always remains constant. This conservation law is written as

$$\Delta E_{tot} = \Delta K + \Delta U + \Delta E_{int} \\ + \left(\begin{array}{c} \text{changes in other} \\ \text{forms of energy} \end{array}\right) = 0. \tag{8-34}$$

Here ΔE_{int} is the change in the internal energy of the bodies within the system.

If, instead, the system is not isolated, then an external force can change the total energy of the system by doing work W. In that case

$$W = \Delta E_{tot} = \Delta K + \Delta U + \Delta E_{int}. \tag{8-35}$$

Power

The **power** of a force is the *rate* at which that force transfers energy. If an amount of energy ΔE is transferred by a force in an amount of time Δt, the **average power** of the force is

$$\overline{P} = \frac{\Delta E}{\Delta t}. \qquad (8\text{-}36)$$

The **instantaneous power** of a force is

$$P = \frac{dE}{dt}. \qquad (8\text{-}37)$$

Mass and Energy

The **mass energy** E of an object is the energy equivalent of its mass m:

$$E = mc^2, \qquad (8\text{-}39)$$

in which c is the speed of light. To convert a mass given in atomic mass units (u) to an energy equivalent in megaelectron-volts, multiply by 931.5 MeV/u. In a nuclear or chemical reaction, the energy released or absorbed is the Q of the reaction, given as

$$Q = -\Delta m\, c^2, \qquad (8\text{-}40)$$

in which Δm is the corresponding decrease or increase in mass. (Q is positive for an energy release, corresponding to a mass decrease.)

Quantized Energy

The energy within a microscopic system, such as an atom, is **quantized** (restricted to certain values) and the system is said to be in a **quantum state** that is characterized by its energy value. The system cannot have any other values of energy. The lowest energy corresponds to the **ground state** of the system. Greater values of energy correspond to **excited states** of the system. If the system gains or loses energy via light, the energy of the light must equal the difference between two of the allowed energy values of the system.

QUESTIONS

1. When a particle moves from f to i and from j to i along the paths shown in Fig. 8-18, and in the indicated directions, a conservative force **F** does the indicated amounts of work on it. How much work is done by **F** when the particle moves from f to j?

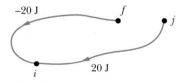

FIGURE 8-18 Question 1.

2. Figure 8-19 shows one direct path and four indirect paths from point i to point f. Along the direct path and three of the indirect paths, only a conservative force F_c acts on a certain particle. Along the fourth indirect path, both F_c and a nonconservative force F_{nc} act on the particle. The work (in joules) done on a particle in going from i to f is indicated along each straight line segment of the indirect paths. (a) How much work is done on the particle in moving from i to f along the direct path? (b) How much work is done on the particle by F_{nc} along the one path where it acts?

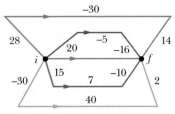

FIGURE 8-19 Question 2.

3. A spring is initially stretched by 3.0 cm from its relaxed length. Here are four choices: the initial stretch is changed to (a) a stretch of 2.0 cm, (b) a compression of 2.0 cm, (c) a compression of 4.0 cm, and (d) a stretch of 4.0 cm. Rank the choices according to the change in the elastic potential energy of the spring, most positive first, most negative last.

4. In Fig. 8-20, a brave skater slides down three slopes of frictionless ice whose vertical heights d are identical. Rank the slopes according to (a) the work done on the skater by her weight during the descent on each slope and (b) the change in her kinetic energy produced along the slope, greatest first.

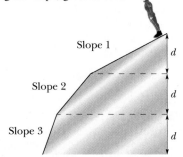

FIGURE 8-20 Question 4.

5. A coconut is thrown from a cliff edge toward a wide, flat valley, with initial speed 8 m/s. Rank the following choices for the launch direction according to (a) the initial kinetic energy of the coconut and (b) its kinetic energy just before hitting the valley bottom, greatest first: (1) **v** almost vertically upward, (2) **v** angled upward by 45°, (3) **v** horizontal, (4) **v** angled downward by 45°, and (5) **v** almost vertically downward.

6. Figure 8-21 shows three plums that are launched from the same level with the same speed. One moves straight upward, one is launched at a small angle to the vertical, and one is launched along a frictionless incline. Rank the plums according to their speed when they reach the level of the dashed line, greatest first.

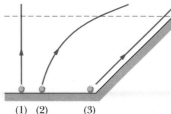

FIGURE 8-21 Question 6. (1) (2) (3)

7. In Fig. 8-22, a horizontally moving block can take three frictionless routes, differing in elevation, to reach the dashed finish line. Rank the routes according to (a) the speed of the block at the finish line and (b) the travel time of the block to the finish line, greatest first.

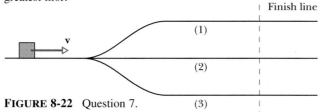

FIGURE 8-22 Question 7. (3)

8. In Fig. 8-23, a small, initially stationary block is released on a frictionless ramp at a height of 3.0 m. Hill heights along the ramp are as shown. The hills have identical circular tops (assume that the block does not fly off any hill). (a) Which hill is the first the block cannot cross? (b) What does it do after failing to cross that hill? On which hilltop is (c) the centripetal acceleration of the block greatest? (d) On which hilltop is the normal force on the block least?

FIGURE 8-23 Question 8.

9. Figure 8-24 shows two arrangements of the same two blocks on a frictionless plane; the blocks are connected with a taut cord that runs over a massless, frictionless pulley. In each arrangement, the hanging block descends when the blocks are released. Consider the total kinetic energy of the two blocks when that block has descended by a certain distance d. Is the total kinetic energy in arrangement (a) more than, less than, or the same as that in arrangement (b)?

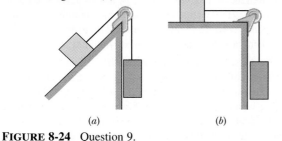

(a) *(b)*

FIGURE 8-24 Question 9.

10. In Fig. 8-25, an initially stationary block is released at time t_1 to slide down a frictionless ramp to a massless spring, which it reaches at time t_2 and then compresses until the maximum compression is reached at time t_3. From t_1 to t_3, what happens to (a) the kinetic energy of the block, (b) the gravitational potential energy of the block–Earth system, (c) the elastic potential energy of the spring, (d) the mechanical energy of the block–Earth system, (e) the mechanical energy of the spring, and (f) the mechanical energy of the block–Earth–spring system?

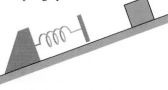

FIGURE 8-25
Question 10.

11. In Checkpoint 4, what values must the mechanical energy E and kinetic energy K of the particle not exceed if the particle is to be (a) trapped in the potential well and (b) able to move to the left of point D but not to the right of it.

12. Figure 8-26 gives the potential energy function of a particle. (a) Rank regions AB, BC, CD, and DE according to the magnitude of the force on the particle, greatest first. What value must the energy E of the particle not exceed if the particle is to be (b) trapped in the potential well at the left, (c) trapped in the potential well at the right, and (d) able to move between the two potential wells but not to the right of point H? For the situation of (d), in what region—BC, DE, or FG—will the particle have (e) the greatest kinetic energy and (f) the least speed?

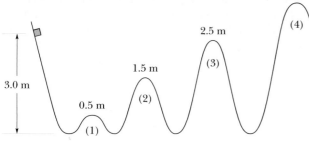

FIGURE 8-26 Question 12.

13. In Fig. 8-27, a block slides from A to C along a frictionless ramp, and then it passes through horizontal region CD, where a frictional force acts on it. Is the block's kinetic energy increasing, decreasing, or constant in (a) region AB, (b) region BC, and (c) region CD? (d) Is the block's mechanical energy increasing, decreasing, or constant in those regions?

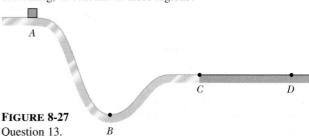

FIGURE 8-27
Question 13. B

EXERCISES & PROBLEMS

SECTION 8-3 Determining Potential Energy Values

1E. The only force acting on a particle is conservative force **F**. If the particle is at point A, the potential energy of the system associated with **F** and the particle is 40 J. If the particle moves from point A to point B, the work done on the particle by **F** is $+25$ J. What is the potential energy of the system with the particle at B?

2E. What is the spring constant of a spring that stores 25 J of elastic potential energy when compressed by 7.5 cm from its relaxed length?

3E. You drop a 2.00 kg textbook to a friend who stands on the ground 10.0 m below the textbook with outstretched hands 1.50 m above the ground (Fig. 8-28). (a) How much work is done on the textbook by its weight as it drops to your friend's hands? (b) What is the change in the gravitational potential energy of the textbook–Earth system during the drop? If the gravitational potential energy of that system is taken to be zero at ground level, what is its potential energy when the textbook (c) is released and (d) reaches the hands?

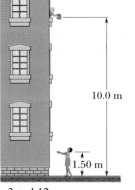

FIGURE 8-28 Exercises 3 and 12.

4E. In Fig. 8-29, a 2.00 g ice flake is released from the edge of a hemispherical bowl whose radius r is 22.0 cm. The flake–bowl contact is frictionless. (a) How much work is done on the flake by its weight during the flake's descent to the bottom of the bowl? (b) What is the change in the potential energy of the flake–Earth system during that descent? (c) If that potential energy is taken to be zero at the bottom of the bowl, what is its value when the flake is released? (d) If, instead, the potential energy is taken to be zero at the release point, what is its value when the flake reaches the bottom of the bowl?

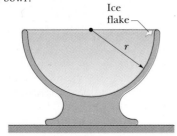

FIGURE 8-29 Exercises 4 and 13.

5E. In Fig. 8-30, a frictionless roller coaster of mass m tops the first hill with speed v_0. How much work does its weight do on it from that point to (a) point A, (b) point B, and (c) point C? If the gravitational potential energy of the coaster–Earth system is taken to be zero at point C, what is its value when the coaster is at (d) point B and (e) point A?

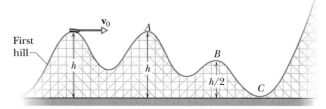

FIGURE 8-30 Exercises 5 and 14.

6E. Figure 8-31 shows a ball with mass m attached to the end of a thin rod with length L and negligible mass. The other end of the rod is pivoted so that the ball can move in a vertical circle. The rod is held in the horizontal position as shown and then given enough of a downward push to cause the ball to swing down and around and just reach the vertically upward position, with zero speed there. How much work is done on the ball by its weight from the initial point to (a) the lowest point, (b) the highest point, and (c) the point on the right at which the ball is level with the initial point? (d) If the gravitational potential energy of the ball–Earth system is taken to be zero at the initial point, what is its value when the ball reaches those three other points, respectively?

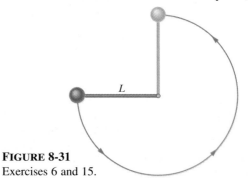

FIGURE 8-31
Exercises 6 and 15.

7P. A spring with a spring constant of 3200 N/m is initially stretched until the elastic potential energy is 1.44 J. ($U = 0$ for no stretch.) What is the change in the elastic potential energy if the initial stretch is changed to (a) a stretch of 2.0 cm, (b) a compression of 2.0 cm, and (c) a compression of 4.0 cm?

8P. A 1.50 kg snowball is fired from a cliff 12.5 m high with an initial velocity of 14.0 m/s, directed 41.0° above the horizontal. (a) How much work is done on the snowball by its weight during its flight to the ground below the cliff? (b) What is the change in the gravitational potential energy of the snowball–Earth system during the flight? (c) If that gravitational potential energy is taken to be zero at the height of the cliff, what is its value when the snowball reaches the ground?

9P. Figure 8-32 shows a thin rod, of length L and negligible mass, that can pivot about one end to rotate in a vertical circle. A heavy ball of mass m is attached to the other end. The rod is pulled aside through an angle θ and released. As the ball descends to its lowest point, (a) how much work does its weight do on it and (b) what is the change in the gravitational potential energy of the ball–Earth system? (c) If the gravitational potential energy is taken to be zero at the lowest point, what is its value just as the ball is released?

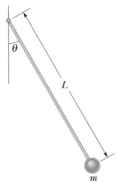

FIGURE 8-32   Problems 9 and 20.

10P. In Fig. 8-33, a small block of mass m can slide along the frictionless loop-the-loop. The block is released from rest at point P, at height $h = 5R$ above the bottom of the loop. How much work does the weight of the block do on the block as the block travels from point P to (a) point Q and (b) the top of the loop? If the gravitational potential energy of the block–Earth system is taken to be zero at the bottom of the loop, what is that potential energy when the block is (c) at point P, (d) at point Q, and (e) at the top of the loop?

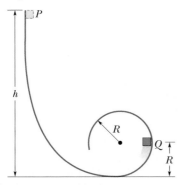

FIGURE 8-33   Problems 10 and 39.

11P. A force $\mathbf{F}$ that is directed along an x axis acts on a particle as the particle moves from $x = 1.0$ m to $x = 4.0$ m and then back to $x = 1.0$ m. What is the net work done on the particle by $\mathbf{F}$ for the round trip if the values of the force during the outward and the return trips are (a) 3.0 N and -3.0 N; (b) 5.0 N and 5.0 N; (c) $2.0x$ and $-2.0x$; (d) $3.0x^2$ and $3.0x^2$? Here, x is in meters and F is in newtons. (e) In which situation(s) is the force conservative?

SECTION 8-4   Conservation of Mechanical Energy

12E. (a) In Exercise 3, what is the speed of the textbook when it reaches the hands? (b) If we substituted a second textbook with twice the mass, what would its speed be?

13E. (a) In Exercise 4, what is the speed of the flake when it reaches the bottom of the bowl? (b) If we substituted a second flake with twice the mass, what would its speed be?

14E. In Exercise 5, what is the speed of the coaster at (a) point A, (b) point B, and (c) point C? (d) How high will it go on the last hill, which is too high for it to cross? (e) If we substitute a second coaster with twice the mass, what then are the answers (a) through (d)?

15E. (a) In Exercise 6, what initial speed is given the ball? What is its speed at (b) the lowest point and (c) the point on the right at which the ball is level with the initial point?

16E. A 70.0 kg man jumping from a window lands in an elevated fire rescue net 11.0 m below the window. He momentarily stops when he has stretched the net by 1.50 m. Assuming that mechanical energy is conserved during this process and that the net functions like an ideal spring, find the elastic potential energy of the net when it is stretched by 1.50 m.

17E. In Fig. 8-34, a runaway truck with failed brakes is moving downgrade at 80 mi/h just before the driver has the truck travel up an emergency escape ramp with an inclination of 15°. What minimum length L must the ramp have if the truck is to stop (momentarily) along it? Why are real escape ramps often covered with a thick layer of sand or gravel?

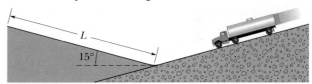

FIGURE 8-34   Exercise 17.

18E. A volcanic ash flow is moving across horizontal ground when it encounters a 10° upslope. The front of the flow then travels 920 m on the upslope before stopping. Assume that the gases entrapped in the flow lift the flow and thus make the frictional force from the ground negligible; assume also that mechanical energy of the front of the flow is conserved. What was the initial speed of the front of the flow?

19P. A 1.50 kg water balloon is shot straight up with an initial speed of 3.00 m/s. (a) What is the kinetic energy of the balloon just as it is launched? (b) How much work does the weight of the balloon do on the balloon during the balloon's full ascent? (c) What is the change in the gravitational potential energy of the balloon–Earth system during the full ascent? (d) If the gravitational potential energy is taken to be zero at the launch point, what is its value when the balloon reaches its maximum height? (e) If, instead, the gravitational potential energy is taken to be zero at the maximum height, what is its value at the launch point? (f) What is the maximum height of the balloon?

20P. In Problem 9, what is the speed of the ball at the lowest point if $L = 2.00$ m and $\theta = 30.0°$?

21P. (a) In Problem 8, using energy techniques, rather than techniques of Chapter 4, find the speed of the snowball as it reaches the ground below the cliff. (b) What is that speed if, instead, the launch angle is 41.0° *below* the horizontal?

22P. Figure 8-35 shows an 8.00 kg stone resting on a spring. The spring is compressed 10.0 cm by the stone. (a) What is the spring constant? (b) The stone is pushed down an additional 30.0 cm and released. What is the elastic potential energy of the compressed spring just before that release? (c) What is the change in the gravitational potential energy of the stone–Earth system when the stone moves from the release point to its maximum height? (d) What is that maximum height, measured from the release point?

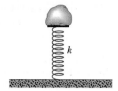

FIGURE 8-35 Problem 22.

23P. A 5.0 g marble is fired vertically upward using a spring gun. The spring must be compressed 8.0 cm if the marble is to just reach a target 20 m above the marble's position on the compressed spring. (a) What is the change in the gravitational potential energy of the marble–Earth system during the 20 m ascent? (b) What is the change in the elastic potential energy of the spring during its launch of the marble? (c) What is the spring constant of the spring?

24P. Figure 8-36a applies to the spring in a cork gun (Fig. 8-36b): it shows the spring force as a function of the stretch or compression of the spring. The spring is compressed by 5.5 cm and used to propel a 3.8 g cork from the gun. (a) What is the speed of the cork if it is released as the spring passes through its relaxed position? (b) Suppose, instead, that the cork sticks to the spring and stretches it 1.5 cm before separation occurs. What now is the speed of the cork at the time of release?

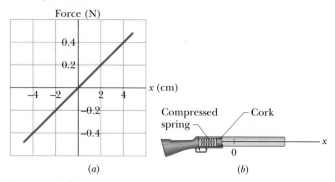

FIGURE 8-36 Problem 24.

25P. A 2.00 kg block is placed against a spring on a frictionless 30.0° incline (Fig. 8-37). The spring, whose spring constant is 19.6 N/cm, is compressed 20.0 cm and then released. (a) What is the elastic potential energy of the compressed spring? (b) What is the change in the gravitational potential energy of the block–Earth system as the block moves from the release point to its

highest point on the incline? (c) How far along the incline is the highest point from the release point?

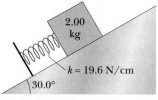

FIGURE 8-37 Problem 25.

26P. In Fig. 8-38, a 12 kg block is released from rest on an incline angled at $\theta = 30°$. Below the block is a spring that can be compressed 2.0 cm by a force of 270 N. The block momentarily stops when it compresses the spring by 5.5 cm. (a) How far has the block moved down the incline to this stopping point? (b) What is the speed of the block just as it touches the spring?

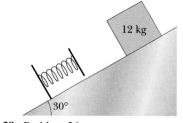

FIGURE 8-38 Problem 26.

27P. A 0.55 kg projectile is launched from the edge of a cliff with an initial kinetic energy of 1550 J and at its highest point is 140 m above the launch point. (a) What is the horizontal component of its velocity? (b) What was the vertical component of its velocity just after launch? (c) At one instant during its flight the vertical component of its velocity is 65 m/s. At that time, how far is it above or below the launch point?

28P. A 50 g ball is thrown from a window with an initial velocity of 8.0 m/s at an angle of 30° above the horizontal. Using energy methods, determine (a) the kinetic energy of the ball at the top of its flight and (b) its speed when it is 3.0 m below the window. Does the answer to (b) depend on either (c) the mass of the ball or (d) the initial angle?

29P. The spring of a child's spring gun has a spring constant of 4.0 lb/in. When the gun is inclined upward by 30° to the horizontal, a 2.0 oz ball is shot to a height of 6.0 ft above the muzzle of the gun. (a) What was the muzzle speed of the ball? (b) By how much must the spring have been compressed initially?

30P. In Fig. 8-39, the pulley is massless, and both it and the inclined plane are frictionless. If the masses are released from rest with the connecting cord taut, what is their total kinetic energy when the 2.0 kg mass has fallen 25 cm?

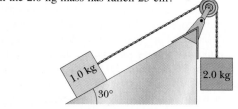

FIGURE 8-39 Problem 30.

31P. A 1.50 kg snowball is shot upward at an angle of 34.0° to the horizontal with an initial speed of 20.0 m/s. (a) What is its initial kinetic energy? (b) By how much does the gravitational potential energy of the snowball–Earth system change as the snowball moves from the launch point to the point of maximum height? (c) What is that maximum height?

32P. A pendulum consists of a 2.0 kg stone swinging on a 4.0 m string of negligible mass. The stone has a speed of 8.0 m/s when it passes its lowest point. (a) What is the speed when the string is at 60° to the vertical? (b) What is the greatest angle with the vertical that the string will reach during the stone's motion? (c) If the potential energy of the pendulum–Earth system is taken to be zero at the stone's lowest point, what is the total mechanical energy of the system?

33P. The string in Fig. 8-40 is $L = 120$ cm long, and the distance d to the fixed peg at point P is 75.0 cm. When the initially stationary ball is released with the string horizontal as shown, it will swing along the dashed arc. What is its speed when it reaches (a) its lowest point and (b) its highest point after the string catches on the peg?

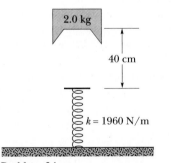

FIGURE 8-40 Problems 33 and 41.

34P. A 2.0 kg block is dropped from a height of 40 cm onto a spring of spring constant $k = 1960$ N/m (Fig. 8-41). Find the maximum distance the spring is compressed.

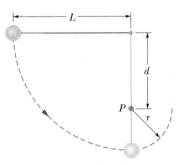

FIGURE 8-41 Problem 34.

35P. Figure 8-42 shows a pendulum of length L. Its bob (which effectively has all the mass) has speed v_0 when the cord makes an angle θ_0 with the vertical. (a) Derive an expression for the speed of the bob when it is in its lowest position. What is the least value that v_0 can have if the pendulum is to swing down and then up (b) to a horizontal position, and (c) to a vertical position with the cord remaining straight?

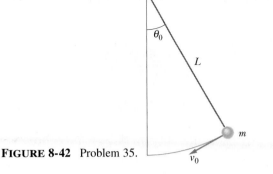

FIGURE 8-42 Problem 35.

36P. Two children are playing a game in which they try to hit a small box on the floor with a marble fired from a spring-loaded gun that is mounted on a table. The target box is 2.20 m horizontally from the edge of the table; see Fig. 8-43. Bobby compresses the spring 1.10 cm, but the center of the marble falls 27.0 cm short of the center of the box. How far should Rhoda compress the spring to score a direct hit?

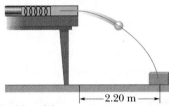

FIGURE 8-43 Problem 36.

37P. The magnitude of the gravitational force between a particle of mass m_1 and one of mass m_2 is given by

$$F(x) = G\frac{m_1 m_2}{x^2},$$

where G is a constant and x is the distance between the particles. (a) What is the corresponding potential energy function $U(x)$? Assume that $U(x) \to 0$ as $x \to \infty$. (b) How much work is required to increase the separation of the particles from $x = x_1$ to $x = x_1 + d$?

38P. A 20 kg object is acted on by a conservative force given by $F = -3.0x - 5.0x^2$, with F in newtons and x in meters. Take the potential energy associated with the force to be zero when the object is at $x = 0$. (a) What is the potential energy of the system associated with the force when the object is at $x = 2.0$ m? (b) If the object has a velocity of 4.0 m/s in the negative direction of the x axis when it is at $x = 5.0$ m, what is its speed when it passes through the origin? (c) What are the answers to (a) and (b) if the potential energy of the system is taken to be -8.0 J when the object is at $x = 0$?

39P. (a) In Problem 10, what is the *net* force acting on the block when it reaches point Q? (b) At what height h should the block be released from rest so that it is on the verge of losing contact with the track at the top of the loop? (*On the verge of losing contact* means that the normal force on the block from the track has just then become zero.)

40P. Tarzan, who weighs 688 N, swings from a cliff at the end of a convenient vine that is 18 m long (Fig. 8-44). From the top of the cliff to the bottom of the swing, he descends by 3.2 m. The vine will break if the force on it exceeds 950 N. (a) Does the vine break? (b) If no, what is the greatest force on it during the swing? If yes, at what angle does it break?

FIGURE 8-44 Problem 40.

41P. In Fig. 8-40 show that, if the ball is to swing completely around the fixed peg, then $d > 3L/5$. (*Hint:* The ball must still be moving at the top of its swing. Do you see why?)

42P. An effectively massless rigid rod of length L has a ball with mass m attached to its end, forming a pendulum. The pendulum is inverted, with the rod straight up, and then released. What are (a) the ball's speed at the lowest point and (b) the tension in the rod at that point? (c) The same pendulum is next put in a horizontal position and released from rest. At what angle from the vertical do the magnitudes of the tension in the rod and the weight of the ball match?

43P*. A chain is held on a frictionless table with one-fourth of its length hanging over the edge, as shown in Fig. 8-45. If the chain has length L and mass m, how much work is required to pull the hanging part back onto the table?

FIGURE 8-45 Problem 43.

44P*. A 3.20 kg block starts at rest and slides a distance d down a frictionless 30.0° incline, where it runs into a spring (Fig. 8-46). The block slides an additional 21.0 cm before it is brought to rest momentarily by compressing the spring, whose spring constant k is 431 N/m. (a) What is the value of d? (b) What is the distance between the point of first contact and the point where the block's speed is greatest?

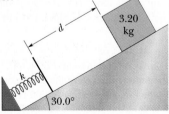

FIGURE 8-46 Problem 44.

45P*. A boy is seated on the top of a hemispherical mound of ice (Fig. 8-47). He is given a very small push and starts sliding down the ice. Show that he leaves the ice at a point whose height is $2R/3$ if the ice is frictionless. (*Hint:* The normal force vanishes as he leaves the ice.)

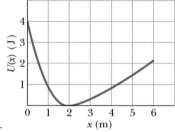

FIGURE 8-47 Problem 45.

SECTION 8-5 Reading a Potential Energy Curve

46E. A conservative force $F(x)$ acts on a particle that moves along the x axis. Figure 8-48 shows how the potential energy $U(x)$ associated with $F(x)$ varies with the particle's position. (a) Plot $F(x)$, using the same x scale as in Fig. 8-48. (b) The mechanical energy E of the system is 4.0 J. Plot the particle's kinetic energy $K(x)$ on Fig. 8-48.

FIGURE 8-48 Exercise 46.

47P. The potential energy of a diatomic molecule (a two-atom system like H_2 or O_2) is given by

$$U = \frac{A}{r^{12}} - \frac{B}{r^6},$$

with r being the separation of the two atoms of the molecule and A and B being positive constants. This potential energy is associated with the force that binds the two atoms together. (a) Find the *equilibrium separation,* that is, the distance between the atoms at which the force on each atom is zero. Is the force repulsive (the atoms are pushed apart) or attractive (they are pulled together) if their separation is (b) smaller and (c) larger than the equilibrium separation?

48P. A conservative force $F(x)$ acts on a 2.0 kg particle that

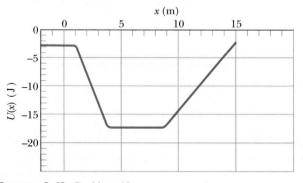

FIGURE 8-49 Problem 48.

moves along the x axis. The potential energy $U(x)$ associated with $F(x)$ is graphed in Fig. 8-49. When the particle is at $x = 2.0$ m, its velocity is -1.5 m/s. (a) What are the magnitude and direction of $F(x)$ at this position? (b) Between what limits of x does the particle move? (c) What is its speed at $x = 7.0$ m?

49P. Figure 8-50*a* shows a molecule consisting of two atoms of masses m and M (with $m \ll M$) and separation r. Figure 8-50*b* shows the potential energy $U(r)$ of the molecule as a function of r. Describe the motion of the atoms (a) if the total mechanical energy E of the two-atom system is greater than zero (as is E_1), and (b) if E is less than zero (as is E_2). For $E_1 = 1 \times 10^{-19}$ J and $r = 0.3$ nm, find (c) the potential energy of the system, (d) the total kinetic energy of the atoms, and (e) the force (magnitude and direction) acting on each atom. For what values of r is the force (f) repulsive, (g) attractive, and (h) zero?

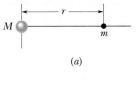

(a)

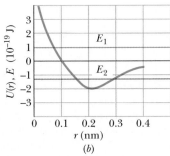

(b)

FIGURE 8-50 Problem 49.

SECTION 8-6 Work Done by Nonconservative Forces

50E. A collie drags its bed box across a floor by applying a horizontal force of 8.0 N. The kinetic frictional force acting on the box has magnitude 5.0 N. As the box is dragged through 0.70 m along the way, (a) what is the work done by the collie's applied force and (b) how much mechanical energy is dissipated by the frictional force?

51E. The temperature of a plastic cube is monitored while the cube is pushed 3.0 m across a floor at constant speed by a horizontal force of 15 N. The monitoring reveals that the thermal energy of the cube increases by 20 J. How much work is done on the cube by the kinetic frictional force acting on it?

52P. A worker pushed a 27 kg block 9.2 m along a level floor at constant speed with a force directed 32° below the horizontal. (a) If the coefficient of kinetic friction is 0.20, how much work was done by the worker's force? (b) How much energy was dissipated by the frictional force?

53P. A 50 kg trunk is pushed 6.0 m at constant speed up a 30° incline by a constant horizontal force. The coefficient of kinetic friction between the trunk and the incline is 0.20. Calculate the work done by (a) the applied force and (b) the weight of the

trunk. (c) How much energy was dissipated by the frictional force acting on the trunk?

54P. A 3.57 kg block is drawn at constant speed 4.06 m along a horizontal floor by a rope exerting a 7.68 N force at an angle of 15.0° above the horizontal. Compute (a) the work done by the rope's force and (b) the coefficient of kinetic friction between block and floor. (c) How much energy is dissipated by the frictional force?

55P. A 1400 kg block of granite is pulled up an incline at a constant speed of 1.34 m/s by a cable and winch (Fig. 8-51). The coefficient of kinetic friction between the block and the incline is 0.40. What is the power due to the force applied to the block by the cable?

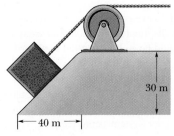

FIGURE 8-51 Problem 55.

SECTION 8-7 Conservation of Energy

56E. If a 70 kg baseball player steals home by sliding into the plate with an initial speed of 10 m/s, (a) how much kinetic energy is dissipated by the frictional force stopping him and (b) what is the change in the thermal energy of his body and the ground along which he slides?

57E. A 75 g Frisbee is thrown from a point 1.1 m above the ground with a speed of 12 m/s. When it has reached a height of 2.1 m, its speed is 10.5 m/s. (a) How much work was done on the Frisbee by its weight? (b) How much of the Frisbee's mechanical energy was dissipated by air drag?

58E. An outfielder throws a baseball with an initial speed of 120 ft/s. Just before an infielder catches the ball at the same level, its speed is 110 ft/s. How much of the ball's mechanical energy is dissipated by air drag? The weight of a baseball is 9.0 oz.

59E. A 0.63 kg ball is thrown up with an initial speed of 14 m/s and reaches a maximum height of 8.1 m. How much energy is dissipated by the air drag acting on the ball during the ascent?

60E. A 9.4 kg projectile is fired vertically upward. Air drag dissipates 68 kJ during its ascent. How much higher would it have gone were air drag negligible?

61E. A river descends 15 m through rapids. The speed of the water is 3.2 m/s upon entering the rapids and 13 m/s upon leaving. What percentage of the gravitational potential energy of the water–Earth system is transferred to kinetic energy during the descent? (*Hint:* Consider the descent of, say, 10 kg of water.)

62E. Approximately 5.5×10^6 kg of water falls 50 m over Niagara Falls each second. (a) What is the decrease in the gravitational potential energy of the water–Earth system each second?

(b) If all this energy could be converted to electrical energy by an electric generating plant (it cannot be), at what rate would electrical energy by supplied? (The mass of 1 m³ of water is 1000 kg.) (c) If the electrical energy were sold at 1 cent/kW·h, what would be the yearly cost?

63E. Each second, 1200 m³ of water passes over a waterfall 100 m high. Three-fourths of the kinetic energy gained by the water in falling is transferred to electrical energy by a hydroelectric generator. At what rate does the generator produce electrical energy? (The mass of 1 m³ of water is 1000 kg.)

64E. The area of the continental United States is about 8 × 10⁶ km², and the average elevation of its land surface is about 500 m (above sea level). The average yearly rainfall is 75 cm. Two-thirds of this rainwater returns to the atmosphere by evaporation, but the rest eventually flows into the ocean. If the associated decrease in gravitational potential energy of the water–Earth system could be fully converted to electrical energy, what would be the average power of the conversion? (The mass of 1 m³ of water is 1000 kg.)

65E. A 68 kg sky diver falls at a constant terminal speed of 59 m/s. (a) At what rate is the gravitational potential energy of the Earth–sky diver system being reduced? (b) At what rate is mechanical energy being dissipated?

66E. A 25 kg bear slides, from rest, 12 m down a lodgepole pine tree, moving with a speed of 5.6 m/s just before hitting the ground. (a) What change occurs in the gravitational potential energy of the bear–Earth system during the slide? (b) What is the kinetic energy of the bear just before hitting the ground? (c) What is the average frictional force that acts on the bear?

67E. During a rockslide, a 520 kg rock slides from rest down a hillside that is 500 m long and 300 m high. The coefficient of kinetic friction between the rock and the hill surface is 0.25. (a) If the gravitational potential energy U of the rock–Earth system is set to zero at the bottom of the hill, what is the value of U just before the slide? (b) How much mechanical energy is dissipated by frictional forces during the slide? (c) What is the kinetic energy of the rock as it reaches the bottom of the hill? (d) What is its speed then?

68E. A 30 g bullet, with a horizontal velocity of 500 m/s, stops 12 cm within a solid wall. (a) What is the change in its mechanical energy? (b) What is the magnitude of the average force from the wall stopping it?

69P. As Fig. 8-52 shows, a 3.5 kg block is accelerated by a compressed spring whose spring constant is 640 N/m. After leaving the spring at the spring's relaxed length, the block travels over a horizontal surface, with a coefficient of kinetic friction of 0.25, for a distance of 7.8 m before stopping. (a) How much mechanical energy was dissipated by the frictional force in stopping the block? (b) What was the maximum kinetic energy of the block? (c) Through what distance was the spring compressed before the block began to move?

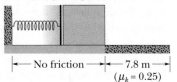

FIGURE 8-52   Problem 69.

—No friction—→ ←—7.8 m—→
($\mu_k = 0.25$)

70P. In Fig. 8-53, a block is moved down an incline a distance of 5.0 m from point A to point B by a force $\mathbf{F}$ that is parallel to the incline and has a magnitude of 2.0 N. The magnitude of the frictional force acting on the block is 10 N. If the kinetic energy of the block increases by 35 J between A and B, how much work is done on the block by its weight as the block moves from A to B?

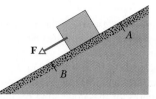

FIGURE 8-53   Problem 70.

71P. Fasten one end of a vertical spring to a ceiling, attach a cabbage to the other end, and then slowly lower the cabbage until the upward force exerted by the spring on the cabbage balances the weight of the cabbage. Show that the loss of gravitational potential energy of the cabbage–Earth system equals twice the gain in the spring's potential energy. Why are these two quantities not equal?

72P. You push a 2.0 kg block against a horizontal spring, compressing the spring by 15 cm. When you release the block, the spring forces it to slide across a tabletop. It stops 75 cm from where you released it. The spring constant is 200 N/m. What is the coefficient of kinetic friction between the block and the table?

73P. A moving 2.5 kg block (Fig. 8-54) collides with a horizontal spring whose spring constant is 320 N/m. The block compresses the spring a maximum distance of 7.5 cm from its rest position. The coefficient of kinetic friction between the block and the horizontal surface is 0.25. (a) How much work is done by the spring in bringing the block to rest? (b) How much mechanical energy is dissipated by the force of friction while the block is being brought to rest by the spring? (c) What was the speed of the block when it hit the spring?

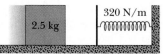

320 N/m

2.5 kg

FIGURE 8-54   Problem 73.

74P. Two snowy peaks are 850 m and 750 m above the valley between them. A ski run extends down from the top of the higher peak and then back up to the top of the lower one, with a total length of 3.2 km and an average slope of 30° (Fig. 8-55). (a) A skier starts from rest on the higher peak. At what speed will he arrive at the top of the lower peak if he just coasts without using the poles? Ignore friction. (b) Approximately what coefficient of kinetic friction between snow and skis would make him stop just at the top of the lower peak?

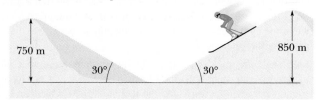

750 m   850 m

30°   30°

FIGURE 8-55   Problem 74.

75P. A factory worker accidentally releases a 400 lb crate that was being held at rest at the top of a 12 ft long ramp inclined at 39° to the horizontal. The coefficient of kinetic friction between the crate and the ramp, and between the crate and the horizontal factory floor, is 0.28. (a) How fast is the crate moving as it reaches the bottom of the ramp? (b) How far will it subsequently slide across the factory floor? (Assume that the crate's kinetic energy does not change as it moves from the ramp onto the floor.) (c) Why don't the answers to (a) and (b) depend on the mass of the crate?

76P. Two blocks are connected by a string, as shown in Fig. 8-56. They are released from rest. Show that, after they have moved a distance L, their common speed is given by

$$v = \sqrt{\frac{2(m_2 - \mu m_1)gL}{m_1 + m_2}},$$

in which μ is the coefficient of kinetic friction between the upper block and the surface. Assume that the pulley is massless and frictionless.

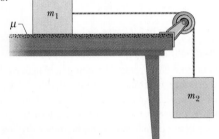

FIGURE 8-56 Problem 76.

77P. A 4.0 kg bundle starts up a 30° incline with 128 J of kinetic energy. How far will it slide up the plane if the coefficient of friction is 0.30?

78P. A cookie jar is moving up a 40° incline. At a point 1.8 ft from the bottom of the incline (measured along the incline), it has a speed of 4.5 ft/s. The coefficient of kinetic friction between jar and incline is 0.15. (a) How much farther up the incline will the jar move? (b) How fast will it be going when it slides back to the bottom of the incline?

79P. A certain spring is found *not* to conform to Hooke's law. The force (in newtons) it exerts when stretched a distance x (in meters) is found to have magnitude $52.8x + 38.4x^2$ in the direction opposing the stretch. (a) Compute the work required to stretch the spring from $x = 0.500$ m to $x = 1.00$ m. (b) With one end of the spring fixed, a particle of mass 2.17 kg is attached to the other end of the spring when it is extended by an amount $x = 1.00$ m. If the particle is then released from rest, compute its speed at the instant the spring has returned to the configuration in which the extension is $x = 0.500$ m. (c) Is the force exerted by the spring conservative or nonconservative? Explain.

80P. A girl whose weight is 267 N slides down a 6.1 m playground slide that makes an angle of 20° with the horizontal. The coefficient of kinetic friction is 0.10. (a) Find the work done on her by her weight. (b) Find the amount of energy dissipated by the frictional force. (c) If she starts at the top with a speed of 0.457 m/s, what is her speed at the bottom?

81P. In Fig. 8-57, a block slides along a track from one level to a higher level, by moving through an intermediate valley. The track is frictionless until the block reaches the higher level. There a frictional force stops the block in a distance d. The block's initial speed v_0 is 6.0 m/s; the height difference h is 1.1 m; and the coefficient of kinetic friction μ is 0.60. Find d.

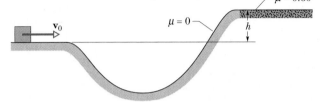

FIGURE 8-57 Problem 81.

82P. In the hydrogen atom, the magnitude of the force of attraction between the positively charged nucleus (a proton) and the negatively charged electron is given by

$$F = k\frac{e^2}{r^2},$$

where e is the magnitude of the charge of the electron and the proton, k is a constant, and r is the separation between electron and nucleus. Assume that the nucleus is fixed in place. Imagine that the electron, which is initially moving in a circle of radius r_1 about the nucleus, suddenly "jumps" into a circular orbit of smaller radius r_2 (Fig. 8-58). (a) Calculate the change in kinetic energy of the electron, using Newton's second law. (b) Using the relation between force and potential energy, calculate the change in potential energy of the atom. (c) By how much has the atom's total energy decreased in this process? (The total energy of the atom must decrease to provide the energy of the light that the atom emits because of the electron's jump.)

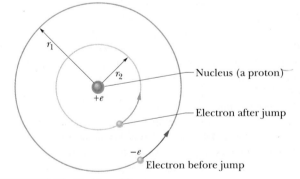

FIGURE 8-58 Problem 82.

83P. A stone with weight w is thrown vertically upward into the air from ground level with initial speed v_0. If a constant force f due to air drag acts on the stone throughout its flight, (a) show that the maximum height reached by the stone is

$$h = \frac{v_0^2}{2g(1 + f/w)}.$$

(b) Show that the speed of the stone just before impact with the ground is

$$v = v_0 \left(\frac{w - f}{w + f}\right)^{1/2}.$$

84P. A playground slide is in the form of an arc of a circle with a maximum height of 4.0 m, with a radius of 12 m, and with the ground tangent to the circle (Fig. 8-59). A 25 kg child starts from rest at the top of the slide and is observed to have a speed of 6.2 m/s at the bottom. (a) What is the length of the slide? (b) What average frictional force acts on the child over this distance? If, instead of the ground, a vertical line through the *top of the slide* is tangent to the circle, what are (c) the length of the slide and (d) the average frictional force on the child?

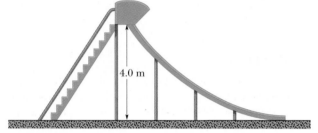

FIGURE 8-59 Problem 84.

85P. A particle can slide along a track with elevated ends and a flat central part, as shown in Fig. 8-60. The flat part has length L. The curved portions of the track are frictionless, but for the flat part the coefficient of kinetic friction is $\mu_k = 0.20$. The particle is released from rest at point A, which is a height $h = L/2$ above the flat part of the track. Where does the particle finally come to rest?

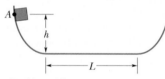

FIGURE 8-60 Problem 85.

86P. A massless rigid rod of length L has a ball of mass m attached to one end (Fig. 8-61). The other end is pivoted in such a way that the ball will move in a vertical circle. The system is launched downward from the horizontal position A with initial speed v_0. The ball just reaches point D and then stops. (a) Derive an expression for v_0 in terms of L, m, and g. (b) What is the tension in the rod when the ball is at B? (c) A little grit is placed on the pivot, after which the ball just reaches C when launched from A with the same speed as before. How much mechanical energy is dissipated by friction during this motion? (d) How much total mechanical energy has been dissipated by friction when the ball finally comes to rest at B after several oscillations?

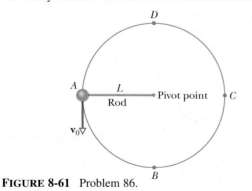

FIGURE 8-61 Problem 86.

87P. The cable of the 4000 lb elevator in Fig. 8-62 snaps when the elevator is at rest at the first floor, where the bottom is a distance $d = 12$ ft above a cushioning spring whose spring constant is $k = 10,000$ lb/ft. A safety device clamps the elevator against guide rails so that a constant frictional force of 1000 lb opposes the motion of the elevator. (a) Find the speed of the elevator just before it hits the spring. (b) Find the maximum distance x that the spring is compressed. (c) Find the distance that the elevator will bounce back up the shaft. (d) Using conservation of energy, find the approximate total distance that the elevator will move before coming to rest. Why is the answer not exact?

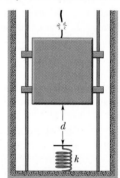

FIGURE 8-62 Problem 87.

88P. A metal tool is sharpened by being held against the rim of a wheel on a grinding machine by a force of 180 N. The wheel has a radius of 20 cm and rotates at 2.5 rev/s. The coefficient of kinetic friction between the wheel and the tool is 0.32. At what rate is energy being transferred from the motor driving the wheel to the thermal energy of the wheel and tool and to the kinetic energy of the material thrown from the tool?

89P*. At a factory, a 300 kg crate is dropped vertically from a packing machine onto a conveyor belt moving at 1.20 m/s (Fig. 8-63). (A motor maintains the belt's constant speed.) The coefficient of kinetic friction between belt and crate is 0.400. After a short time, slipping between the belt and the crate ceases and the crate then moves along with the belt. For the period of time during which the crate is being brought to rest relative to the belt, calculate, for a coordinate system at rest in the factory, (a) the kinetic energy supplied to the crate, (b) the magnitude of the kinetic frictional force acting on the crate, and (c) the energy supplied by the motor. (d) Explain why the answers to (a) and (c) are different.

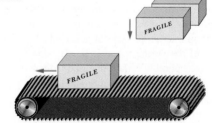

FIGURE 8-63 Problem 89.

SECTION 8-8 Mass and Energy

90E. (a) How much energy in joules is represented by a mass of 102 g? (b) For how many years would this supply the energy needs of a one-family home consuming energy at the average rate of 1.00 kW?

91E. The magnitude M of an earthquake on the so-called Richter scale is related to the released energy E in joules by the equation

$$\log E = 5.24 + 1.44M.$$

(a) The 1906 San Francisco earthquake was of magnitude 8.2 (Fig. 8-64). How much energy was released? (b) How much mass is equivalent to this amount of energy?

92E. The United States generated about 2.31×10^{12} kW·h of electrical energy in 1983. What is the mass equivalent of this electrical energy?

93P. What is the minimum energy that is required to break a nucleus of ^{12}C (of mass 11.99671 u) into three nuclei of ^{4}He (of mass 4.00151 u each)?

94P. The nucleus of a gold atom contains 79 protons and 118 neutrons and has a mass of 196.9232 u. What is its binding energy? See Sample Problem 8-10 for additional data.

95P. In the nuclear fusion reaction d + t → ^{4}He + n, a deuteron (d) combines with the nucleus of a tritium atom (t, with one proton and two neutrons) to produce the nucleus of a helium atom (with two protons and two neutrons) and a free neutron (n). The masses involved are

d: 2.01355 u ^{4}He: 4.00151 u

t: 3.01550 u   n: 1.00867 u

(a) Does the reaction release or absorb energy? (b) How much energy is absorbed or released?

SECTION 8-9 Quantized Energy

96E. Here are the lowest five energy levels, in electron-volts, of atoms of three types:

FIGURE 8-64 Exercise 91. Destruction on Nob Hill in San Francisco due to the earthquake of 1906. Over 400 km of the San Andreas fault line ruptured during the earthquake.

TYPE A	TYPE B	TYPE C
3.8	3.2	3.1
3.0	2.7	2.9
2.4	2.0	2.2
1.4	1.2	1.5
0	0	0

Two emissions of light from atoms are detected at the energies of (a) 1.4 eV and (b) 1.5 eV. For each emission, identify all transitions that could have produced the light.

97E. An atom that is being monitored emits light five times without being re-excited. The energies detected for four of those emissions are 0.7 eV, 0.8 eV, 0.9 eV, and 2.0 eV. The energy information for the other emission was lost by the computer controlling the detectors, as was information about the sequence of detection. The lowest 12 energy levels of the atom, in electron-volts, are

6.5	4.1	2.6
5.3	3.8	2.0
4.9	3.4	1.5
4.5	2.9	0

(a) From which energy level did the atom begin its quantum jumping downward in energy? (b) What value of energy was lost by the computer?

ELECTRONIC COMPUTATION

98. A nerfball™ has a mass of 9.8 g and a terminal velocity in air of 7.3 m/s. It is thrown straight up with an initial speed of 15 m/s. Use numerical integration to calculate the coordinate and velocity of the ball every 0.1 s from the time it is thrown to the time it returns to its initial position. For each of those times calculate the potential energy, kinetic energy, and total mechanical energy. On the way up how much of the mechanical energy of the ball-Earth system is dissipated? How much is dissipated on the way down? Take the potential energy to be zero initially.

99. A 700 g block is released from rest at height h_0 above a vertical spring with negligible mass and a spring constant $k = 400$ N/m. The block sticks to the spring and momentarily comes to rest after compressing the spring 19.0 cm. How much work is done (a) by the block on the spring and (b) by the spring on the block? (c) What was the value of h_0? (d) If the block was released from a height $2h_0$ above the spring, what would be the maximum compression of the spring?

100. To make a pendulum, a 300 g mass is tied to the end of a string 1.4 m long. The mass is pulled to one side until the string makes an angle of 30.0° with the vertical; then (with the string taut) the mass is released from rest. Find (a) the speed of the mass when the string makes an angle of 20.0° with the vertical and (b) the maximum speed of the mass. (c) What is the angle between the string and the vertical when the speed of the mass is one third its maximum value?

9
Systems of Particles

If you leap forward, chances are that your head and torso will follow a parabolic path, like a baseball thrown in from the outfield. But when a skilled ballet dancer leaps across the stage in a <u>grand jeté</u> as shown in the photograph, the path taken by her head and torso is nearly horizontal during much of the jump. She seems to be floating across the stage. The audience may not know about gravitational attraction, but they still sense that something unusual has happened. How does the ballerina seemingly "turn off" gravity?

9-1 A SPECIAL POINT

Physicists love to look at something complicated and find in it something simple and familiar. Here is an example. If you flip a baseball bat into the air, its motion as it turns is clearly more complicated than that of, say, a nonspinning tossed ball (Fig. 9-1a), which moves like a particle. Every part of the bat moves in a different way from every other part, so you cannot represent the bat as a particle that is tossed into the air; instead, it is a system of particles.

However, if you look closely, you will find that one special point of the bat moves in a simple parabolic path, just as a particle would if tossed into the air (Fig. 9-1b). In fact, that special point moves as though (1) the bat's total mass were concentrated there and (2) the weight of the bat acted only there. This special point is said to be the **center of mass** of the bat. In general:

> The center of mass of a body or a system of bodies is the point that moves as though all of the mass were concentrated there and all external forces were applied there.

The center of mass of a baseball bat lies along the central axis. You can locate it by balancing the bat horizontally on an outstretched finger: the center of mass is on the bat's axis just above your finger.

9-2 THE CENTER OF MASS

We shall now spend some time determining how to find the center of mass in various systems. We start with a system of a few particles, and then we consider a system of a great many particles (as in a baseball bat).

Systems of Particles

Figure 9-2a shows two particles of masses m_1 and m_2 separated by a distance d. We have arbitrarily chosen the origin of the x axis to coincide with m_1. We *define* the position of the center of mass of this two-particle system to be

$$x_{\text{cm}} = \frac{m_2}{m_1 + m_2} d. \qquad (9\text{-}1)$$

Suppose, as an example, that $m_2 = 0$. Then there is only one particle (m_1), and the center of mass must lie at the position of that particle; Eq. 9-1 dutifully reduces to $x_{\text{cm}} = 0$. If $m_1 = 0$, there is again only one particle (m_2), and we have, as we expect, $x_{\text{cm}} = d$. If $m_1 = m_2$, the masses of the particles are equal and the center of mass

(a)

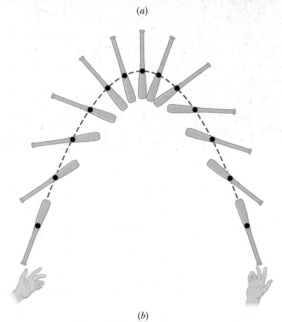

(b)

FIGURE 9-1 (a) A ball tossed into the air follows a parabolic path. (b) The center of mass (the black dot) of a baseball bat that is flipped into the air does also, but all other points of the bat follow more complicated curved paths.

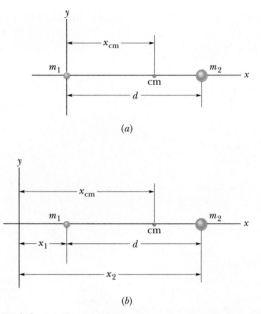

(a)

(b)

FIGURE 9-2 (a) Two particles of masses m_1 and m_2 are separated by a distance d. The dot labeled cm shows the position of the center of mass, calculated from Eq. 9-1. (b) The same as (a) except that the origin is located farther from the particles. The position of the center of mass is calculated from Eq. 9-2. The relative location of the center of mass (with respect to the particles) is the same in both cases.

should be halfway between them; Eq. 9-1 reduces to $x_{cm} = \frac{1}{2}d$, again as we expect. Finally, Eq. 9-1 tells us that if neither m_1 nor m_2 is zero, x_{cm} can have only values that lie between zero and d; that is, the center of mass must lie somewhere between the two particles.

Figure 9-2b shows a more generalized situation, in which the coordinate system has been shifted leftward. The position of the center of mass is now defined as

$$x_{cm} = \frac{m_1 x_1 + m_2 x_2}{m_1 + m_2}. \tag{9-2}$$

Note that if we put $x_1 = 0$, then x_2 becomes d and Eq. 9-2 reduces to Eq. 9-1, as it must. Note also that in spite of the shift of the coordinate system, the center of mass is still the same distance from each particle.

We can rewrite Eq. 9-2 as

$$x_{cm} = \frac{m_1 x_1 + m_2 x_2}{M}, \tag{9-3}$$

in which M is the total mass of the system. Here, $M = m_1 + m_2$. We can extend this equation to a more general situation in which n particles are strung out along the x axis. Then the total mass is $M = m_1 + m_2 + \cdots + m_n$,

and the location of the center of mass is

$$x_{cm} = \frac{m_1 x_1 + m_2 x_2 + m_3 x_3 + \cdots + m_n x_n}{M}$$

$$= \frac{1}{M} \sum_{i=1}^{n} m_i x_i. \tag{9-4}$$

Here the subscript i is a running number, or index, that takes on all integer values from 1 to n. It identifies the various particles, their masses, and their x coordinates.

If the particles are distributed in three dimensions, the center of mass must be identified by three coordinates. By extension of Eq. 9-4, they are

$$x_{cm} = \frac{1}{M} \sum_{i=1}^{n} m_i x_i,$$

$$y_{cm} = \frac{1}{M} \sum_{i=1}^{n} m_i y_i, \tag{9-5}$$

$$z_{cm} = \frac{1}{M} \sum_{i=1}^{n} m_i z_i.$$

We can also define the center of mass with the language of vectors. The position of a particle whose coordinates are x_i, y_i, and z_i is given by a position vector:

$$\mathbf{r}_i = x_i \mathbf{i} + y_i \mathbf{j} + z_i \mathbf{k}. \tag{9-6}$$

Here the index identifies the particle, and $\mathbf{i}$, $\mathbf{j}$, and $\mathbf{k}$ are unit vectors pointing, respectively, in the direction of the x, y, and z axes. Similarly, the position of the center of mass of a system of particles is given by a position vector:

$$\mathbf{r}_{cm} = x_{cm} \mathbf{i} + y_{cm} \mathbf{j} + z_{cm} \mathbf{k}. \tag{9-7}$$

The three scalar equations of Eq. 9-5 can now be replaced by a single vector equation,

$$\mathbf{r}_{cm} = \frac{1}{M} \sum_{i=1}^{n} m_i \mathbf{r}_i, \tag{9-8}$$

where again M is the total mass of the system. You can check that this equation is correct by substituting Eqs. 9-6 and 9-7 into it, and then separating out the x, y, and z components. The scalar relations of Eq. 9-5 result.

Rigid Bodies

An ordinary object, such as a baseball bat, contains so many particles (atoms) that we can best treat it as a continuous distribution of matter. The "particles" then become differential mass elements dm, the sums of Eq. 9-5 become integrals, and the coordinates of the center of mass are defined as

$$x_{cm} = \frac{1}{M} \int x \, dm,$$

$$y_{cm} = \frac{1}{M} \int y \, dm, \qquad (9\text{-}9)$$

$$z_{cm} = \frac{1}{M} \int z \, dm,$$

where M is now the mass of the object. The integrals are to be evaluated over all the mass elements in the object. In practice, however, we rewrite them in terms of the coordinates of the mass elements. If the object has uniform density ρ (mass per volume), then we can write

$$\rho = \frac{dm}{dV} = \frac{M}{V}, \qquad (9\text{-}10)$$

where dV is the volume occupied by a mass element dm, and V is the total volume of the object (ρ is ''rho''). We next substitute dm from Eq. 9-10 into Eq. 9-9, finding

$$x_{cm} = \frac{1}{V} \int x \, dV,$$

$$y_{cm} = \frac{1}{V} \int y \, dV, \qquad (9\text{-}11)$$

$$z_{cm} = \frac{1}{V} \int z \, dV.$$

These integrals are evaluated over the volume of the object. An example is given in Sample Problem 9-4.

Many objects have a point, a line, or a plane of symmetry. The center of mass of such an object then lies at that point, on that line, or in that plane. For example, the center of mass of a homogeneous sphere (which has a point of symmetry) is at the center of the sphere. The center of mass of a homogeneous cone (whose axis is a line of symmetry) lies on the axis of the cone. The center of mass of a banana (which has a plane of symmetry that splits it into two equal parts) lies somewhere in that plane.

The center of mass of an object need not lie within the object. There is no dough at the center of mass of a doughnut, and no iron at the center of mass of a horseshoe.

SAMPLE PROBLEM 9-1

Figure 9-3 shows three particles of masses $m_1 = 1.2$ kg, $m_2 = 2.5$ kg, and $m_3 = 3.4$ kg located at the corners of an equilateral triangle of edge $a = 140$ cm. Where is the center of mass?

SOLUTION: We choose our x and y coordinate axes so that one of the particles is located at the origin and the x axis coincides with one of the sides of the triangle. The three particles then have the following coordinates:

PARTICLE	MASS (kg)	x (cm)	y (cm)
m_1	1.2	0	0
m_2	2.5	140	0
m_3	3.4	70	121

Because of our wise choice of coordinate axes, three of the coordinates in the table are zero, simplifying the calculations. The total mass M of the system is 7.1 kg.

From Eq. 9-5, the coordinates of the center of mass are

$$x_{cm} = \frac{1}{M} \sum_{i=1}^{3} m_i x_i = \frac{m_1 x_1 + m_2 x_2 + m_3 x_3}{M}$$

$$= \frac{(1.2 \text{ kg})(0) + (2.5 \text{ kg})(140 \text{ cm}) + (3.4 \text{ kg})(70 \text{ cm})}{7.1 \text{ kg}}$$

$$= 83 \text{ cm} \qquad \text{(Answer)}$$

and

$$y_{cm} = \frac{1}{M} \sum_{i=1}^{3} m_i y_i = \frac{m_1 y_1 + m_2 y_2 + m_3 y_3}{M}$$

$$= \frac{(1.2 \text{ kg})(0) + (2.5 \text{ kg})(0) + (3.4 \text{ kg})(121 \text{ cm})}{7.1 \text{ kg}}$$

$$= 58 \text{ cm}. \qquad \text{(Answer)}$$

In Fig. 9-3, the center of mass is located by the position vector $\mathbf{r}_{cm}$, with components x_{cm} and y_{cm}.

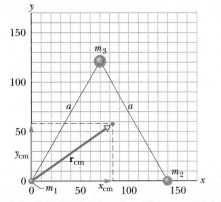

FIGURE 9-3 Sample Problem 9-1. Three particles having different masses form an equilateral triangle of side a. The center of mass is located by the position vector $\mathbf{r}_{cm}$.

SAMPLE PROBLEM 9-2

Find the center of mass of the uniform triangular plate that is shown in each part of Fig. 9-4.

SOLUTION: Figure 9-4a shows the plate divided into thin slats, parallel to one side of the triangle. From symmetry, the center of mass of a thin, uniform slat is at its midpoint. The center of mass of the triangular plate must then lie somewhere along the line that connects the midpoints of all the slats. That bisecting line also connects the upper vertex with the midpoint of the opposite side. The plate would balance if it were placed on a knife-edge coinciding with this line of symmetry.

In Figs. 9-4b and 9-4c, we subdivide the plate into slats parallel to the other two sides of the triangle. Again, the center of mass must lie somewhere along each of the bisecting lines shown. Hence the center of mass of the plate must lie at the intersection of these three symmetry lines, as Fig. 9-4d shows. It is the only point that the three lines have in common.

You can check this conclusion experimentally by taking advantage of the (correct) intuitive notion that an object suspended from a point will orient itself so that its center of mass lies vertically below that point. Suspend the triangle from each vertex in turn, and draw a line vertically downward from the suspension point, as in Fig. 9-4e. The center of mass of the triangle will be at the intersection of the three lines.

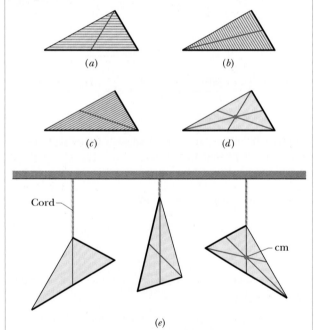

FIGURE 9-4 Sample Problem 9-2. In (a), (b), and (c), the triangular plate is divided into thin slats, parallel to a side. The center of mass must lie along the bisecting lines shown. (d) The dot, the only point common to all three lines, is the position of the center of mass. (e) Finding the center of mass by suspending the triangle from each vertex in turn.

CHECKPOINT 1: The figure shows a uniform square plate from which four identical squares at the corners will be removed. (a) Where is the center of mass of the

plate originally? Where is it after the removal of (b) square 1; (c) squares 1 and 2; (d) squares 1 and 3; (e) squares 1, 2, and 3; (f) all four squares? Answer in terms of quadrants, axes, or points (without calculation, of course).

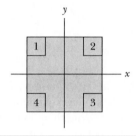

SAMPLE PROBLEM 9-3

Figure 9-5a shows a uniform circular metal plate of radius $2R$ from which a disk of radius R has been removed. Let us call this plate with a hole object X. Its center of mass is shown as a dot on the x axis. Locate this point.

SOLUTION: Figure 9-5b shows object X before the disk was removed. Call the disk object D, and the original composite plate object C. From symmetry, the center of mass of object C is at the origin of a coordinate system placed as shown.

In finding the center of mass of a composite object, we can assume that the masses of its components are concentrated at their individual centers of mass. Thus object C can be treated as equivalent to two point masses, representing objects X and D. Figure 9-5c shows the positions of the centers of mass of these three objects.

The position of the center of mass of object C is given, from Eq. 9-2, as

$$x_C = \frac{m_D x_D + m_X x_X}{m_D + m_X},$$

in which x_D and x_X are the positions of the centers of mass of objects D and X, respectively. Noting that $x_C = 0$ and solving for x_X, we obtain

$$x_X = -\frac{x_D m_D}{m_X}. \tag{9-12}$$

If ρ is the density (mass per volume) of the plate material and t is the uniform thickness of the plate, we have

$$m_D = \pi R^2 \rho t \quad \text{and} \quad m_X = \pi (2R)^2 \rho t - \pi R^2 \rho t.$$

With these substitutions and with $x_D = -R$, Eq. 9-12 becomes

$$x_X = -\frac{(-R)(\pi R^2 \rho t)}{\pi (2R)^2 \rho t - \pi R^2 \rho t} = \tfrac{1}{3}R. \quad \text{(Answer)}$$

Note that the uniform density and the uniform thickness of the plate cancel out and thus do not determine x_X.

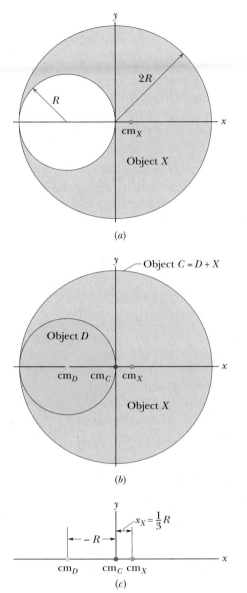

(a)

(b)

$x_X = \frac{1}{3}R$

$-R-$

cm_D cm_C cm_X

(c)

FIGURE 9-5 Sample Problem 9-3. (a) Object X is a metal disk of radius $2R$ with a hole of radius R cut in it; its center of mass is at point cm_X. (b) Object D is a metal disk that fills the hole in object X; its center of mass is at point cm_D, at $x = -R$. Object C is the composite object made up of objects X and D; its center of mass is at the origin of coordinates. (c) The centers of mass of the three objects.

SAMPLE PROBLEM 9-4

Silbury Hill (Fig. 9-6a), a mound on the plains near Stonehenge, was built 4600 years ago for unknown reasons, perhaps as a burial site. It is an incomplete right-circular cone (see Fig. 9-6b), with a flattened top of radius $r_2 = 16$ m, a base radius $r_1 = 88$ m, a height $h = 40$ m, and a volume V of $4.09 \times$

(a)

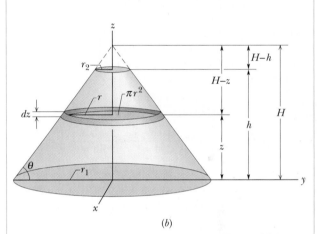

(b)

FIGURE 9-6 Sample Problem 9-4. (a) Silbury Hill in England, built by Neolithic people, required an estimated 1.8×10^7 hours of labor. (b) An incomplete, right-circular cone that resembles Silbury Hill. A ''wafer'' of radius r and thickness dz is shown at height z from the base.

10^5 m^3. The sides of the cone make an angle $\theta = 30°$ with the horizontal.

(a) Where is the center of mass of the mound?

SOLUTION: Because of the circular symmetry of the mound, the center of mass lies on the central vertical axis of the cone, at height z_{cm} above the base. To find z_{cm}, we use the last part of Eq. 9-11. We can simplify the integral by using the symmetry of the mound. To do this, we consider a thin, horizontal ''wafer,'' as shown in Fig. 9-6b. The wafer has radius r, thickness dz, and horizontal area πr^2 and is at height z from the base. Its volume dV is

$$dV = \pi r^2 \, dz. \qquad (9\text{-}13)$$

The mound consists of a stack of such wafers, with radii ranging from r_1 at the bottom of the stack to r_2 at the top. If the cone were complete, it would have a height that we call H in

Fig. 9-6*b*. The radius *r* of each wafer is related to *H* by

$$\tan \theta = \frac{H}{r_1} = \frac{H - z}{r},$$

or

$$r = (H - z)\frac{r_1}{H}. \tag{9-14}$$

Substituting Eqs. 9-13 and 9-14 into the last part of Eq. 9-11, we have

$$z_{cm} = \frac{1}{V} \int z \, dV = \frac{\pi r_1^2}{VH^2} \int_0^h z(H - z)^2 \, dz$$

$$= \frac{\pi r_1^2}{VH^2} \int_0^h (z^3 - 2z^2 H + zH^2) \, dz$$

$$= \frac{\pi r_1^2}{VH^2} \left[\frac{z^4}{4} - \frac{2z^3 H}{3} + \frac{z^2 H^2}{2} \right]_0^h$$

$$= \frac{\pi r_1^2 h^4}{VH^2} \left[\frac{1}{4} - \frac{2H}{3h} + \frac{H^2}{2h^2} \right].$$

Substituting known values, we find

$$z_{cm} = \frac{\pi (88 \text{ m})^2 (40 \text{ m})^4}{(4.09 \times 10^5 \text{ m}^3)(50.8 \text{ m})^2}$$

$$\times \left[\frac{1}{4} - \frac{2(50.8 \text{ m})}{3(40 \text{ m})} + \frac{(50.8 \text{ m})^2}{2(40 \text{ m})^2} \right]$$

$$= 12.37 \text{ m} \approx 12 \text{ m}. \qquad \text{(Answer)}$$

(b) If Silbury Hill has density $\rho = 1.5 \times 10^3$ kg/m³, then how much work was required to lift the dirt from the level of the base to build the mound?

SOLUTION: To find the work dW required to lift a mass element dm to height z, we use Eq. 7-21, with $\phi = 180°$:

$$dW = -dm \, gz \cos 180° = dm \, gz.$$

With Eq. 9-10, we substitute $\rho \, dV$ for dm, finding

$$dW = \rho gz \, dV.$$

To find the total work required to lift all the mass of Silbury Hill into place, we sum, via integration, the work dW associated with each volume element dV:

$$W = \int dW = \int \rho gz \, dV = \rho g \int z \, dV.$$

Using Eq. 9-11, we replace the integral with Vz_{cm}, finding

$$W = \rho V g z_{cm}. \tag{9-15}$$

This tells us that the work required to lift all the mass of Silbury Hill into place is the same as if all the mass were lifted to (and somehow concentrated at) the center of mass of the hill. Substituting the known data in Eq. 9-15 yields

$$W = (1.5 \times 10^3 \text{ kg/m}^3)(4.09 \times 10^5 \text{ m}^3)$$

$$\times (9.8 \text{ m/s}^2)(12.37 \text{ m})$$

$$= 7.4 \times 10^{10} \text{ J}. \qquad \text{(Answer)}$$

PROBLEM SOLVING TACTICS

TACTIC 1: *Center-of-Mass Problems*

Sample Problems 9-1 to 9-3 provide three strategies for simplifying center-of-mass problems. (1) Make full use of the symmetry of the object, be it point, line, or plane. (2) If the object can be divided into several parts, treat each of these parts as a particle, located at its own center of mass. (3) Choose your axes wisely: if your system is a group of particles, choose one of the particles as your origin. If your system is a body with a line of symmetry, there is your *x* axis! The choice of origin is completely arbitrary; the location of the center of mass is the same regardless of the origin from which it is measured.

9-3 NEWTON'S SECOND LAW FOR A SYSTEM OF PARTICLES

If you roll a cue ball at a second billiard ball that is at rest, you expect that the two-ball system will continue to have some forward motion after impact. You would be surprised, for example, if both balls came back toward you or if both moved to the right or to the left.

What continues to move forward, its steady motion completely unaffected by the collision, is the center of mass of the two balls. If you focus on this point—which is always halfway between these bodies of identical mass—you can easily convince yourself by trial at a pool table that this is so. No matter whether the collision is glancing, head on, or somewhere in between, the center of mass moves majestically forward, as if the collision had never occurred. Let us look into this in more detail.

To do so, we replace the pair of billiard balls with an assemblage of *n* particles of (possibly) different masses. We are interested not in the individual motions of these particles but *only* in the motion of their center of mass. Although the center of mass is just a point, it moves like a particle whose mass is equal to the total mass of the system; we can assign a position, a velocity, and an acceleration to it. We state (and shall prove below) that the (vector) equation that governs the motion of the center of mass of such a system of particles is

$$\sum \mathbf{F}_{ext} = M\mathbf{a}_{cm} \quad \text{(system of particles)}. \tag{9-16}$$

Equation 9-16 is Newton's second law governing the motion of the center of mass of a system of particles. Remarkably, it retains the same form ($\Sigma \mathbf{F} = m\mathbf{a}$) that holds for the motion of a single particle. In using Eq. 9-16, the three quantities that appear in it must be evaluated with some care:

1. $\Sigma\mathbf{F}_{ext}$ is the *vector* sum of *all* the *external forces* that act on the system. Forces exerted by one part of the system on another are called *internal forces,* and you must be careful not to include them when using Eq. 9-16.

2. M is the *total mass* of the system. We assume that no mass enters or leaves the system as it moves, so that M remains constant. The system is said to be **closed.**

3. $\mathbf{a}_{cm}$ is the acceleration of the *center of mass* of the system. Equation 9-16 gives no information about the acceleration of any other point of the system.

Equation 9-16, like all other vector equations, is equivalent to three equations involving the components of $\Sigma\mathbf{F}_{ext}$ and $\mathbf{a}_{cm}$ along the three coordinate axes. These equations are

$$\sum F_{ext,x} = Ma_{cm,x},$$
$$\sum F_{ext,y} = Ma_{cm,y}, \qquad (9\text{-}17)$$
$$\sum F_{ext,z} = Ma_{cm,z}.$$

Now we can go back and examine the behavior of the billiard balls. Once the cue ball has begun to roll, no net external force acts on the (two-ball) system. And thus, because $\Sigma\mathbf{F}_{ext} = 0$, Eq. 9-16 tells us that $\mathbf{a}_{cm} = 0$ also. Because acceleration is the rate of change of velocity, we conclude that the velocity of the center of mass of the system of two balls does not change. When the two balls collide, the forces that come into play are *internal* forces, exerted by one ball on the other. Such forces do not contribute to $\Sigma\mathbf{F}_{ext}$, which remains zero. Thus the center of mass of the system, which was moving forward before the collision, must continue to move forward after the collision, with the same speed and in the same direction.

Equation 9-16 applies not only to a system of particles but also to a solid body, such as the bat of Fig. 9-1*b*. In that case, M in Eq. 9-16 is the mass of the bat and $\Sigma\mathbf{F}_{ext}$ is the weight $M\mathbf{g}$ of the bat. Equation 9-16 then tells us that $\mathbf{a}_{cm} = \mathbf{g}$. In other words, the center of mass of the bat moves as if the bat were a single particle of mass M, with force $M\mathbf{g}$ acting on it.

Figure 9-7 shows another interesting case. Suppose that at a fireworks display, a rocket is launched on a parabolic path. At a certain point, it explodes into fragments. If the explosion had not occurred, the rocket would have continued along the trajectory shown in the figure. The forces of the explosion are *internal* to the system (the rocket or its fragments); that is, they are forces exerted by one part of the system on another part. If we ignore air resistance, the total *external* force $\Sigma\mathbf{F}_{ext}$ acting on the system is the weight $M\mathbf{g}$ of the system, regardless of whether the rocket explodes. Thus, from Eq. 9-16, the acceleration

FIGURE 9-7 A fireworks rocket explodes in flight. In the absence of air drag, the center of mass of the fragments would continue to follow the original parabolic path, until fragments began to hit the ground.

of the center of mass of the fragments (while they are in flight) remains equal to $\mathbf{g}$, and the center of mass of the fragments follows the same parabolic trajectory that the unexploded rocket would have followed.

When a ballet dancer leaps across the stage in a grand jeté, she raises her arms and stretches her legs out horizontally as soon as her feet leave the stage (Fig. 9-8). These actions shift her center of mass upward through her body. Although the shifting center of mass faithfully follows a parabolic path across the stage, its movement relative to the body decreases the height that would be attained by the head and torso in a normal jump. The result is that the head and torso follow a nearly horizontal path.

Proof of Equation 9-16

Now let us prove this important equation. From Eq. 9-8 we have, for a system of n particles,

$$M\mathbf{r}_{cm} = m_1\mathbf{r}_1 + m_2\mathbf{r}_2 + m_3\mathbf{r}_3 + \cdots + m_n\mathbf{r}_n, \quad (9\text{-}18)$$

in which M is the system's total mass and $\mathbf{r}_{cm}$ is the vector locating the position of the system's center of mass.

Differentiating Eq. 9-18 with respect to time gives

$$M\mathbf{v}_{cm} = m_1\mathbf{v}_1 + m_2\mathbf{v}_2 + m_3\mathbf{v}_3 + \cdots + m_n\mathbf{v}_n. \quad (9\text{-}19)$$

Here $\mathbf{v}_i (= d\mathbf{r}_i/dt)$ is the velocity of the ith particle, and $\mathbf{v}_{cm}$ $(= d\mathbf{r}_{cm}/dt)$ is the velocity of the center of mass.

Differentiating Eq. 9-19 with respect to time leads to

$$M\mathbf{a}_{cm} = m_1\mathbf{a}_1 + m_2\mathbf{a}_2 + m_3\mathbf{a}_3 + \cdots + m_n\mathbf{a}_n. \quad (9\text{-}20)$$

Here $a_i (= d\mathbf{v}_i/dt)$ is the acceleration of the ith particle, and $\mathbf{a}_{cm}$ $(= d\mathbf{v}_{cm}/dt)$ is the acceleration of the center of mass. Although the center of mass is just a geometrical point, it has a position, a velocity, and an acceleration, as if it were a particle.

From Newton's second law, $m_i\mathbf{a}_i$ is equal to the result-

Path of head

Path of center of mass

FIGURE 9-8 A grand jeté. (Adapted from *The Physics of Dance,* by Kenneth Laws, Schirmer Books, 1984.)

ant force $\mathbf{F}_i$ that acts on the ith particle. Thus we can rewrite Eq. 9-20 as

$$M\mathbf{a}_{cm} = \mathbf{F}_1 + \mathbf{F}_2 + \mathbf{F}_3 + \cdots + \mathbf{F}_n. \quad (9\text{-}21)$$

Among the forces that contribute to the right side of Eq. 9-21 will be forces that the particles of the system exert on each other (internal forces) and forces exerted on the particles from outside the system (external forces). By Newton's third law, the internal forces form action–reaction pairs and cancel out in the sum that appears on the right side of Eq. 9-21. What remains is the vector sum of all the *external* forces that act on the system. Equation 9-21 then reduces to Eq. 9-16, the relation that we set out to prove.

$\mathbb{C}$HECKPOINT **2:** Two skaters on frictionless ice hold opposite ends of a pole of negligible mass. An axis runs along the pole, and the origin of the axis is at the center of mass of the two-skater system. One skater, Fred, weighs twice a much as the other skater, Ethel. Where do the skaters meet if (a) Fred pulls hand over hand along the pole so as to draw himself to Ethel, (b) Ethel pulls hand over hand to draw herself to Fred, and (c) both skaters pull hand over hand?

SAMPLE PROBLEM 9-5

Figure 9-9a shows a system of three particles, each particle acted on by a different external force and all initially at rest. What is the acceleration of the center of mass of this system?

SOLUTION: The position of the center of mass, calculated by the method of Sample Problem 9-1, is marked by a dot in the figure. As Fig. 9-9b suggests, we treat this point as if it were a real particle, assigning to it a mass M equal to the total mass of the system (16 kg) and assuming that all external forces are applied at that point.

The x component of the net external force $\Sigma\mathbf{F}_{ext}$ acting on the center of mass is

$$\sum F_{ext,x} = 14 \text{ N} - 6.0 \text{ N} + (12 \text{ N})(\cos 45°)$$
$$= 16.5 \text{ N},$$

and the y component is

$$\sum F_{ext,y} = (12 \text{ N})(\sin 45°) = 8.49 \text{ N}.$$

The net external force thus has the magnitude

$$\sum F_{ext} = \sqrt{(16.5 \text{ N})^2 + (8.49 \text{ N})^2} = 18.6 \text{ N}$$

and makes an angle with the x axis given by

$$\theta = \tan^{-1} \frac{8.49 \text{ N}}{16.5 \text{ N}} = \tan^{-1} 0.515$$
$$= 27°. \qquad \text{(Answer)}$$

This is also the direction of the acceleration $\mathbf{a}_{cm}$ of the center of mass. From Eq. 9-16, the magnitude of $\mathbf{a}_{cm}$ is given by

$$a_{cm} = \frac{\sum F_{ext}}{M} = \frac{18.6 \text{ N}}{16 \text{ kg}} = 1.16 \text{ m/s}^2$$
$$\approx 1.2 \text{ m/s}^2. \qquad \text{(Answer)}$$

The three particles of Fig. 9-9a and their center of mass move with (different) constant accelerations. Since the particles start from rest, each will move, with ever-increasing speed, along a straight line in the direction of the force acting *on it*. The center of mass will move in the directon of $\mathbf{a}_{cm}$.

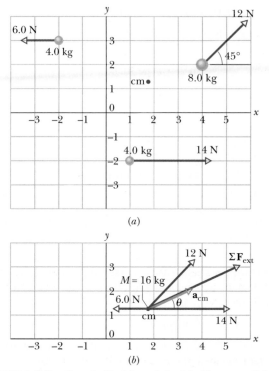

FIGURE 9-9 Sample Problem 9-5. (*a*) Three particles, placed at rest in the positions shown, are acted on by the external forces shown. The center of mass of the system is marked. (*b*) The forces are now transferred to the center of mass of the system, which behaves like a particle whose mass *M* is equal to the total mass of the system. The net force and the acceleration of the center of mass are shown.

9-4 LINEAR MOMENTUM

Momentum is another word that has several meanings in everyday language but only a single precise meaning in physics. The **linear momentum** of a particle is a vector **p**, defined as

$$\mathbf{p} = m\mathbf{v} \qquad \begin{array}{c}\text{(linear momentum}\\ \text{of a particle),}\end{array} \qquad (9\text{-}22)$$

in which *m* is the mass of the particle and **v** is its velocity. (The adjective *linear* is often dropped, but it serves to distinguish **p** from *angular* momentum, which is introduced in Chapter 12 and which is associated with rotation.) Note that since *m* is always a positive scalar quantity, Eq. 9-22 tells us that **p** and **v** are in the same direction. Equation 9-22 also tells us that the SI unit for momentum is the kilogram-meter per second.

Newton actually expressed his second law of motion in terms of momentum:

The rate of change of the momentum of a particle is proportional to the net force acting on the particle and is in the direction of that force.

In equation form this becomes

$$\sum\mathbf{F} = \frac{d\mathbf{p}}{dt}. \qquad (9\text{-}23)$$

Substituting for **p** from Eq. 9-22 gives

$$\sum\mathbf{F} = \frac{d\mathbf{p}}{dt} = \frac{d}{dt}\,(m\mathbf{v}) = m\,\frac{d\mathbf{v}}{dt} = m\mathbf{a}.$$

Thus the relations $\sum\mathbf{F} = d\mathbf{p}/dt$ and $\sum\mathbf{F} = m\mathbf{a}$ are completely equivalent expressions of Newton's second law of motion as it applies to the motion of single particles in classical mechanics.

Momentum at Very High Speeds

For particles moving with speeds that are near the speed of light, Newtonian mechanics predicts results that do not agree with experiment. In such cases, we must use Einstein's theory of special relativity. In relativity, the formulation $\mathbf{F} = d\mathbf{p}/dt$ is correct, *provided* we define the momentum of a particle not as $m\mathbf{v}$ but as

$$\mathbf{p} = \frac{m\mathbf{v}}{\sqrt{1 - (v/c)^2}}, \qquad (9\text{-}24)$$

in which *c*, the speed of light, is a sure indicator of a relativistic equation.

The speeds of common macroscopic objects such as baseballs, bullets, or space probes are so much lower than the speed of light that the quantity $(v/c)^2$ in Eq. 9-24 is very close to zero. Under these conditions, Eq. 9-24 reduces to Eq. 9-22 and Einstein's relativity theory reduces to Newtonian mechanics. For electrons and other subatomic particles, however, speeds very close to that of light are easily obtained and the definition in Eq. 9-24 *must* be used, often as a matter of routine engineering practice.

9-5 THE LINEAR MOMENTUM OF A SYSTEM OF PARTICLES

Consider now a system of *n* particles, each with its own mass, velocity, and linear momentum. The particles may interact with each other, and external forces may act on them as well. The system as a whole has a total linear momentum **P**, which is defined to be the vector sum of the individual particles' linear momenta. Thus

TABLE 9-1   SOME DEFINITIONS AND LAWS IN CLASSICAL MECHANICS

LAW OR DEFINITION	SINGLE PARTICLE	SYSTEM OF PARTICLES
Newton's second law	$\Sigma\mathbf{F} = m\mathbf{a}$	$\Sigma\mathbf{F}_{ext} = M\mathbf{a}_{cm}$ (9-16)
Linear momentum	$\mathbf{p} = m\mathbf{v}$ (9-22)	$\mathbf{P} = M\mathbf{v}_{cm}$ (9-26)
Newton's second law	$\Sigma\mathbf{F} = \dfrac{d\mathbf{p}}{dt}$ (9-23)	$\Sigma\mathbf{F}_{ext} = \dfrac{d\mathbf{P}}{dt}$ (9-28)
Work–kinetic energy theorem	$W = \Delta K$	

$$\mathbf{P} = \mathbf{p}_1 + \mathbf{p}_2 + \mathbf{p}_3 + \cdots + \mathbf{p}_n$$
$$= m_1\mathbf{v}_1 + m_2\mathbf{v}_2 + m_3\mathbf{v}_3 + \cdots + m_n\mathbf{v}_n. \qquad (9\text{-}25)$$

If we compare this equation with Eq. 9-19, we see that

$$\mathbf{P} = M\mathbf{v}_{cm} \qquad \begin{array}{l}\text{(linear momentum,}\\ \text{system of particles),}\end{array} \qquad (9\text{-}26)$$

which gives us another way to define the linear momentum of a system of particles:

The linear momentum of a system of particles is equal to the product of the total mass M of the system and the velocity of the center of mass.

If we take the time derivative of Eq. 9-26, we find

$$\frac{d\mathbf{P}}{dt} = M\,\frac{d\mathbf{v}_{cm}}{dt} = M\mathbf{a}_{cm}. \qquad (9\text{-}27)$$

Comparing Eqs. 9-16 and 9-27 allows us to write Newton's second law for a system of particles in the equivalent form

$$\sum\mathbf{F}_{ext} = \frac{d\mathbf{P}}{dt}. \qquad (9\text{-}28)$$

This equation is the generalization of the single-particle equation $\Sigma\mathbf{F} = d\mathbf{p}/dt$ to a system of many particles. Table 9-1 displays the important relations that we have derived for single particles and for systems of particles.

CHECKPOINT 3: The figure gives the linear momentum versus time for a particle moving along an axis. A force directed along the axis acts on the particle. (a) Rank the four regions indicated according to the magnitude of the force, greatest first. (b) In which region is the particle slowing?

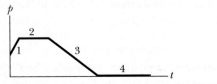

SAMPLE PROBLEM 9-6

Figure 9-10a shows a 2.0 kg toy race car before and after taking a turn on a track. Its speed is 0.50 m/s before the turn and 0.40 m/s after the turn. What is the change $\Delta\mathbf{P}$ in the linear momentum of the car?

SOLUTION: Before we can use Eq. 9-26 to find the car's linear momenta before and after the turn, we need its velocity $\mathbf{v}_i$ before the turn and its velocity $\mathbf{v}_f$ after the turn. Using the coordinate system of Fig. 9-10a, we write $\mathbf{v}_i$ and $\mathbf{v}_f$ as

$$\mathbf{v}_i = -(0.50 \text{ m/s})\mathbf{j} \quad \text{and} \quad \mathbf{v}_f = (0.40 \text{ m/s})\mathbf{i}.$$

Equation 9-26 then gives us the linear momentum $\mathbf{P}_i$ before the turn and the linear momentum $\mathbf{P}_f$ after the turn:

$$\mathbf{P}_i = M\mathbf{v}_i = (2.0 \text{ kg})(-0.50 \text{ m/s})\mathbf{j} = (-1.0 \text{ kg}\cdot\text{m/s})\mathbf{j}$$

and

$$\mathbf{P}_f = M\mathbf{v}_f = (2.0 \text{ kg})(0.40 \text{ m/s})\mathbf{i} = (0.80 \text{ kg}\cdot\text{m/s})\mathbf{i}.$$

Because these linear momenta are not along the same axis, we *cannot* find the change in linear momentum $\Delta\mathbf{P}$ by merely subtracting the magnitude of $\mathbf{P}_i$ from the magnitude of $\mathbf{P}_f$. Instead, we write the change in linear momentum as

$$\Delta\mathbf{P} = \mathbf{P}_f - \mathbf{P}_i \qquad (9\text{-}29)$$

and then as

$$\Delta\mathbf{P} = (0.80 \text{ kg}\cdot\text{m/s})\mathbf{i} - (-1.0 \text{ kg}\cdot\text{m/s})\mathbf{j}$$
$$= (0.8\mathbf{i} + 1.0\mathbf{j}) \text{ kg}\cdot\text{m/s}. \qquad \text{(Answer)}$$

Figure 9-10b shows $\Delta\mathbf{P}$, $\mathbf{P}_f$, and $-\mathbf{P}_i$. Note that we subtract $\mathbf{P}_i$ from $\mathbf{P}_f$ by adding $-\mathbf{P}_i$ to $\mathbf{P}_f$.

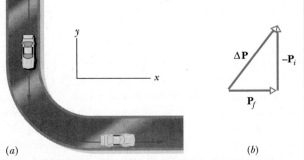

(a)   (b)

FIGURE 9-10   Sample Problem 9-6. (a) A toy car takes a turn. (b) The change $\Delta\mathbf{P}$ in the car's linear momentum is the vector sum of its final linear momentum $\mathbf{P}_f$ and the negative of its initial linear momentum $\mathbf{P}_i$.

9-6 CONSERVATION OF LINEAR MOMENTUM

Suppose that the sum of the external forces acting on a system of particles is zero (the system is isolated) and that no particles leave or enter the system (the system is closed). Putting $\Sigma\mathbf{F}_{ext} = 0$ in Eq. 9-28 then yields $d\mathbf{P}/dt = 0$, or

$$\mathbf{P} = \text{constant} \qquad \text{(closed, isolated system).} \qquad (9\text{-}30)$$

This important result, called the **law of conservation of linear momentum,** can also be written as

$$\mathbf{P}_i = \mathbf{P}_f \qquad \text{(closed, isolated system),} \qquad (9\text{-}31)$$

where the subscripts refer to the values of $\mathbf{P}$ at some initial time i and some later time f. Equations 9-30 and 9-31 tell us that if no net external force acts on a system of particles, the total linear momentum of the system remains constant.

Like the law of conservation of energy that you met in Chapter 8, the law of conservation of linear momentum is a more general law than Newtonian mechanics itself. It holds in the subatomic realm, where Newton's laws fail. It holds for the highest particle speeds, where Einstein's relativity theory prevails, that is, where Eq. 9-24, rather than Eq. 9-22, must be used to find the linear momentum.

From Eq. 9-26 ($\mathbf{P} = M\mathbf{v}_{cm}$) we see that if $\mathbf{P}$ is constant, then $\mathbf{v}_{cm}$, the velocity of the center of mass, is also constant. This in turn means that $\mathbf{a}_{cm}$, the acceleration of the center of mass, must be zero, which is consistent with Eq. 9-16 ($\Sigma\mathbf{F}_{ext} = M\mathbf{a}_{cm}$) when $\Sigma\mathbf{F}_{ext} = 0$. Thus the law of conservation of linear momentum is consistent with both forms of Newton's second law for systems of particles, as displayed in Table 9-1.

Equations 9-30 and 9-31 are vector equations and, as such, both are equivalent to three equations corresponding to the conservation of linear momentum in three mutually perpendicular directions. Depending on the forces acting on a system, linear momentum might be conserved in one or two directions but not in all directions:

If a component of the net *external* force on a closed system is zero along an axis, then the component of the linear momentum of the system along that axis cannot change.

As an example, suppose that you toss a grapefruit across a room. During its flight, the only external force acting on the grapefruit (which we take as the system) is its weight $m\mathbf{g}$, which is directed vertically downward. Thus, the vertical component of the linear momentum of the grapefruit changes, but since no horizontal external force acts on the grapefruit, the horizontal component of the linear momentum cannot change.

Note we focus on external forces acting on a closed system. Although internal forces can change the linear momentum of portions of the system, they cannot change the total linear momentum of the system itself.

CHECKPOINT 4: An initially stationary device lying on a frictionless floor explodes into two pieces, which then slide across the floor. One piece slides in the positive direction of an x axis. (a) What is the sum of the momenta of the two pieces after the explosion? (b) Can the second piece move at an angle to the x axis? (c) What is the direction of the momentum of the second piece?

SAMPLE PROBLEM 9-7

A box with mass $m = 6.0$ kg slides with speed $v = 4.0$ m/s across a frictionless floor in the positive direction of an x axis. It suddenly explodes into two pieces: one piece, with mass $m_1 = 2.0$ kg, moves in the positive direction of the x axis with speed $v_1 = 8.0$ m/s. What is the velocity of the second piece, with mass m_2?

SOLUTION: Our reference frame will be that of the floor. Our system, which initially consists of the original box and then of the two pieces, is closed but is not isolated, since the box and pieces have weight and each experiences a normal force from the floor. However, those forces are all vertical and thus cannot change the horizontal component of the momentum of the system. Neither can the forces produced by the explosion, because those forces are internal to the system. Thus, the horizontal component of the momentum of the system is conserved, and we can apply Eq. 9-31 along the x axis.

The initial momentum of the system was that of the box:

$$\mathbf{P}_i = m\mathbf{v}.$$

Similarly, we can write the final momenta of the two pieces as

$$\mathbf{P}_{f1} = m_1\mathbf{v}_1 \quad \text{and} \quad \mathbf{P}_{f2} = m_2\mathbf{v}_2.$$

The final total momentum $\mathbf{P}_f$ of the system is the vector sum of the momenta of the two pieces:

$$\mathbf{P}_f = \mathbf{P}_{f1} + \mathbf{P}_{f2} = m_1\mathbf{v}_1 + m_2\mathbf{v}_2.$$

Since all the velocities and momenta in this problem are vectors along a single axis, we can represent them by their magnitudes, along with an implied plus sign or an explicit minus sign. Doing so while applying Eq. 9-31, we now obtain

$$P_i = P_f$$

or

$$mv = m_1v_1 + m_2v_2.$$

Inserting the given data, and using the fact that the mass of the second piece is $m_2 = m - m_1 = 4.0$ kg, we find

$$(6.0 \text{ kg})(4.0 \text{ m/s}) = (2.0 \text{ kg})(8.0 \text{ m/s}) + (4.0 \text{ kg})v_2$$

and thus

$$v_2 = 2.0 \text{ m/s}. \qquad \text{(Answer)}$$

Since the result is positive, the second piece moves in the positive direction of the x axis.

SAMPLE PROBLEM 9-8

As Fig. 9-11 shows, a cannon whose mass M is 1300 kg fires a ball whose mass m is 72 kg in a horizontal direction with a velocity $\mathbf{v}$ relative to the cannon, which recoils (freely) with velocity $\mathbf{V}$ relative to the Earth. The magnitude of $\mathbf{v}$ is 55 m/s.

(a) What is $\mathbf{V}$?

SOLUTION: We choose the cannon plus the ball as our system. By doing so, we ensure that the forces involved in the firing of the cannon are internal to the system, and we do not have to deal with them. The external forces acting on the system have no components in the horizontal direction. Thus the horizontal component of the total linear momentum of the system must remain unchanged as the cannon is fired.

We choose the Earth as a reference frame. Because all the velocities here are horizontal, they are positive if they point rightward in Fig. 9-11 and negative if they point leftward. Although $\mathbf{V}$ is shown pointing leftward, we actually do not yet know its direction.

The velocity of the ball relative to the Earth is $\mathbf{v}_E$, represented by v_E. The ball's velocity relative to the cannon is the difference between its velocity relative to the Earth and the cannon's velocity relative to the Earth. We represent this as

$$v = v_E - V,$$

and thus

$$v_E = v + V. \qquad \text{(9-32)}$$

Before the cannon is fired, the system has an initial linear momentum P_i of zero. While the ball is in flight, the system has a horizontal linear momentum P_f which, with the aid of Eq. 9-32, is given by

$$P_f = MV + mv_E = MV + m(v + V),$$

in which the first term on the right is the linear momentum of the cannon and the second term that of the ball.

Conservation of linear momentum in the horizontal direction requires that $P_i = P_f$, or

$$0 = MV + m(v + V).$$

Solving for V yields

$$V = -\frac{mv}{M + m} = -\frac{(72 \text{ kg})(55 \text{ m/s})}{1300 \text{ kg} + 72 \text{ kg}}$$

$$= -2.9 \text{ m/s}. \qquad \text{(Answer)}$$

The minus sign tells us that the cannon recoils to the left in Fig. 9-11, as we know it does.

(b) What is v_E?

SOLUTION: From Eq. 9-32, we find

$$v_E = v + V = 55 \text{ m/s} + (-2.9 \text{ m/s})$$

$$= 52 \text{ m/s}. \qquad \text{(Answer)}$$

Because of the recoil, the ball is moving a little slower relative to the Earth than it would were there no recoil.

Note the importance in this problem of choosing the system (*cannon + ball*) wisely and of being clear about the reference frame (Earth or recoiling cannon) to which the various measurements are referenced.

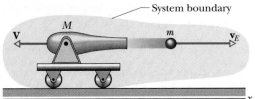

FIGURE 9-11 Sample Problem 9-8. A cannon of mass M fires a ball of mass m. The ball has velocity $\mathbf{v}_E$ relative to the Earth and velocity $\mathbf{v}$ relative to the cannon. The recoiling cannon has velocity $\mathbf{V}$ relative to the Earth.

SAMPLE PROBLEM 9-9

Imagine a spaceship and cargo module, of total mass M, traveling in deep space with velocity $v_i = 2100$ km/h relative to the Sun (Fig. 9-12a). With a small explosion, the ship ejects the cargo module, of mass 0.20M (Fig. 9-12b). The ship then travels 500 km/h faster than the module; that is, the relative speed v_{rel} between the module and the ship is 500 km/h. What then is the velocity v_f of the ship relative to the Sun?

SOLUTION: Because the ship–module system is closed and isolated, its total linear momentum is conserved; that is,

$$P_i = P_f, \qquad \text{(9-33)}$$

where the subscripts i and f refer to values before and after the ejection, respectively. Before the ejection, we have

$$P_i = Mv_i. \qquad \text{(9-34)}$$

Let U be the velocity of the ejected module relative to the Sun. The total linear momentum of the system after the ejection is then

$$P_f = (0.20M)U + (0.80M)v_f, \qquad \text{(9-35)}$$

where the first term on the right is the linear momentum of the module and the second term is that of the ship.

The relative speed v_{rel} between the module and the ship is the difference in their velocities

$$v_{\text{rel}} = v_f - U,$$

from which we write

$$U = v_f - v_{\text{rel}}.$$

Substituting this expression for U into Eq. 9-35, and then substituting Eqs. 9-34 and 9-35 into Eq. 9-33, we find

$$Mv_i = 0.20M(v_f - v_{rel}) + 0.80Mv_f,$$

which gives us

$$v_f = v_i + 0.20v_{rel},$$

or $\qquad v_f = 2100 \text{ km/h} + (0.20)(500 \text{ km/h})$

$$= 2200 \text{ km/h}. \qquad \text{(Answer)}$$

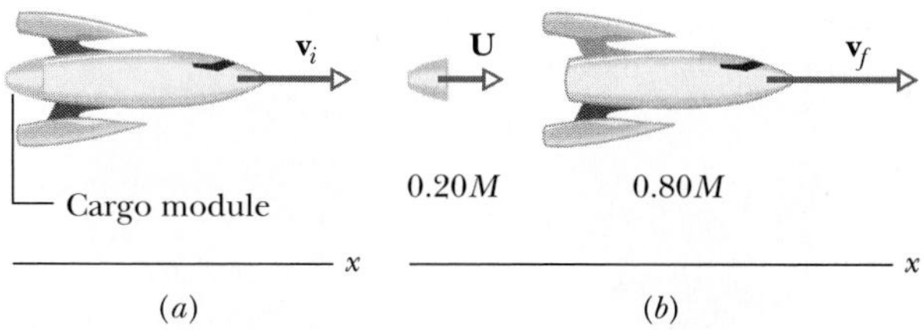

FIGURE 9-12 Sample Problem 9-9. (a) A spaceship, with a cargo module, moving at velocity $\mathbf{v}_i$. (b) The spaceship has ejected the cargo module. The spaceship now moves at velocity $\mathbf{v}_f$ and the cargo module moves at velocity $\mathbf{U}$.

CHECKPOINT **5:** The table gives velocities of the ship and module in Sample Problem 9-9 (after ejection and relative to the Sun), and the relative speed between the ship and the module for three situations. What are the missing values?

	VELOCITIES (km/h)		RELATIVE SPEED (km/h)
	MODULE	SHIP	
(a)	1500	2000	
(b)		3000	400
(c)	1000		600

SAMPLE PROBLEM 9-10

Figure 9-13 shows two blocks that are connected by an ideal spring and are free to slide on a frictionless horizontal surface. Block 1 has mass m_1 and block 2 has mass m_2. The blocks are pulled in opposite directions (stretching the spring) and then released from rest.

(a) What is the ratio v_1/v_2 of the velocity of block 1 to the velocity of block 2 as the separation between the blocks decreases?

SOLUTION: We take the two blocks and the spring as our system, and the horizontal surface on which they slide as our reference frame. The initial momentum P_i of the system before the blocks are released is zero. The momentum of the system at any instant as the blocks move toward each other is

$$P_f = m_1v_1 + m_2v_2.$$

Conservation of momentum requires that $P_i = P_f$, or

$$0 = m_1v_1 + m_2v_2. \qquad \text{(9-36)}$$

Thus we have

$$\frac{v_1}{v_2} = -\frac{m_2}{m_1}. \qquad \text{(9-37)}$$

The minus sign tells us that the two velocities always have opposite signs. Equation 9-37 holds at every instant after release, no matter what the actual speeds of the blocks are.

(b) What is the ratio K_1/K_2 of the kinetic energies of the blocks as their separation decreases?

SOLUTION: We can write the ratio K_1/K_2 as

$$\frac{K_1}{K_2} = \frac{\frac{1}{2}m_1v_1^2}{\frac{1}{2}m_2v_2^2} = \frac{m_1}{m_2}\left(\frac{v_1}{v_2}\right)^2.$$

Substituting for v_1/v_2 from Eq. 9-37 and simplifying, we find

$$\frac{K_1}{K_2} = \frac{m_2}{m_1}. \qquad \text{(9-38)}$$

As the blocks move toward each other and the stretch of the spring decreases, energy is transferred from the elastic potential energy of the spring to the kinetic energies K_1 and K_2 of the blocks. Although K_1 and K_2 then increase, Eq. 9-38 tells us that their ratio does not change but is preset by the ratio of the masses. After the spring reaches its rest length and begins to be compressed by the blocks, the energy transfer is reversed, but Eq. 9-38 still holds.

Equations 9-36 through 9-38 apply to other situations in which two bodies attract (or repel) each other. For example, they apply to a stone falling toward the Earth. The stone corresponds to, say, block 1 in Fig. 9-13 and the Earth to block 2. The gravitational force between the stone and the Earth corresponds to the mutual force between the blocks that is provided by the spring in Fig. 9-13. Our reference frame is the frame in which the center of mass of the stone–Earth system is stationary. In this frame, Eq. 9-36 tells us that the magnitudes of the linear momenta of the stone and the Earth remain equal throughout the fall. Equations 9-37 and 9-38 tell us that, because $m_2 \gg m_1$, the stone has much greater speed and kinetic energy than the Earth during the fall.

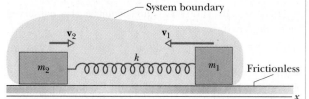

FIGURE 9-13 Sample Problem 9-10. Two blocks, resting on a frictionless surface and connected by a spring, have been pulled apart and then released from rest. The vector sum of their linear momenta remains zero during their subsequent motions. The system boundary is shown.

SAMPLE PROBLEM 9-11

A firecracker placed inside a coconut of mass M, initially at rest on a frictionless floor, blows the fruit into three pieces and sends them sliding across the floor. An overhead view is shown in Fig. 9-14a. Piece C, with mass $0.30M$, has final speed $v_{fC} = 5.0$ m/s.

(a) What is the speed of piece B, with mass $0.20M$?

SOLUTION: We superimpose an xy coordinate system as shown in Fig. 9-14b, with the direction of decreasing x coinciding with the direction of $\mathbf{v}_{fA}$. The x axis makes an angle of $80°$ with $\mathbf{v}_{fC}$ and an angle of $50°$ with $\mathbf{v}_{fB}$.

The linear momentum of the coconut (and its pieces) is conserved along both the x and y axes, because the forces involved in the explosion are internal forces, and because no external force acts on the coconut along the x or y axis. Along the y axis the conservation of linear momentum is written as

$$P_{iy} = P_{fy}, \tag{9-39}$$

where the subscript i refers to the initial value (before the explosion), and the subscript y refers to the y component of the vector $\mathbf{P}_i$ or $\mathbf{P}_f$.

The component P_{iy} of the initial linear momentum is zero, because the coconut is initially at rest. To get an expression for P_{fy}, we find the y component of the final linear momentum of each piece:

$$p_{fA,y} = 0,$$

$$p_{fB,y} = -0.20Mv_{fB,y} = -0.20Mv_{fB}\sin 50°,$$

$$p_{fC,y} = 0.30Mv_{fC,y} = 0.30Mv_{fC}\sin 80°.$$

(Note that $p_{fA,y} = 0$ because of our choice of axes.) Equation 9-39 can now be written as

$$P_{iy} = P_{fy} = p_{fA,y} + p_{fB,y} + p_{fC,y}.$$

Then, with $v_{fC} = 5.0$ m/s, we have

$$0 = 0 - 0.20Mv_{fB}\sin 50° + (0.30M)(5.0 \text{ m/s})\sin 80°,$$

from which we find

$$v_{fB} = 9.64 \text{ m/s} \approx 9.6 \text{ m/s.} \tag{Answer}$$

(b) What is the speed of piece A?

SOLUTION: Because linear momentum is also conserved along the x axis, we have

$$P_{ix} = P_{fx}, \tag{9-40}$$

where $P_{ix} = 0$ because the coconut is initially at rest. To get P_{fx}, we find the x components of the final momenta, using the fact that piece A must have a mass of $0.50M$:

$$p_{fA,x} = -0.50Mv_{fA},$$

$$p_{fB,x} = 0.20Mv_{fB,x} = 0.20Mv_{fB}\cos 50°,$$

$$p_{fC,x} = 0.30Mv_{fC,x} = 0.30Mv_{fC}\cos 80°.$$

Equation 9-40 can now be written as

$$P_{ix} = P_{fx} = p_{fA,x} + p_{fB,x} + p_{fC,x}.$$

Then, with $v_{fC} = 5.0$ m/s and $v_{fB} = 9.64$ m/s, we have

$$0 = -0.50Mv_{fA} + 0.20M(9.64 \text{ m/s})\cos 50°$$
$$+ 0.30M(5.0 \text{ m/s})\cos 80°,$$

from which we find

$$v_{fA} = 3.0 \text{ m/s.} \tag{Answer}$$

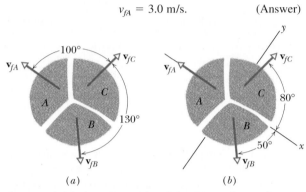

FIGURE 9-14   Sample Problem 9-11. Three pieces of an exploded coconut move off in three directions along a frictionless floor. (a) An overhead view of the event. (b) The same with a two-dimensional axis system imposed.

CHECKPOINT **6:** Suppose that the exploding coconut in Sample Problem 9-11 is accelerating in the negative direction of y in Fig. 9-14 (say it is on a ramp that slants downward in that direction). Is linear momentum conserved along (a) the x axis (as stated in Eq. 9-40) and (b) the y axis (as stated in Eq. 9-39)?

PROBLEM SOLVING TACTICS

TACTIC 2: *Conservation of Linear Momentum*

This is a good time to reread Tactic 2 of Chapter 8, which deals with problems involving the conservation of mechanical energy. The same broad suggestions hold here for problems involving the conservation of linear momentum.

First, make sure that you have chosen a closed, isolated system. *Closed* means that no matter (no particles) passes through the system boundary in any direction. *Isolated* means that the net external force acting on the system is zero. If the system is *not* isolated or *not* closed, then Eqs. 9-30 and 9-31 do not hold.

Remember that linear momentum is a vector quantity so that each component can be conserved separately, provided only that the corresponding component of the net external force is zero. In Sample Problem 9-8, no net external force acts horizontally on the *cannon + ball* system, so the horizontal component of linear momentum is conserved. However, the net external force acting vertically on this system is not zero; weight acts on the cannonball while it is in flight.

Thus the vertical component of linear momentum for this system is not conserved.

Select two appropriate states of the system (which you may choose to call the initial state and the final state) and write expressions for the linear momentum of the system in each of these two states. In writing these expressions, make sure that you know what inertial reference frame you are using, and make sure also that you include the entire system, not missing any part of it and not including objects that do not belong to your system. In Sample Problem 9-8 we must be particularly clear about the reference frame (Earth or recoiling cannon) to which the various measurements are referenced.

Finally, set your expressions for P_i and $\mathbf{P}_f$ equal to each other and solve for what is requested.

9-7 SYSTEMS WITH VARYING MASS: A ROCKET (OPTIONAL)

In the systems we have dealt with so far, we have assumed that the total mass of the system remains constant. Sometimes, as in a rocket (Fig. 9-15), it does not. Most of the mass of a rocket on its launching pad is fuel, all of which will eventually be burned and ejected from the nozzle of the rocket engine.

We handle the variation of the mass of the rocket as the rocket accelerates by applying Newton's second law, not to the rocket alone but to the rocket and its ejected

FIGURE 9-15 Liftoff of Project Mercury spacecraft.

combustion products taken together. The mass of *this* system does *not* change as the rocket accelerates.

Finding the Acceleration

Assume that you are in an inertial reference frame, watching a rocket accelerate through deep space with no gravitational or atmospheric drag forces acting on it. For this one-dimensional motion, let M be the mass of the rocket and v its velocity at an arbitrary time t (see Fig. 9-16a).

Figure 9-16b shows how things stand a time interval dt later. The rocket now has velocity $v + dv$ and mass $M + dM$, where the change in mass dM is a *negative quantity*. The exhaust products released by the rocket during interval dt have mass $-dM$ and velocity U relative to our inertial reference frame.

Our system consists of the rocket and the exhaust products released during interval dt. The system is closed and isolated, so the linear momentum of the system must be conserved during dt; that is,

$$P_i = P_f, \tag{9-41}$$

where the subscripts i and f indicate the values at the beginning and end of time interval dt. We can rewrite Eq. 9-41 as

$$Mv = -dM\,U + (M + dM)(v + dv), \tag{9-42}$$

where the first term on the right is the linear momentum of the exhaust products released during interval dt and the second term is the linear momentum of the rocket at the end of interval dt.

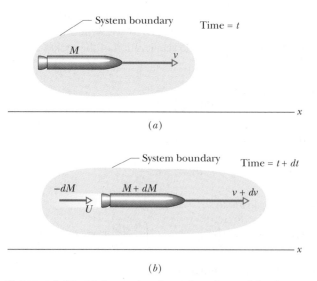

FIGURE 9-16 (a) An accelerating rocket of mass M at time t, as seen from an inertial reference frame. (b) The same but at time $t + dt$. The exhaust products released during interval dt are shown.

We can simplify Eq. 9-42 by using the speed u of the exhaust products relative to the rocket. We find this relative speed u by subtracting the velocity U of the exhaust products from the velocity $v + dv$ of the rocket at the end of interval dt:

$$u = (v + dv) - U.$$

This gives us

$$U = v + dv - u. \tag{9-43}$$

Substituting this result for U into Eq. 9-42 yields, with a little algebra,

$$-dM\, u = M\, dv. \tag{9-44}$$

Dividing each side by dt yields

$$-\frac{dM}{dt}\, u = M\, \frac{dv}{dt}. \tag{9-45}$$

We replace dM/dt (the rate at which the rocket loses mass) by $-R$, where R is the (positive) rate of fuel consumption. And we recognize that dv/dt is the acceleration of the rocket. With these changes, Eq. 9-45 becomes

$$Ru = Ma \quad \text{(first rocket equation).} \tag{9-46}$$

Equation 9-46 holds at any instant, with the mass M, the fuel consumption rate R, and the acceleration a evaluated at that instant.

The left side of Eq. 9-46 has the dimensions of a force ($\text{kg} \cdot \text{m/s}^2 = \text{N}$) and depends only on design characteristics of the rocket engine, namely, the rate R at which it consumes fuel mass and the speed u with which that mass is ejected relative to the rocket. We call this term Ru the **thrust** of the rocket engine and represent it with T. Newton's second law emerges clearly if we write Eq. 9-46 as $T = Ma$, in which a is the acceleration of the rocket at the time that its mass is M.

Finding the Velocity

How will the velocity of a rocket change as it consumes its fuel? From Eq. 9-44 we have

$$dv = -u\, \frac{dM}{M}.$$

Integrating leads to

$$\int_{v_i}^{v_f} dv = -u \int_{M_i}^{M_f} \frac{dM}{M},$$

in which M_i is the initial mass of the rocket and M_f its final mass. Evaluating the integral then gives

$$v_f - v_i = u \ln \frac{M_i}{M_f} \quad \begin{array}{l}\text{(second rocket}\\ \text{equation)}\end{array} \tag{9-47}$$

for the increase in the speed of the rocket during the change in mass from M_i to M_f.* We see here the advantage of multistage rockets, in which M_f is reduced by discarding successive stages when their fuel is depleted. An ideal rocket would reach its destination with only its payload remaining.

SAMPLE PROBLEM 9-12

A rocket whose initial mass M_i is 850 kg consumes fuel at the rate $R = 2.3$ kg/s. The speed u of the exhaust gases relative to the rocket engine is 2800 m/s.

(a) What thrust does the rocket engine provide?

SOLUTION: The thrust is

$$T = Ru = (2.3 \text{ kg/s})(2800 \text{ m/s})$$
$$= 6440 \text{ N} \approx 6400 \text{ N}. \quad \text{(Answer)}$$

(b) What is the initial acceleration of the rocket?

SOLUTION: From Newton's second law, we have

$$a = \frac{T}{M_i} = \frac{6440 \text{ N}}{850 \text{ kg}} = 7.6 \text{ m/s}^2. \quad \text{(Answer)}$$

To be launched from the Earth's surface, a rocket must have an initial acceleration greater than $g = 9.8$ m/s^2. Put another way, the thrust T of the rocket engine must exceed the initial weight of the rocket, which here is $M_i g = 850$ kg $\times$ 9.8 m/s^2 = 8300 N. Because the acceleration or thrust requirement is not met (here $T = 6400$ N), our rocket could not be launched from the Earth's surface. It would have to be launched into space by another, more powerful, rocket.

(c) Suppose, instead, that the rocket is launched from a spacecraft already in deep space, where we can neglect any gravitational force acting on it. The mass M_f of the rocket when its fuel is exhausted is 180 kg. What is its speed relative to the spacecraft at that time? Assume that the spacecraft is so massive that the launch does not alter its speed.

SOLUTION: The rocket begins with a speed of $v_i = 0$ relative to the spacecraft. From Eq. 9-47 we have

$$v_f = u \ln \frac{M_i}{M_f}$$
$$= (2800 \text{ m/s}) \ln \frac{850 \text{ kg}}{180 \text{ kg}}$$
$$= (2800 \text{ m/s}) \ln 4.72 \approx 4300 \text{ m/s}. \quad \text{(Answer)}$$

Note that the ultimate speed of the rocket can exceed the exhaust speed u.

---

*The symbol "ln" in Eq. 9-47 means the *natural logarithm*, which is the logarithm taken to the base e ($= 2.718 \ldots$); punch "ln," not "log," on your calculator.

9-8 EXTERNAL FORCES AND INTERNAL ENERGY CHANGES

As the ice skater in Fig. 9-17a pushes herself away from a railing, the railing exerts a force $\mathbf{F}_{ext}$ on her at angle ϕ to the horizontal. This force accelerates her from an initial velocity of zero to some final velocity with which she leaves the railing (Fig. 9-17b). Thus the force increases her kinetic energy. This example differs in two ways from earlier examples in which an external force changes an object's kinetic energy:

1. Previously, each part of an object moved rigidly in the same direction; here the skater's arm does not move like the rest of her body.

2. Previously, the external force transferred energy between the object and its environment; here the external force transfers energy from an internal biochemical energy of the skater's muscles to the kinetic energy of her body as a whole.

In spite of these two differences, we can relate the external force to the change in kinetic energy much as we did for a particle in Section 7-3. To do so, we treat the skater as a particle located at her center of mass, with all of her mass M concentrated there. Then we treat the external force $\mathbf{F}_{ext}$ as acting on the center of mass (Fig. 9-17c). The horizontal force component $F_{ext} \cos \phi$ gives the center of mass a horizontal acceleration $a_{cm,x}$, changing the velocity of the center of mass from an initial value $\mathbf{v}_{cm,0}$.

Suppose that with displacement $\mathbf{d}_{cm}$ of the center of mass, the velocity of the center of mass becomes $\mathbf{v}_{cm}$. Then the magnitude of $\mathbf{v}_{cm}$ is, by Eq. 2-16,

$$v_{cm}^2 = v_{cm,0}^2 + 2a_{cm,x}d_{cm}. \qquad (9\text{-}48)$$

Multiplying both sides of Eq. 9-48 by mass M and rearranging yield

$$\tfrac{1}{2}Mv_{cm}^2 - \tfrac{1}{2}Mv_{cm,0}^2 = Ma_{cm,x}d_{cm}. \qquad (9\text{-}49)$$

The left side of Eq. 9-49 is the difference between the initial kinetic energy $K_{cm,i}$ of the center of mass and the final kinetic energy $K_{cm,f}$. This difference is the change ΔK_{cm} in the kinetic energy of the center of mass due to the force $\mathbf{F}_{ext}$. Making that substitution, and substituting the product $F_{ext} \cos \phi$ for the product $Ma_{cm,x}$ according to Newton's second law, we get

$$\Delta K_{cm} = F_{ext}d_{cm} \cos \phi. \qquad (9\text{-}50)$$

Because the skater is an isolated system in the sense that no energy is transferred to or from her, the change

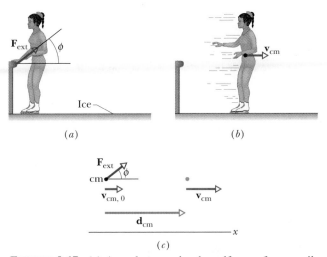

FIGURE 9-17 (a) As a skater pushes herself away from a railing, the railing exerts a force $\mathbf{F}_{ext}$ on her. (b) After the skater leaves the railing, her center of mass has velocity $\mathbf{v}_{cm}$. (c) External force $\mathbf{F}_{ext}$ is taken to act at the skater's center of mass; the vector is at angle ϕ with a horizontal x axis. When the center of mass goes through displacement $\mathbf{d}_{cm}$, its velocity is changed from $\mathbf{v}_{cm,0}$ to $\mathbf{v}_{cm}$ by the horizontal component of $\mathbf{F}_{ext}$.

ΔK_{cm} in her kinetic energy must be accompanied by a change ΔE_{int} in an internal energy. (Let us assume that the only internal energy changed is muscular biochemical energy.) From the principle of the conservation of energy, we write

$$\Delta K_{cm} + \Delta E_{int} = 0,$$

or

$$\Delta E_{int} = -\Delta K_{cm}. \qquad (9\text{-}51)$$

Substituting Eq. 9-50 into Eq. 9-51 yields the change in her internal energy:

$$\Delta E_{int} = -F_{ext}d_{cm} \cos \phi. \qquad (9\text{-}52)$$

Taken together, Eqs. 9-51 and 9-52 say that the external force transfers an amount of energy equal to $F_{ext}d_{cm} \cos \phi$ from internal energy to kinetic energy.

If the height of the skater's center of mass changes while the external force is applied, a change ΔU_{cm} in the gravitational potential energy of the skater–Earth system occurs. Then the isolated system to which the principle of conservation of energy applies is the skater–Earth system, and that principle is written as

$$\Delta K_{cm} + \Delta U_{cm} + \Delta E_{int} = 0. \qquad (9\text{-}53)$$

Substituting Eq. 9-52 into Eq. 9-53, we find

$$\Delta K_{cm} + \Delta U_{cm} - F_{ext}d_{cm} \cos \phi = 0,$$

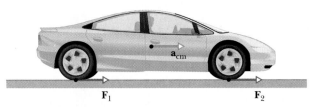

FIGURE 9-18 A car, initially at rest, accelerates to the right. The road exerts four frictional forces (two of them shown) on the bottom surfaces of the tires. Taken together, these four forces makeup the net external force $\mathbf{F}_{ext}$ acting on the car.

or

$$\Delta E = \Delta K_{cm} + \Delta U_{cm} = F_{ext}d_{cm} \cos \phi, \quad (9\text{-}54)$$

where ΔE is the change in the mechanical energy of the skater–Earth system. In words, Eq. 9-54 tells us that the external force transfers an amount of energy equal to the product $F_{ext}d_{cm} \cos \phi$ from internal energy to the mechanical energy of the system, increasing either the kinetic energy of the skater or the height of the skater, or both.

If the external force $\mathbf{F}_{ext}$ is not constant in magnitude, we can replace the symbol F_{ext} in Eqs. 9-52 and 9-54 with the symbol $\overline{F}_{ext}$ for the average magnitude.

Although we have used an ice skater to derive Eqs. 9-52 and 9-54, the equations hold for other objects on which an external force causes a change ΔE_{int} in an internal energy of the object. For example, consider a car that is increasing in speed. During the acceleration, the engine causes the tires to push backward on the road's surface. This push produces frictional forces that act on each tire in the forward direction (Fig. 9-18). The net external force $\mathbf{F}_{ext}$, which is the sum of these frictional forces, gives the car's center of mass an acceleration $\mathbf{a}_{cm}$. It also transfers energy from an internal energy (released by the combustion of fuel) to the kinetic energy of the car. If $\mathbf{F}_{ext}$ is constant, then for a given displacement $\mathbf{d}_{cm}$ of the car's center of mass along level road, we can relate the change ΔK_{cm} in the kinetic energy to the external force $\mathbf{F}_{ext}$ with Eq. 9-54, setting $\Delta U_{cm} = 0$ and $\phi = 0°$.

If the driver applies the brakes, Eq. 9-54 still applies. Now the net external force $\mathbf{F}_{ext}$ due to the frictional forces is toward the rear, and $\phi = 180°$. And now energy is transferred from kinetic energy of the car's center of mass to thermal energy of the brakes.

SAMPLE PROBLEM 9-13

When a click beetle is upside down on its back, it jumps upward by suddenly arching its back, transferring energy stored in a muscle to kinetic energy of its center of mass. This launching mechanism produces an audible click, giving the beetle its name. Videotape of a certain jump shows that the center of mass of a click beetle of mass $m = 4.0 \times 10^{-6}$ kg moved upward by displacement $d_{cm} = 0.77$ mm during the launch and then to a maximum height of $h = 0.30$ m. What was the average magnitude of the external force $\mathbf{F}_{ext}$ on the beetle's back from the floor during the launch?

SOLUTION: The beetle–Earth system is an isolated system in the sense that no energy is transferred to or from it. Thus we can apply the principle of conservation of energy to the system during the time interval T from the beginning of the launch to the moment the center of mass reaches maximum height h. During interval T, the changes in energy are the following: (1) the change ΔK_{cm} in the kinetic energy of the center of mass is zero (the beetle is stationary both initially and at maximum height h); (2) the change ΔU_{cm} in the gravitational potential energy of the beetle–Earth system is equal to mgh; (3) during the launch, the external force $\mathbf{F}_{ext}$ on the beetle causes a change ΔE_{int} in an internal energy of the beetle.

We write the conservation of energy during time interval T as

$$\Delta K_{cm} + \Delta U_{cm} + \Delta E_{int} = 0.$$

Substituting for ΔK_{cm} and ΔU_{cm}, we have

$$0 + mgh + \Delta E_{int} = 0,$$

or

$$\Delta E_{int} = -mgh. \quad (9\text{-}55)$$

Substituting Eq. 9-52 into Eq. 9-55 yields

$$\overline{F}_{ext}d_{cm} \cos \phi = mgh,$$

or

$$\overline{F}_{ext} = \frac{mgh}{d_{cm} \cos \phi}. \quad (9\text{-}56)$$

The angle ϕ between upward force $\mathbf{F}_{ext}$ and upward displacement $\mathbf{d}_{cm}$ is 0°. Substituting this and given data into Eq. 9-56, we find

$$\overline{F}_{ext} = \frac{(4.0 \times 10^{-6} \text{ kg})(9.8 \text{ m/s}^2)(0.30 \text{ m})}{(7.7 \times 10^{-4} \text{ m})(\cos 0°)}$$

$$= 1.5 \times 10^{-2} \text{ N}. \quad \text{(Answer)}$$

This force magnitude may seem small, but to the click beetle it is enormous because, as you can show, it gives the beetle an acceleration of over $380g$ during the launch.

REVIEW & SUMMARY

Center of Mass

The **center of mass** of a system of particles is defined to be the point whose coordinates are given by

$$x_{cm} = \frac{1}{M} \sum_{i=1}^{n} m_i x_i, \quad y_{cm} = \frac{1}{M} \sum_{i=1}^{n} m_i y_i, \quad z_{cm} = \frac{1}{M} \sum_{i=1}^{n} m_i z_i, \tag{9-5}$$

or

$$\mathbf{r}_{cm} = \frac{1}{M} \sum_{i=1}^{n} m_i \mathbf{r}_i, \tag{9-8}$$

where M is the total mass of the system. If the mass is continuously distributed, the center of mass is given by

$$x_{cm} = \frac{1}{M} \int x \, dm, \quad y_{cm} = \frac{1}{M} \int y \, dm, \quad z_{cm} = \frac{1}{M} \int z \, dm. \tag{9-9}$$

If the density (mass per volume) is uniform, then Eq. 9-9 can be rewritten as

$$x_{cm} = \frac{1}{V} \int x \, dV, \quad y_{cm} = \frac{1}{V} \int y \, dV, \quad z_{cm} = \frac{1}{V} \int z \, dV, \tag{9-11}$$

where V is the volume occupied by M.

Newton's Second Law for a System of Particles

The motion of the center of mass of any system of particles is governed by **Newton's second law for a system of particles,** which is written as

$$\sum \mathbf{F}_{ext} = M \mathbf{a}_{cm}. \tag{9-16}$$

Here $\sum \mathbf{F}_{ext}$ is the resultant of the *external* forces acting on the system, M is the total mass of the system, and $\mathbf{a}_{cm}$ is the acceleration of the system's center of mass.

Linear Momentum and Newton's Second Law

For a single particle, we define a vector quantity $\mathbf{p}$ called the **linear momentum** as

$$\mathbf{p} = m\mathbf{v}, \tag{9-22}$$

and write Newton's second law in terms of momentum:

$$\sum \mathbf{F} = \frac{d\mathbf{p}}{dt}. \tag{9-23}$$

For a system of particles these relations become

$$\mathbf{P} = M\mathbf{v}_{cm} \quad \text{and} \quad \sum \mathbf{F}_{ext} = \frac{d\mathbf{P}}{dt}. \tag{9-26, 9-28}$$

Relativistic Momentum

A more complete (relativistic) definition of linear momentum is

$$\mathbf{p} = \frac{m\mathbf{v}}{\sqrt{1 - (v/c)^2}}. \tag{9-24}$$

Equation 9-24 must be used whenever a particle's speed is near the speed of light c. It is valid under all circumstances and is equivalent to Eq. 9-22 whenever $v \ll c$.

Conservation of Linear Momentum

If a system is isolated so that no net *external* force acts on the system, the linear momentum $\mathbf{P}$ of the system remains constant:

$$\mathbf{P} = \text{constant} \quad \text{(closed, isolated system),} \tag{9-30}$$

which can also be written as

$$\mathbf{P}_i = \mathbf{P}_f \quad \text{(closed, isolated system),} \tag{9-31}$$

where the subscripts refer to the values of $\mathbf{P}$ at some initial time and at a later time. Equations 9-30 and 9-31 are equivalent statements of the **law of conservation of linear momentum.**

Variable-Mass Systems

For a system of varying mass, we redefine the system, enlarging its boundaries until it encompasses a larger system whose mass *does* remain constant; then we apply the law of conservation of linear momentum. For a rocket, this means that the system includes both the rocket and its exhaust gases. Such analysis shows that in the absence of external forces a rocket accelerates at an instantaneous rate given by

$$Ru = Ma \quad \text{(first rocket equation),} \tag{9-46}$$

in which M is the rocket's instantaneous mass (including unexpended fuel), R is the fuel consumption rate, and u is the fuel's exhaust speed relative to the rocket. The term Ru is the **thrust** of the rocket engine. For a rocket with constant R and u, whose speed changes from v_i to v_f when its mass changes from M_i to M_f,

$$v_f - v_i = u \ln \frac{M_i}{M_f} \quad \text{(second rocket equation).} \tag{9-47}$$

External Forces and Internal Energy Changes

An external force $\mathbf{F}_{ext}$ acting on an object can cause a change ΔE_{int} in an internal energy of the object:

$$\Delta E_{int} = -F_{ext} d_{cm} \cos \phi. \tag{9-52}$$

Here $\mathbf{d}_{cm}$ is the displacement of the object's center of mass and ϕ is the angle between $\mathbf{d}_{cm}$ and $\mathbf{F}_{ext}$. If the external force transfers energy from an internal energy to the kinetic energy of the object's center of mass, the change ΔK_{cm} in that kinetic energy is

$$\Delta K_{cm} = F_{ext} d_{cm} \cos \phi. \tag{9-50}$$

If the transfer is to the mechanical energy of a system containing the object and having potential energy U, then the change ΔE in the mechanical energy is

$$\Delta E = \Delta K_{cm} + \Delta U_{cm} = F_{ext} d_{cm} \cos \phi. \tag{9-54}$$

QUESTIONS

1. An amateur sculptor cuts the shape of a bird from a single sheet of metal of uniform thickness (Fig. 9-19). Of the numbered points, which is most likely to be the center of mass of the "sculpture"?

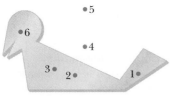

FIGURE 9-19 Question 1.

2. Figure 9-20 shows four situations in which a uniform square metal plate has had a section removed. The origin of the *x* and *y* axes is at the center of the original plate, and in each situation the center of mass of the removed section was at the origin. In each situation, where is the center of mass of the remaining section of the plate? Give quadrant, line, or point.

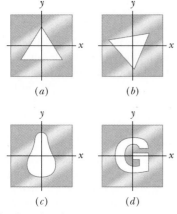

FIGURE 9-20 Question 2.

3. In Fig. 9-21, a penguin stands at the left edge of a uniform sled of length *L*, which lies on frictionless ice. The sled and penguin have equal masses. (a) Where is the center of mass of the sled? (b) How far and in what direction is the center of the sled from the center of mass of sled–penguin system? The penguin then waddles to the right edge of the sled and the sled slides on the ice. (c) Does the center of mass of the sled–penguin system move leftward, rightward, or not at all? (d) Now how far and in what direction is the center of the sled from the center of mass of the sled–penguin system? (e) How far does the penguin move relative to the sled? Relative to the center of mass of the sled–penguin system, how far does (f) the center of the sled move and (g) the penguin move? (Warmup for Problem 23)

FIGURE 9-21 Questions 3 and 4.

4. In Question 3 and Fig. 9-21, suppose that the sled and penguin are initially moving rightward at speed v_0. (a) As the penguin waddles to the right edge of the sled, is the speed *v* of the sled less than, greater than, or equal to v_0? (b) If the penguin then waddles back to the left edge, during that motion, is the speed *v* of the sled less than, greater than, or equal to v_0?

5. Figure 9-22 shows an overhead view of four particles of equal mass sliding over a frictionless surface at constant velocity. The directions of the velocities are indicated; their magnitudes are equal. Consider pairing the particles. Which pairs form a system with a center of mass that (a) is stationary, (b) is stationary and at the origin, and (c) passes through the origin?

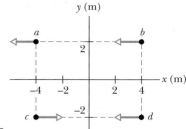

FIGURE 9-22 Question 5.

6. If you drop two watermelons side by side from a bridge, what is the acceleration of the center of mass of the two melon system? (b) If you delay dropping one of the melons, what is the acceleration of the two-melon system while both are falling?

7. Figure 9-23 shows an overhead view of three particles on which external forces act. The magnitudes and directions of the forces on two of the particles are indicated. What are the magnitude and direction of the force acting on the third particle if the center of mass of the three-particle system is (a) stationary, (b) moving at a constant velocity rightward, (c) accelerating rightward?

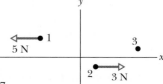

FIGURE 9-23 Question 7.

8. A container sliding along an *x* axis on a frictionless surface explodes into three pieces. The pieces then move along the *x* axis in the directions indicated in Figure 9-24. The following table gives four sets of magnitudes (in kg·m/s) for the momenta $\mathbf{p}_1$, $\mathbf{p}_2$, and $\mathbf{p}_3$ of the pieces. Rank the sets according to the initial speed of the container, greatest first.

	p_1	p_2	p_3		p_1	p_2	p_3
(a)	10	2	6	(b)	10	6	2
(c)	2	10	6	(d)	6	2	10

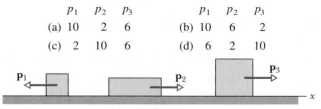

FIGURE 9-24 Question 8.

9. Consider a box, like that in Sample Problem 9-7, which explodes into two pieces while moving with a constant positive velocity along an x axis. If one piece, with mass m_1, ends up with positive velocity $\mathbf{v}_1$, then the second piece, with mass m_2, could end up with (a) a positive velocity $\mathbf{v}_2$ (Fig. 9-25a) or (b) a negative velocity $\mathbf{v}_2$ (Fig. 9-25b), or (c) zero velocity (Fig. 9-25c). Rank those three possible results for the second piece according to the corresponding magnitude of $\mathbf{v}_1$, greatest first.

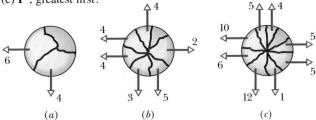

FIGURE 9-25 Question 9.

10. Figure 9-26 shows overhead views of three two-dimensional explosions in which a stationary grapefruit is blown by a firecracker into three pieces, seven pieces, and nine pieces. The pieces then slide over a frictionless floor. For each situation, Fig. 9-26 shows the momentum vectors of all but one piece; that piece has momentum $\mathbf{P}'$. The numbers next to the vectors are the magnitudes of the momenta (in kilogram-meter per second). Rank the three situations according to the magnitudes of (a) P'_x, (b) P'_y, and (c) $\mathbf{P}'$, greatest first.

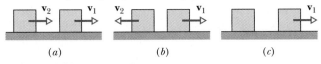

FIGURE 9-26 Question 10.

11. Figure 9-27 shows an overhead view showing six particles that have emerged from a region in which a two-dimensional explosion took place. The explosion was of an object that had been stationary on a frictionless floor. The directions of the momenta of the particles are indicated by the vectors; the numbers give the magnitudes of the momenta (in kg·m/s). (a) Will more particles be emerging from the explosion region? (b) If so, give (b) their net momentum and (c) their direction of travel.

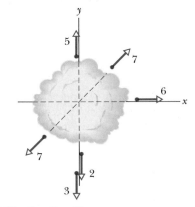

FIGURE 9-27 Question 11.

12. For three pairs of particles, here are the masses of the particles and their initial separation d:

	m_1	m_2	INITIAL SEPARATION d
Pair 1	$2m$	$8m$	1.0 m
Pair 2	$3m$	$6m$	2.0 m
Pair 3	$4m$	$9m$	0.5 m

The particles in each pair attract each other. They are released from rest at the given initial separation d. When the separation has decreased to $d/2$, the speed of m_1 is v_1 and the speed of m_2 is v_2. Without written calculation, rank the pairs according to the ratio v_1/v_2, greatest first. (*Hint:* See Sample Problem 9-10.)

EXERCISES & PROBLEMS

SECTION 9-2 The Center of Mass

1E. (a) How far is the center of mass of the Earth–Moon system from the center of the Earth? (Appendix C gives the masses of the Earth and the Moon and the distance between the two.) (b) Express the answer to (a) as a fraction of the Earth's radius.

2E. The distance between the centers of the carbon (C) and oxygen (O) atoms in a carbon monoxide (CO) gas molecule is 1.131×10^{-10} m. Locate the center of mass of a CO molecule relative to the carbon atom. (Find the masses of C and O in Appendix D.)

3E. (a) What are the coordinates of the center of mass of the three-particle system shown in Fig. 9-28? (b) What happens to the center of mass as the mass of the topmost particle is gradually increased?

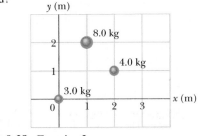

FIGURE 9-28 Exercise 3.

4E. Three thin rods each of length L are arranged in an inverted U, as shown in Fig. 9-29. The two rods on the arms of the U each have mass M; the third rod has mass $3M$. Where is the center of mass of the assembly?

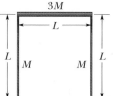

FIGURE 9-29   Exercise 4.

5E. A uniform square plate 6 m on a side has had a square piece 2 m on a side cut out of it (Fig. 9-30). The center of that piece is at $x = 2$ m, $y = 0$. The center of the square plate is at $x = y = 0$. Find the coordinates of the center of mass of the remaining piece.

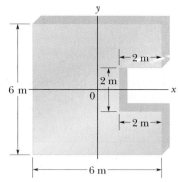

FIGURE 9-30   Exercise 5.

6P. Show that the ratio of the distances of two particles from the center of mass of the two-particle system is the inverse ratio of their masses.

7P. Figure 9-31 shows the dimensions of a composite slab; half the slab is made of aluminum (density = 2.70 g/cm³) and half is made of iron (density = 7.85 g/cm³). Where is the center of mass of the slab?

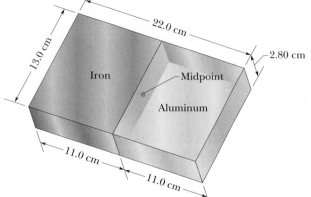

FIGURE 9-31   Problem 7.

8P. In the ammonia (NH_3) molecule (see Fig. 9-32), the three hydrogen (H) atoms form an equilateral triangle; the center of the triangle is 9.40×10^{-11} m from each hydrogen atom. The nitrogen (N) atom is at the apex of a pyramid, with the three hydrogen atoms forming the base. The nitrogen-to-hydrogen distance is 10.14×10^{-11} m, and the nitrogen-to-hydrogen atomic mass ratio is 13.9. Locate the center of mass of the molecule relative to the nitrogen atom.

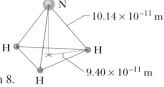

FIGURE 9-32   Problem 8.

9P. A cubical box, open at the top, with edge length 40 cm, is constructed from metal plate of uniform density and negligible thickness (Fig. 9-33). Find the coordinates of the center of mass of the box.

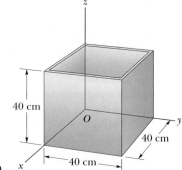

FIGURE 9-33   Problem 9.

10P. The Great Pyramid of Cheops at El Gizeh, Egypt (Fig. 9-34a), had height $H = 147$ m before its topmost stone fell. Its base is a square with edge length $L = 230$ m (see Fig. 9-34b). Its volume V is equal to $L^2H/3$. Assuming that it has uniform density $\rho = 1.8 \times 10^3$ kg/m³, find (a) the original height of its center of mass above the base and (b) the work required to lift the blocks into place from the base level.

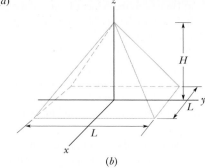

(a)

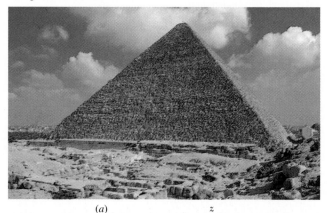

FIGURE 9-34
Problem 10.

(b)

11P*. A right cylindrical can with mass M, height H, and uniform density is initially filled with pop of mass m (Fig. 9-35). We punch small holes in the top and bottom to drain the pop; we then consider the height h of the center of mass of the can and any pop within it. What is h (a) initially, and (b) when all the pop has drained? (c) What happens to h during the draining of the pop? (d) If x is the height of the remaining pop at any given instant, find x (in terms of M, H, and m) when the center of mass reaches its lowest point.

FIGURE 9-35 Problem 11.

SECTION 9-3 Newton's Second Law for a System of Particles

12E. Two skaters, one with mass 65 kg and the other with mass 40 kg, stand on an ice rink holding a pole with a length of 10 m and a negligible mass. Starting from the ends of the pole, the skaters pull themselves along the pole until they meet. How far will the 40 kg skater move?

13E. An old Chrysler with mass 2400 kg is moving along a straight stretch of road at 80 km/h. It is followed by a Ford with mass 1600 kg moving at 60 km/h. How fast is the center of mass of the two cars moving?

14E. A man of mass m clings to a rope ladder suspended below a balloon of mass M; see Fig. 9-36. The balloon is stationary with respect to the ground. (a) If the man begins to climb the ladder at speed v (with respect to the ladder), in what direction and with what speed (with respect to the Earth) will the balloon move? (b) What is the state of the motion after the man stops climbing?

FIGURE 9-36 Exercise 14.

15E. Two particles P and Q are initially at rest 1.0 m apart. P has a mass of 0.10 kg and Q a mass of 0.30 kg. P and Q attract each other with a constant force of 1.0×10^{-2} N. No external forces act on the system. (a) Describe the motion of the center of mass. (b) At what distance from P's original position do the particles collide?

16E. A cannon and a supply of cannonballs are inside a sealed railroad car of length L, as in Fig. 9-37. The cannon fires to the right; the car recoils to the left. Fired cannonballs remain in the car after hitting the far wall and landing on the floor there. (a) After all the cannonballs have been fired, what is the greatest distance the car can have moved from its original position? (b) Under what circumstance would the car move that distance? (c) What is the speed of the car just after all the cannonballs have been fired?

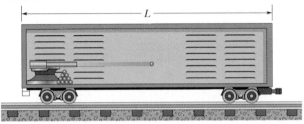

FIGURE 9-37 Exercise 16.

17P. At the instant a 3.0 kg particle has a velocity of 6.0 m/s in the direction of decreasing y, a 4.0 kg particle has a velocity of 7.0 m/s in the direction of increasing x. What is the speed of the center of mass of the two-particle system?

18P. A stone is dropped at $t = 0$. A second stone, with twice the mass of the first, is dropped from the same point at $t = 100$ ms. (a) Where is the center of mass of the two stones at $t = 300$ ms? Neither stone has yet reached the ground. (b) How fast is the center of mass of the two-stone system moving at that time?

19P. A 1000 kg automobile is at rest at a traffic signal. At the instant the light turns green, the automobile starts to move with a constant acceleration of 4.0 m/s². At the same instant a 2000 kg truck, traveling at a constant speed of 8.0 m/s, overtakes and passes the automobile. (a) How far is the center of mass of the automobile–truck system from the traffic light at $t = 3.0$ s? (b) What is the speed of the center of mass of the automobile–truck system then?

20P. Ricardo, mass 80 kg, and Carmelita, who is lighter, are enjoying Lake Merced at dusk in a 30 kg canoe. When the canoe is at rest in the placid water, they exchange seats, which are 3.0 m apart and symmetrically located with respect to the canoe's center. Ricardo notices that the canoe moved 40 cm relative to a submerged log during the exchange and calculates Carmelita's mass, which she has not told him. What is it?

21P. A shell is fired from a gun with a muzzle velocity of 20 m/s, at an angle of 60° with the horizontal. At the top of the trajectory, the shell explodes into two fragments of equal mass (Fig. 9-38). One fragment, whose speed immediately after the explosion is zero, falls vertically. How far from the gun does the other fragment land, assuming that the terrain is level and that the air drag is negligible?

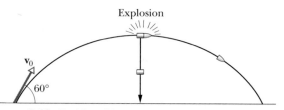

FIGURE 9-38 Problem 21.

22P. Two identical containers of sugar are connected by a massless cord that passes over a massless, frictionless pulley with a diameter of 50 mm (Fig. 9-39). The two containers are at the same level. Each originally has a mass of 500 g. (a) Locate the horizontal position of their center of mass. (b) Now 20 g of sugar is transferred from one container to the other, but the containers are prevented from moving. Locate the new horizontal position of their center of mass. (c) The two containers are now released. In what direction does the center of mass move? (d) What is its acceleration?

FIGURE 9-39 Problem 22.

23P. A dog, weighing 10.0 lb, is standing on a flatboat so that he is 20.0 ft from the shore (to the left in Fig. 9-40a). He walks 8.00 ft on the boat toward shore and then halts. The boat weighs 40.0 lb, and one can assume there is no friction between it and the water. How far is the dog then from the shore? (*Hint:* See Fig. 9-40b. The dog moves leftward; the boat moves rightward; but does the center of mass of the *boat + dog* system move?)

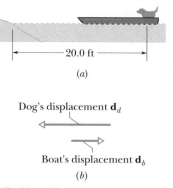

(a)

Dog's displacement $\mathbf{d}_d$

Boat's displacement $\mathbf{d}_b$

(b)

FIGURE 9-40 Problem 23.

SECTION 9-5 The Linear Momentum of a System of Particles

24E. Suppose that your mass is 80 kg. How fast would you have to run to have the same linear momentum as a 1600 kg car moving at 1.2 km/h?

25E. How fast must an 816 kg Geo travel (a) to have the same linear momentum as a 2650 kg Cadillac going 16 km/h and (b) to have the same kinetic energy?

26E. What is the linear momentum of an electron of speed $0.99c$ ($= 2.97 \times 10^8$ m/s)?

27E. The linear momentum of a particle moving at 1.5×10^8 m/s is measured to be 2.9×10^{-19} kg·m/s. By finding its mass, determine whether the particle is an electron or a proton.

28E. A 0.70 kg mass is moving horizontally with a speed of 5.0 m/s when it strikes a vertical wall. The mass rebounds with a speed of 2.0 m/s. What is the magnitude of the change in linear momentum of the mass?

29P. A 0.165 kg cue ball with an initial speed of 2.00 m/s bounces off the rail in a game of pool, as shown from overhead in Fig. 9-41. For x and y axes located as shown, the bounce reverses the y component of the ball's velocity but does not alter the x component. (a) What is θ in Fig. 9-41? (b) What is the change in the ball's linear momentum in unit-vector notation? (The fact that the ball rolls is not relevant to either question.)

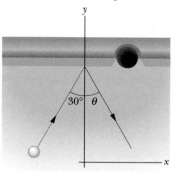

FIGURE 9-41 Problem 29.

30P. A 2100 kg truck traveling north at 41 km/h turns east and accelerates to 51 km/h. (a) What is the change in the kinetic energy of the truck? (b) What are the magnitude and direction of the change in the linear momentum of the truck?

31P. An object is tracked by a radar station and found to have a position vector given by $\mathbf{r} = (3500 - 160t)\mathbf{i} + 2700\mathbf{j} + 300\mathbf{k}$, with $\mathbf{r}$ in meters and t in seconds. The radar station's x axis points east, its y axis north, and its z axis vertically up. If the object is a 250 kg meteorological missile, what are (a) its linear momentum, (b) its direction of motion, and (c) the net force on it?

32P. A 50 g ball is thrown from ground level into the air with an initial speed of 16 m/s at an angle of 30° above the horizontal. (a) What are the values of the kinetic energy of the ball initially and just before it hits the ground? (b) Find the corresponding values of the linear momentum (magnitude and direction). (c) Show that the change in linear momentum is just equal to the weight of the ball multiplied by the time of flight.

33P. A particle with mass m has linear momentum $\mathbf{p}$ of magnitude mc. What is its speed in terms of c, the speed of light?

SECTION 9-6 Conservation of Linear Momentum

34E. A 200 lb man standing on a surface of negligible friction kicks forward a 0.15 lb stone lying at his feet, giving it a speed of 13 ft/s. What velocity does the man acquire as a result?

35E. Two blocks of masses 1.0 kg and 3.0 kg are connected by a spring and rest on a frictionless surface. They are given velocities toward each other such that the 1.0 kg block travels initially at 1.7 m/s toward the center of mass, which remains at rest. What is the initial velocity of the other block?

36E. A space vehicle is traveling at 4300 km/h relative to the Earth when the exhausted rocket motor is disengaged and sent backward with a speed of 82 km/h relative to the command module. The mass of the motor is four times the mass of the module. What is the speed of the command module relative to Earth after the separation?

37E. A 75 kg man is riding on a 39 kg cart traveling at a speed of 2.3 m/s. He jumps off with zero horizontal speed. What is the resulting change in the speed of the cart?

38E. A railroad flatcar of weight W can roll without friction along a straight horizontal track. Initially, a man of weight w is standing on the car, which is moving to the right with speed v_0; see Fig. 9-42. What is the change in velocity of the car if the man runs to the left so that his speed relative to the car is v_{rel}?

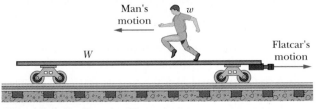

FIGURE 9-42 Exercise 38.

39P. The last stage of a rocket is traveling at a speed of 7600 m/s. This last stage is made up of two parts that are clamped together, namely, a rocket case with a mass of 290.0 kg and a payload capsule with a mass of 150.0 kg. When the clamp is released, a compressed spring causes the two parts to separate with a relative speed of 910.0 m/s. (a) What are the speeds of the two parts after they have separated? Assume that all velocities are along the same line. (b) Find the total kinetic energy of the two parts before and after they separate; account for any difference.

40P. A radioactive nucleus, initially at rest, decays by emitting an electron and a neutrino perpendicular to each other. (A *neutrino* is one of the fundamental particles of physics.) The linear momentum of the electron is 1.2×10^{-22} kg·m/s and that of the neutrino is 6.4×10^{-23} kg·m/s. (a) Find the direction and magnitude of the linear momentum of the nucleus as it recoils from the decay. (b) The mass of the residual nucleus is 5.8×10^{-26} kg. What is its kinetic energy as it recoils?

41E. An electron (mass $m_1 = 9.11 \times 10^{-31}$ kg) and a proton (mass $m_2 = 1.67 \times 10^{-27}$ kg) attract each other via an electrical force. Suppose that an electron and a proton are released from rest with an initial separation $d = 3.0 \times 10^{-6}$ m. When their separation has decreased to 1.0×10^{-6} m, what is the ratio of (a) the electron's linear momentum to the proton's linear momentum, (b) the electron's speed to the proton's speed, and (c) the electron's kinetic energy to the proton's kinetic energy? (d) As the separation continues to decrease, do the answers to (a) through (c) increase, decrease, or remain the same?

42P. A 4.0 kg mass sliding on a frictionless surface explodes into two 2.0 kg masses. After the explosion the velocities of the 2.0 kg masses are 3.0 m/s, due north, and 5.0 m/s, 30° north of east. What was the original speed of the 4.0 kg mass?

43P. A vessel at rest explodes, breaking into three pieces. Two pieces, having equal mass, fly off perpendicular to one another with the same speed of 30 m/s. The third piece has three times the mass of each other piece. What are the direction and magnitude of its velocity immediately after the explosion?

44P. A 2140 kg railroad flatcar, which can move with negligible friction, is motionless next to a platform. A 242 kg sumo wrestler runs at 5.3 m/s along the platform (parallel to the track) and then jumps onto the flatcar. What is the speed of the flatcar if he then (a) stands on it, (b) runs at 5.3 m/s relative to it in his original direction, and (c) turns and runs at 5.3 m/s relative to the flatcar opposite his original direction?

45P. A rocket sled with a mass of 2900 kg moves at 250 m/s on a set of rails. At a certain point, a scoop on the sled dips into a trough of water located between the tracks and scoops water into an empty tank on the sled. By applying the principle of conservation of linear momentum, determine the speed of the sled after 920 kg of water has been scooped up. Ignore any retarding force on the scoop.

46P. A body of mass 8 kg is traveling at 2 m/s with no external force acting. At a certain instant an internal explosion occurs, splitting the body into two chunks of 4 kg mass each. The explosion gives the chunks an additional 16 J of kinetic energy. Neither chunk leaves the line of original motion. Determine the speed and direction of motion of each of the chunks after the explosion.

47P. You are on an iceboat on frictionless, flat ice; you and the boat have a combined mass M. Along with you are two stones with masses m_1 and m_2 such that $M = 6.00m_1 = 12.0m_2$. To get the boat moving, you throw the stones rearward, either in succession or together, but in each case with a certain speed v_{rel} relative to the boat. What is the resulting speed of the boat if you throw the stones (a) simultaneously, (b) m_1 and then m_2, and (c) m_2 and then m_1?

48P. A 1400 kg cannon, which fires a 70.0 kg shell with a speed of 556 m/s relative to the muzzle, is set at an elevation angle of 39.0° above the horizontal. The cannon is mounted on frictionless rails, so that it recoils freely. (a) What is the speed of the shell with respect to the Earth? (b) At what angle with the ground is the shell projected? (*Hint:* The horizontal component of the linear momentum of the system remains unchanged as the gun is fired.)

SECTION 9-7 Systems with Varying Mass: A Rocket

49E. A rocket at rest in space, where there is virtually no gravitational force, has a mass of 2.55×10^5 kg, of which 1.81×10^5 kg is fuel. The engine consumes fuel at the rate of 480 kg/s, and the exhaust speed is 3.27 km/s. The engine is fired for 250 s. (a) Find the thrust of the rocket engine. (b) What is the mass of the rocket after the engine burn? (c) What is the final speed attained?

50E. A rocket is moving away from the solar system at a speed of 6.0×10^3 m/s. It fires its engine, which ejects exhaust with a velocity of 3.0×10^3 m/s relative to the rocket. The mass of the rocket at this time is 4.0×10^4 kg, and its acceleration is 2.0 m/s². (a) What is the thrust of the engine? (b) At what rate, in kilograms per second, was exhaust ejected during the firing?

51E. A 6090 kg space probe, moving nose-first toward Jupiter at 105 m/s relative to the Sun, fires its rocket engine, ejecting 80.0 kg of exhaust at a speed of 253 m/s relative to the space probe. What is the final velocity of the probe?

52E. Consider a rocket at rest in deep space. For the rocket's speed to be (a) equal to the exhaust speed and (b) equal to twice the exhaust speed after firing of the rocket engine, what must be the rocket's *mass ratio* (ratio of initial to final mass)?

53E. During a lunar mission, it is necessary to increase the speed of a spacecraft by 2.2 m/s when it is moving at 400 m/s. The exhaust speed of rocket engine is 1000 m/s. What fraction of the initial mass of the spacecraft must be burned and ejected for the increase?

54E. A railroad car moves at a constant speed of 3.20 m/s under a grain elevator. Grain drops into it at the rate of 540 kg/min. What force must be applied to the railroad car, in the absence of friction, to keep it moving at constant speed?

55P. A single-stage rocket, at rest in a certain inertial reference frame, has mass M when the rocket engine is ignited. Show that when the mass has decreased to $0.368M$, the gases streaming out of the rocket engine at that time will be at rest in the original reference frame.

56P. A 6100 kg rocket is set for vertical firing from the Earth's surface. If the exhaust speed is 1200 m/s, how much gas must be ejected each second if the thrust (a) is to equal the weight of the rocket and (b) is to give the rocket an initial upward acceleration of 21 m/s²?

57P. Two long barges are moving in the same direction in still water, one with a speed of 10 km/h and the other with a speed of 20 km/h. While they are passing each other, coal is shoveled from the slower to the faster one at a rate of 1000 kg/min; see Fig. 9-43. How much additional force must be provided by the driving engines of each barge if neither is to change speed? Assume that the shoveling is always perfectly sideways and that the frictional forces between the barges and the water do not depend on the weight of the barges.

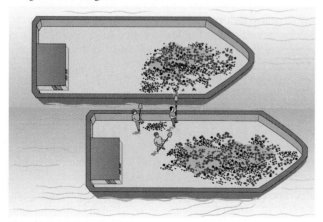

FIGURE 9-43 Problem 57.

58P. A jet airplane is traveling 180 m/s. Each second, the engine takes in 68 m³ of air, which has a mass of 70 kg. The air is used to burn 2.9 kg of fuel each second. The energy is used to compress the products of combustion and to eject them at the rear of the plane at 490 m/s relative to the plane. Find (a) the thrust of the jet engine and (b) its power in watts.

SECTION 9-8 External Forces and Internal Energy Changes

59E. The summit of Mount Everest is 8850 m above sea level. (a) How much energy would a 90 kg climber expend against the gravitational force (his weight) in climbing to the summit from sea level? (b) How many candy bars, at 300 kcal per bar, would supply an energy equivalent to this? Your answer should suggest that work done against the gravitational force is a very small part of the energy expended in climbing a mountain.

60E. A 51 kg boy climbs, with constant speed, a vertical rope 6.0 m long in 10 s. (a) What is the increase in the boy–Earth gravitational potential energy? (b) What is the boy's average power during the climb?

61E. A 55 kg woman runs up a flight of stairs with a height of 4.5 m in 3.5 s. What average power is required of her?

62E. A sprinter who weighs 670 N runs the first 7.0 m of a race in 1.6 s, starting from rest and accelerating uniformly. What are the sprinter's (a) speed and (b) kinetic energy at the end of that 1.6 s? (c) What average power does the sprinter generate during the 1.6 s interval?

63E. The luxury liner *Queen Elizabeth 2* has a diesel-electric powerplant with a maximum power of 92 MW at a cruising speed of 32.5 knots. What forward force is exerted on the ship at this highest attainable speed? (1 knot = 6076 ft/h.)

64E. What power, in horsepower, is required of the engine of a 1600 kg car moving at 25.1 m/s on a level road if the forces of resistance total 703 N?

65E. A swimmer moves through the water at an average speed of 0.22 m/s. The average drag force opposing this motion is 110 N. What average power is required of the swimmer?

66E. The energy required for a person to run is about 335 J/m, regardless of speed. What is a runner's average power during (a) a 100 m dash (time = 10 s) and (b) a marathon (distance = 26.2 mi; time = 2 h 10 min)?

67E. An automobile with passengers has weight 16,400 N and is moving at 113 km/h when the driver brakes to a stop. The road exerts a frictional force of 8230 N on the wheels. Find the stopping distance.

68E. You crouch from a standing position, lowering your center of mass 18 cm in the process. Then you jump vertically. The average force exerted on you by the floor while you jump is three times your weight. What is your upward speed as you pass through your standing position in leaving the floor?

69E. A 55-kg woman leaps vertically from a crouching position in which her center of mass is 40 cm above the ground. As her feet leave the floor her center of mass is 90 cm above the ground;

it rises to 120 cm at the top of her leap. (a) What average force was exerted on her by the ground during the jump? (b) What maximum speed does she attain?

70P. A 110 kg ice hockey player skates at 3.0 m/s toward a railing at the edge of the ice and then stops himself by grasping the railing with his outstretched arms. During this stopping process his center of mass moves 30 cm toward the railing. (a) What is the change in the kinetic energy of his center of mass during this process? (b) What average force must he exert on the railing?

71P. A 1500 kg automobile starts from rest on a horizontal road and gains a speed of 72 km/h in 30 s. (a) What is the kinetic energy of the auto at the end of the 30 s? (b) What is the average power required of the car during the 30 s interval? (c) What is the instantaneous power at the end of the 30 s interval, assuming that the acceleration was constant?

72P. While a 1710 kg automobile is moving at a constant speed of 15.0 m/s, the engine supplies 16.0 kW of power to overcome friction, wind resistance, and so on. (a) What is the effective retarding force associated with all the frictional forces combined? (b) What power must the engine supply if the car is to move up an 8.00% grade (8.00 m vertically for each 100 m horizontally) at 15.0 m/s? (c) On what downgrade, expressed as a percentage, would the car coast at 15.0 m/s?

73P. If a locomotive with a power capability of 1.5 MW can accelerate a train from a speed of 10 m/s to 25 m/s in 6.0 min, (a) calculate the mass of the train. Find (b) the speed of the train and (c) the force accelerating the train as functions of time (in seconds) during the 6.0 min interval. (d) Find the distance moved by the train during the interval.

74P. Resistance to the motion of an automobile consists of road friction, which is almost independent of speed, and air drag, which is proportional to speed squared. For a car with a weight of 12,000 N, the total resistant force F is given by $F = 300 + 1.8v^2$, where F is in newtons and v is in meters per second. Calculate the power required to accelerate the car at 0.92 m/s^2 when the speed is 80 km/h.

75P*. A dragster with mass m races another dragster through a distance d from an initial dead stop. Assume that its engine provides a constant instantaneous power P for the entire race and that the dragster can be treated as a particle. Find the elapsed time for the race.

Electronic Computation

76. The following gives the masses of three objects and, for a certain instant, the (xy) coordinates and the velocities of the objects. At that instant, what are (a) the position and (b) the velocity of the center of mass of the three particle system, and (c) what is the net linear momentum of the system?

OBJECT	MASS (kg)	COORDINATES (m)	VELOCITY (m/s)
1	4.00	(0, 0)	$1.50\mathbf{i} - 2.50\mathbf{j}$
2	3.00	(7.00, 3.00)	0
3	5.00	(3.00, 2.00)	$2.00\mathbf{i} - 1.00\mathbf{j}$

77. A 2.00 kg block is released from rest over the side of a very tall building ($y = 0$) at time $t = 0$. At time $t = 1.00$ s, a 3.00 kg block is released from rest at the same point. The first block hits the ground at $t = 5.00$ s. Plot, for the time interval $t = 0$ to $t = 6.00$ s, (a) the position and (b) the speed of the center of mass of the two block system.

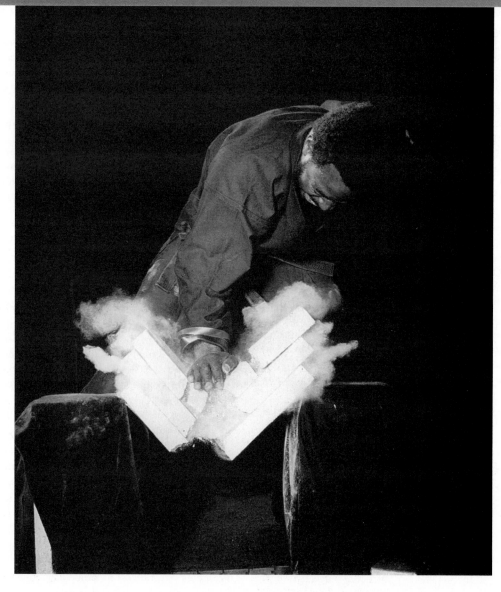

Ronald McNair, a physicist and one of the astronauts killed in the explosion of the Challenger *space shuttle, held a black belt in karate. Here he breaks several concrete slabs with one blow. In such karate demonstrations, a pine board or a concrete "patio block" is typically used. When struck, the board or block bends, storing energy like a stretched spring does, until a critical energy is reached. Then the object breaks. Surprisingly, the energy necessary to break the block is about one-third of that for the board, yet the board is considerably easier to break. Why?*

10-1 WHAT IS A COLLISION?

In everyday language, a *collision* occurs when objects crash into each other. Although we will refine that definition, it conveys the meaning well enough and covers common collisions, such as those between billiard balls, a hammer and a nail, and—too commonly—automobiles. Figure 10-1*a* shows the lasting result of a collision (an impressive crash) that occurred about 20,000 years ago. Collisions range from the microscopic scale of subatomic particles (Fig. 10-1*b*) to the astronomic scale of colliding stars and colliding galaxies. Even when they are on a normal scale, they are often too brief to be visible, although they involve significant distortion of the *colliding bodies* (Fig. 10-1*c*).

We shall use the following more formal definition of collision:

> A **collision** is an isolated event in which two or more bodies (the colliding bodies) exert relatively strong forces on each other for a relatively short time.

We must be able to distinguish times that are *before, during,* and *after* a collision, as suggested in Fig. 10-2. In the illustration, a system boundary surrounds the two colliding bodies. The forces that the bodies exert on each other are internal to the system.

Note that our formal definition of collision does not require the ''crash'' of our informal definition. When a space probe swings around a large planet to pick up speed (a *slingshot* encounter), that too is a collision. The probe and planet do not actually ''touch,'' but a collision does not require contact, and a collision force does not have to be a force of contact; it can just as easily be a gravitational force, as in this case.

Many physicists today spend their time playing what we can call ''the collision game.'' A principal goal of this game is to find out as much as possible about the forces that act during a collision, knowing the state of the particles both before and after the collision. Virtually all our knowledge of the subatomic world—electrons, protons, neutrons, muons, quarks, and the like—comes from experiments involving collisions. The rules of the game are the laws of conservation of momentum and of energy.

(a)

(b)

(c)

FIGURE 10-1 Collisions range widely in scale. (*a*) Meteor Crater in Arizona is about 1200 m wide and 200 m deep. (*b*) An alpha particle coming in from the left (whose trail is colored yellow in this false-color photograph) bounces off a nitrogen nucleus that had been stationary and that now moves toward the bottom right (red trail). (*c*) In a tennis match, the ball is in contact with the racquet for about 4 ms in each collision (for a cumulative time of only 1 s per set).

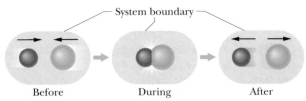

FIGURE 10-2 A flowchart showing the system in which a collision occurs.

10-2 IMPULSE AND LINEAR MOMENTUM

Single Collision

Figure 10-3 shows the equal but opposite forces, $\mathbf{F}(t)$ and $-\mathbf{F}(t)$, that act during a simple head-on collision between two particle-like bodies of different masses. These forces will change the linear momentum of both bodies; the amount of the change will depend not only on the average values of the forces, but also on the time Δt during which they act. To see this quantitatively, let us apply Newton's second law in the form $\mathbf{F} = d\mathbf{p}/dt$ to body R on the right in Fig. 10-3. We have

$$d\mathbf{p} = \mathbf{F}(t)\, dt, \qquad (10\text{-}1)$$

in which $\mathbf{F}(t)$ is the time-varying force, displayed as the curve in Fig. 10-4a. Let us integrate Eq. 10-1 over the collision interval Δt, that is, from an initial time t_i (just before the collision) to a final time t_f (just after the collision). We obtain

$$\int_{\mathbf{p}_i}^{\mathbf{p}_f} d\mathbf{p} = \int_{t_i}^{t_f} \mathbf{F}(t)\, dt. \qquad (10\text{-}2)$$

The left side of this equation is $\mathbf{p}_f - \mathbf{p}_i$, the change in linear momentum of the body R. The right side, which is a measure of both the strength and the duration of the collision force, is called the **impulse J** of the collision. Thus

$$\mathbf{J} = \int_{t_i}^{t_f} \mathbf{F}(t)\, dt \qquad \text{(impulse defined)}. \quad (10\text{-}3)$$

Equation 10-3 tells us that the impulse is equal to the area under the $F(t)$ curve of Fig. 10-4a.

From Eqs. 10-2 and 10-3 we see that the change in the linear momentum of each body in a collision is equal to the impulse that acts on that body:

$$\mathbf{p}_f - \mathbf{p}_i = \Delta \mathbf{p} = \mathbf{J} \qquad \begin{array}{l}\text{(impulse–linear}\\ \text{momentum theorem)}.\end{array} \quad (10\text{-}4)$$

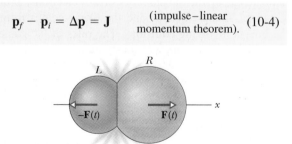

FIGURE 10-3 Two particle-like bodies L and R collide with each other. During the collision, body L exerts force $\mathbf{F}(t)$ on body R, and body R exerts force $-\mathbf{F}(t)$ on body L. Forces $\mathbf{F}(t)$ and $-\mathbf{F}(t)$ are an action–reaction pair. Their magnitudes vary with time during the collision, but at any given instant these magnitudes are equal.

(Furthermore, from the conservation of linear momentum, we know that $\Delta \mathbf{p}$ for body R in Fig. 10-3 equals $-\Delta \mathbf{p}$ for body L.) Equation 10-4 can also be written in component form as

$$p_{fx} - p_{ix} = \Delta p_x = J_x, \qquad (10\text{-}5)$$

$$p_{fy} - p_{iy} = \Delta p_y = J_y, \qquad (10\text{-}6)$$

and

$$p_{fz} - p_{iz} = \Delta p_z = J_z. \qquad (10\text{-}7)$$

Impulse and linear momentum are both vectors, and they have the same units and dimensions. The **impulse–linear momentum theorem** of Eq. 10-4, like the work–kinetic energy theorem, is not a new and independent theorem but is a direct consequence of Newton's second law. Both theorems are special forms of this law, useful for special purposes.

If $\overline{F}$ is the average magnitude of the force in Fig. 10-4a, we can write the magnitude of the impulse as

$$J = \overline{F}\, \Delta t, \qquad (10\text{-}8)$$

where Δt is the duration of the collision. The value of $\overline{F}$ is chosen so that the area within the rectangle of Fig. 10-4b is equal to the area under the $F(t)$ curve of Fig. 10-4a.

CHECKPOINT 1: A paratrooper whose chute fails to open lands in snow; he is hurt slightly. Had he landed on bare ground, the stopping time would have been 10 times shorter and the collision lethal. Does the presence of the snow increase, decrease, or leave unchanged the values of (a) the paratrooper's change in momentum, (b) the impulse stopping the paratrooper, and (c) the force stopping the paratrooper?

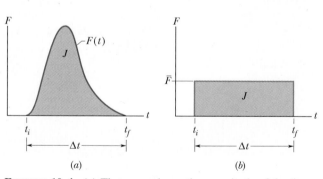

FIGURE 10-4 (a) The curve shows the magnitude of the time-varying force $F(t)$ that acts on body R during the collision of Fig. 10-3. The area under the curve for $F(t)$ is equal to the magnitude of the impulse $\mathbf{J}$ on body R in the collision. (b) The height of the rectangle represents the average force $\overline{F}$ acting on body R over the time interval Δt. The area within the rectangle is equal to the area under the curve in (a) and thus is also equal to the magnitude of the impulse $\mathbf{J}$ in the collision.

Series of Collisions

In Fig. 10-5, a steady stream of bodies, with identical linear momenta mv, collides with body R, which is fixed in place. In each collision of this one-dimensional situation, the impulse J acting on body R and the change Δp in linear momentum of the incident body have the same magnitude but opposite directions; that is, $J = -\Delta p$. If n bodies collide with R in time interval Δt, then by Eq. 10-4 the total impulse J acting on body R during Δt is

$$J = -n\,\Delta p. \qquad (10\text{-}9)$$

By rearranging Eq. 10-8 and substituting Eq. 10-9, we find the average force $\overline{F}$ acting on body R during the collisions:

$$\overline{F} = \frac{J}{\Delta t} = -\frac{n}{\Delta t}\,\Delta p = -\frac{n}{\Delta t}\,m\,\Delta v. \qquad (10\text{-}10)$$

This equation gives us $\overline{F}$ in terms of $n/\Delta t$, the rate at which the bodies collide with body R, and Δv, the change in the velocity of those bodies.

If the colliding bodies stop upon impact, then in Eq. 10-10 we substitute

$$\Delta v = v_f - v_i = 0 - v = -v, \qquad (10\text{-}11)$$

where $v_i\ (= v)$ and $v_f\ (= 0)$ are the velocities before and after the collision, respectively. If, instead, the colliding bodies bounce directly backward from body R with no change in speed, then $v_f = -v$ and we substitute

$$\Delta v = v_f - v_i = -v - v = -2v. \qquad (10\text{-}12)$$

In time interval Δt, an amount of mass $\Delta m = nm$ collides with body R. With this result, we can rewrite Eq. 10-10 as

$$\overline{F} = -\frac{\Delta m}{\Delta t}\,\Delta v. \qquad (10\text{-}13)$$

This equation gives the average force $\overline{F}$ in terms of $\Delta m/\Delta t$, the rate at which mass collides with body R. Here again we can substitute Eq. 10-11 or 10-12 for Δv, depending on what the colliding bodies do.

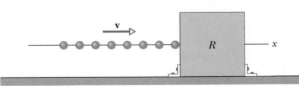

FIGURE 10-5 A steady stream of bodies, with identical linear momenta, collides with body R, which is fixed in place. The average force $\overline{F}$ on body R is to the right and has a magnitude that depends on the rate at which the bodies collide or, equivalently, the rate at which mass collides.

SAMPLE PROBLEM 10-1

A pitched 140 g baseball, in horizontal flight with a speed v_i of 39 m/s, is struck by a batter. After leaving the bat, the ball travels in the opposite direction with speed v_f, also 39 m/s.

(a) What impulse J acts on the ball while it is in contact with the bat?

SOLUTION: We can calculate the impulse from the change it produces in the ball's linear momentum, using Eq. 10-4 for one-dimensional motion. Let us choose the direction in which the bat is moving to be positive. From Eq. 10-4 we have

$$\begin{aligned} J = p_f - p_i &= mv_f - mv_i \\ &= (0.14\text{ kg})(39\text{ m/s}) - (0.14\text{ kg})(-39\text{ m/s}) \\ &= 10.9\text{ kg·m/s} \approx 11\text{ kg·m/s}. \qquad \text{(Answer)} \end{aligned}$$

With our sign convention, the initial velocity of the ball is negative and the final velocity is positive. The impulse turned out to be positive, which tells us that the direction of the impulse vector acting on the ball is the direction in which the bat was swinging, which makes sense.

(b) The impact time Δt for the baseball–bat collision is 1.2 ms. What average force acts on the baseball?

SOLUTION: From Eq. 10-8 we have

$$\begin{aligned} \overline{F} = \frac{J}{\Delta t} &= \frac{10.9\text{ kg·m/s}}{0.0012\text{ s}} \\ &= 9100\text{ N}, \qquad \text{(Answer)} \end{aligned}$$

which is about a ton. The *maximum* force will be larger than this. The sign of the average force exerted on the ball is positive, which means that the direction of the force vector is the same as that of the impulse vector.

(c) What is the average acceleration a of the baseball?

SOLUTION: We find this with

$$\overline{a} = \frac{\overline{F}}{m} = \frac{9100\text{ N}}{0.14\text{ kg}} = 6.5 \times 10^4\text{ m/s}^2, \qquad \text{(Answer)}$$

or about $6600g$.

In defining a collision, we assumed that no net external force acts on the colliding bodies. That is not true in this case, because the weight $m\mathbf{g}$ of the ball always acts on the ball, whether the ball is in flight or in contact with the bat. However, this force, with a magnitude of 1.4 N, is negligible compared to the average force exerted by the bat, which has a magnitude of 9100 N. We are quite safe in treating the collision as "isolated."

SAMPLE PROBLEM 10-2

As in Sample Problem 10-1, the baseball approaches the bat horizontally at a speed v_i of 39 m/s, but now the collision is not head-on, and the ball leaves the bat with a speed v_f of 45 m/s at an upward angle of 30° from the horizontal (Fig. 10-6). What is the average force $\overline{\mathbf{F}}$ exerted on the ball if the collision lasts 1.2 ms?

SOLUTION: We find the components J_x and J_y of the impulse from Eqs. 10-5 and 10-6:

$$J_x = p_{fx} - p_{ix} = m(v_{fx} - v_{ix})$$
$$= (0.14 \text{ kg})[(45 \text{ m/s})(\cos 30°) - (-39 \text{ m/s})]$$
$$= 10.92 \text{ kg} \cdot \text{m/s},$$

and
$$J_y = p_{fy} - p_{iy} = m(v_{fy} - v_{iy})$$
$$= (0.14 \text{ kg})[(45 \text{ m/s})(\sin 30°) - 0]$$
$$= 3.150 \text{ kg} \cdot \text{m/s}.$$

The magnitude of the impulse $\mathbf{J}$ is given by

$$J = \sqrt{J_x^2 + J_y^2}$$
$$= \sqrt{(10.92 \text{ kg} \cdot \text{m/s})^2 + (3.150 \text{ kg} \cdot \text{m/s})^2}$$
$$= 11.37 \text{ kg} \cdot \text{m/s}.$$

From Eq. 10-8, the magnitude $\overline{F}$ of the average force is

$$\overline{F} = \frac{J}{\Delta t} = \frac{11.37 \text{ kg} \cdot \text{m/s}}{0.0012 \text{ s}}$$
$$= 9475 \text{ N} \approx 9500 \text{ N}. \qquad \text{(Answer)}$$

The impulse $\mathbf{J}$ is angled upward from the horizontal by θ, where θ is given by

$$\tan \theta = \frac{J_y}{J_x} = \frac{3.150 \text{ kg} \cdot \text{m/s}}{10.92 \text{ kg} \cdot \text{m/s}} = 0.288,$$

or
$$\theta = 16°. \qquad \text{(Answer)}$$

The average force $\overline{\mathbf{F}}$ is in the same direction as $\mathbf{J}$. Note that, unlike Sample Problem 10-1, the direction of $\overline{\mathbf{F}}$ and $\mathbf{J}$ in this two-dimensional situation is different from the direction taken by the ball as it leaves the bat.

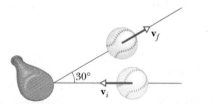

FIGURE 10-6 Sample Problem 10-2. A bat collides with a pitched baseball, sending the ball off at an angle of 30° from the horizontal.

CHECKPOINT 2: The figure shows an overhead view of a ball bouncing from a wall without any change in its speed. Consider the change $\Delta\mathbf{p}$ in the ball's linear momentum. (a) Is Δp_x positive, negative, or zero? (b) Is Δp_y positive, negative, or zero? (c) What is the direction of $\Delta\mathbf{p}$?

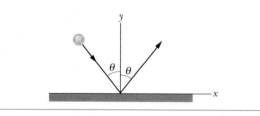

10-3 ELASTIC COLLISIONS IN ONE DIMENSION

Stationary Target

Consider a simple head-on collision of two bodies of masses m_1 and m_2 as illustrated in Fig. 10-7. For convenience, we take one of the bodies to be stationary, with velocity $\mathbf{v}_{2i} = 0$ before the collision. That body will be the "target," and the other body will be the "projectile,"* with velocity $\mathbf{v}_{1i}$ before the collision. We assume that this two-body system is closed (no mass enters or leaves it) and isolated (no net external force acts on it). Let us also make another, special assumption: the kinetic energy of the system is the same before and after the collision. The collision is then said to be of a special type called **elastic collisions.**

> In an elastic collision, the kinetic energy of each colliding body can change, but the total kinetic energy of the system does not change.

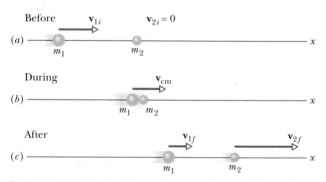

FIGURE 10-7 Two bodies undergo an elastic collision. One of them (the target body with mass m_2) is initially at rest before the collision. The velocities are shown (a) before, (b) during, and (c) after the collision. (The velocity during the collision is the velocity of the center of mass of the two bodies, which are momentarily touching.) The velocities are drawn to scale for the case in which $m_1 = 3m_2$.

---

*If the target body is moving with respect to our laboratory frame, we can always find another inertial reference frame in which the target body is initially stationary.

Note carefully that the linear momentum of a closed, isolated system is *always* conserved in a collision, regardless of whether the collision is elastic, because the forces involved in the collision are all internal forces.

In a closed, isolated system, the linear momentum of each colliding body can change, but the net linear momentum cannot change, regardless of whether the collision is elastic.

The conservations of linear momentum and of kinetic energy for the collision of Fig. 10-7 give us

$$m_1 v_{1i} = m_1 v_{1f} + m_2 v_{2f} \quad \text{(linear momentum)} \quad (10\text{-}14)$$

and

$$\tfrac{1}{2} m_1 v_{1i}^2 = \tfrac{1}{2} m_1 v_{1f}^2 + \tfrac{1}{2} m_2 v_{2f}^2 \quad \text{(kinetic energy).} \quad (10\text{-}15)$$

In each of these equations, the subscript i identifies the initial velocities and the subscript f the final velocities of the bodies. If we know the masses of the bodies and if we also know v_{1i}, the initial velocity of body 1, the only unknown quantities are v_{1f} and v_{2f}, the final velocities of the two bodies. With two equations at our disposal, we should be able to find these two unknowns.

To do so, we rewrite Eq. 10-14 as

$$m_1(v_{1i} - v_{1f}) = m_2 v_{2f} \quad (10\text{-}16)$$

and Eq. 10-15 as*

$$m_1(v_{1i} - v_{1f})(v_{1i} + v_{1f}) = m_2 v_{2f}^2. \quad (10\text{-}17)$$

After dividing Eq. 10-17 by Eq. 10-16 and doing some more algebra, we obtain

$$v_{1f} = \frac{m_1 - m_2}{m_1 + m_2} v_{1i} \quad (10\text{-}18)$$

and

$$v_{2f} = \frac{2m_1}{m_1 + m_2} v_{1i}. \quad (10\text{-}19)$$

We note from Eq. 10-19 that v_{2f} is always positive (the target body with mass m_2 always moves forward). From Eq. 10-18 we see that v_{1f} may be of either sign (the projectile body with mass m_1 moves forward if $m_1 > m_2$ but rebounds if $m_1 < m_2$).

CHECKPOINT **3:** What is the final linear momentum of the target in Fig. 10-7 if the initial linear momentum of the projectile is 6 kg·m/s and the final linear momentum of the projectile is (a) 2 kg·m/s and (b) −2 kg·m/s? (c) What is the final kinetic energy of the target if the initial and final kinetic energies of the projectile are, respectively, 5 J and 2 J?

Let us look at a few special situations.

1. *Equal masses.* If $m_1 = m_2$, Eqs. 10-18 and 10-19 reduce to

$$v_{1f} = 0 \quad \text{and} \quad v_{2f} = v_{1i},$$

which we might call a pool player's result. It predicts that after a head-on collision of bodies with equal masses, body 1 (initially moving) stops dead in its tracks and body 2 (initially at rest) takes off with the initial speed of body 1. In head-on collisions, bodies of equal mass simply exchange velocities. This is true even if the target particle (body 2) is not initially at rest.

2. *A massive target.* In terms of Fig. 10-7, a massive target means that $m_2 \gg m_1$. For example, we might fire a golf ball at a cannonball. Equations 10-18 and 10-19 then reduce to

$$v_{1f} \approx -v_{1i} \quad \text{and} \quad v_{2f} \approx \left(\frac{2m_1}{m_2}\right) v_{1i}. \quad (10\text{-}20)$$

This tells us that body 1 (the golf ball) simply bounces back in the same direction from which it came, its speed essentially unchanged. Body 2 (the cannonball) moves forward at a low speed, because the quantity in parentheses in Eq. 10-20 is much less than unity. All this is what we should expect.

3. *A massive projectile.* This is the opposite case; that is, $m_1 \gg m_2$. This time, we fire a cannonball at a golf ball. Equations 10-18 and 10-19 reduce to

$$v_{1f} \approx v_{1i} \quad \text{and} \quad v_{2f} \approx 2v_{1i}. \quad (10\text{-}21)$$

Equation 10-21 tells us that body 1 (the cannonball) simply keeps on going, scarcely slowed by the collision. Body 2 (the golf ball) charges ahead at twice the speed of the cannonball.

You may wonder: "Why twice the speed?" As a starting point in thinking about the matter, recall the collision described by Eq. 10-20, in which the velocity of the incident light body (the golf ball) changed from $+v$ to $-v$, a velocity *change* of $2v$. The same *change* in velocity (from zero to $2v$) occurs in this example also.

4. *Motion of the center of mass.* The center of mass of two colliding bodies continues to move, totally uninfluenced by the collision. This follows from conservation

---

*In this step, we use the identity $a^2 - b^2 = (a - b)(a + b)$. It reduces the amount of algebra needed to solve the simultaneous equations, Eqs. 10-16 and 10-17.

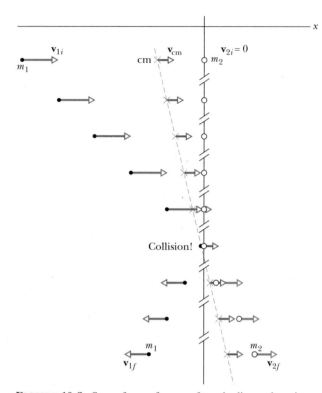

FIGURE 10-8 Some freeze-frames of two bodies undergoing an elastic collision. Body 2 is initially at rest, and $m_2 = 3m_1$. The velocity of the center of mass is also shown. Note that it is unaffected by the collision.

of linear momentum and from Eq. 9-26,

$$P = Mv_{\text{cm}} = (m_1 + m_2)v_{\text{cm}}, \quad (10\text{-}22)$$

which relates the linear momentum P of the system of two bodies to v_{cm}, the velocity of their center of mass. Because the momentum P is unchanged by the collision, v_{cm} must also be unchanged. So the center of mass continues to move in the same direction and at the same speed. From Eq. 10-22, the velocity of the center of mass for the collision shown in Fig. 10-7 (target initially at rest) is

$$v_{\text{cm}} = \frac{P}{m_1 + m_2} = \frac{m_1}{m_1 + m_2} v_{1i}. \quad (10\text{-}23)$$

Figure 10-8, a series of freeze-frames of a typical elastic collision, shows that the center of mass does indeed move steadily forward, unaffected in any way by the collision.

Moving Target

Now that we have examined the elastic collision of a projectile and a stationary target, let us examine the situation in which both bodies are moving before they undergo an elastic collision.

FIGURE 10-9 Two bodies headed for an elastic collision.

For the situation of Fig. 10-9, the conservation of linear momentum is written as

$$m_1v_{1i} + m_2v_{2i} = m_1v_{1f} + m_2v_{2f}, \quad (10\text{-}24)$$

and the conservation of kinetic energy is written as

$$\tfrac{1}{2}m_1v_{1i}^2 + \tfrac{1}{2}m_2v_{2i}^2 = \tfrac{1}{2}m_1v_{1f}^2 + \tfrac{1}{2}m_2v_{2f}^2. \quad (10\text{-}25)$$

To solve these simultaneous equations for v_{1f} and v_{2f}, we first rewrite Eq. 10-24 as

$$m_1(v_{1i} - v_{1f}) = -m_2(v_{2i} - v_{2f}), \quad (10\text{-}26)$$

and Eq. 10-25 as

$$m_1(v_{1i} - v_{1f})(v_{1i} + v_{1f})$$
$$= -m_2(v_{2i} - v_{2f})(v_{2i} + v_{2f}). \quad (10\text{-}27)$$

After dividing Eq. 10-27 by Eq. 10-26 and doing some more algebra, we obtain

$$v_{1f} = \frac{m_1 - m_2}{m_1 + m_2} v_{1i} + \frac{2m_2}{m_1 + m_2} v_{2i} \quad (10\text{-}28)$$

and

$$v_{2f} = \frac{2m_1}{m_1 + m_2} v_{1i} + \frac{m_2 - m_1}{m_1 + m_2} v_{2i}. \quad (10\text{-}29)$$

Note that the assignment of subscripts 1 and 2 to the bodies is arbitrary. If we exchange those subscripts in Fig. 10-9 and in Eqs. 10-28 and 10-29, we end up with the same set of equations. Note also that if we set $v_{2i} = 0$, body 2 becomes a stationary target, and Eqs. 10-28 and 10-29 reduce to Eqs. 10-18 and 10-19, respectively.

From Eq. 10-22, the constant velocity v_{cm} of the center of mass of the two-body system of Fig. 10-9 is

$$v_{\text{cm}} = \frac{P}{m_1 + m_2} = \frac{m_1v_{1i} + m_2v_{2i}}{m_1 + m_2}. \quad (10\text{-}30)$$

Because the two-body system is closed and isolated, its linear momentum P is unchanged by the collision. Thus, the velocity v_{cm} of its center of mass is also unchanged.

$\mathcal{C}$HECKPOINT **4:** The initial momenta of body 1 and body 2 in Fig. 10-9 are, respectively, 10 kg·m/s and −8 kg·m/s. What is the final momentum of body 2 if the final momentum of body 1 is (a) 2 kg·m/s, and (b) −2 kg·m/s?

SAMPLE PROBLEM 10-3

Two metal spheres, suspended by vertical cords, initially just touch, as shown in Fig. 10-10. Sphere 1, with mass $m_1 = 30$ g, is pulled to the left to height $h_1 = 8.0$ cm, and then released. After swinging down, it undergoes an elastic collision with sphere 2, whose mass $m_2 = 75$ g.

(a) What is the velocity v_{1f} of sphere 1 just after the collision?

SOLUTION: We let v_{1i} represent the speed of sphere 1 just before the collision. When the sphere begins its descent, its kinetic energy is zero and the gravitational potential energy is m_1gh_1. Just before the collision, sphere 1 has kinetic energy $\frac{1}{2}m_1v_{1i}^2$ and the gravitational potential energy is zero. For the descent, the conservation of mechanical energy gives us

$$\tfrac{1}{2}m_1v_{1i}^2 = m_1gh_1,$$

which we solve for the speed v_{1i} of sphere 1 just before the collision:

$$v_{1i} = \sqrt{2gh_1} = \sqrt{(2)(9.8 \text{ m/s}^2)(0.080 \text{ m})} = 1.252 \text{ m/s}.$$

Although sphere 1 swings down in a two-dimensional arc, its velocity is horizontal when it collides with sphere 2. Thus the collision is one-dimensional, and we can represent the velocity of sphere 1 just before that collision as v_{1i}.

To find the velocity v_{1f} of sphere 1 just after the collision, we use Eq. 10-18:

$$v_{1f} = \frac{m_1 - m_2}{m_1 + m_2}v_{1i} = \frac{0.030 \text{ kg} - 0.075 \text{ kg}}{0.030 \text{ kg} + 0.075 \text{ kg}}(1.252 \text{ m/s})$$
$$= -0.537 \text{ m/s} \approx -0.54 \text{ m/s}. \qquad \text{(Answer)}$$

The minus sign tells us that sphere 1 moves to the left just after the collision.

(b) To what height h_1' does sphere 1 swing to the left after the collision?

SOLUTION: When sphere 1 begins its upward swing, its kinetic energy is $\frac{1}{2}m_1v_{1f}^2$ and the gravitational potential energy is zero. When it momentarily stops at height h_1', its kinetic energy is zero and the gravitational potential energy is m_1gh_1'. Conserving mechanical energy for the upward swing, we find

$$m_1gh_1' = \tfrac{1}{2}m_1v_{1f}^2,$$

or
$$h_1' = \frac{v_{1f}^2}{2g} = \frac{(-0.537 \text{ m/s})^2}{(2)(9.8 \text{ m/s}^2)}$$
$$= 0.0147 \text{ m} \approx 1.5 \text{ cm}. \qquad \text{(Answer)}$$

(c) What is the velocity v_{2f} of sphere 2 just after the collision?

SOLUTION: From Eq. 10-19, we have

$$v_{2f} = \frac{2m_1}{m_1 + m_2}v_{1i} = \frac{(2)(0.030 \text{ kg})}{0.030 \text{ kg} + 0.075 \text{ kg}}(1.252 \text{ m/s})$$
$$= 0.715 \text{ m/s} \approx 0.72 \text{ m/s}. \qquad \text{(Answer)}$$

(d) To what height h_2 does sphere 2 swing after the collision?

SOLUTION: Sphere 2 begins its ascent with kinetic energy $\frac{1}{2}m_2v_{2f}^2$. When it momentarily stops at height h_2, the gravitational potential energy is m_2gh_2. Conservation of mechanical energy for the ascent gives us

$$m_2gh_2 = \tfrac{1}{2}m_2v_{2f}^2,$$

or
$$h_2 = \frac{v_{2f}^2}{2g} = \frac{(0.715 \text{ m/s})^2}{(2)(9.8 \text{ m/s}^2)}$$
$$= 0.0261 \text{ m} \approx 2.6 \text{ cm}. \qquad \text{(Answer)}$$

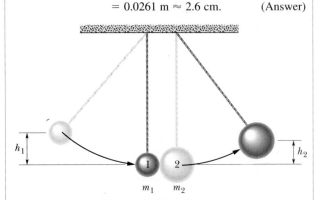

FIGURE 10-10 Sample Problem 10-3. Two metal spheres suspended by cords just touch when they are at rest. Sphere 1, with mass m_1, is pulled to the left to height h_1 and then released. The subsequent elastic collision with sphere 2 sends sphere 2 to height h_2.

SAMPLE PROBLEM 10-4

In a nuclear reactor, newly produced fast neutrons must be slowed down before they can participate effectively in the chain-reaction process. This is done by allowing them to collide with the nuclei of atoms in a *moderator*.

(a) By what fraction is the kinetic energy of a neutron (of mass m_1) reduced in a head-on elastic collision with a nucleus of mass m_2, initially at rest?

SOLUTION: The initial and final kinetic energies of the neutron are

$$K_i = \tfrac{1}{2}m_1v_{1i}^2 \quad \text{and} \quad K_f = \tfrac{1}{2}m_1v_{1f}^2.$$

The fraction we seek is then

$$\text{frac} = \frac{K_i - K_f}{K_i} = \frac{v_{1i}^2 - v_{1f}^2}{v_{1i}^2} = 1 - \frac{v_{1f}^2}{v_{1i}^2}. \qquad (10\text{-}31)$$

For such a collision we have, from Eq. 10-18,

$$\frac{v_{1f}}{v_{1i}} = \frac{m_1 - m_2}{m_1 + m_2}. \qquad (10\text{-}32)$$

Substituting Eq. 10-32 into Eq. 10-31 yields, after a little algebra,

$$\text{frac} = \frac{4m_1m_2}{(m_1 + m_2)^2}. \quad \text{(Answer)} \quad \text{(10-33)}$$

(b) Evaluate the fraction for lead, carbon, and hydrogen. The ratios of the mass of a nucleus to the mass of a neutron ($= m_2/m_1$) for these nuclei are 206 for lead, 12 for carbon, and about 1 for hydrogen.

SOLUTION: The following values of the fraction can be calculated with Eq. 10-33: for lead ($m_2 = 206m_1$),

$$\text{frac} = \frac{(4)(206)}{(1 + 206)^2} = 0.019 \text{ or } 1.9\%; \quad \text{(Answer)}$$

for carbon ($m_2 = 12m_1$),

$$\text{frac} = \frac{(4)(12)}{(1 + 12)^2} = 0.28 \text{ or } 28\%; \quad \text{(Answer)}$$

and for hydrogen ($m_2 \approx m_1$),

$$\text{frac} = \frac{(4)(1)}{(1 + 1)^2} = 1 \text{ or } 100\%. \quad \text{(Answer)}$$

These results partially explain why water, which contains lots of hydrogen, is a much better moderator of neutrons than lead.

10-4 INELASTIC COLLISIONS IN ONE DIMENSION

An **inelastic collision** is one in which the kinetic energy of the system of colliding bodies is not conserved. If you drop a Superball onto a hard floor, it loses very little of its kinetic energy on impact and rebounds to almost its original height. If the ball did regain the original height, its collision with the floor would be elastic. However, the small loss of kinetic energy in the collision lowers the rebound height, and therefore the collision is somewhat inelastic.

A dropped golf ball will lose more of its kinetic energy and will rebound to only 60% of its original height. This collision is noticeably inelastic. If you drop a ball of wet putty onto the floor, it sticks to the floor and does not rebound at all. Because the putty sticks and does not rebound, this collision is said to be a **completely inelastic collision.**

The kinetic energy that is lost in any inelastic collision is transferred to some other form of energy, perhaps thermal energy. Nonetheless, the linear momentum of the system is always conserved (provided the system is isolated and closed). Since kinetic energy and linear momentum both involve the speeds of the colliding bodies, the conservation of linear momentum limits just how much kinetic energy is lost by a system in an inelastic collision. When the bodies stick together in a completely inelastic collision, the amount of kinetic energy that is lost is the maximum

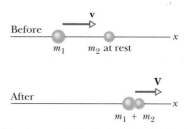

FIGURE 10-11 A completely inelastic collision between two bodies. Before the collision, the body of mass m_2 is at rest. Afterward, the bodies stick together, which is the criterion for a *completely* inelastic collision. The velocities are drawn to scale for the case in which $m_1 = 3m_2$.

allowed by the conservation of linear momentum. In some situations, this maximum loss is all the kinetic energy of the system.

In this section, we restrict ourselves to completely inelastic collisions. Figure 10-11 shows a one-dimensional inelastic collision in which one body is initially stationary. The law of conservation of linear momentum holds, so

$$m_1v = (m_1 + m_2)V, \quad \text{(10-34)}$$

or

$$V = v \frac{m_1}{m_1 + m_2}, \quad \text{(10-35)}$$

where V represents the final velocity of the stuck-together bodies. Equation 10-35 tells us that the final speed is always less than that of the incoming body.

Figure 10-12 (compare it with Fig. 10-8) shows that the motion of the center of mass in a completely inelastic collision is unaffected by the collision. Here, even though the bodies stick together, the kinetic energy associated with the motion of the center of mass is still present. Kinetic energy cannot vanish *totally* in an inelastic collision unless the reference frame is fixed with respect to the center of mass of the colliding bodies. In such cases, all motion relative to the reference frame ceases when the bodies stick together. If the target body m_2 happens to be exceedingly massive, the center of mass of the system essentially coincides with that of the target. This is the situation in which a putty ball is dropped onto the floor, the target body being the Earth. As the ball hits, all the kinetic energy is dissipated, being transferred to some other form of energy.

If both bodies are moving prior to a completely inelastic collision, we replace Eq. 10-34 with

$$m_1v_1 + m_2v_2 = (m_1 + m_2)V, \quad \text{(10-36)}$$

where m_1v_1 is the initial linear momentum of one body, and m_2v_2 is that of the other body. Here again, all the kinetic energy is dissipated only if the reference frame is fixed with respect to the center of mass of the bodies. Figure

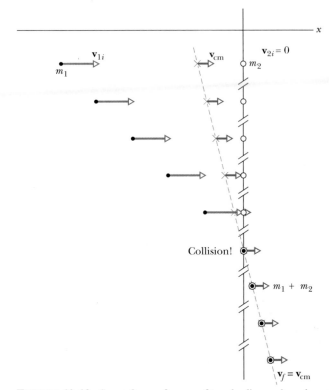

FIGURE 10-12 Some freeze-frames of two bodies undergoing a completely inelastic collision. Body 2 is initially at rest, and $m_2 = 3m_1$. The bodies stick together after the collision and move forward together. The velocity of the center of mass is also shown. Note that it is unaffected by the collision and that it is equal to the final velocity of the stuck-together bodies.

FIGURE 10-13 Two cars after an almost head-on, almost completely inelastic collision.

10-13 shows an example: identical cars driven at identical speeds underwent an almost head-on and almost completely inelastic collision. Their center of mass was stationary relative to the Earth. So a bystander, who would have been stationary relative to the Earth, would have seen these cars stop dead because of the collision, rather than move afterward in one direction or another.

CHECKPOINT 5: Body 1 and body 2 are in a completely inelastic one-dimensional collision. What is their final momentum if their initial momenta are, respectively, (a) 10 kg·m/s and 0; (b) 10 kg·m/s and 4 kg·m/s; (c) 10 kg·m/s and −4 kg·m/s?

SAMPLE PROBLEM 10-5

The *ballistic pendulum* was used to measure the speeds of bullets before electronic timing devices were developed. The version shown in Fig. 10-14 consists of a large block of wood of mass $M = 5.4$ kg, hanging from two long cords. A bullet of mass $m = 9.5$ g is fired into the block, coming quickly to rest. The *block + bullet* then swing upward, their center of mass rising a vertical distance $h = 6.3$ cm before the pendulum comes momentarily to rest at the end of its arc.

(a) What was the speed v of the bullet just prior to the collision?

SOLUTION: Just after the collision, the *bullet + block* have speed V. Applying the conservation of linear momentum to

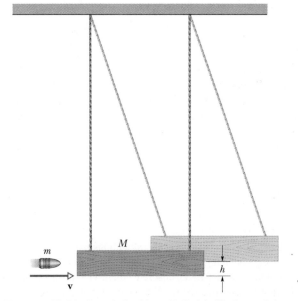

FIGURE 10-14 Sample Problem 10-5. A ballistic pendulum, used to measure the speeds of bullets.

the collision, we have

$$mv = (M + m)V.$$

Since the bullet and block stick together, the collision is completely inelastic, and kinetic energy is not conserved during it. However, *after* the collision, mechanical energy *is* conserved, because no force then acts to dissipate that energy. So the kinetic energy of the system when the block is at the bottom of its arc must equal the potential energy of the system when the block is at the top:

$$\tfrac{1}{2}(M + m)V^2 = (M + m)gh.$$

Eliminating V between these two equations leads to

$$
\begin{aligned}
v &= \frac{M + m}{m}\sqrt{2gh} \\
&= \left(\frac{5.4 \text{ kg} + 0.0095 \text{ kg}}{0.0095 \text{ kg}}\right)\sqrt{(2)(9.8 \text{ m/s}^2)(0.063 \text{ m})} \\
&= 630 \text{ m/s}. \qquad\qquad \text{(Answer)}
\end{aligned}
$$

The ballistic pendulum is a kind of "transformer," exchanging the high speed of a light object (the bullet) for the low — and thus more easily measurable — speed of a massive object (the block).

(b) What is the initial kinetic energy of the bullet? How much of this energy remains as mechanical energy of the swinging pendulum?

SOLUTION: The kinetic energy of the bullet is

$$
\begin{aligned}
K_b &= \tfrac{1}{2}mv^2 = (\tfrac{1}{2})(0.0095 \text{ kg})(630 \text{ m/s})^2 \\
&= 1900 \text{ J}. \qquad\qquad \text{(Answer)}
\end{aligned}
$$

The mechanical energy of the swinging pendulum is equal to the gravitational potential energy when the block is at the top of its swing:

$$
\begin{aligned}
E &= (M + m)gh \\
&= (5.4 \text{ kg} + 0.0095 \text{ kg})(9.8 \text{ m/s}^2)(0.063 \text{ m}) \\
&= 3.3 \text{ J}. \qquad\qquad \text{(Answer)}
\end{aligned}
$$

Thus only 3.3/1900, or 0.2%, of the original kinetic energy of the bullet is transferred to mechanical energy of the pendulum. The rest is transferred to thermal energy of the block and bullet, or goes into the breaking of wood fibers as the bullet bores into the wood.

SAMPLE PROBLEM 10-6

A karate expert strikes downward with his fist (of mass $m_1 = 0.70$ kg), breaking a 0.14 kg board (Fig. 10-15a). He then does the same to a 3.2 kg concrete block. The spring constants k for bending are 4.1×10^4 N/m for the board and 2.6×10^6 N/m for the block. Breaking occurs at a deflection d of 16 mm for the board and 1.1 mm for the block (Fig. 10-15c).*

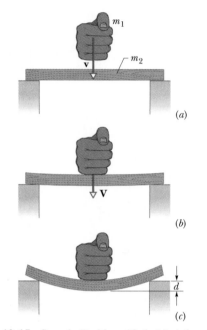

FIGURE 10-15 Sample Problem 10-6. (*a*) A karate expert strikes at a flat object with speed v. (*b*) Fist and object undergo a completely inelastic collision, and bending begins. The *fist + object* then have speed V. (*c*) The object breaks when its center has been deflected by an amount d.

(a) Just before the board and block break, what is the energy stored in each?

SOLUTION: We treat the bending as the compression of a spring for which Hooke's law applies. The stored energy is then, from Eq. 8-11, $U = \tfrac{1}{2}kd^2$. For the board,

$$
\begin{aligned}
U &= \tfrac{1}{2}(4.1 \times 10^4 \text{ N/m})(0.016 \text{ m})^2 \\
&= 5.248 \text{ J} \approx 5.2 \text{ J}. \qquad \text{(Answer)}
\end{aligned}
$$

For the block,

$$
\begin{aligned}
U &= \tfrac{1}{2}(2.6 \times 10^6 \text{ N/m})(0.0011 \text{ m})^2 \\
&= 1.573 \text{ J} \approx 1.6 \text{ J}. \qquad \text{(Answer)}
\end{aligned}
$$

(b) What fist speed v is required to break the board and the block? Assume that mechanical energy is conserved during the bending, that the fist and struck object stop just before the break, and that the fist–object collision at the onset of bending (Fig. 10-15b) is completely inelastic.

SOLUTION: If mechanical energy is conserved during the bending, then the kinetic energy K of the fist–object system at the onset of bending must be equal to the stored energy U just at breaking: 5.2 J for the board and 1.6 J for the block. The fist

---

*The data are taken from "The Physics of Karate," by S. R. Wilk, R. E. McNair, and M. S. Feld, *American Journal of Physics,* September 1983.

speed required to break the object is the speed required to produce that kinetic energy K. To find it, we must first find the speed V of *fist + object* at the onset of bending. We write the equality between K and U as

$$K = \tfrac{1}{2}(m_1 + m_2)V^2 = U,$$

from which we find

$$V = \sqrt{\frac{2U}{m_1 + m_2}},$$

where V is the speed of *fist + object* at the onset of bending, $m_1 = 0.70$ kg, and m_2 is 0.14 kg for the board or 3.2 kg for the block. For the board we have

$$V = \sqrt{\frac{2(5.248 \text{ J})}{0.70 \text{ kg} + 0.14 \text{ kg}}} = 3.534 \text{ m/s}.$$

For the block we have

$$V = \sqrt{\frac{2(1.573 \text{ J})}{0.70 \text{ kg} + 3.2 \text{ kg}}} = 0.8981 \text{ m/s}.$$

Now let the fist have speed v just before hitting the board or block. Then Eq. 10-35 holds for the collision. By rearranging that equation, we get

$$v = \left(\frac{m_1 + m_2}{m_1}\right)V.$$

For the board we find

$$v = \left(\frac{0.70 \text{ kg} + 0.14 \text{ kg}}{0.70 \text{ kg}}\right)(3.534 \text{ m/s})$$

$$\approx 4.2 \text{ m/s}, \qquad \text{(Answer)}$$

and for the block we find

$$v = \left(\frac{0.70 \text{ kg} + 3.2 \text{ kg}}{0.70 \text{ kg}}\right)(0.8981 \text{ m/s})$$

$$\approx 5.0 \text{ m/s}. \qquad \text{(Answer)}$$

The fist speed must be about 20% faster for the fist to break the block, because the larger mass of the block makes the transfer of energy to the block more difficult.

10-5 COLLISIONS IN TWO DIMENSIONS

Here we consider a *glancing collision* (it is not head-on) between a projectile body and a target body at rest. Figure 10-16 shows a typical situation. After the collision, the two bodies fly off at angles θ_1 and θ_2, as the figure shows.

From the conservation of linear momentum (a vector relation), we can write two scalar equations:

$$m_1v_{1i} = m_1v_{1f}\cos\theta_1 + m_2v_{2f}\cos\theta_2 \qquad \text{(x component)}$$
$$\text{(10-37)}$$

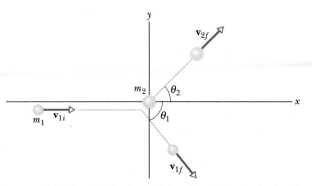

FIGURE 10-16 An elastic collision between two bodies in which the collision is not head-on. The body with mass m_2 (the target) is initially at rest.

and

$$0 = -m_1v_{1f}\sin\theta_1 + m_2v_{2f}\sin\theta_2 \qquad \text{(y component)}.$$
$$\text{(10-38)}$$

If the collision is elastic, kinetic energy is also conserved:

$$\tfrac{1}{2}m_1v_{1i}^2 = \tfrac{1}{2}m_1v_{1f}^2 + \tfrac{1}{2}m_2v_{2f}^2 \qquad \text{(kinetic energy)}. \quad \text{(10-39)}$$

These three equations contain seven variables: two masses, m_1 and m_2; three speeds, v_{1i}, v_{1f}, and v_{2f}; and two angles, θ_1 and θ_2. If we know any four of these quantities, we can solve the three equations for the remaining three quantities. Often the known quantities are the two masses, the initial speed, and one of the angles. The unknowns to be solved for are then the two final speeds and the remaining angle.

CHECKPOINT **6:** In Fig. 10-16, suppose that the projectile has an initial momentum of 6 kg·m/s, a final x component of momentum of 4 kg·m/s, and a final y component of momentum of −3 kg·m/s. For the target, what then are (a) the final x component of momentum and (b) the final y component of momentum?

SAMPLE PROBLEM 10-7

Two particles of equal masses have an elastic collision, the target particle being initially at rest. Show that (unless the collision is head-on) the two particles will always move off perpendicular to each other after the collision.

SOLUTION: You might be tempted to jump into the problem and solve it in a straightforward way, by applying Eqs. 10-37, 10-38, and 10-39. There is a neater way.

Figure 10-17a shows the situation before and after the collision, each particle with its linear momentum vector attached. Because linear momentum is conserved in the collision, these vectors must form a closed triangle, as Fig. 10-17b shows. (The vector $m\mathbf{v}_{1i}$ must be the vector sum of $m\mathbf{v}_{1f}$ and

$m\mathbf{v}_{2f}$.) Because the masses of the particles are equal, the closed linear momentum triangle of Fig. 10-17b is also a closed velocity triangle, because dividing by the scalar m does not change the relation of the vectors. That is, we may draw the velocity vectors as in Fig. 10-17c, because

$$\mathbf{v}_{1i} = \mathbf{v}_{1f} + \mathbf{v}_{2f}. \qquad (10\text{-}40)$$

Because kinetic energy is conserved, Eq. 10-39 holds. With the equal terms $\frac{1}{2}m$ canceled out, this equation becomes

$$v_{1i}^2 = v_{1f}^2 + v_{2f}^2. \qquad (10\text{-}41)$$

Equation 10-41 applies to the lengths of the sides in the triangle of Fig. 10-17c. For it to hold, the triangle must be a right triangle (and Eq. 10-41 is then the Pythagorean theorem). Therefore the angle ϕ between the vectors $\mathbf{v}_{1f}$ and $\mathbf{v}_{2f}$ in Fig. 10-17 must be 90°, which is what we set out to prove.

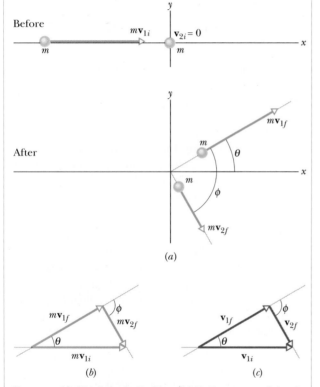

(a)

(b)   (c)

FIGURE 10-17 Sample Problem 10-7. A neat proof that in an elastic collision between two particles of the same mass, the particles fly off at 90° to each other afterward. For this to hold true, the target particle must be initially at rest and the collision must not be head-on.

SAMPLE PROBLEM 10-8

Two skaters collide and embrace, in a completely inelastic collision. That is, they stick together after impact, as suggested by Fig. 10-18, where the origin is placed at the point of collision. Alfred, whose mass m_A is 83 kg, is originally moving east with speed $v_A = 6.2$ km/h. Barbara, whose mass m_B is 55 kg, is originally moving north with speed $v_B = 7.8$ km/h.

(a) What is the velocity $\mathbf{V}$ of the couple after impact?

SOLUTION: Linear momentum is conserved during the collision. We can write, for the linear momentum components in the x and y directions,

$$m_A v_A = MV \cos \theta \qquad (x \text{ component}) \qquad (10\text{-}42)$$

and

$$m_B v_B = MV \sin \theta \qquad (y \text{ component}), \qquad (10\text{-}43)$$

in which $M = m_A + m_B$. Dividing Eq. 10-43 by Eq. 10-42 yields

$$\tan \theta = \frac{m_B v_B}{m_A v_A} = \frac{(55 \text{ kg})(7.8 \text{ km/h})}{(83 \text{ kg})(6.2 \text{ km/h})} = 0.834.$$

Thus

$$\theta = \tan^{-1} 0.834 = 39.8° \approx 40°. \qquad \text{(Answer)}$$

From Eq. 10-43 we then have

$$V = \frac{m_B v_B}{M \sin \theta} = \frac{(55 \text{ kg})(7.8 \text{ km/h})}{(83 \text{ kg} + 55 \text{ kg})(\sin 39.8°)}$$
$$= 4.86 \text{ km/h} \approx 4.9 \text{ km/h}. \qquad \text{(Answer)}$$

(b) What is the velocity of the center of mass of the two skaters before and after the collision?

SOLUTION: We can answer this without further calculation. After the collision, the velocity of the center of mass is the same as the velocity that we calculated in part (a), namely, 4.9 km/h at 40° north of east. Because the velocity of the center of mass is not changed by the collision, the same value must prevail before the collision.

(c) What is the fractional change in the kinetic energy of the skaters because of the collision?

SOLUTION: The initial kinetic energy is

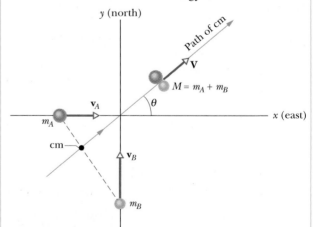

FIGURE 10-18 Sample Problem 10-8. Two skaters, Alfred (A) and Barbara (B), represented with spheres in this simplified overhead view, have a completely inelastic collision. Afterward, they move off together at angle θ, with speed V. The path of their center of mass is shown. The position of the center of mass for the indicated positions of the skaters before the collision is also shown.

$$K_i = \tfrac{1}{2}m_A v_A^2 + \tfrac{1}{2}m_B v_B^2$$
$$= (\tfrac{1}{2})(83 \text{ kg})(6.2 \text{ km/h})^2 + (\tfrac{1}{2})(55 \text{ kg})(7.8 \text{ km/h})^2$$
$$= 3270 \text{ kg}\cdot\text{km}^2/\text{h}^2.$$

The final kinetic energy is

$$K_f = \tfrac{1}{2}MV^2$$
$$= (\tfrac{1}{2})(83 \text{ kg} + 55 \text{ kg})(4.86 \text{ km/h})^2$$
$$= 1630 \text{ kg}\cdot\text{km}^2/\text{h}^2.$$

The fractional change is then

$$\text{frac} = \frac{K_f - K_i}{K_i}$$
$$= \frac{1630 \text{ kg}\cdot\text{km}^2/\text{h}^2 - 3270 \text{ kg}\cdot\text{km}^2/\text{h}^2}{3270 \text{ kg}\cdot\text{km}^2/\text{h}^2}$$
$$= -0.50. \qquad \text{(Answer)}$$

Thus 50% of the initial kinetic energy is lost as a result of the collision.

CHECKPOINT 7: In Sample Problem 10-8 and Fig. 10-18, does θ increase, decrease, or remain the same if we increase (a) the speed of Barbara and (b) the mass of Barbara?

PROBLEM SOLVING TACTICS

TACTIC 1: *Switch to SI Units?*

More often than not, it is wise to express all physical quantities in their basic SI units; thus all speeds in meters per second, all masses in kilograms, and so on. Sometimes, however, it is not necessary to do this. In Sample Problem 10-8a, for example, the units cancel out when we calculate the angle θ. In Sample Problem 10-8c, they cancel when we calculate the dimensionless quantity frac. In this latter case, for example, there is no need to change the kinetic energy units to joules, the basic SI energy unit; we can leave them in the units kg·km²/h² because we can look ahead and see that they are going to cancel when we compute the ratio frac.

10-6 REACTIONS AND DECAY PROCESSES (OPTIONAL)

Here we consider certain collisions (called **nuclear reactions**) in which the identity and even the number of interacting nuclear particles may change because of the collisions. We consider too the *spontaneous decay* of unstable particles, in which one particle is transformed to two other particles. For both kinds of event, there is a clear distinction between times ''before the event'' and times ''after

the event,'' and both linear momentum and *total* energy are conserved. In short, we can treat these events by the methods we have already developed for collisions.

SAMPLE PROBLEM 10-9

A radioactive nucleus of uranium-235 decays spontaneously to thorium-231 by emitting an alpha particle (the nucleus of a helium atom, symbolized as α, helium-4, or ⁴He):

$$^{235}\text{U} \rightarrow \alpha + ^{231}\text{Th}.$$

The alpha particle ($m_\alpha = 4.00$ u) has a kinetic energy K_α of 4.60 MeV. What is the kinetic energy of the thorium-231 nucleus ($m_{\text{Th}} = 231$ u)?

SOLUTION: The ²³⁵U nucleus is initially at rest in the laboratory. After decay, the alpha particle flies off and the ²³¹Th moves in the opposite direction, with kinetic energies K_α and K_{Th}, respectively. Applying the law of conservation of linear momentum leads to

$$0 = m_{\text{Th}}v_{\text{Th}} + m_\alpha v_\alpha,$$

which we can recast as

$$m_{\text{Th}}v_{\text{Th}} = -m_\alpha v_\alpha. \qquad (10\text{-}44)$$

Squaring both sides of Eq. 10-44 yields

$$m_{\text{TH}}^2 v_{\text{Th}}^2 = m_\alpha^2 v_\alpha^2. \qquad (10\text{-}45)$$

Because $K = \tfrac{1}{2}mv^2$, we can write Eq. 10-45 as

$$m_{\text{Th}}K_{\text{Th}} = m_\alpha K_\alpha.$$

Thus

$$K_{\text{Th}} = K_\alpha \frac{m_\alpha}{m_{\text{Th}}} = (4.60 \text{ MeV})\frac{4.00 \text{ u}}{231 \text{ u}}$$
$$= 7.97 \times 10^{-2} \text{ MeV} = 79.7 \text{ keV}. \quad \text{(Answer)}$$

We see that, of the total amount of kinetic energy made available during the decay (4.60 MeV + 0.0797 MeV = 4.68 MeV), the heavy thorium-231 nucleus receives only about 0.0797/4.68, or 1.7%.

SAMPLE PROBLEM 10-10

A nuclear reaction of great importance for the generation of energy by nuclear fusion is the so-called d-d reaction, one form of which is

$$d + d \rightarrow t + p. \qquad (10\text{-}46)$$

The particles represented by these letters are all isotopes of hydrogen, with the following properties:

SYMBOLS		NAME	MASS
p	¹H	Proton	$m_p = 1.00783$ u
d	²H	Deuteron	$m_d = 2.01410$ u
t	³H	Triton	$m_t = 3.01605$ u

(a) How much energy appears because of the mass change Δm that occurs in this reaction?

SOLUTION: From Eq. 8-40, the Q of a reaction (the energy released or absorbed in the reaction) is $Q = -\Delta m \, c^2$. Here the change Δm in mass is $\Delta m = m_p + m_t - 2m_d$. So, here Q is

$$Q = -\Delta m \, c^2 = (2m_d - m_p - m_t)c^2$$
$$= (2 \times 2.01410 \text{ u} - 1.00783 \text{ u} - 3.01605 \text{ u})$$
$$\times (931.5 \text{ MeV/u})$$
$$= (0.00432 \text{ u})(931.5 \text{ MeV/u})$$
$$= 4.02 \text{ MeV}. \qquad \text{(Answer)}$$

We have used the value of 931.5 MeV/u for c^2 (Eq. 8-43).

A positive value for Q (as in this case) means that the reaction is **exothermic**; in this reaction mass energy is transferred to kinetic energy of the products. Only $0.00432/(2 \times 2.01410)$, or about 0.1%, of the mass energy originally present has been so transferred. A negative Q signals an **endothermic** reaction, in which the transfer is from kinetic energy to mass energy. And $Q = 0$ means an *elastic encounter,* with no mass change and kinetic energy conserved.

(b) A deuteron of kinetic energy $K_d = 1.50$ MeV strikes a stationary deuteron, initiating the reaction of Eq. 10-46. A proton is observed to move off at an angle of 90° to the incident direction, with a kinetic energy of 3.39 MeV; see Fig. 10-19. What is the kinetic energy of the triton?

SOLUTION: The energy Q released due to the decrease in mass appears as an increase in the kinetic energies of the particles. Thus we can write

$$Q = \Delta K = K_p + K_t - K_d,$$

where Q is the energy computed for this reaction in (a). Solving for K_t gives us

$$K_t = Q + K_d - K_p$$
$$= 4.02 \text{ MeV} + 1.50 \text{ MeV} - 3.39 \text{ MeV}$$

$$= 2.13 \text{ MeV.} \qquad \text{(Answer)}$$

(c) At what angle ϕ with the incident direction (see Fig. 10-19) does the triton emerge?

SOLUTION: We have not yet made use of the fact that linear momentum is conserved in the reaction of Eq. 10-46. The law of conservation of linear momentum yields two scalar equations:

$$m_d v_d = m_t v_t \cos \phi \qquad \text{(x component)} \qquad \text{(10-47)}$$

and

$$0 = m_p v_p + m_t v_t \sin \phi \qquad \text{(y component).} \qquad \text{(10-48)}$$

From Eq. 10-48 we have

$$\sin \phi = -\frac{m_p v_p}{m_t v_t}. \qquad \text{(10-49)}$$

Using the relation $K = \frac{1}{2}mv^2$, we can rewrite each linear momentum mv as $\sqrt{2mK}$ and thus recast Eq. 10-49 as

$$\phi = \sin^{-1}\left(-\sqrt{\frac{m_p K_p}{m_t K_t}}\right)$$
$$= \sin^{-1}\left(-\sqrt{\frac{(1.01 \text{ u})(3.39 \text{ MeV})}{(3.02 \text{ u})(2.13 \text{ MeV})}}\right)$$
$$= \sin^{-1}(-0.730) = -46.9°. \qquad \text{(Answer)}$$

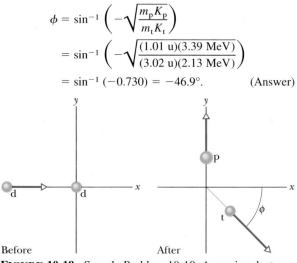

FIGURE 10-19 Sample Problem 10-10. A moving deuteron (d) strikes a stationary deuteron, producing a proton (p) and a triton (t).

REVIEW & SUMMARY

Collisions

In a **collision**, two bodies exert strong forces on each other for a relatively short time. These forces are internal to the two-body system and are significantly larger than any external force during the collision. The laws of conservation of energy and of linear momentum, applied immediately before and after a collision, allow us to predict the outcome of the collision and to understand the interactions between the colliding bodies.

Impulse and Linear Momentum

Applying Newton's second law in momentum form to a particle-like body involved in a collision leads to the **impulse–linear momentum theorem**:

$$\mathbf{p}_f - \mathbf{p}_i = \Delta \mathbf{p} = \mathbf{J}, \qquad \text{(10-4)}$$

where $\mathbf{p}_f - \mathbf{p}_i = \Delta \mathbf{p}$ is the change in the body's linear momentum, and $\mathbf{J}$ is the **impulse** due to the force $\mathbf{F}(t)$ exerted on the body by the other body in the collision:

$$\mathbf{J} = \int_{t_i}^{t_f} \mathbf{F}(t) \, dt. \qquad \text{(10-3)}$$

If $\overline{F}$ is the average of $\mathbf{F}(t)$ during the collision and Δt is the duration of the collision, then for one-dimensional motion

$$J = \overline{F} \, \Delta t. \qquad \text{(10-8)}$$

When a steady stream of bodies, each with mass m and speed v, collides with a body fixed in position, the average force on the fixed body is

$$\overline{F} = -\frac{n}{\Delta t}\, \Delta p = -\frac{n}{\Delta t}\, m\, \Delta v, \qquad (10\text{-}10)$$

where $n/\Delta t$ is the rate at which the bodies collide with the fixed body, and Δv is the change in velocity of each colliding body. This average force can also be written as

$$\overline{F} = -\frac{\Delta m}{\Delta t}\, \Delta v, \qquad (10\text{-}13)$$

where $\Delta m/\Delta t$ is the rate at which mass collides with the fixed body. In Eqs. 10-10 and 10-13, $\Delta v = -v$ if the bodies stop upon impact or $\Delta v = -2v$ if they bounce directly backward with no change in speed.

Elastic Collision — One Dimension

An *elastic collision* is one in which the kinetic energy of a system of two colliding bodies is conserved. For a one-dimensional situation in which one body (the target) is stationary and the other body (the projectile) is initially moving, we get the following relations from conservation of kinetic energy and linear momentum:

$$v_{1f} = \frac{m_1 - m_2}{m_1 + m_2}\, v_{1i} \qquad (10\text{-}18)$$

and

$$v_{2f} = \frac{2m_1}{m_1 + m_2}\, v_{1i}. \qquad (10\text{-}19)$$

Here subscripts i and f refer to the velocities immediately before and after the collision, respectively. If both bodies are moving prior to the collision, their velocities immediately after the collision are given by

$$v_{1f} = \frac{m_1 - m_2}{m_1 + m_2}\, v_{1i} + \frac{2m_2}{m_1 + m_2}\, v_{2i} \qquad (10\text{-}28)$$

and

$$v_{2f} = \frac{2m_1}{m_1 + m_2}\, v_{1i} + \frac{m_2 - m_1}{m_1 + m_2}\, v_{2i}. \qquad (10\text{-}29)$$

Inelastic Collision — One Dimension

An *inelastic collision* is one in which the kinetic energy of a system of two colliding bodies is not conserved. The total linear momentum of the system must, however, still be conserved. If the colliding bodies stick together, the collision is **completely inelastic,** and the reduction in kinetic energy is maximum (but not necessarily to zero). For one-dimensional motion, with one body initially stationary and the other having initial velocity v, the velocity V of the stuck-together bodies is found by applying conservation of linear momentum to the system:

$$m_1 v = (m_1 + m_2)\, V. \qquad (10\text{-}34)$$

If both bodies are moving prior to the collision, conservation of linear momentum is written as

$$m_1 v_1 + m_2 v_2 = (m_1 + m_2)\, V. \qquad (10\text{-}36)$$

Motion of the Center of Mass

The center of mass of a closed, isolated system of colliding bodies is unaffected by the collision, whether the collision is elastic or inelastic. For a one-dimensional collision, the constant velocity v_{cm} of the center of mass is given by

$$v_{\mathrm{cm}} = \frac{P}{m_1 + m_2} = \frac{m_1 v_{1i} + m_2 v_{2i}}{m_1 + m_2} \qquad (10\text{-}30)$$

before and after the collision.

Collisions in Two Dimensions

Collisions in two dimensions are governed by the conservation of vector linear momentum, a condition that leads to two-component equations. These determine the final motion if the collision is completely inelastic. Otherwise, the laws of conservation of linear momentum and of energy generally lead to equations that cannot be solved completely unless other data, such as the final direction of one of the velocities, are available.

Nuclear Reactions and Decay

In a *nuclear reaction* or *decay* of nuclear particles, linear momentum and *total* energy are conserved. If the mass of a system of such particles changes by Δm, the mass energy of the system changes by $\Delta m\, c^2$. The Q of the reaction or decay is defined as

$$Q = -\Delta m\, c^2.$$

This process is said to be **exothermic,** and Q is a positive quantity, if mass energy is transferred to kinetic energy of particles in the system. It is said to be **endothermic,** and Q is a negative quantity, if kinetic energy of particles in the system is transferred to mass energy.

QUESTIONS

1. Figure 10-20 shows three graphs of force magnitude versus time for a body involved in a collision. Rank the graphs according to the magnitude of the impulse on the body, greatest first.

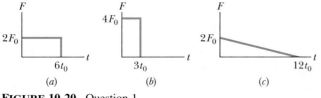

FIGURE 10-20 Question 1.

2. Figure 10-21 shows an overhead view of a golf ball bouncing off a tree trunk with no change in its speed. During the collision the force $\mathbf{F}$ on the ball by the trunk causes a change $\Delta \mathbf{p}$ in the linear momentum of the ball. If angle θ is increased (and assuming that the duration of the collision is unchanged), do the following increase, decrease, or remain the same: (a) Δp_x, (b) Δp_y, (c) magnitude of $\Delta \mathbf{p}$, (d) F_x, (e) F_y, and (f) magnitude of $\mathbf{F}$?

3. The following table gives, for three situations, the masses (kilograms) and velocities (meters per second) for the two parti-

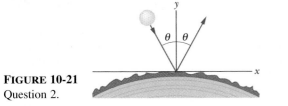

FIGURE 10-21
Question 2.

cles in Fig. 10-9. For which situations is the center of mass of the two-particle system stationary?

SITUATION	m_1	v_1	m_2	v_2
a	2	3	4	−3
b	6	2	3	−4
c	4	3	4	−3

4. (a) In Sample Problem 10-3, does the rebound height of sphere 1 increase, decrease, or remain the same if we increase the mass of sphere 2? If, instead, we set $m_2 = m_1$, (b) what is the rebound height of sphere 1, and (c) what is the height h_2 reached by sphere 2?

5. Two bodies undergo an elastic one-dimensional collision along an x axis. Figure 10-22 is a graph of position versus time for those bodies and for their center of mass. (a) Were both bodies initially moving, or was one initially stationary? Which line segment corresponds to the motion of the center of mass (b) before the collision and (c) after the collision? (d) Is the mass of the body that was moving faster before the collision greater than, less than, or equal to that of the other body?

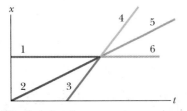

FIGURE 10-22 Question 5.

6. One body catches up with a second body and they then undergo a one-dimensional collision. Figure 10-23 is a graph of position versus time for those bodies and for their center of mass. Which line segment corresponds to (a) the faster body before the collision, (b) the center of mass before the collision, (c) the center of mass after the collision, and (d) the initially faster body after the collision? (e) Is the mass of the initially faster body greater than, less than, or equal to that of the other body?

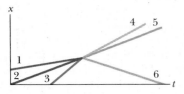

FIGURE 10-23 Question 6.

7. Figure 10-24 shows, for four situations, three identical blocks that undergo elastic collisions on a frictionless surface. In situations 1 and 2, two of the blocks are glued together. In all four situations, the initially moving blocks have the same velocity **v**. Rank the situations according to (a) the total linear momentum of the blocks after the collisions and (b) the speed of the rightmost block after the collisions, greatest first.

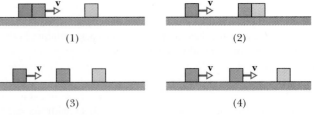

FIGURE 10-24 Question 7.

8. Figure 10-25 shows seven identical blocks on a frictionless floor. Initially, blocks a and b are moving rightward and block g is moving leftward, each with speed $v = 3$ m/s. The other blocks are stationary. A series of elastic collisions occurs. After the last collision, what are the speeds and directions of motion of each of the seven blocks?

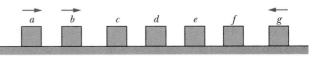

FIGURE 10-25 Question 8.

9. In Fig. 10-26, blocks A and B have linear momenta with directions as shown and with magnitudes of 9 kg·m/s and 4 kg·m/s, respectively. (a) What is the direction of motion of the center of mass of the two-block system over the frictionless floor? (b) If the blocks stick together during their collision, in what direction will they move? (c) If, instead, block A ends up moving to the left, will the magnitude of its momentum then be smaller than, more than, or the same as that of block B?

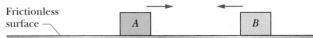

FIGURE 10-26 Question 9.

10. Figure 10-27 shows four graphs of position versus time for two bodies and their center of mass. The two bodies undergo a completely inelastic, one-dimensional collision while moving along an x axis. In graph 1, are (a) the two bodies and (b) the center of mass moving in the positive or negative direction of the x axis? (c) Which graphs correspond to a physically impossible situation?

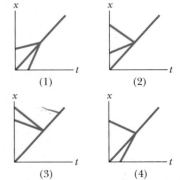

FIGURE 10-27
Question 10.

11. Body Q, with linear momentum $\mathbf{p}_Q = (2\mathbf{i} - 3\mathbf{j})$ kg·m/s, collides with and also sticks to body R, with linear momentum $\mathbf{p}_R = (8\mathbf{i} + 3\mathbf{j})$ kg·m/s. In what direction do the bodies move after the collision?

12. Drop, in succession, a baseball and a basketball from about shoulder height above a hard floor, and note how high each rebounds. Then align the baseball above the basketball (with a small separation as in Fig. 10-28a) and drop them simultaneously. (Be prepared to duck, and guard your face.) (a) Is the rebound height of the basketball now higher or lower than before (Fig. 10-28b)? (b) Is the rebound height of the baseball less than or greater than the sum of the individual baseball and basketball rebound heights? (See also Problem 37.)

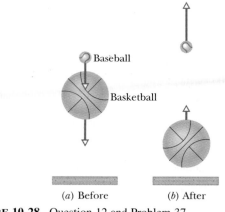

FIGURE 10-28 Question 12 and Problem 37.

EXERCISES & PROBLEMS

SECTION 10-2 Impulse and Linear Momentum

1E. The linear momentum of a 1500 kg car increased by 9.0×10^3 kg·m/s in 12 s. (a) What is the magnitude of the constant force that accelerated the car? (b) By how much did the speed of the car increase?

2E. A cue stick strikes a stationary pool ball, exerting an average force of 50 N over a time of 10 ms. If the ball has mass 0.20 kg, what speed does it have after impact?

3E. The National Transportation Safety Board is testing the crash-worthiness of a new car. The 2300 kg vehicle, moving at 15 m/s, is allowed to collide with a bridge abutment, being brought to rest in a time of 0.56 s. What force, assumed constant, acted on the car during impact?

4E. A ball of mass m and speed v strikes a wall perpendicularly and rebounds in the opposite direction with the same speed. (a) If the time of collision is Δt, what is the average force exerted by the ball on the wall? (b) Evaluate this average force numerically for a rubber ball with mass 140 g moving at 7.8 m/s; the duration of the collision is 3.8 ms.

5E. A 150 g (weight ≈ 5.3 oz) baseball pitched at a speed of 40 m/s (≈ 130 ft/s) is hit straight back to the pitcher at a speed of 60 m/s (≈ 200 ft/s). What average force was exerted by the bat if it was in contact with the ball for 5.0 ms?

6E. Until he was in his seventies, Henri LaMothe excited audiences by belly-flopping from a height of 40 ft into 12 in. of water (Fig. 10-29). Assuming that he stops just as he reaches the bottom of the water, what is the average force on him from the water? Assume his weight is 160 lb.

7E. In February 1955, a paratrooper fell 1200 ft from an airplane without being able to open his chute but happened to land in snow, suffering only minor injuries. Assume that his speed at impact was 56 m/s (terminal speed), that his mass (including gear) was 85 kg, and that the force on him from the snow was at the survivable limit of 1.2×10^5 N. What is the minimum depth of snow that would have stopped him safely?

8E. A force that averages 1200 N is applied to a 0.40 kg steel ball moving at 14 m/s in a collision lasting 27 ms. If the force is in a direction opposite the initial velocity of the ball, find the final speed and direction of the ball.

9E. A 1.2 kg medicine ball drops vertically onto a floor, hitting with a speed of 25 m/s. It rebounds with an initial speed of 10 m/s. (a) What impulse acts on the ball during the contact? (b) If the ball is in contact with the floor for 0.020 s, what is the average force exerted on the floor?

10E. A golfer hits a golf ball, giving it an initial velocity of magnitude 50 m/s directed 30° above the horizontal. Assuming that the mass of the ball is 46 g and the club and ball are in contact

FIGURE 10-29 Exercise 6.

for 1.7 ms, find (a) the impulse on the ball, (b) the impulse on the club, (c) the average force exerted on the ball by the club, and (d) the work done on the ball.

11P. A 1400 kg car moving at 5.3 m/s is initially traveling north in the positive y direction. After completing a 90° right-hand turn to the positive x direction in 4.6 s, the inattentive operator drives into a tree, which stops the car in 350 ms. In unit-vector notation, what is the impulse on the car (a) during the turn and (b) during the collision? What is the magnitude of the average force that acts on the car (c) during the turn and (d) during the collision? (e) What is the angle between the average force in (c) and the positive x direction?

12P. The force on a 10 kg object increases uniformly from zero to 50 N in 4.0 s. What is the object's speed at the end of the 4.0 s interval if it started from rest?

13P. A pellet gun fires ten 2.0 g pellets per second with a speed of 500 m/s. The pellets are stopped by a rigid wall. (a) What is the momentum of each pellet? (b) What is the kinetic energy of each pellet? (c) What is the average force exerted by the stream of pellets on the wall? (d) If each pellet is in contact with the wall for 0.6 ms, what is the average force exerted on the wall by each pellet during contact? Why is this average force so different from the average force calculated in (c)?

14P. A movie set machine gun fires 50 g bullets at a speed of 1000 m/s. An actor, holding the machine gun in his hands, can exert an average force of 180 N against the gun. Determine the maximum number of bullets he can fire per minute while still holding the gun steady.

15P. It is well known that bullets and other missiles fired at Superman simply bounce off his chest (Fig. 10-30). Suppose that a gangster sprays Superman's chest with 3 g bullets at the rate of 100 bullets/min, the speed of each bullet being 500 m/s. Suppose too that the bullets rebound straight back with no change in speed. What is the average force exerted by the stream of bullets on Superman's chest?

16P. During a violent thunderstorm, hail of diameter 1.0 cm falls at a speed of 25 m/s. There are estimated to be 120 hailstones per cubic meter of air. Ignore the bounce of the hail on impact. (a) What is the mass of each hailstone (density = 0.92 g/cm³)? (b) What average force is exerted by hail on a 10 m × 20 m flat roof during the storm?

17P. A stream of water impinges on a stationary "dished" turbine blade, as shown in Fig. 10-31. The speed of the water is v, both before and after it strikes the curved surface of the blade, and the mass of water striking the blade per unit time is constant at the value μ. Find the force exerted by the water on the blade.

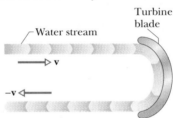

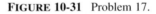

FIGURE 10-31 Problem 17.

18P. A stream of water from a hose is sprayed on a wall. If the speed of the water is 5.0 m/s and the hose sprays 300 cm³/s, what is the average force exerted on the wall by the stream of water? Assume that the water does not spatter back appreciably. Each cubic centimeter of water has a mass of 1.0 g.

19P. Figure 10-32 shows an approximate plot of force versus time during the collision of a 58 g tennis ball with a wall. The initial velocity of the ball is 34 m/s perpendicular to the wall; it rebounds with the same speed, also perpendicular to the wall. What is F_{max}, the maximum value of the contact force during the collision?

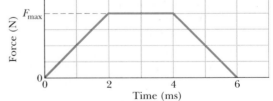

FIGURE 10-32 Problem 19.

20P. A ball having a mass of 150 g strikes a wall with a speed of 5.2 m/s and rebounds with only 50% of its initial kinetic energy.

FIGURE 10-30 Problem 15.

(a) What is the speed of the ball immediately after rebounding? (b) What was the magnitude of the impulse of the ball on the wall? (c) If the ball was in contact with the wall for 7.6 ms, what was the magnitude of the average force exerted by the wall on the ball during this time interval?

21P. In the overhead view of Fig. 10-33, a 300 g ball with a speed v of 6.0 m/s strikes a wall at an angle θ of 30° and then rebounds with the same speed and angle. It is in contact with the wall for 10 ms. (a) What is the impulse on the ball? (b) What is the average force exerted by the ball on the wall?

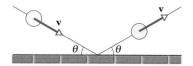

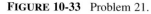

FIGURE 10-33 Problem 21.

22P. A 2500 kg unmanned space probe is moving in a straight line at a constant speed of 300 m/s. Control rockets on the space probe execute a burn in which a thrust of 3000 N acts for 65.0 s. (a) What is the change in linear momentum (magnitude only) of the probe if the thrust is backward, forward, or directly sideways? (b) What is the change in kinetic energy under the same three conditions? Assume that the mass of the ejected fuel is negligible compared to the mass of the space probe.

23P. A force exerts an impulse J on an object of mass m, changing its speed from v to u. The force and the object's motion are along the same straight line. Show that the work done by the force is $\frac{1}{2}J(u + v)$.

24P. A spacecraft is separated into two parts by detonating the explosive bolts that hold them together. The masses of the parts are 1200 kg and 1800 kg; the magnitude of the impulse on each part is 300 N·s. With what relative speed do the two parts separate because of the detonation?

25P. A stationary croquet ball with mass 0.50 kg is struck by a mallet, receiving the impulse shown in Fig. 10-34. What is the ball's velocity just after the force has become zero?

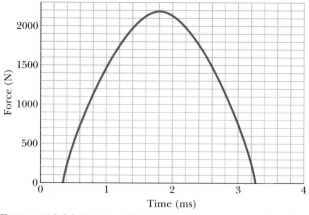

FIGURE 10-34 Problem 25.

26P. A soccer player kicks a soccer ball of mass 0.45 kg that is initially at rest. The player's foot is in contact with the ball for 3.0×10^{-3} s, and the force of the kick is given by

$$F(t) = [(6.0 \times 10^6)t - (2.0 \times 10^9)t^2] \text{ N},$$

for $0 \le t \le 3.0 \times 10^{-3}$ s, where t is in seconds. Find the magnitudes of the following: (a) the impulse imparted to the ball, (b) the average force exerted by the player's foot on the ball during the period of contact, (c) the maximum force exerted by the player's foot on the ball during the period of contact, and (d) the ball's velocity immediately after it loses contact with the player's foot.

27P. Spacecraft *Voyager 2* (of mass m and speed v relative to the Sun) approaches the planet Jupiter (of mass M and speed V relative to the Sun) as shown in Fig. 10-35. The spacecraft rounds the planet and departs in the opposite direction. What is its speed, relative to the Sun, after this slingshot encounter, which can be analyzed as a collision? Assume $v = 12$ km/s and $V = 13$ km/s (the orbital speed of Jupiter). The mass of Jupiter is very much greater than the mass of the spacecraft; $M \gg m$.

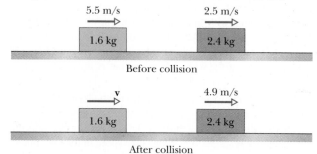

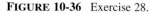

FIGURE 10-35 Problem 27.

SECTION 10-3 Elastic Collisions in One Dimension

28E. The blocks in Fig. 10-36 slide without friction. (a) What is the velocity **v** of the 1.6 kg block after the collision? (b) Is the collision elastic? (c) Suppose the initial velocity of the 2.4 kg block is the reverse of what is shown. Can the velocity **v** of the 1.6 kg block after the collision be in the direction shown?

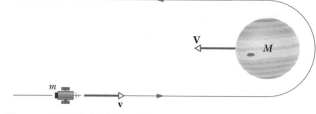

FIGURE 10-36 Exercise 28.

29E. A hovering fly is approached by an enraged elephant charging at 2.1 m/s. Assuming that the collision is elastic, at what speed does the fly rebound? Note that the projectile (the elephant) is much more massive than the stationary target (the fly).

30E. An electron collides elastically with a hydrogen atom initially at rest. (All motions are along the same straight line.) What percentage of the electron's initial kinetic energy is transferred to the hydrogen atom? The mass of the hydrogen atom is 1840 times the mass of the electron.

31E. A cart with mass 340 g moving on a frictionless linear air track at an initial speed of 1.2 m/s strikes a second cart of unknown mass at rest. The collision between the carts is elastic.

After the collision, the first cart continues in its original direction at 0.66 m/s. (a) What is the mass of the second cart? (b) What is its speed after impact? (c) What is the speed of the two-cart center of mass?

32E. An alpha particle (mass 4 u) experiences an elastic head-on collision with a gold nucleus (mass 197 u) that is originally at rest. What percentage of its original kinetic energy does the alpha particle lose?

33E. A body of mass 2.0 kg makes an elastic collision with another body at rest and continues to move in the original direction but with one-fourth of its original speed. (a) What is the mass of the struck body? (b) What is the speed of the two-body center of mass if the initial speed of the 2.0 kg body was 4.0 m/s?

34P. A steel ball of mass 0.500 kg is fastened to a cord 70.0 cm long and fixed at the far end, and is released when the cord is horizontal (Fig. 10-37). At the bottom of its path, the ball strikes a 2.50 kg steel block initially at rest on a frictionless surface. The collision is elastic. Find (a) the speed of the ball and (b) the speed of the block, both just after the collision.

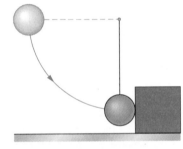

FIGURE 10-37 Problem 34.

35P. A platform scale is calibrated to indicate the mass in kilograms of an object placed on it. Particles fall from a height of 3.5 m and collide with the platform of the scale. The collisions are elastic; the particles rebound upward with the same speed they had before hitting the pan. If each particle has a mass of 110 g and collisions occur at the rate of 42 s^{-1}, what is the average scale reading?

36P. Two titanium spheres approach each other head-on with the same speed and collide elastically. After the collision, one of the spheres, whose mass is 300 g, remains at rest. (a) What is the mass of the other sphere? (b) What is the speed of the two-sphere center of mass if the initial speed of each sphere was 2.0 m/s?

37P. A ball of mass m is aligned above a ball of mass M (with a slight separation, as in Fig. 10-28a), and the two are dropped simultaneously from height h. (Assume the radius of each ball is negligible compared to h.) (a) If M rebounds elastically from the floor and then m rebounds elastically from M, what ratio m/M results in M stopping upon its collision with m? (The answer is approximately the mass ratio of a baseball to a basketball, as in Question 12.) (b) What height does m then reach?

38P. A block of mass m_1 is at rest on a long frictionless table, one end of which terminates at a wall. A block of mass m_2 is placed between the first block and the wall and set in motion to the left, toward m_1, with constant speed v_{2i}, as in Fig. 10-38. Assuming that all collisions are elastic, find the value of m_2 (in terms of m_1) for which both blocks move with the same velocity after m_2 has collided once with m_1 and once with the wall. Assume the wall to have infinite mass.

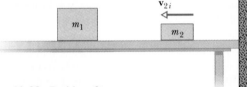

FIGURE 10-38 Problem 38.

39P. A target glider, whose mass m_2 is 350 g, is at rest on an air track, a distance $d = 53$ cm from the end of the track. A projectile glider, whose mass m_1 is 590 g, approaches the target glider with velocity $v_{1i} = -75$ cm/s and collides elastically with it (Fig. 10-39). The target glider rebounds elastically from a short spring at the end of the track and meets the projectile glider for a second time. How far from the end of the track does this second collision occur?

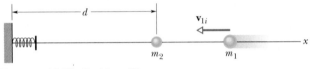

FIGURE 10-39 Problem 39.

SECTION 10-4 Inelastic Collisions in One Dimension

40E. Meteor Crater in Arizona (Fig. 10-1a) is thought to have been formed by the impact of a meteor with the Earth some 20,000 years ago. The mass of the meteor is estimated at 5 × 10^{10} kg, and its speed at 7200 m/s. What speed would such a meteor impart to the Earth in a head-on collision?

41E. A 6.0 kg box sled is coasting across frictionless ice at a speed of 9.0 m/s when a 12 kg package is dropped into it from above. What is the new speed of the sled?

42E. A 5.20 g bullet moving at 672 m/s strikes a 700 g wooden block at rest on a frictionless surface. The bullet emerges with its speed reduced to 428 m/s. (a) What is the resulting speed of the block? (b) What is the speed of the bullet–block center of mass?

43E. Two 2.0 kg masses, A and B, collide. The velocities before the collision are $\mathbf{v}_A = 15\mathbf{i} + 30\mathbf{j}$ and $\mathbf{v}_B = -10\mathbf{i} + 5.0\mathbf{j}$. After the collision, $\mathbf{v}'_A = -5.0\mathbf{i} + 20\mathbf{j}$. All speeds are given in meters per second. (a) What is the final velocity of B? (b) How much kinetic energy was gained or lost in the collision?

44E. A bullet of mass 10 g strikes a ballistic pendulum of mass 2.0 kg. The center of mass of the pendulum rises a vertical distance of 12 cm. Assuming that the bullet remains embedded in the pendulum, calculate the bullet's initial speed.

45E. A bullet of mass 4.5 g is fired horizontally into a 2.4 kg wooden block at rest on a horizontal surface. The coefficient of kinetic friction between block and surface is 0.20. The bullet comes to rest in the block, which moves 1.8 m. (a) What is the speed of the block immediately after the bullet comes to rest within it? (b) At what speed is the bullet fired?

46E. A 5.0 kg block with a speed of 3.0 m/s collides with a 10 kg block that has a speed of 2.0 m/s in the same direction. After the collision, the 10 kg block is observed to be traveling in the original direction with a speed of 2.5 m/s. (a) What is the speed of the 5.0 kg block immediately after the collision? (b) By how much does the total kinetic energy of the system of two blocks change because of the collision? (c) Suppose, instead, that the 10 kg block ends up with a speed of 4.0 m/s. What then is the change in the total kinetic energy? (d) Account for the result you obtained in (c).

47P. Two cars A and B slide on an icy road as they attempt to stop at a traffic light. The mass of A is 1100 kg, and the mass of B is 1400 kg. The coefficient of kinetic friction between the locked wheels of either car and the road is 0.13. Car A succeeds in coming to rest at the light, but car B cannot stop and rear-ends car A. After the collision, A comes to rest 8.2 m ahead of the impact point and B 6.1 m ahead; see Fig. 10-40. Both drivers had their brakes locked throughout the incident. (a) From the distance each car moved after the collision, find the speed of each car immediately after impact. (b) Use conservation of linear momentum to find the speed at which car B struck car A. On what grounds can the use of linear momentum conservation be criticized here?

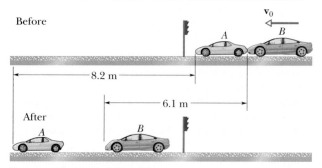

FIGURE 10-40 Problem 47.

48P. A 3000 kg weight falls vertically through 6.0 m and then collides with a 500 kg pile, driving it 3.0 cm into bedrock. Assuming that the weight–pile collision is completely inelastic, find the magnitude of the average force on the pile by the bedrock during the 3.0 cm descent.

49P. Two particles, one having twice the mass of the other, are held together with a compressed spring between them. The energy stored in the spring is 60 J. How much kinetic energy does each particle have after the two are released? Assume that all the stored energy is transferred to the particles and that neither particle is attached to the spring after the release.

50P. A 3.50 g bullet is fired horizontally at two blocks resting on a smooth tabletop, as shown in Fig. 10-41a. The bullet passes

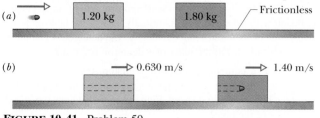

FIGURE 10-41 Problem 50.

through the first block, with mass 1.20 kg, and embeds itself in the second, with mass 1.80 kg. Speeds of 0.630 m/s and 1.40 m/s, respectively, are thereby imparted to the blocks, as shown in Fig. 10-41b. Neglecting the mass removed from the first block by the bullet, find (a) the speed of the bullet immediately after it emerges from the first block and (b) the bullet's original speed.

51P. An object with mass m and speed v explodes into two pieces, one three times as massive as the other; the explosion takes place in gravity-free space. The less massive piece comes to rest. How much kinetic energy was added to the system in the explosion?

52P. A box is put on a scale that is marked in units of mass and adjusted to read zero when the box is empty. A stream of marbles is then poured into the box from a height h above its bottom at a rate of R (marbles per second). Each marble has mass m. If the collisions between the marbles and the box are completely inelastic, find the scale reading at time t after the marbles begin to fill the box. Determine a numerical answer when $R = 100$ s^{-1}, $h = 7.60$ m, $m = 4.50$ g, and $t = 10.0$ s.

53P. A 35-ton railroad freight car collides with a stationary caboose car. They couple together, and 27% of the initial kinetic energy is dissipated as heat, sound, vibrations, and so on. Find the weight of the caboose.

54P. A ball of mass m is projected with speed v_i into the barrel of a spring gun of mass M initially at rest on a frictionless surface; see Fig. 10-42. The ball sticks in the barrel at the point of maximum compression of the spring. No mechanical energy is dissipated by friction. (a) What is the speed of the spring gun after the ball comes to rest in the barrel? (b) What fraction of the initial kinetic energy of the ball is stored in the spring?

FIGURE 10-42 Problem 54.

55P. A block of mass $m_1 = 2.0$ kg slides along a frictionless table with a speed of 10 m/s. Directly in front of it, and moving in the same direction, is a block of mass $m_2 = 5.0$ kg moving at 3.0 m/s. A massless spring with spring constant $k = 1120$ N/m is attached to the near side of m_2, as shown in Fig. 10-43. When the blocks collide, what is the maximum compression of the spring? (*Hint:* At the moment of maximum compression of the spring, the two blocks move as one. Find the velocity by noting that the collision is completely inelastic to this point.)

FIGURE 10-43 Problem 55.

56P. A 1.0 kg block at rest on a horizontal frictionless surface is connected to an unstretched spring ($k = 200$ N/m) whose other end is fixed (Fig. 10-44). A 2.0 kg block whose speed is 4.0 m/s collides with the 1.0 kg block. If the two blocks stick together after the one-dimensional collision, what maximum compression of the spring occurs when the blocks momentarily stop?

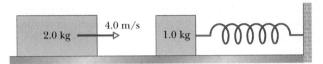

FIGURE 10-44 Problem 56.

57P. Two 22.7 kg ice sleds are placed a short distance apart, one directly behind the other, as shown in Fig. 10-45. A 3.63 kg cat, standing on one sled, jumps across to the other and immediately back to the first. Both jumps are made at a speed of 3.05 m/s relative to the ice. Find the final speeds of the two sleds.

FIGURE 10-45 Problem 57.

58P. The bumper of a 1200 kg car is designed so that it can just absorb all the energy when the car runs head-on into a solid wall at 5.00 km/h. The car is involved in a collision in which it runs at 70.0 km/h into the rear of a 900 kg car moving at 60.0 km/h in the same direction. The 900 kg car is accelerated to 70.0 km/h as a result of the collision. (a) What is the speed of the 1200 kg car immediately after impact? (b) What is the ratio of the kinetic energy absorbed in the collision to that which can be absorbed by the bumper of the 1200 kg car?

59P. A railroad freight car weighing 32 tons and traveling at 5.0 ft/s overtakes one weighing 24 tons and traveling at 3.0 ft/s in the same direction. If the cars couple together, find (a) the speed of the cars after collision and (b) the loss of kinetic energy during collision. (c) If instead, as is very unlikely, the collision is elastic, find the speeds of the cars after collision.

SECTION 10-5 Collisions in Two Dimensions

60E. An alpha particle collides with an oxygen nucleus, initially at rest. The alpha particle is scattered at an angle of 64.0° above its initial direction of motion, and the oxygen nucleus recoils at an angle of 51.0° below this initial direction. The final speed of the nucleus is 1.20×10^5 m/s. Find (a) the final speed and (b) the initial speed of the alpha particle. (The mass of an alpha particle is 4.0 u; the mass of an oxygen nucleus is 16 u.)

61E. A proton (atomic mass 1 u) with a speed of 500 m/s collides elastically with another proton at rest. The projectile proton is scattered 60° from its initial direction. (a) What is the direction of the velocity of the target proton after the collision? (b) What are the speeds of the two protons after the collision?

62E. A certain nucleus, at rest, spontaneously disintegrates into three particles. Two of them are detected; their masses and velocities are as shown in Fig. 10-46. (a) In unit-vector notation, what is the linear momentum of the third particle, which is known to have a mass of 11.7×10^{-27} kg? (b) How much kinetic energy appears in the disintegration process?

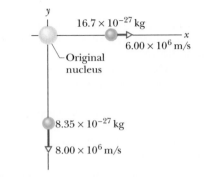

FIGURE 10-46 Exercise 62.

63E. In a game of pool, the cue ball strikes another ball initially at rest. After the collision, the cue ball moves at 3.50 m/s along a line making an angle of 22.0° with its original direction of motion, and the second ball has a speed of 2.00 m/s. Find (a) the angle between the direction of motion of the second ball and the original direction of motion of the cue ball and (b) the original speed of the cue ball. (c) Is kinetic energy conserved?

64E. Two vehicles A and B are traveling west and south, respectively, toward the same intersection, where they collide and lock together. Before the collision, A (total weight 2700 lb) is moving with a speed of 40 mi/h and B (total weight 3600 lb) has a speed of 60 mi/h. Find the magnitude and direction of the velocity of the (interlocked) vehicles immediately after the collision.

65E. In a game of billiards, the cue ball is given an initial speed V and strikes the pack of 15 stationary balls. All 16 balls then engage in numerous ball–ball and ball–cushion collisions. Some time later, it is observed that (by some accident) all 16 balls have the same speed v. Assuming that all collisions are elastic and ignoring the rotational aspect of the balls' motion, calculate v in terms of V.

66P. A 20.0 kg body is moving in the direction of the positive x axis with a speed of 200 m/s when, owing to an internal explosion, it breaks into three parts. One part, whose mass is 10.0 kg, moves away from the point of explosion with a speed of 100 m/s in the direction of the positive y axis. A second fragment, with a mass of 4.00 kg, moves in the direction of the negative x axis with a speed of 500 m/s. (a) What is the velocity of the third (6.00 kg) fragment? (b) How much energy was released in the explosion? Ignore effects due to gravity.

67P. Two balls A and B, having different but unknown masses, collide. A is initially at rest, and B has speed v. After collision, B has speed v/2 and moves perpendicularly to its original motion. (a) Find the direction in which ball A moves after collision. (b) Can you determine the speed of A from the information given? Explain.

68P. Show that if a neutron is scattered through 90° in an elastic collision with a deuteron that is initially at rest, the neutron loses two-thirds of its initial kinetic energy to the deuteron. (The mass of a neutron is 1.0 u; the mass of a deuteron is 2.0 u.)

69P. After a completely inelastic collision, two objects of the same mass and same initial speed are found to move away together at half their initial speed. Find the angle between the initial velocities of the objects.

70P. Two pendulums, both of length l, are initially situated as in Fig. 10-47. The left pendulum is released and strikes the other. Assume that the collision is completely inelastic, and neglect the mass of the strings and any frictional effects. How high does the center of mass of the pendulum system rise after the collision?

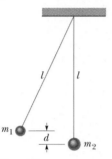

FIGURE 10-47 Problem 70.

71P. A billiard ball moving at a speed of 2.2 m/s strikes an identical stationary ball a glancing blow. After the collision, one ball is found to be moving at a speed of 1.1 m/s in a direction making a 60° angle with the original line of motion. (a) Find the velocity of the other ball. (b) Can the collision be inelastic, given these data?

72P. A ball with an initial speed of 10 m/s collides elastically with two identical stationary balls whose centers are on a line perpendicular to the initial velocity and that are initially in contact with each other (Fig. 10-48). The first ball is aimed directly at the contact point, and all motion is frictionless. Find the velocities of all three balls after the collision. (*Hint:* With friction absent, each impulse is directed along the line connecting the centers of the colliding balls, normal to the colliding surfaces.)

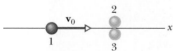

FIGURE 10-48 Problem 72.

73P. A barge with mass 1.50×10^5 kg is proceeding down river at 6.2 m/s in heavy fog when it collides broadside with a barge heading directly across the river (see Fig. 10-49). The second

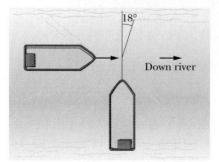

FIGURE 10-49 Problem 73.

barge has mass 2.78×10^5 kg and was moving at 4.3 m/s. Immediately after impact, the second barge finds its course deflected by 18° in the downriver direction and its speed increased to 5.1 m/s. The river current was practically zero at the time of the

accident. (a) What are the speed and direction of motion of the first barge immediately after the collision? (b) How much kinetic energy is lost in the collision?

SECTION 10-6 Reactions and Decay Processes

74E. The precise masses in the reaction

$$p + {}^{19}F \rightarrow \alpha + {}^{16}O$$

have been determined to be

$$m_p = 1.007825 \text{ u}, \quad m_\alpha = 4.002603 \text{ u},$$
$$m_F = 18.998405 \text{ u}, \quad m_O = 15.994915 \text{ u}.$$

Calculate the Q of the reaction from these data.

75E. A particle called Σ^- (sigma minus) is initially at rest and decays spontaneously into two other particles according to

$$\Sigma^- \rightarrow \pi^- + n.$$

The masses are

$$m_\Sigma = 2340.5m_e, \quad m_\pi = 273.2m_e, \quad m_n = 1838.65m_e,$$

where m_e (9.11×10^{-31} kg) is the electron mass. (a) How much energy is transferred to kinetic energy in this process? (b) How do the linear momenta of the decay products (π^- and n) compare? (c) Which product gets the larger share of the kinetic energy?

76P*. An alpha particle with kinetic energy 7.70 MeV strikes an ^{14}N nucleus at rest. An ^{17}O nucleus and a proton are produced; the proton is emitted at 90° to the direction of the incident alpha particle and has a kinetic energy of 4.44 MeV. The masses of the various particles are: alpha particle, 4.00260 u; ^{14}N, 14.00307 u; proton, 1.007825 u; and ^{17}O, 16.99914 u. (a) What is the kinetic energy of the oxygen nucleus? (b) What is the Q of the reaction?

77P*. Consider the alpha decay of radium (Ra) to radon (Rn), according to the reaction

$$^{226}Ra \rightarrow \alpha + {}^{222}Rn.$$

The masses of the various nuclei are: ^{226}Ra, 226.0254 u; alpha, 4.0026 u; ^{222}Rn, 222.0175 u. (a) Calculate the Q of the reaction. (b) What value of Q would be obtained if the accurate masses given above were rounded off to three significant figures? What is the kinetic energy of (c) the alpha particle and (d) the radon nucleus? (For this calculation the rounded-off values of the masses *can* be used. Why?)

ELECTRONIC COMPUTATION

78. A 6.00 kg model rocket is traveling horizontally and due south with a speed of 20.0 m/s when it explodes into two pieces. The velocity of one piece, with a mass of 2.00 kg, is

$$\mathbf{v}_1 = (-12.0 \text{ m/s})\mathbf{i} + (30.0 \text{ m/s})\mathbf{j} - (15.0 \text{ m/s})\mathbf{k},$$

with $\mathbf{i}$ pointing due east, $\mathbf{j}$ pointing due north, and $\mathbf{k}$ pointing vertically upward. (a) What is the linear momentum of the other piece, in unit-vector notation? (b) What is the kinetic energy of the other piece? (c) How much kinetic energy is produced by the explosion?

11
Rotation

In judo, a weaker and smaller fighter who understands physics can defeat a stronger and larger fighter who does not. This fact is demonstrated by the basic "hip throw," in which a fighter rotates the fighter's opponent around his hip and—if the throw is successful—onto the mat. Without the proper use of physics, the throw requires considerable strength and can easily fail. What is the advantage offered by physics?

11-1 TRANSLATION AND ROTATION

The graceful movement of a figure skater can be used to illustrate, in an aesthetically pleasing way, two kinds of pure, or unmixed, motion. Figure 11-1a shows a skater gliding across the ice in a straight line with constant speed. Her motion is one of pure **translation.** Figure 11-1b shows her spinning at a constant rate about a fixed vertical axis, in

FIGURE 11-1 Figure skater Kristi Yamaguchi in motion of (a) pure translation along a fixed direction and (b) pure rotation about a fixed axis.

(a)

(b)

a motion of pure **rotation.** This second kind of motion is our focus in this chapter.

Translation is motion along a straight line, the motion we have discussed almost exclusively so far. Rotation is the motion of wheels, gears, motors, the hands of clocks, the rotors of jet engines, and the blades of helicopters. It is the motion of hurricanes, planets, stars, and galaxies.

11-2 THE ROTATIONAL VARIABLES

In this chapter, we deal with the rotation of a *rigid* body about a *fixed* axis. The first of these restrictions means that we shall not examine the rotation of such objects as the Sun, because the Sun—a ball of gas—is not a rigid body. Our second restriction rules out objects like a bowling ball rolling down a bowling lane. Such a ball is in *rolling* motion, rotating about a *moving* axis.

Figure 11-2 shows a rigid body of arbitrary shape in pure *rotation around a fixed axis,* called the **axis of rotation** or the **rotation axis.** Every point of the body moves in a circle whose center lies on the axis of rotation, and every point moves through the same angle during a particular time interval. (In pure translation, every point of the body moves in a straight line, and every point moves through the same *linear distance* during a particular time interval. Comparisons between linear and angular motion will be a constant part of what follows.)

We deal now—one at a time—with the angular equivalents of the linear quantities position, displacement, velocity, and acceleration.

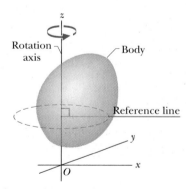

FIGURE 11-2 A rigid body of arbitrary shape in pure rotation about the *z* axis of a coordinate system. The position of the *reference line* with respect to the rigid body is arbitrary, but it is perpendicular to the rotation axis. It is fixed in the body and rotates with the body.

Angular Position

Figure 11-2 also shows a reference line, fixed in the body, perpendicular to the rotation axis, and rotating with the body. We can describe the motion of the rotating body by specifying the **angular position** of this line, that is, the angle of the line relative to a fixed direction. In Fig. 11-3, the angular position θ is measured relative to the positive direction of the x axis, and θ is given by

$$\theta = \frac{s}{r} \quad \text{(radian measure).} \quad (11\text{-}1)$$

Here s is the length of arc (or the arc distance) along a circle and between the x axis and the reference line, and r is the radius of that circle.

An angle defined in this way is measured in **radians** (rad) rather than in revolutions (rev) or degrees. The radian, being the ratio of two lengths, is a pure number and thus has no dimension. Because the circumference of a circle of radius r is $2\pi r$, there are 2π radians in a complete circle:

$$1 \text{ rev} = 360° = \frac{2\pi r}{r} = 2\pi \text{ rad,} \quad (11\text{-}2)$$

and thus

$$1 \text{ rad} = 57.3° = 0.159 \text{ rev.} \quad (11\text{-}3)$$

We do *not* reset θ to zero with each complete rotation of the reference line about the rotation axis. If the reference line completes two revolutions, then the angular position θ is $\theta = 4\pi$ rad.

For pure translational motion along the x direction, we know all there is to know about a moving body if we know $x(t)$, its position as a function of time. Similarly, for pure rotation, we know all there is to know about a rotating body if we know $\theta(t)$, the angular position of the body's reference line as a function of time.

FIGURE 11-3 The rotating rigid body of Fig. 11-2 in cross section, viewed from above. The plane of the cross section is perpendicular to the rotation axis, which now extends out of the page, toward you. In this position of the body, the reference line makes an angle θ with the x axis.

Angular Displacement

If the body of Fig. 11-3 rotates about the rotation axis as in Fig. 11-4, changing the angular position of the reference line from θ_1 to θ_2, the body undergoes an **angular displacement** $\Delta\theta$ given by

$$\Delta\theta = \theta_2 - \theta_1. \quad (11\text{-}4)$$

This definition of angular displacement holds not only for the rigid body as a whole but also for *every particle within that body.*

If a body is in translational motion along an x axis, its displacement Δx is either positive or negative, depending on whether the body is moving in the direction of increasing x or decreasing x. Similarly, the angular displacement $\Delta\theta$ of a rotating body can be either positive or negative, depending on whether the body is rotating in the direction of increasing θ (counterclockwise, as in Figs. 11-3 and 11-4) or decreasing θ (clockwise).

$\mathbb{C}$HECKPOINT **1:** The disk in the figure can rotate about its central axis like a merry-go-round. Which of the following pairs of values for its initial and final angular positions, respectively, give a negative angular displacement: (a) -3 rad, $+5$ rad, (b) -3 rad, -7 rad, (c) 7 rad, -3 rad?

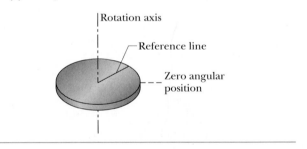

FIGURE 11-4 The reference line of the rigid body of Figs. 11-2 and 11-3 is at angular position θ_1 at time t_1 and at angular position θ_2 at a later time t_2. The quantity $\Delta\theta (= \theta_2 - \theta_1)$ is the angular displacement that occurs during the interval $\Delta t (= t_2 - t_1)$. The body itself is not shown.

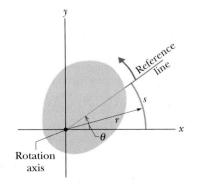

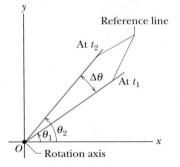

Angular Velocity

Suppose (see Fig. 11-4) that our rotating body is at angular position θ_1 at time t_1 and at angular position θ_2 at time t_2. We define the **average angular velocity** of the body in the time interval Δt from t_1 to t_2 to be

$$\bar{\omega} = \frac{\theta_2 - \theta_1}{t_2 - t_1} = \frac{\Delta \theta}{\Delta t}, \qquad (11\text{-}5)$$

in which $\Delta \theta$ is the angular displacement that occurs during Δt (ω is the lowercase Greek letter omega).

The **(instantaneous) angular velocity** ω, with which we shall be most concerned, is the limit of the ratio in Eq. 11-5 as Δt is made to approach zero. Thus

$$\omega = \lim_{\Delta t \to 0} \frac{\Delta \theta}{\Delta t} = \frac{d\theta}{dt}. \qquad (11\text{-}6)$$

If we know $\theta(t)$, we can find the angular velocity ω by differentiation.

Equations 11-5 and 11-6 hold not only for the rotating rigid body as a whole but also for *every particle of that body*. The unit of angular velocity is commonly the radian per second (rad/s) or the revolution per second (rev/s).

If a particle moves in translation along an x axis, its linear velocity v can be either positive or negative, depending on whether the particle is moving in the direction of increasing x or decreasing x. Similarly, the angular velocity ω of a rotating rigid body can be either positive or negative, depending on whether the body is rotating in the direction of increasing θ (counterclockwise) or of decreasing θ (clockwise). The magnitude of an angular velocity is called the **angular speed,** which is also represented with ω.

Angular Acceleration

If the angular velocity of a rotating body is not constant, then the body has an angular acceleration. Let ω_2 and ω_1 be the angular velocities at times t_2 and t_1, respectively. The **average angular acceleration** of the rotating body in the interval from t_1 to t_2 is defined as

$$\bar{\alpha} = \frac{\omega_2 - \omega_1}{t_2 - t_1} = \frac{\Delta \omega}{\Delta t}, \qquad (11\text{-}7)$$

in which $\Delta \omega$ is the change in the angular velocity that occurs during the time interval Δt. The **(instantaneous) angular acceleration** α, with which we shall be most concerned, is the limit of this quantity as Δt is made to approach zero. Thus

$$\alpha = \lim_{\Delta t \to 0} \frac{\Delta \omega}{\Delta t} = \frac{d\omega}{dt}. \qquad (11\text{-}8)$$

Equations 11-7 and 11-8 hold not only for the rotating rigid body as a whole but also for *every particle of that body*. The unit of angular acceleration is commonly the radian per second-squared (rad/s²) or the revolution per second-squared (rev/s²).

SAMPLE PROBLEM 11-1

The angular position of a reference line on a spinning wheel is given by

$$\theta = t^3 - 27t + 4,$$

where t is in seconds and θ is in radians.

(a) Find $\omega(t)$ and $\alpha(t)$.

SOLUTION: To get $\omega(t)$, we differentiate $\theta(t)$ with respect to t:

$$\omega = \frac{d\theta(t)}{dt} = 3t^2 - 27. \qquad \text{(Answer)}$$

To get $\alpha(t)$, we differentiate $\omega(t)$ with respect to t:

$$\alpha = \frac{d\omega(t)}{dt} = \frac{d(3t^2 - 27)}{dt} = 6t. \qquad \text{(Answer)}$$

(b) Do we ever find $\omega = 0$?

SOLUTION: Setting $\omega(t) = 0$ yields

$$0 = 3t^2 - 27,$$

which we solve, finding

$$t = \pm 3 \text{ s}. \qquad \text{(Answer)}$$

That is, the angular velocity is momentarily zero 3 s before and 3 s after our clock reads zero.

(c) Describe the wheel's motion for $t \geq 0$.

SOLUTION: To answer, we examine the expressions for $\theta(t)$, $\omega(t)$, and $\alpha(t)$.

At $t = 0$, the reference line on the wheel is at $\theta = 4$ rad, and the wheel is rotating with an angular velocity of -27 rad/s (that is, *clockwise* at an angular speed of 27 rad/s) and an angular acceleration of zero.

For $0 < t < 3$ s, the wheel continues to rotate clockwise, but at decreasing angular speed, because it now has a positive (counterclockwise) angular acceleration. (Check $\omega(t)$ and $\alpha(t)$ for, say, $t = 2$ s.)

At $t = 3$ s, the wheel stops momentarily ($\omega = 0$) and has rotated as far clockwise as it will ever get (the reference line is now at $\theta = -50$ rad).

For $t > 3$ s, the wheel's angular acceleration continues to increase. Its angular velocity, which is now also counterclockwise, increases rapidly because the signs of ω and α are the same.

SAMPLE PROBLEM 11-2

A child's top is spun with angular acceleration

$$\alpha = 5t^3 - 4t,$$

where the coefficients are in units compatible with seconds and radians. At $t = 0$, the top has angular velocity 5 rad/s, and a reference line on it is at angular position $\theta = 2$ rad.

(a) Obtain an expression for the angular velocity $\omega(t)$ of the top.

SOLUTION: From Eq. 11-8 we have

$$d\omega = \alpha \, dt,$$

which we integrate to get

$$\omega = \int \alpha \, dt = \int (5t^3 - 4t) \, dt$$
$$= \tfrac{5}{4}t^4 - \tfrac{4}{2}t^2 + C.$$

To evaluate the constant of integration C, we note that $\omega = 5$ rad/s at $t = 0$. Substituting these values in our expression for ω yields

$$5 \text{ rad/s} = 0 - 0 + C,$$

so $C = 5$ rad/s. Then

$$\omega = \tfrac{5}{4}t^4 - 2t^2 + 5. \qquad \text{(Answer)}$$

(b) Obtain an expression for the angular position $\theta(t)$ of the top.

SOLUTION: From Eq. 11-6 we have

$$d\theta = \omega \, dt,$$

which we integrate to get

$$\theta = \int \omega \, dt = \int (\tfrac{5}{4}t^4 - 2t^2 + 5) \, dt,$$
$$= \tfrac{1}{4}t^5 - \tfrac{2}{3}t^3 + 5t + C'$$
$$= \tfrac{1}{4}t^5 - \tfrac{2}{3}t^3 + 5t + 2, \qquad \text{(Answer)}$$

where C' has been evaluated by noting that $\theta = 2$ rad at $t = 0$.

11-3 ARE ANGULAR QUANTITIES VECTORS?

We can describe the position, velocity, and acceleration of a single particle by means of vectors. If the particle is confined to a straight line, however, we do not really need the power of vectors. Such a particle has only two directions available to it, and we can designate these with plus and minus signs.

In the same way, a rigid body rotating about a fixed axis can rotate only clockwise or counterclockwise about this axis, and again we can select between them by means of plus and minus signs. The question arises: "Can we treat the angular displacement, velocity, and acceleration of a rotating body as vectors?" The answer is a qualified "yes" (see caution below, in connection with angular displacements).

Consider the angular velocity. Figure 11-5a shows a phonograph record rotating about a fixed spindle. The record has a fixed rotation rate $\omega \, (= 33\tfrac{1}{3} \text{ rev/min})$ and a fixed direction of rotation (clockwise as viewed from above). By convention, we represent its angular velocity as a vector $\boldsymbol{\omega}$ pointing along the axis of rotation, as in Fig. 11-5b. We choose the length of this vector according to some convenient scale, for example, with 1 cm corresponding to 10 rev/min.

We establish a direction for the vector $\boldsymbol{\omega}$ by using a **right-hand rule,** as Fig. 11-5c shows. Curl your right hand about the rotating record, your fingers pointing *in the direction of rotation.* Your extended thumb will then point in the direction of the angular velocity vector. If the record were to rotate in the opposite sense, the right-hand rule would tell you that the angular velocity vector then points in the opposite direction.

It is not easy to get used to representing angular quantities as vectors. We instinctively expect that something should be moving *along* the direction of a vector. That is not the case here. Instead, something (the rigid body) is rotating *around* the direction of the vector. In the world of pure rotation, a vector defines an axis of rotation, not a direction in which something moves. Nonetheless, the

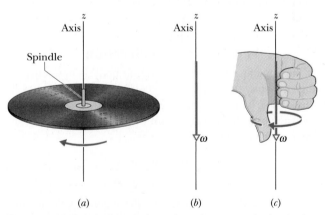

FIGURE 11-5 (a) A record rotating about a vertical axis that coincides with the axis of the spindle. (b) The angular velocity of the rotating record can be represented by the vector $\boldsymbol{\omega}$, lying along the axis and pointing down, as shown. (c) We establish the direction of the angular velocity vector as downward by using a right-hand rule. When the fingers of the right hand curl around the record and point the way it is moving, the extended thumb points in the direction of $\boldsymbol{\omega}$.

vector also defines the motion. Furthermore, it obeys all the rules for vector manipulation discussed in Chapter 3. The angular acceleration $\boldsymbol{\alpha}$ is another vector, and it too obeys those rules.

In this chapter we consider only rotations that are about a fixed axis. For such situations, we need not consider vectors—we can represent angular velocity with ω and angular acceleration with α, and we can indicate a direction with an implied plus sign for counterclockwise or an explicit minus sign for clockwise. In more complicated situations, however, we would use vectors $\boldsymbol{\omega}$ and $\boldsymbol{\alpha}$.

Now for the caution. Angular *displacements* (unless they are very small) *cannot* be treated as vectors. Why not? We can certainly give them both magnitude and direction, just as we did for the angular velocity vector in Fig. 11-5. However, that is (as the mathematicians say) a necessary condition but not sufficient. To be represented as a vector, a quantity must *also* obey the rules of vector addition, one of which says that if you add two vectors, the order in which you add them does not matter. Angular displacements fail this test.

To see this, place a book flat on a table, as in Fig. 11-6*a*. Now give the book two successive 90° angular displacements, *first* about the (horizontal) x axis and *then* about the (vertical) y axis, using the right-hand rule as a guide to positive rotation in each case.

Now, with a book in the same initial position (Fig. 11-6*b*), carry out these two angular displacements in the reverse order (that is, *first* about the y axis and *then* about the x axis). As the diagrams show, the book ends up in a very different orientation.

Thus the same two operations produce different results, depending on the order in which you carry them out. Addition of angular displacements is thus not commutative, so angular displacements are not vector quantities. With some practice, you should be able to show that the final positions of the book are much closer together if you use displacements much smaller than 90°. In the limiting case of differential angular displacements (such as $d\theta$ in Eq. 11-6), angular displacements *can* be treated as vectors.

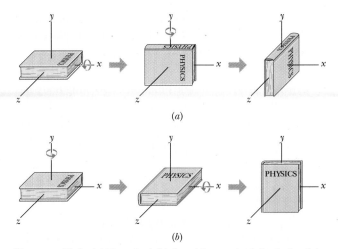

FIGURE 11-6 (*a*) From its initial position on the left, the book is given two successive 90° rotations, first about the (horizontal) x axis and then about the (vertical) y axis. (*b*) The book is given the same rotations, but in the reverse order. If angular displacement were truly a vector quantity, the order of these displacements would not matter. It clearly does matter, so (large) angular displacements are not vector quantities, even though we can assign magnitude and direction to them.

11-4 ROTATION WITH CONSTANT ANGULAR ACCELERATION

In pure translation, motion with a *constant linear acceleration* (for example, that of a falling body) is an important special case. In Table 2-1, we displayed a series of equations that hold for such motion.

In pure rotation, the case of *constant angular acceleration* is also important, and a parallel set of equations holds for this case also. We shall not derive them here, but simply write them from the corresponding linear equations, substituting equivalent angular quantities for the linear ones. This is done in Table 11-1, which displays both sets of equations (Eqs. 2-11 and 2-15 to 2-18; 11-9 to 11-13). For simplicity, we let $x_0 = 0$ and $\theta_0 = 0$ in these equa-

TABLE 11-1 EQUATIONS OF MOTION FOR CONSTANT LINEAR AND FOR CONSTANT ANGULAR ACCELERATION

EQUATION NUMBER	LINEAR FORMULA	MISSING VARIABLE		ANGULAR FORMULA	EQUATION NUMBER
(2-11)	$v = v_0 + at$	x	θ	$\omega = \omega_0 + \alpha t$	(11-9)
(2-15)	$x = v_0 t + \frac{1}{2}at^2$	v	ω	$\theta = \omega_0 t + \frac{1}{2}\alpha t^2$	(11-10)
(2-16)	$v^2 = v_0^2 + 2ax$	t	t	$\omega^2 = \omega_0^2 + 2\alpha\theta$	(11-11)
(2-17)	$x = \frac{1}{2}(v_0 + v)t$	a	α	$\theta = \frac{1}{2}(\omega_0 + \omega)t$	(11-12)
(2-18)	$x = vt - \frac{1}{2}at^2$	v_0	ω_0	$\theta = \omega t - \frac{1}{2}\alpha t^2$	(11-13)

tions. With those *initial conditions,* a linear displacement $\Delta x\ (= x - x_0)$ is equal to x, and an angular displacement $\Delta\theta\ (= \theta - \theta_0)$ is equal to θ.

CHECKPOINT 2: In four situations, a rotating body has angular position $\theta(t)$ given by (a) $\theta = 3t - 4$, (b) $\theta = -5t^3 + 4t^2 + 6$, (c) $\theta = 2/t^2 - 4/t$, and (d) $\theta = 5t^2 - 3$. To which situations do the angular equations of Table 11-1 apply?

SAMPLE PROBLEM 11-3

A grindstone (Fig. 11-7) has a constant angular acceleration $\alpha = 0.35$ rad/s^2. It starts from rest (that is, $\omega_0 = 0$) with an arbitrary reference line horizontal, at angular position $\theta_0 = 0$.

(a) What is the angular displacement θ of the reference line (hence of the wheel) at $t = 18$ s?

SOLUTION: From Eq. 11-10 of Table 11-1 ($\theta = \omega_0 t + \frac{1}{2}\alpha t^2$), we obtain:

$$\theta = (0)(18 \text{ s}) + (\tfrac{1}{2})(0.35 \text{ rad/s}^2)(18 \text{ s})^2$$
$$= 56.7 \text{ rad} \approx 57 \text{ rad} \approx 3200° \approx 9.0 \text{ rev}. \quad \text{(Answer)}$$

(b) What is the wheel's angular velocity at $t = 18$ s?

SOLUTION: From Eq. 11-9 of Table 11-1 ($\omega = \omega_0 + \alpha t$) we now get:

$$\omega = 0 + (0.35 \text{ rad/s}^2)(18 \text{ s})$$
$$= 6.3 \text{ rad/s} = 360°/\text{s} = 1.0 \text{ rev/s}. \quad \text{(Answer)}$$

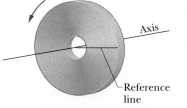

FIGURE 11-7 Sample Problems 11-3 and 11-4. A grindstone. At $t = 0$ the reference line (which we imagine to be marked on the stone) is horizontal.

SAMPLE PROBLEM 11-4

For the grindstone of Sample Problem 11-3, let us assume the same angular acceleration ($\alpha = 0.35$ rad/s^2), but let us now assume that the wheel does not start from rest but has an initial angular velocity ω_0 of -4.6 rad/s; that is, the angular acceleration acts initially to slow the wheel (because the signs of α and ω_0 are opposite).

(a) At what time t will the grindstone momentarily stop?

SOLUTION: Solving Eq. 11-9 ($\omega = \omega_0 + \alpha t$) for t yields

$$t = \frac{\omega - \omega_0}{\alpha} = \frac{0 - (-4.6 \text{ rad/s})}{0.35 \text{ rad/s}^2} = 13 \text{ s}. \quad \text{(Answer)}$$

(b) At what time will the grindstone have rotated such that its angular displacement is five revolutions in the positive direction of rotation? (The angular displacement of the reference line will then be $\theta = 5$ rev.)

SOLUTION: The wheel is initially rotating in the negative (clockwise) direction with $\omega_0 = -4.6$ rad/s, but its angular acceleration α is positive (counterclockwise). This initial opposition of the signs of angular velocity and angular acceleration means that the wheel slows in its rotation in the negative direction, stops, and then reverses to rotate in the positive direction. After the reference line comes back through its initial orientation of $\theta = 0$, the wheel must turn an additional five revolutions to reach the angular displacement we want. This is all "taken care of" if we use Eq. 11-10:

$$\theta = \omega_0 t + \tfrac{1}{2}\alpha t^2.$$

Substituting known values and setting $\theta = 5$ rev $= 10\pi$ rad give us

$$10\pi \text{ rad} = (-4.6 \text{ rad/s})t + (\tfrac{1}{2})(0.35 \text{ rad/s}^2)t^2.$$

Note that for t in seconds, the units in this equation are consistent. Dropping units (for convenience) and rearranging give

$$t^2 - 26.3t - 180 = 0. \quad (11\text{-}14)$$

Solving this quadratic equation for t and discarding the negative root, we obtain

$$t = 32 \text{ s}. \quad \text{(Answer)}$$

PROBLEM SOLVING TACTICS

TACTIC 1: *Unexpected Answers*

Do not hasten to throw away one root of a quadratic equation as meaningless. Often, as in Sample Problem 11-4, a discarded root has physical meaning.

The two solutions to Eq. 11-14 are $t = 32$ s and $t = -5.6$ s. We chose the first (positive) solution and ignored the second as perhaps meaningless. But is it? A negative time in this problem simply means a time before $t = 0$, that is, a time before we started to pay attention to what was going on.

Figure 11-8 is a plot of the angular position θ of the reference line on the grindstone of Sample Problem 11-4 as a function of time, for both positive and negative times. It is a plot of Eq. 11-10 ($\theta = \omega_0 t + \frac{1}{2}\alpha t^2$), using as constants $\omega_0 = -4.6$ rad/s and $\alpha = +0.35$ rad/s^2. Point a corresponds to $t = 0$, at which time we arbitrarily took the angular position of the reference line to be zero. The wheel is moving in the direction of decreasing θ at that time, and continues to do so until coming to rest at point b at $t = 13$ s. It then reverses, with the reference line returning to its $\theta = 0$ position at point c and

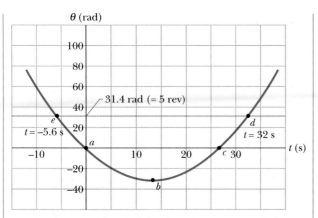

FIGURE 11-8 A plot of angular position versus time for the grindstone of Sample Problem 11-4. Negative times (that is, times before $t = 0$) have been included. The two roots of Eq. 11-14 are indicated by points d and e.

going on for five additional revolutions ($= 31.4$ rad) to point d. This latter point (with $t = 32$ s) represents the root that we accepted as the answer to our problem.

Note, however, that the reference line was at this same angular position at $t = -5.6$ s, before the "official start" of the problem. This root (point e) is just as valid as the root at point d. More important, by asking if this negative root can have any physical meaning, you learn a little more about the rotation of the wheel.

SAMPLE PROBLEM 11-5

During an analysis of a helicopter engine, you determine that the rotor's velocity changes from 320 rev/min to 225 rev/min in 1.50 min as the rotor is slowing to a stop.

(a) What is the average angular acceleration of the rotor blades during this interval?

SOLUTION: From Eq. 11-7,

$$\bar{\alpha} = \frac{\omega - \omega_0}{\Delta t} = \frac{225 \text{ rev/min} - 320 \text{ rev/min}}{1.50 \text{ min}}$$

$$= -63.3 \text{ rev/min}^2. \qquad \text{(Answer)}$$

(b) Assume the angular acceleration α is constant at this average value. How long will the rotor blades take to stop from their initial angular velocity of 320 rev/min?

SOLUTION: Solving Eq. 11-9 ($\omega = \omega_0 + \alpha t$) for t gives

$$t = \frac{\omega - \omega_0}{\alpha} = \frac{0 - 320 \text{ rev/min}}{-63.3 \text{ rev/min}^2}$$

$$= 5.1 \text{ min}. \qquad \text{(Answer)}$$

(c) How many revolutions will the rotor blades make in coming to rest from their initial angular velocity of 320 rev/min?

SOLUTION: Solving Eq. 11-11 ($\omega^2 = \omega_0^2 + 2\alpha\theta$) for θ gives us

$$\theta = \frac{\omega^2 - \omega_0^2}{2\alpha} = \frac{0 - (320 \text{ rev/min})^2}{(2)(-63.3 \text{ rev/min}^2)}$$

$$= 809 \text{ rev}. \qquad \text{(Answer)}$$

11-5 RELATING THE LINEAR AND ANGULAR VARIABLES

In Section 4-7, we discussed uniform circular motion, in which a particle travels at constant linear speed v along a circle and around an axis of rotation. When a rigid body, such as a merry-go-round, turns around an axis of rotation, each particle in the body moves in its own circle around that axis. Since the body is rigid, all the particles make one revolution in the same amount of time; that is, they all have the same angular speed ω.

However, the farther a particle is from the axis, the greater the circumference of its circle is, and so the faster its linear speed v must be. You can notice this on a merry-go-round. You turn with the same angular speed ω regardless of your distance from the center, but your linear speed v increases noticeably if you move to the outside edge of the merry-go-round.

We often need to relate the linear variables s, v, and a for a particular point in a rotating body to the angular variables θ, ω, and α for that body. The two sets of variables are related by r, the *perpendicular distance* of the point from the rotation axis. This perpendicular distance is the distance between the point and the rotation axis, measured along a perpendicular to the axis. It is also the radius r of the circle traveled by the point around the axis of rotation.

The Position

If a reference line on a rigid body is rotated through an angle θ, a point within the body is moved a distance s along a circular arc, where s is given by Eq. 11-1:

$$s = \theta r \qquad \text{(radian measure)}. \qquad (11\text{-}15)$$

This is the first of our linear–angular relations. The angle θ must be measured in radians because Eq. 11-15 is itself the definition of angular measure in radians.

The Speed

Differentiating Eq. 11-15 with respect to time—with r held constant—leads to

$$\frac{ds}{dt} = \frac{d\theta}{dt} r.$$

But ds/dt is the linear speed (the magnitude of the linear velocity) of the point in question and $d\theta/dt$ is the angular speed ω of the rotating body, so

$$v = \omega r \qquad \text{(radian measure)}. \qquad (11\text{-}16)$$

Again, the angular speed ω must be expressed in radian measure. Equation 11-16 tells us that since all points within the rigid body have the same angular speed ω, points with greater radius r have greater linear speed v. Figure 11-9a shows that the linear velocity is always tangent to the circular path of the point in question.

If the angular speed ω of the rigid body is constant, then Eq. 11-16 tells us that the linear speed v for any point within it is also constant. Thus each point within the body undergoes uniform circular motion. The period of revolution T for the motion of each point and for the rigid body itself is given by Eq. 4-23:

$$T = \frac{2\pi r}{v}. \qquad (11\text{-}17)$$

This equation tells us that the time for one revolution is the distance $2\pi r$ traveled in one revolution divided by the speed at which that distance is traveled. Substituting for v from Eq. 11-16 and canceling r, we find also that

$$T = \frac{2\pi}{\omega} \qquad \text{(radian measure)}. \qquad (11\text{-}18)$$

This equivalent equation says that the time for one revolution is the angle 2π rad traversed in one revolution divided by the angular speed at which that angle is traversed.

The Acceleration

Differentiating Eq. 11-16 with respect to time—again with r held constant—leads to

$$\frac{dv}{dt} = \frac{d\omega}{dt} r. \qquad (11\text{-}19)$$

Here we run up against a complication. In Eq. 11-19, dv/dt represents only the part of the linear acceleration that is responsible for changes in the *magnitude* v of the linear velocity **v**. Like **v**, that part of the linear acceleration is tangent to the path of the point in question. We call it the *tangential component* a_t of the linear acceleration of the point, and we write

$$a_t = \alpha r \qquad \text{(radian measure)}, \qquad (11\text{-}20)$$

where $\alpha = d\omega/dt$.

In addition, as Eq. 4-22 tells us, a particle (or point) moving in a circular path has a *radial component* of linear acceleration, $a_r = v^2/r$ (radially inward), that is responsi-

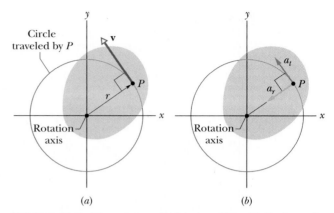

FIGURE 11-9   The rotating rigid body of Fig. 11-2, shown in cross section. Every point of the body (such as P) moves in a circle around the rotation axis. (a) The linear velocity **v** of every point is tangent to the circle in which the point moves. (b) The linear acceleration **a** of the point has (in general) two components: a tangential component a_t and a radial component a_r.

ble for changes in the *direction* of the linear velocity **v**. By substituting for v from Eq. 11-16, we can write this component as

$$a_r = \frac{v^2}{r} = \omega^2 r \qquad \text{(radian measure)}. \qquad (11\text{-}21)$$

Thus, as Fig. 11-9b shows, the linear acceleration of a point on a rotating rigid body has, in general, two components. The radially inward component a_r (given by Eq. 11-21) is present unless the angular velocity of the body is zero. The tangential component a_t (given by Eq. 11-20) is present unless the angular acceleration is zero.

CHECKPOINT **3:** A cockroach rides the rim of a rotating merry-go-round. If the angular speed of this system *(merry-go-round + cockroach)* is constant, does the cockroach have (a) radial acceleration and (b) tangential acceleration? If the angular speed is decreasing, does the cockroach have (c) radial and (d) tangential acceleration?

SAMPLE PROBLEM 11-6

Figure 11-10 shows a centrifuge used to accustom astronaut trainees to high accelerations. The radius r of the circle traveled by an astronaut is 15 m.

(a) At what constant angular velocity must the centrifuge rotate if the astronaut is to be subject to a linear acceleration that is $11g$?

SOLUTION: Because the angular velocity is constant, the

angular acceleration α ($= d\omega/dt$) is zero and so is the tangential component of the linear acceleration (see Eq. 11-20). This leaves only the radial component. From Eq. 11-21 ($a_r = \omega^2 r$), with $a_r = 11g$, we have

$$\omega = \sqrt{\frac{a_r}{r}} = \sqrt{\frac{(11)(9.8 \text{ m/s}^2)}{15 \text{ m}}}$$

$$= 2.68 \text{ rad/s} \approx 26 \text{ rev/min}. \qquad \text{(Answer)}$$

(b) What is the tangential acceleration of the astronaut if the centrifuge accelerates uniformly from rest to the angular velocity of (a) in 120 s?

SOLUTION: Since the angular acceleration is constant during the speed-up of the centrifuge, Eq. 11-9 applies and we get

$$\alpha = \frac{\omega - \omega_0}{t} = \frac{2.68 \text{ rad/s} - 0}{120 \text{ s}} = 0.0223 \text{ rad/s}^2.$$

With Eq. 11-20, we then find

$$a_t = \alpha r = (0.0223 \text{ rad/s}^2)(15 \text{ m})$$

$$= 0.33 \text{ m/s}^2. \qquad \text{(Answer)}$$

Although the final radial acceleration a_r ($= 11g$) is large (and alarming), the tangential acceleration a_t ($= 0.034g$) during the speed-up is not.

FIGURE 11-10 Sample Problem 11-6. A centrifuge in Cologne, Germany, is used to accustom astronauts to the large acceleration experienced during a liftoff.

PROBLEM SOLVING TACTICS

TACTIC 2: *Units for Angular Variables*

In Eq. 11-1 ($\theta = s/r$), we committed ourselves to the use of radian measure for all angular variables whenever we are using equations that contain both angular and linear variables. That is, we must express angular displacements in radians, angular velocities in rad/s and rad/min, and angular accelerations in rad/s^2 and rad/min^2. Equations 11-15, 11-16, 11-18, 11-20, and 11-21 are marked to emphasize this. The only exceptions to this rule are equations that involve *only* angular

variables, such as the angular equations listed in Table 11-1. Here you are free to use any unit you wish for the angular variables. That is, you may use radians, degrees, or revolutions, as long as you use them consistently.

In equations where radian measure must be used, you need not keep track of the unit "radian" (rad) algebraically, as you must do for other units. You can add or delete it at will, to suit the context. In Sample Problem 11-6a the unit was added to the answer; in Sample Problem 11-6b it was omitted from the answer.

11-6 KINETIC ENERGY OF ROTATION

The rapidly rotating blade of a table saw certainly has kinetic energy. How can we express it? We cannot use the familiar formula $K = \frac{1}{2}mv^2$ directly because it applies only to particles, and we would not know what to use for v.

Instead, we shall treat the table saw (and any other rotating rigid body) as a collection of particles —all with different speeds. We can then add up the kinetic energies of these particles to find the kinetic energy of the body as a whole. In this way we obtain, for the kinetic energy of a rotating body,

$$K = \frac{1}{2}m_1 v_1^2 + \frac{1}{2}m_2 v_2^2 + \frac{1}{2}m_3 v_3^2 + \cdots$$
$$= \sum \frac{1}{2}m_i v_i^2, \qquad (11\text{-}22)$$

in which m_i is the mass of the ith particle and v_i is its speed. The sum is taken over all the particles in the body.

The problem with Eq. 11-22 is that v_i is not the same for all particles. We solve this problem by substituting for v from Eq. 11-16 ($v = \omega r$), so that we have

$$K = \sum \frac{1}{2}m_i(\omega r_i)^2 = \frac{1}{2}\left(\sum m_i r_i^2\right)\omega^2, \quad (11\text{-}23)$$

in which ω *is* the same for all particles.

The quantity in parentheses on the right side of Eq. 11-23 tells us how the mass of the rotating body is distributed about its axis of rotation. We call that quantity the **rotational inertia** (or **moment of inertia**) I of the body with respect to the axis of rotation. It is a constant for a particular rigid body and for a particular rotation axis. (That axis must always be specified if the value of I is to be meaningful.)

We may now write

$$I = \sum m_i r_i^2 \qquad \text{(rotational inertia)} \quad (11\text{-}24)$$

and substitute into Eq. 11-23, obtaining

$$K = \frac{1}{2}I\omega^2 \qquad \text{(radian measure)} \quad (11\text{-}25)$$

as the expression we seek. Because we have used the relation $v = \omega r$ in deriving Eq. 11-25, ω must be expressed in radian measure. The SI unit for I is the kilogram–square meter ($kg \cdot m^2$).

Equation 11-25, which gives the kinetic energy of a rigid body in pure rotation, is the angular equivalent of the formula $K = \frac{1}{2}Mv_{cm}^2$, which gives the kinetic energy of a rigid body in pure translation. In each case, there is a factor of $\frac{1}{2}$. Where mass M (which can be called the *translational inertia*) appears in one formula, I (the *rotational inertia*) appears in the other. Finally, each equation contains as a factor the square of a speed—translational or rotational as appropriate. The kinetic energies of translation and of rotation are not different kinds of energy. They are both kinetic energy, expressed in ways that are appropriate to the motion at hand.

The rotational inertia of a rotating body depends not only on its mass but also on how that mass is distributed with respect to the rotation axis. Figure 11-11 suggests a convincing way to develop a physical feeling for rotational inertia. Figure 11-11a shows the exterior of either of two plastic rods that outwardly appear to be identical. The dimensions and the weights of the two rods are the same, and both rods balance at their midpoints. If you grasp each rod at its center and move it back and forth rapidly in translational motion, you still cannot tell them apart.

However, a truly striking difference appears if, with a twisting wrist motion, you twist the rods (like a baton) rapidly in back-and-forth angular motion. One rod twists quite easily; the other does not. As Figs. 11-11b and 11-11c show, the "easy" rod has two internal weights near its

center and the "hard" rod has one at each end. Although the masses of the rods are equal, their rotational inertias about a central axis are quite different because their mass distributions are different.

$\mathbb{C}$HECKPOINT 4: The figure shows three masses that rotate about a vertical axis. The perpendicular distance between the axis and the center of each mass is given. Rank the three masses according to their rotational inertia about that axis, greatest first.

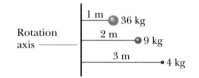

11-7 CALCULATING THE ROTATIONAL INERTIA

If a rigid body is made up of discrete particles, we can calculate its rotational inertia from Eq. 11-24. If the body is continuous, we can replace the sum in Eq. 11-24 with an integral, and the definition of rotational inertia becomes

$$I = \int r^2 \, dm \qquad \text{(rotational inertia,} \atop \text{continuous).} \qquad (11\text{-}26)$$

In the sample problems that follow this section, we calculate I for bodies of both kinds. In general, the rotational inertia of any rigid body with respect to a rotation axis depends on (1) the shape of the body, (2) the perpendicular distance from the axis to the body's center of mass, and (3) the orientation of the body with respect to the axis.

Table 11-2 gives the rotational inertias of several common bodies, about various axes. Note how the distribution of mass relative to the rotational axis affects the value of the rotational inertia I. For example, the rod in (f) has more mass far from the rotational axis than does the equally long rod in (e). So, if the rods have the same mass M, the rod in (f) has the greater rotational inertia.

The Parallel-Axis Theorem

If you know the rotational inertia of a body about any axis that passes through its center of mass, you can find its rotational inertia about any other axis parallel to that axis with the **parallel-axis theorem**:

$$I = I_{cm} + Mh^2 \qquad \text{(parallel-axis} \atop \text{theorem).} \qquad (11\text{-}27)$$

Here M is the mass of the body and h is the perpendicular

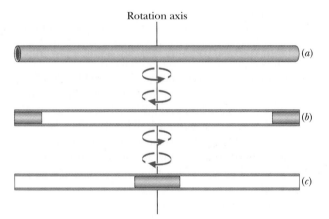

FIGURE 11-11 (a) Two plastic rods like this seem identical until you try to twist them back and forth rapidly about their midpoints. Rod (c) twists readily; rod (b) does not. Although both rods have the same mass, the rotational inertia of rod (b) about an axis through its midpoint is considerably greater than that of rod (c) because of the different internal distributions of mass.

TABLE 11-2 SOME ROTATIONAL INERTIAS

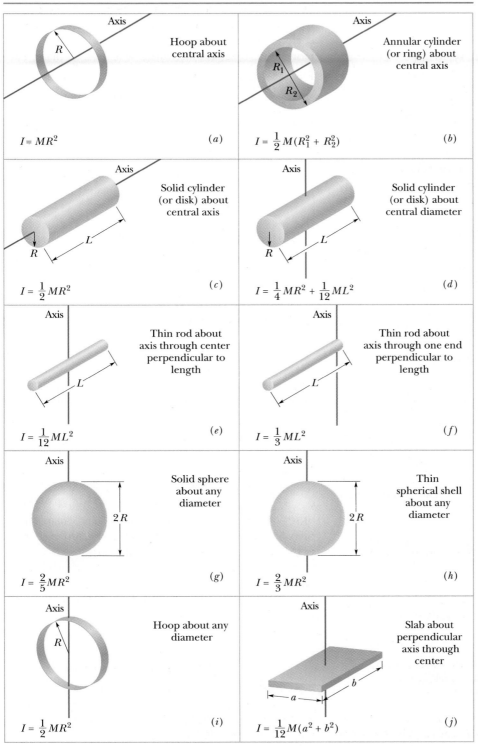

Axis

Hoop about central axis

$I = MR^2$ (a)

Axis

Annular cylinder (or ring) about central axis

$I = \frac{1}{2} M(R_1^2 + R_2^2)$ (b)

Axis

Solid cylinder (or disk) about central axis

$I = \frac{1}{2} MR^2$ (c)

Axis

Solid cylinder (or disk) about central diameter

$I = \frac{1}{4} MR^2 + \frac{1}{12} ML^2$ (d)

Axis

Thin rod about axis through center perpendicular to length

$I = \frac{1}{12} ML^2$ (e)

Axis

Thin rod about axis through one end perpendicular to length

$I = \frac{1}{3} ML^2$ (f)

Axis

Solid sphere about any diameter

$I = \frac{2}{5} MR^2$ (g)

Axis

Thin spherical shell about any diameter

$I = \frac{2}{3} MR^2$ (h)

Axis

Hoop about any diameter

$I = \frac{1}{2} MR^2$ (i)

Axis

Slab about perpendicular axis through center

$I = \frac{1}{12} M(a^2 + b^2)$ (j)

distance between the two (parallel) axes. In words, this theorem can be stated as follows:

> The rotational inertia of a body about any axis is equal to the rotational inertia ($= Mh^2$) it would have about that axis if all its mass were concentrated at its center of mass *plus* its rotational inertia ($= I_{cm}$) about a parallel axis through its center of mass.

Proof of the Parallel-Axis Theorem

Let O be the center of mass of the arbitrarily shaped body shown in cross section in Fig. 11-12. Place the origin of coordinates at O. Consider an axis through O perpendicular to the plane of the figure, and another axis through point P parallel to the first axis. Let the coordinates of P be a and b.

Let dm be a mass element with coordinates x and y. The rotational inertia of the body about the axis through P is then, from Eq. 11-26,

$$I = \int r^2 \, dm = \int [(x - a)^2 + (y - b)^2] \, dm,$$

which we can rearrange as

$$I = \int (x^2 + y^2) \, dm - 2a \int x \, dm$$
$$- 2b \int y \, dm + \int (a^2 + b^2) \, dm. \quad (11\text{-}28)$$

From the definition of the center of mass (Eq. 9-9), the middle two integrals of Eq. 11-28 give the coordinates of the center of mass (multiplied by a constant) and thus must

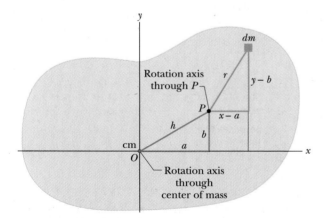

FIGURE 11-12   A rigid body in cross section, with its center of mass at O. The parallel-axis theorem (Eq. 11-27) relates the rotational inertia of the body about an axis through O to that about a parallel axis through a point such as P, a distance h from the body's center of mass. Both axes are perpendicular to the plane of the figure.

each be zero. Because $x^2 + y^2$ is equal to R^2, where R is the distance from O to dm, the first integral is simply I_{cm}, the rotational inertia of the body about an axis through its center of mass. Inspection of Fig. 11-12 shows that the last term in Eq. 11-28 is Mh^2, where M is the total mass. Thus Eq. 11-28 reduces to Eq. 11-27, which is the relation that we set out to prove.

CHECKPOINT **5:** The figure shows a booklike object (one side is longer than the other) and four choices of rotation axes, all perpendicular to the face of the object. Rank the choices according to the rotational inertia of the object about the axis, greatest first.

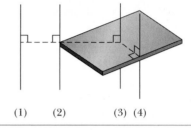

(1)   (2)   (3) (4)

SAMPLE PROBLEM 11-7

Figure 11-13 shows a rigid body consisting of two particles of mass m connected by a rod of length L and negligible mass.

(a) What is the rotational inertia of this body about an axis through its center, perpendicular to the rod (see Fig. 11-13a)?

SOLUTION: From Eq. 11-24 we have

$$I = \sum m_i r_i^2 = (m)(\tfrac{1}{2}L)^2 + (m)(\tfrac{1}{2}L)^2$$
$$= \tfrac{1}{2}mL^2. \qquad \text{(Answer)}$$

(b) What is the rotational inertia of the body about an axis through one end of the rod and parallel to the first axis, as in Fig. 11-13b?

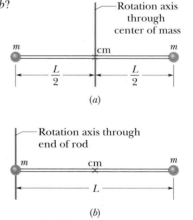

(a)

(b)

FIGURE 11-13   Sample Problem 11-7. A rigid body that consists of two particles of mass m which are joined by a rod of negligible mass.

SOLUTION: We can use the parallel-axis theorem of Eq. 11-27. We have just calculated I_{cm} in (a), and the distance h between the parallel axes is half the length of the rod. Thus, from Eq. 11-27,

$$I = I_{cm} + Mh^2 = \tfrac{1}{2}mL^2 + (2m)(\tfrac{1}{2}L)^2$$
$$= mL^2. \qquad \text{(Answer)}$$

We can check this result by direct calculation, using Eq. 11-24:

$$I = \sum m_i r_i^2 = (m)(0)^2 + (m)(L)^2$$
$$= mL^2. \qquad \text{(Answer)}$$

SAMPLE PROBLEM 11-8

Figure 11-14 shows a thin, uniform rod of mass M and length L.

(a) What is its rotational inertia about an axis perpendicular to the rod, through its center of mass?

SOLUTION: We place the rod on an x axis, with its center of mass at the origin, and we choose as a mass element a slice dx of the rod. The center of the slice is a distance x from the rotation axis. The mass per unit length of the rod is M/L, so the mass dm of the element dx is

$$dm = \left(\frac{M}{L}\right) dx.$$

From Eq. 11-26 we have

$$I = \int r^2\, dm = \int_{x=-L/2}^{x=+L/2} x^2 \left(\frac{M}{L}\right) dx$$
$$= \frac{M}{3L}\left[x^3 \right]_{-L/2}^{+L/2} = \frac{M}{3L}\left[\left(\frac{L}{2}\right)^3 - \left(-\frac{L}{2}\right)^3 \right]$$
$$= \tfrac{1}{12}ML^2. \qquad \text{(Answer)}$$

This agrees with the result given in Table 11-2(e).

(b) What is the rotational inertia of the rod about an axis perpendicular to the rod through one end?

SOLUTION: We combine the result in (a) with the parallel-axis theorem (Eq. 11-27), obtaining

$$I = I_{cm} + Mh^2$$
$$= \tfrac{1}{12}ML^2 + (M)(\tfrac{1}{2}L)^2 = \tfrac{1}{3}ML^2, \qquad \text{(Answer)}$$

which agrees with the result given in Table 11-2(f).

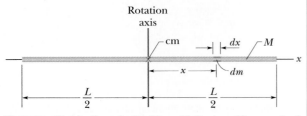

FIGURE 11-14 Sample Problem 11-8. A uniform rod of length L and mass M.

SAMPLE PROBLEM 11-9

A hydrogen chloride molecule consists of a hydrogen atom whose mass m_H is 1.01 u and a chlorine atom whose mass m_{Cl} is 35.0 u. The centers of the two atoms are a distance $d = 1.27 \times 10^{-10}$ m = 127 pm apart (Fig. 11-15). What is the rotational inertia of the molecule about an axis perpendicular to the line joining the two atoms and passing through the center of mass of the molecule?

SOLUTION: Let x be the distance from the center of mass of the molecule to the chlorine atom. Then from Fig. 11-15 and Eq. 9-3, we see that

$$0 = \frac{-m_{Cl}x + m_H(d-x)}{m_{Cl} + m_H},$$

or

$$m_{Cl}x = m_H(d-x).$$

This yields

$$x = \frac{m_H}{m_{Cl} + m_H}\, d. \qquad (11\text{-}29)$$

From Eq. 11-24, the rotational inertia about an axis through the center of mass is

$$I = \sum m_i r_i^2 = m_H(d-x)^2 + m_{Cl}x^2.$$

Substituting for x from Eq. 11-29 leads, after some algebra, to

$$I = d^2\, \frac{m_H m_{Cl}}{m_{Cl} + m_H} = (127 \text{ pm})^2 \frac{(1.01 \text{ u})(35.0 \text{ u})}{35.0 \text{ u} + 1.01 \text{ u}}$$
$$= 15{,}800 \text{ u} \cdot \text{pm}^2. \qquad \text{(Answer)}$$

These units are convenient enough when dealing with the rotational inertia of molecules. If we use the angstrom as a length unit (1 Å = 10^{-10} m), the answer above becomes $I = 1.58$ u·Å², an even more convenient unit.

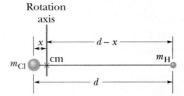

FIGURE 11-15 Sample Problem 11-9. A hydrogen chloride molecule, shown schematically. A rotation axis extends through its center of mass and is perpendicular to the line joining the two atomic centers.

SAMPLE PROBLEM 11-10

With modern technology, it is possible to construct a flywheel that stores enough energy to run an automobile. The energy is stored as rotational kinetic energy when the flywheel is initially made to spin by a machine. The stored energy is then gradually transferred to the automobile by a gear system as the automobile is being driven. Suppose that such a wheel is a solid cylinder whose mass M is 75 kg and whose radius R is

25 cm. If the wheel is spun at 85,000 rev/min, how much rotational kinetic energy can it store?

SOLUTION: The rotational inertia of the cylindrical wheel follows from Table 11-2(c):

$$I = \tfrac{1}{2}MR^2 = (\tfrac{1}{2})(75 \text{ kg})(0.25 \text{ m})^2 = 2.34 \text{ kg} \cdot \text{m}^2.$$

The angular velocity of the wheel is

$$\omega = (85{,}000 \text{ rev/min})(2\pi \text{ rad/rev})(1 \text{ min/60 s})$$
$$= 8900 \text{ rad/s}.$$

From Eq. 11-25, the kinetic energy of rotation is then

$$K = \tfrac{1}{2}I\omega^2 = (\tfrac{1}{2})(2.34 \text{ kg} \cdot \text{m}^2)(8900 \text{ rad/s})^2$$
$$= 9.3 \times 10^7 \text{ J} = 26 \text{ kW} \cdot \text{h}. \qquad \text{(Answer)}$$

This amount of energy, used with the expected efficiency, would take a small car about 200 mi.

11-8 TORQUE

A doorknob is located as far as possible from the door's hinge line for a good reason. If you want to open a heavy door you must certainly apply a force; that alone, however, is not enough. Where you apply that force and in what direction you push are also important. If you apply your force nearer to the hinge line than the knob, or at any angle other than 90° to the plane of the door, you must use a greater force to move the door than if you apply the force at the knob and perpendicular to the door's plane.

Figure 11-16a shows a cross section of a body that is free to rotate about an axis passing through O and perpendicular to the cross section. A force **F** is applied at point P, whose position relative to O is defined by a position vector **r**. Vectors **F** and **r** make an angle φ with each other. (For simplicity, we consider only forces that have no compo-

nent parallel to the rotation axis; thus, **F** is in the plane of the page.)

To determine how **F** results in a rotation of the body around the rotation axis, we resolve **F** into two components (Fig. 11-16b). One component, called the *radial component* F_r, points along **r**. This component does not cause rotation, because it acts along a line that extends through O. (If you pull on a door parallel to the plane of the door, you do not rotate the door.) The other component of **F**, called the *tangential component* F_t, is perpendicular to **r** and has magnitude $F_t = F \sin \phi$. This component *does* cause rotation. (If you pull on a door perpendicular to its plane, you rotate the door.)

The ability of **F** to rotate the body depends not only on the magnitude of its tangential component F_t, but also on just how far from O it is applied. To include both these factors, we define a quantity called **torque** τ as the product of the two factors and write it as

$$\tau = (r)(F \sin \phi). \qquad (11\text{-}30)$$

Two equivalent ways of computing the torque are

$$\tau = (r)(F \sin \phi) = rF_t \qquad (11\text{-}31)$$

and

$$\tau = (r \sin \phi)(F) = r_\perp F, \qquad (11\text{-}32)$$

where $r_\perp$ is the perpendicular distance between the rotation axis at O and an extended line running through the vector **F** (Fig. 11-16c). This extended line is called the **line of action** of **F**, and $r_\perp$ is called the **moment arm** of **F**. Figure 11-16b shows that we can describe r, the magnitude of **r**, as being the moment arm of the force component F_t.

Torque, which comes from the Latin word meaning "to twist," may be loosely identified as the turning or twisting action of the force **F**. When you apply a force to an object—such as a screwdriver or pipe wrench—with the

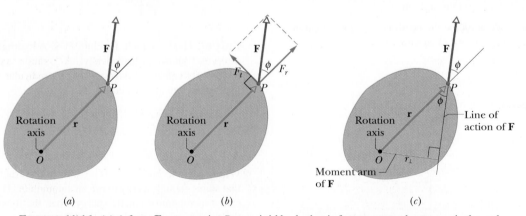

(a)   (b)   (c)

FIGURE 11-16 (a) A force **F** acts at point P on a rigid body that is free to rotate about an axis through O; the axis is perpendicular to the plane of the cross section shown here. (b) The torque exerted by this force is (r) (F sin φ). We can also write it as rF_t, where F_t is the tangential component of **F**. (c) The torque can also be written as $r_\perp F$, where $r_\perp$ is the moment arm of **F**.

purpose of turning that object, you are applying a torque. The SI unit of torque is the newton-meter (N·m).* A torque τ is positive if it tends to rotate the body counterclockwise, in the direction of increasing θ, as in Fig. 11-16. It is negative if it tends to rotate the body clockwise.

The definition of torque in Eq. 11-30 can be rewritten as a *vector cross product:*

$$\tau = \mathbf{r} \times \mathbf{F}. \qquad (11\text{-}33)$$

Thus torque τ is a vector that is directed perpendicular to the plane containing $\mathbf{r}$ and $\mathbf{F}$. The magnitude of τ is given by Eqs. 11-30, 11-31, and 11-32, and its direction is given by the right-hand rule. We shall make use of Eq. 11-33 in Chapter 12.

CHECKPOINT 6: The figure shows an overhead view of a meter stick that can pivot about the dot at the position marked 20 (for 20 cm). All five horizontal forces on the stick have the same magnitude. Rank those forces according to the magnitude of the torque that they produce, greatest first.

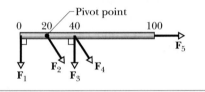

11-9 NEWTON'S SECOND LAW FOR ROTATION

Figure 11-17 shows a simple case of rotation about a fixed axis. The rotating rigid body consists of a single particle of mass m fastened to the end of a massless rod of length r. A force $\mathbf{F}$ acts as shown, causing the particle to move in a circle about the axis. The particle has a tangential component of acceleration a_t governed by Newton's second law:

$$F_t = ma_t.$$

The torque acting on the particle is, from Eq. 11-31,

$$\tau = F_t r = ma_t r.$$

From Eq. 11-20 ($a_t = \alpha r$) we can write this as

$$\tau = m(\alpha r)r = (mr^2)\alpha. \qquad (11\text{-}34)$$

The quantity in parentheses on the right side of Eq. 11-34 is the rotational inertia of the particle about the rotation axis

---

*The newton-meter is also the unit of work. Torque and work, however, are quite different quantities and must not be confused. Work is often expressed in joules (1 J = 1 N·m), but torque never is.

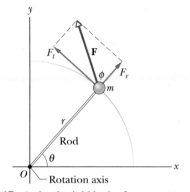

FIGURE 11-17 A simple rigid body, free to rotate about an axis through O, consists of a particle of mass m fastened to the end of a rod of length r and negligible mass. An applied force $\mathbf{F}$ causes the body to rotate.

(see Eq. 11-24). So, Eq. 11-34 reduces to

$$\tau = I\alpha \qquad \text{(radian measure)}. \qquad (11\text{-}35)$$

For the situation in which more than one force is applied to the particle, we can extend Eq. 11-35 as

$$\sum \tau = I\alpha \qquad \text{(radian measure)}, \qquad (11\text{-}36)$$

where $\sum \tau$ is the net torque (the sum of all external torques) acting on the particle. Equation 11-36 is the angular (or rotational) form of Newton's second law.

Although we derived Eq. 11-35 for the special case of a single particle rotating about a fixed axis, it holds for any rigid body rotating about a fixed axis, because any such body can be analyzed as an assembly of single particles.

CHECKPOINT 7: The figure shows an overhead view of a meter stick that can pivot about the point indicated, which is to the left of the stick's midpoint. Two horizontal forces, $\mathbf{F}_1$ and $\mathbf{F}_2$, are applied to the stick. Only $\mathbf{F}_1$ is shown. $\mathbf{F}_2$ is perpendicular to the stick and is applied at the right end. If the stick is not to turn, (a) what should be the direction of $\mathbf{F}_2$, and (b) should F_2 be greater than, less than, or equal to F_1?

SAMPLE PROBLEM 11-11

Figure 11-18*a* shows a uniform disk, whose mass M is 2.5 kg and whose radius R is 20 cm, mounted on a fixed horizontal axle. A block whose mass m is 1.2 kg hangs from a massless cord that is wrapped around the rim of the disk. Find the acceleration of the falling block (assuming that it does fall), the angular acceleration of the disk, and the tension in the cord. The cord does not slip, and there is no friction at the axle.

SOLUTION: Figure 11-18b is a free-body diagram for the block. We assume the block accelerates downward, so the magnitude mg of its weight must exceed the tension T in the cord. From Newton's second law, we have

$$T - mg = ma. \qquad (11\text{-}37)$$

Figure 11-18c is a free-body diagram for the disk. The torque acting on the disk is $-TR$, negative because it rotates the disk clockwise. From Table 11-2(c), we know that the rotational inertia I of the disk is $\frac{1}{2}MR^2$. (Two other forces also act on the disk, its weight $M\mathbf{g}$ and the normal force $\mathbf{N}$ exerted on the disk by its support. Since, however, both these forces act at the axis of the disk, they exert no torque on the disk.) Applying Newton's second law in angular form ($\tau = I\alpha$) to the disk, we obtain

$$-TR = \tfrac{1}{2}MR^2\alpha.$$

Because the cord does not slip, we assume that the linear acceleration a of the block and the (tangential) linear acceleration a_t of the rim of the disk are equal. Then, by Eq. 11-20, $\alpha = a/R$, and the equation above reduces to

$$T = -\tfrac{1}{2}Ma. \qquad (11\text{-}38)$$

Combining Eqs. 11-37 and 11-38 leads to

$$a = -g\,\frac{2m}{M + 2m} = -(9.8 \text{ m/s}^2)\,\frac{(2)(1.2 \text{ kg})}{2.5 \text{ kg} + (2)(1.2 \text{ kg})}$$

$$= -4.8 \text{ m/s}^2. \qquad \text{(Answer)}$$

We then use Eq. 11-38 to find T:

$$T = -\tfrac{1}{2}Ma = -\tfrac{1}{2}(2.5 \text{ kg})(-4.8 \text{ m/s}^2)$$

$$= 6.0 \text{ N}. \qquad \text{(Answer)}$$

As we should expect, the acceleration of the falling block is

less than g, and the tension in the cord ($= 6.0$ N) is less than the weight of the hanging block ($= mg = 11.8$ N). We see also that the acceleration of the block and the tension depend on the mass of the disk but not on its radius. As a check, we note that the formulas derived above predict $a = -g$ and $T = 0$ for the case of a massless disk ($M = 0$). This is what we would expect; the block simply falls as a free body, trailing the string behind it.

From Eq. 11-20, the angular acceleration of the disk is

$$\alpha = \frac{a}{R} = \frac{-4.8 \text{ m/s}^2}{0.20 \text{ m}} = -24 \text{ rad/s}^2. \qquad \text{(Answer)}$$

SAMPLE PROBLEM 11-12

To throw an 80 kg opponent with a basic judo hip throw, you intend to pull his uniform with a force $\mathbf{F}$ and a moment arm $d_1 = 0.30$ m from a pivot point (rotation axis) on your right hip, about which you wish to rotate him with an angular acceleration of -6.0 rad/s^2, that is, with a clockwise acceleration (Fig. 11-19). Assume that his rotational inertia I relative to the pivot point is 15 kg·m^2.

(a) What must the magnitude of $\mathbf{F}$ be if you initially bend your opponent forward to bring his center of mass to your hip (Fig. 11-19a)?

SOLUTION: If your opponent's center of mass is at the rotation axis, his weight vector produces no torque about that axis. The only torque on him then is due to your pull $\mathbf{F}$. From Eqs. 11-31 and 11-35, we have, for the clockwise torque,

$$\tau = -d_1F = I\alpha,$$

which gives

$$F = \frac{-I\alpha}{d_1} = \frac{-(15 \text{ kg·m}^2)(-6.0 \text{ rad/s}^2)}{0.30 \text{ m}}$$

$$= 300 \text{ N}. \qquad \text{(Answer)}$$

FIGURE 11-19 Sample Problem 11-12. A judo hip throw (a) correctly executed and (b) incorrectly executed.

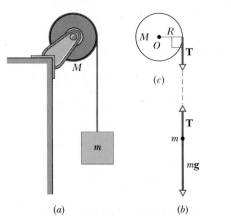

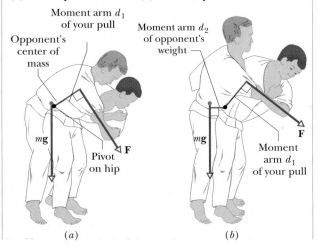

(a)   (b)

FIGURE 11-18 Sample Problems 11-11 and 11-13. (a) The falling block causes the disk to rotate. (b) A free-body diagram for the block. (c) A free-body diagram for the disk.

(b) What must the magnitude of **F** be if he remains upright and his weight $m\mathbf{g}$ has a moment arm $d_2 = 0.12$ m from the pivot point (Fig. 11-19b)?

SOLUTION: In this situation, your opponent's weight $m\mathbf{g}$ provides a positive (counterclockwise) torque that counters your torque. From Eqs. 11-31 and 11-36, we have

$$\sum \tau = -d_1 F + d_2 mg = I\alpha,$$

which gives

$$F = -\frac{I\alpha}{d_1} + \frac{d_2 mg}{d_1}.$$

From (a), we know that the first term on the right is equal to 300 N. Substituting this and the given data, we have

$$F = 300 \text{ N} + \frac{(0.12 \text{ m})(80 \text{ kg})(9.8 \text{ m/s}^2)}{0.30 \text{ m}}$$

$$= 613.6 \text{ N} \approx 610 \text{ N}. \qquad \text{(Answer)}$$

The results indicate that you will have to pull much harder if you do not initially bend your opponent to bring his center of mass to your hip. A good judo fighter knows this lesson from physics. (An analysis of judo and aikido is given in "The Amateur Scientist," *Scientific American,* July 1980.)

11-10 WORK AND ROTATIONAL KINETIC ENERGY

Let us again consider the situation of Fig. 11-17, in which force **F** rotates a rigid body consisting of a single particle of mass m fastened to the end of a massless rod. During the rotation, force **F** does work on the body. Let us assume that the only energy of the body that is changed by **F** is the kinetic energy. Then we can apply the work–kinetic energy theorem of Eq. 7-4:

$$\Delta K = K_f - K_i = W. \qquad (11\text{-}39)$$

Using $K = \frac{1}{2}mv^2$ and Eq. 11-16 ($v = \omega r$), we can rewrite Eq. 11-39 as

$$\Delta K = \tfrac{1}{2}mr^2\omega_f^2 - \tfrac{1}{2}mr^2\omega_i^2 = W. \qquad (11\text{-}40)$$

From Eq. 11-24, the rotational inertia for a body with one particle is $I = mr^2$. Substituting this into Eq. 11-40 yields

$$\Delta K = \tfrac{1}{2}I\omega_f^2 - \tfrac{1}{2}I\omega_i^2 = W. \qquad (11\text{-}41)$$

Equation 11-41 says that the work W done by **F** on the body of Fig. 11-17 changes the rotational kinetic energy K ($= \frac{1}{2}I\omega^2$) of the body. Equation 11-41 is the angular equivalent of the work–kinetic energy theorem for translational motion. We derived it for a rigid body with one particle, but it holds for any rigid body rotated about a fixed axis.

We next relate the work W done on the body in Fig. 11-17 to the torque τ on the body due to force **F**. If the particle in Fig. 11-17 were to move a differential distance ds along its circular path, the body would rotate through differential angle $d\theta$, with $ds = r\,d\theta$. We would use Eq. 7-11 to show the work dW done on the body by force **F** during this motion:

$$dW = \mathbf{F} \cdot d\mathbf{s} = F_t\,ds = F_t r\,d\theta. \qquad (11\text{-}42)$$

(Recall that F_t is the component of **F** along the particle's path.) From Eq. 11-31, we see that the product $F_t r$ is equal to the torque τ. So, we can rewrite Eq. 11-42 as

$$dW = \tau\,d\theta. \qquad (11\text{-}43)$$

The work done during a finite angular displacement from θ_i to θ_f is then

$$W = \int_{\theta_i}^{\theta_f} \tau\,d\theta. \qquad (11\text{-}44)$$

TABLE 11-3 **SOME CORRESPONDING RELATIONS FOR TRANSLATIONAL AND ROTATIONAL MOTION**

PURE TRANSLATION (FIXED DIRECTION)		PURE ROTATION (FIXED AXIS)	
Position	x	Angular position	θ
Velocity	$v = dx/dt$	Angular velocity	$\omega = d\theta/dt$
Acceleration	$a = dv/dt$	Angular acceleration	$\alpha = d\omega/dt$
Mass	m	Rotational inertia	I
Newton's second law	$F = ma$	Newton's second law	$\tau = I\alpha$
Work	$W = \int F\,dx$	Work	$W = \int \tau\,d\theta$
Kinetic energy	$K = \frac{1}{2}mv^2$	Kinetic energy	$K = \frac{1}{2}I\omega^2$
Power	$P = Fv$	Power	$P = \tau\omega$
Work–kinetic energy theorem	$W = \Delta K$	Work–kinetic energy theorem	$W = \Delta K$

Equation 11-44, which holds for any rigid body rotating about a fixed axis, is the rotational equivalent of Eq. 7-27, which is

$$W = \int_{x_i}^{x_f} F \, dx.$$

We can find the power P for rotational motion from Eq. 11-43:

$$P = \frac{dW}{dt} = \tau \frac{d\theta}{dt} = \tau\omega. \qquad (11\text{-}45)$$

This is the rotational analog of $P = Fv$ (from Eq. 7-49), which gives the rate at which a force F does work on a particle moving with speed v, with **F** and **v** parallel.

Table 11-3 summarizes the equations that apply to the rotation of a rigid body about a fixed axis and the equivalent relations for translational motion.

SAMPLE PROBLEM 11-13

(a) In the arrangement of Fig. 11-18, through what angle does the disk rotate in 2.5 s, starting from rest.

SOLUTION: From Eq. 11-10 ($\theta = \omega_0 t + \frac{1}{2}\alpha t^2$), we have, putting $\omega_0 = 0$ and using the value of α calculated in Sample Problem 11-11,

$$\theta = 0 + (\tfrac{1}{2})(-24 \text{ rad/s}^2)(2.5 \text{ s})^2$$
$$= -75 \text{ rad}. \qquad \text{(Answer)}$$

(b) What is the angular velocity of the disk at $t = 2.5$ s?

SOLUTION: We can find this with Eq. 11-9 ($\omega = \omega_0 + \alpha t$). Putting $\omega_0 = 0$ and using the value of α calculated in Sample Problem 11-11, we obtain

$$\omega = 0 + (-24 \text{ rad/s}^2)(2.5 \text{ s})$$
$$= -60 \text{ rad/s}. \qquad \text{(Answer)}$$

(c) What is the kinetic energy K of the disk at $t = 2.5$ s?

SOLUTION: From Eq. 11-25, the kinetic energy of the disk is $\frac{1}{2}I\omega^2$, in which $I = \frac{1}{2}MR^2$. Thus, using the value of ω found in (b), we have

$$K = \tfrac{1}{2}I\omega^2 = \tfrac{1}{2}(\tfrac{1}{2}MR^2)\omega^2$$
$$= (\tfrac{1}{4})(2.5 \text{ kg})(0.20 \text{ m})^2(-60 \text{ rad/s})^2$$
$$= 90 \text{ J}. \qquad \text{(Answer)}$$

Another approach to this problem is to calculate the change in the rotational kinetic energy using the work–kinetic energy theorem. To do so, we must calculate the work W done on the disk by the torque that acts on it. This torque, exerted by the tension T in the cord, is constant. Equation 11-44 then gives us,

$$W = \int_{\theta_i}^{\theta_f} \tau \, d\theta = \tau \int_{\theta_i}^{\theta_f} d\theta = \tau(\theta_f - \theta_i).$$

For the torque τ we use $-TR$, in which T is the tension in the cord ($= 6.0$ N; see Sample Problem 11-11) and R ($= 0.20$ m) is the radius of the disk. The quantity $\theta_f - \theta_i$ is just the angular displacement that we calculated in (a). Thus

$$W = \tau(\theta_f - \theta_i) = -TR(\theta_f - \theta_i)$$
$$= -(6.0 \text{ N})(0.20 \text{ m})(-75 \text{ rad})$$
$$= 90 \text{ J}. \qquad \text{(Answer)}$$

Because K_i is zero (the system starts from rest), Eq. 11-41 tells us that this answer is equal to K.

SAMPLE PROBLEM 11-14

A rigid sculpture, consisting of a thin hoop (of mass m and radius $R = 0.15$ m) and two thin rods (each of mass m and length $L = 2.0R$), is arranged as shown in Fig. 11-20. The sculpture can pivot around a horizontal axis in the plane of the hoop, passing through its center.

(a) In terms of m and R, what is the sculpture's rotational inertia I about the rotation axis?

SOLUTION: From Table 11-2(i), the hoop has rotational inertia $I_{\text{hoop}} = \frac{1}{2}mR^2$ about its diameter. From Table 11-2(e), rod A has rotational inertia $I_{\text{cm},A} = mL^2/12$ about an axis through its center of mass and parallel to the rotation axis. To find the rotational inertia I_A of rod A about the rotation axis, we use Eq. 11-27, the parallel-axis theorem:

$$I_A = I_{\text{cm},A} + mh_{\text{cm},A}^2 = \frac{mL^2}{12} + m\left(R + \frac{L}{2}\right)^2$$
$$= 4.33mR^2,$$

where we have used the fact that $L = 2.0R$ and where $h_{\text{cm},A}$ ($= R + L/2$) is the perpendicular distance between the center of rod A and the rotation axis.

We treat rod B similarly: it has zero rotational inertia $I_{\text{cm},B}$ about an axis along its length, and its rotational inertia I_B about the rotation axis is

$$I_B = I_{\text{cm},B} + mh_{\text{cm},B}^2 = 0 + mR^2 = mR^2.$$

FIGURE 11-20 Sample Problem 11-14. A rigid sculpture consisting of a hoop and two rods can rotate around a horizontal axis.

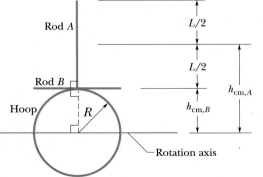

Here $h_{cm, B} (= R)$ is the perpendicular distance between rod B and the rotation axis. Thus the rotational inertia I of the sculpture about the rotation axis is

$$I = I_{hoop} + I_A + I_B = \tfrac{1}{2}mR^2 + 4.33mR^2 + mR^2$$
$$= 5.83mR^2 \approx 5.8mR^2. \qquad \text{(Answer)}$$

(b) Starting from rest, the sculpture rotates around the rotation axis from the initial upright orientation of Fig. 11-20. What is its angular speed ω about the axis when it is inverted?

SOLUTION: Before the sculpture moves, its center of mass is a distance y_{cm} above the rotation axis, with y_{cm} given by Eq. 9-5:

$$y_{cm} = \frac{m(0) + mR + m(R + L/2)}{3m} = R,$$

where $3m$ is the mass of the sculpture. As the sculpture rotates, its center of mass descends to the same distance *below* the rotation axis. Thus the center of mass undergoes a vertical displacement of $\Delta y_{cm} = -2R$. During the descent, gravita-

tional potential energy U of the sculpture is transferred to kinetic energy K of its rotation. The change ΔU in potential energy is the product of the magnitude of the sculpture's weight $(3mg)$ and the vertical displacement Δy_{cm}. Thus

$$\Delta U = 3mg \, \Delta y_{cm} = 3mg(-2R) = -6mgR.$$

From Eq. 11-25, the associated change in kinetic energy is

$$\Delta K = \tfrac{1}{2}I\omega^2 - 0 = \tfrac{1}{2}I\omega^2.$$

So by conservation of mechanical energy

$$\Delta K + \Delta U = 0,$$

and

$$\tfrac{1}{2}I\omega^2 - 6mgR = 0.$$

Substituting $I = 5.83mR^2$ and solving for ω give

$$\omega = \sqrt{\frac{12g}{5.83R}} = \sqrt{\frac{(12)(9.8 \text{ m/s}^2)}{(5.83)(0.15 \text{ m})}}$$
$$= 12 \text{ rad/s.} \qquad \text{(Answer)}$$

REVIEW & SUMMARY

Angular Position

To describe the rotation of a rigid body about a fixed axis, called the **rotation axis,** we assume a **reference line** is fixed in the body, perpendicular to that axis and rotating with the body. We measure the **angular position** θ of this line relative to a fixed direction. When θ is measured in **radians,**

$$\theta = \frac{s}{r} \qquad \text{(radian measure)}, \qquad (11\text{-}1)$$

where s is the arc length of a circular path of radius r and angle θ. Radian measure is related to angle measure in revolutions and degrees by

$$1 \text{ rev} = 360° = 2\pi \text{ rad.} \qquad (11\text{-}2)$$

Angular Displacement

A body that rotates about a rotation axis, changing its angular position from θ_1 to θ_2, undergoes an **angular displacement**

$$\Delta\theta = \theta_2 - \theta_1, \qquad (11\text{-}4)$$

where $\Delta\theta$ is positive for counterclockwise rotation and negative for clockwise rotation.

Angular Velocity and Speed

If a body rotates through an angular displacement $\Delta\theta$ in a time interval Δt, its **average angular velocity** $\overline{\omega}$ is

$$\overline{\omega} = \frac{\Delta\theta}{\Delta t}. \qquad (11\text{-}5)$$

The **(instantaneous) angular velocity** ω of the body is

$$\omega = \frac{d\theta}{dt}. \qquad (11\text{-}6)$$

Both $\overline{\omega}$ and ω are vectors, with directions given by the **right-hand rule** of Fig. 11-5. They are positive for counterclockwise rotation and negative for clockwise rotation. The magnitude of the body's angular velocity is the **angular speed.**

Angular Acceleration

If the angular velocity of a body changes from ω_1 to ω_2 in a time interval $\Delta t = t_2 - t_1$, the **average angular acceleration** $\overline{\alpha}$ of the body is

$$\overline{\alpha} = \frac{\omega_2 - \omega_1}{t_2 - t_1} = \frac{\Delta\omega}{\Delta t}. \qquad (11\text{-}7)$$

The **(instantaneous) angular acceleration** α of a body is

$$\alpha = \frac{d\omega}{dt}. \qquad (11\text{-}8)$$

Both $\overline{\alpha}$ and α are vectors.

The Kinematic Equations for Constant Angular Acceleration

Constant angular acceleration $(\alpha = \text{constant})$ is an important special case of rotational motion. The appropriate kinematic equations, given in Table 11-1, are

$$\omega = \omega_0 + \alpha t, \qquad (11\text{-}9) \qquad \theta = \tfrac{1}{2}(\omega_0 + \omega)t, \qquad (11\text{-}12)$$
$$\theta = \omega_0 t + \tfrac{1}{2}\alpha t^2, \qquad (11\text{-}10) \qquad \theta = \omega t - \tfrac{1}{2}\alpha t^2. \qquad (11\text{-}13)$$
$$\omega^2 = \omega_0^2 + 2\alpha\theta, \qquad (11\text{-}11)$$

Linear and Angular Variables Related

A point in a rigid rotating body, at a *perpendicular distance r* from the rotation axis, moves in a circle with radius r. If the body rotates through an angle θ, the point moves along an arc with length s given by

$$s = \theta r \quad \text{(radian measure)}, \quad (11\text{-}15)$$

where θ is in radians.

The linear velocity $\mathbf{v}$ of the point is tangent to the circle; the point's linear speed v is given by

$$v = \omega r \quad \text{(radian measure)}, \quad (11\text{-}16)$$

where ω is the angular speed (in radians per second) of the body.

The linear acceleration $\mathbf{a}$ of the point has both *tangential* and *radial* components. The tangential component is

$$a_t = \alpha r \quad \text{(radian measure)}, \quad (11\text{-}20)$$

where α is the magnitude of the angular acceleration (in radians per second-squared) of the body. The radial component of $\mathbf{a}$ is

$$a_r = \frac{v^2}{r} = \omega^2 r \quad \text{(radian measure)}. \quad (11\text{-}21)$$

If the point moves in uniform circular motion, the period T of the motion for the point and the body is

$$T = \frac{2\pi r}{v} = \frac{2\pi}{\omega} \quad \text{(radian measure)}. \quad (11\text{-}17, 11\text{-}18)$$

Rotational Kinetic Energy and Rotational Inertia

The kinetic energy K of a rigid body rotating about a fixed axis is given by

$$K = \tfrac{1}{2} I \omega^2 \quad \text{(radian measure)}, \quad (11\text{-}25)$$

in which I is the **rotational inertia** of the body, defined as

$$I = \sum m_i r_i^2 \quad (11\text{-}24)$$

for a system of discrete particles and as

$$I = \int r^2 \, dm \quad (11\text{-}26)$$

for a body with continuously distributed mass. The r and r_i in these expressions represent the perpendicular distance from the axis of rotation to each mass element in the body.

The Parallel-Axis Theorem

The *parallel-axis theorem* relates the rotational inertia I of a body about any axis to that of the same body about a parallel axis through the center of mass:

$$I = I_{\text{cm}} + Mh^2. \quad (11\text{-}27)$$

Here h is the perpendicular distance between the two axes.

Torque

Torque is a turning or twisting action on a body about a rotation axis due to a force $\mathbf{F}$. If $\mathbf{F}$ is exerted at a point given by the position vector $\mathbf{r}$ relative to the axis, then the torque $\boldsymbol{\tau}$ (a vector quantity) is

$$\boldsymbol{\tau} = \mathbf{r} \times \mathbf{F}. \quad (11\text{-}33)$$

The magnitude of the torque is

$$\tau = r F_t = r_\perp F = r F \sin \phi, \quad (11\text{-}31, 11\text{-}32, 11\text{-}30)$$

where F_t is the component of $\mathbf{F}$ perpendicular to $\mathbf{r}$, and ϕ is the angle between $\mathbf{r}$ and $\mathbf{F}$. The quantity $r_\perp$ is the perpendicular distance between the rotation axis and an extended line running through the $\mathbf{F}$ vector. This line is called the **line of action** of $\mathbf{F}$, and $r_\perp$ is called the **moment arm** of $\mathbf{F}$. Similarly, r is the moment arm of F_t.

The SI unit of torque is the newton-meter (N·m). A torque τ is positive if it tends to rotate the body counterclockwise and negative if it tends to rotate the body in the clockwise direction.

Newton's Second Law in Angular Form

The rotational analog of Newton's second law is

$$\sum \tau = I \alpha, \quad (11\text{-}36)$$

where $\Sigma \tau$ is the net torque acting on a particle or rigid body, I is the rotational inertia of the particle or body about the rotation axis, and α is the resulting angular acceleration about that axis.

Work and Rotational Kinetic Energy

The equations used for calculating work and power in rotational motion are analogs of the corresponding equations governing translational motion and are

$$W = \int_{\theta_i}^{\theta_f} \tau \, d\theta \quad (11\text{-}44)$$

and

$$P = \frac{dW}{dt} = \tau \frac{d\theta}{dt} = \tau \omega. \quad (11\text{-}45)$$

The form of the work–kinetic energy theorem used for rotating bodies is

$$\Delta K = \tfrac{1}{2} I \omega_f^2 - \tfrac{1}{2} I \omega_i^2 = W. \quad (11\text{-}41)$$

QUESTIONS

1. Figure 11-21*b* is a graph of the angular position of the rotating disk of Fig. 11-21*a*. Is the angular velocity of the disk positive, negative, or zero at (a) $t = 1$ s, (b) $t = 2$ s, and (c) $t = 3$ s? (d) Is the angular acceleration positive or negative?

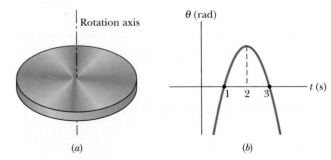

FIGURE 11-21 Question 1.

(a)   (b)

2. Figure 11-22 is a graph of the angular velocity of the rotating disk of Fig. 11-21*a*. What are the (a) initial and (b) final directions of rotation? (c) Does the disk momentarily stop? (d) Is the angular acceleration positive or negative? (e) Is the angular acceleration constant or varying?

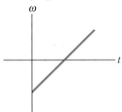

FIGURE 11-22
Question 2.

3. At $t = 0$, the rotating disk of Fig. 11-21*a* is at angular position $\theta_0 = -2$ rad. Here are the signs of the disk's initial angular velocity and constant angular acceleration, respectively, for four situations: (1) +, +; (2) +, −; (3) −, +; (4) −, −. For which situations will the disk (a) undergo a momentary stop, (b) definitely pass through angular position $\theta = 0$ (given enough time), and (c) definitely not pass through that angular position?

4. For the rotating disk of Fig. 11-21*a*, here are the initial and final angular velocities, respectively, in four situations: (a) 2 rad/s, 3 rad/s; (b) −2 rad/s, 3 rad/s; (c) −2 rad/s, −3 rad/s; (d) 2 rad/s, −3 rad/s. The magnitude of the angular acceleration of the disk has the same constant value in all four situations. Rank the situations according to the magnitude of the disk's displacement, greatest first, during the change from the initial to the final angular velocity.

5. For which of the following expressions for $\omega(t)$ of a rotating object do the angular equations of Table 11-1 apply? (a) $\omega = 3$; (b) $\omega = 4t^2 + 2t - 6$; (c) $\omega = 3t - 4$; (d) $\omega = 5t^2 - 3$.

6. Figure 11-23 gives the angular velocity versus time for the rotating disk of Fig. 11-21*a*. For a point on the rim of the disk, rank the four instants *a*, *b*, *c*, and *d* according to the magnitude of (a) the tangential acceleration and (b) the radial acceleration, greatest first.

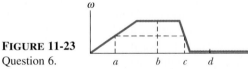

FIGURE 11-23
Question 6.

7. Figure 11-24 shows four gears that rotate together because of friction between them. Gear 2 has radius R; gear 3 has radius $2R$; and gears 1 and 4 have radius $3R$. Gear 2 is forced to rotate by a motor. Rank the four gears according to (a) the linear speed of their rims and (b) their angular speed of rotation, greatest first.

8. Circular disks *A* and *B* have the same weight and thickness, but the density of *A* is greater than that of *B*. Is the rotational inertia of *A* about its central axis greater than, less than, or the same as that of *B*?

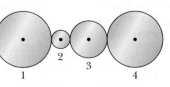

FIGURE 11-24
Question 7.

9. Figure 11-25 shows three uniform disks, along with their radii R and masses M. Rank the disks according to their rotational inertias about their central axes, greatest first.

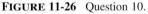

R:	1 m	2 m	3 m
M:	26 kg	7 kg	3 kg
	(a)	(b)	(c)

FIGURE 11-25
Question 9.

10. Five solids with identical masses are shown in cross section in Fig. 11-26. The cross sections have equal widths at the widest part and equal heights (but not necessarily equal thicknesses). (a) Which one has the greatest rotational inertia about an axis through its center of mass and perpendicular to its cross section? (b) Which has the least?

Hoop   Cube   Cylinder   Prism   Sphere

FIGURE 11-26 Question 10.

11. Figure 11-27*a* shows a meter stick, half wood and half steel, that is pivoted at the wood end at *O*. A force **F** is applied to the steel end at *a*. In Fig. 11-27*b*, the stick is reversed and pivoted at the steel end at *O*′, and the same force is applied at the wood end at *a*′. Is the resulting angular acceleration of Fig. 11-27*a* greater than, less than, or the same as that of Fig. 11-27*b*?

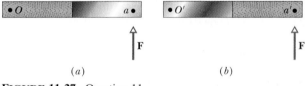

(*a*)   (*b*)

FIGURE 11-27 Question 11.

12. Figure 11-28*a* shows an overhead view of a horizontal bar that can pivot about the point indicated. Two horizontal forces act on the bar, but the bar is stationary. If the angle between the bar and force **F**$_2$ is now decreased from the initial 90°, and the bar is still not to turn, should the magnitude of **F**$_2$ be made larger, be made smaller, or left the same?

13. Figure 11-28*b* shows an overhead view of a horizontal bar that is rotated about the pivot point by two horizontal forces, **F**$_1$ and **F**$_2$, at opposite ends of the bar. The direction of **F**$_2$ is at angle ϕ to the bar. Rank the following values of ϕ according to the magnitude of the angular acceleration of the bar, greatest first: 90°, 70°, and 110°.

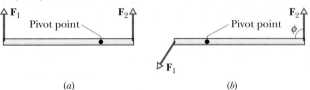

(*a*)   (*b*)

FIGURE 11-28 Questions 12 and 13.

14. A force is applied to the rim of the disk in Fig. 11-21*a* so as to change its angular velocity. Its initial and final angular velocities, respectively, for four situations are: (a) −2 rad/s, 5 rad/s; (b) 2 rad/s, 5 rad/s; (c) −2 rad/s, −5 rad/s; and (d) 2 rad/s, −5 rad/s. Rank the situations according to the work done by the torque due to the force, greatest first.

15. Hold your right arm downward, palm toward your thigh. Keeping your wrist rigid, (1) lift the arm until it is horizontal and forward, (2) move it horizontally until it is pointed toward the right, and (3) then bring it down to your side. Your palm faces forward. If you start over, but reverse the steps, why does your palm *not* face forward?

EXERCISES & PROBLEMS

SECTION 11-2 The Rotational Variables

1E. (a) What angle in radians is subtended by an arc that has length 1.80 m and is part of a circle of radius 1.20 m? (b) Express the same angle in degrees. (c) The angle between two radii of a circle is 0.620 rad. What arc length is subtended if the radius is 2.40 m?

2E. During a time interval *t* the flywheel of a generator turns through the angle $\theta = at + bt^3 - ct^4$, where *a, b,* and *c* are constants. Write expressions for the wheel's (a) angular velocity and (b) angular acceleration.

3E. What is the angular speed of (a) the second hand, (b) the minute hand, and (c) the hour hand of a watch? Answer in radians per second.

4E. Our Sun is 2.3×10^4 ly (light-years) from the center of our Milky Way galaxy and is moving in a circle around that center at a speed of 250 km/s. (a) How long does it take the Sun to make one revolution about the galactic center? (b) How many revolutions has the Sun completed since it was formed about 4.5×10^9 years ago?

5E. The angular position of a point on the rim of a rotating wheel is given by $\theta = 4.0t - 3.0t^2 + t^3$, where θ is in radians if *t* is given in seconds. (a) What are the angular velocities at $t = 2.0$ s and $t = 4.0$ s? (b) What is the average angular acceleration for the time interval that begins at $t = 2.0$ s and ends at $t = 4.0$ s? (c) What are the instantaneous angular accelerations at the beginning and end of this time interval?

6E. The angular position of a point on a rotating wheel is given by $\theta = 2 + 4t^2 + 2t^3$, where θ is in radians and *t* is in seconds. At $t = 0$, what are (a) the angular position and (b) the angular velocity? (c) What is the angular velocity at $t = 4.0$ s? (d) Calculate the angular acceleration at $t = 2.0$ s. (e) Is the angular acceleration constant?

7P. A wheel rotates with an angular acceleration α given by $\alpha = 4at^3 - 3bt^2$, where *t* is the time and *a* and *b* are constants. If the wheel has an initial angular speed ω_0, write equations for (a) the angular speed and (b) the angular displacement of the wheel as functions of time.

8P. A good baseball pitcher can throw a baseball toward home plate at 85 mi/h with a spin of 1800 rev/min. How many revolutions does the baseball make on its way to home plate? For simplicity, assume that the 60 ft trajectory is a straight line.

9P. A diver makes 2.5 complete revolutions on the way from a 10 m high platform to the water. Assuming zero initial vertical velocity, find the diver's average angular velocity during a dive.

10P. The wheel in Fig. 11-29 has eight spokes and a radius of 30 cm. It is mounted on a fixed axle and is spinning at 2.5 rev/s. You want to shoot a 20 cm long arrow parallel to this axle and through the wheel without hitting any of the spokes. Assume that the arrow and the spokes are very thin. (a) What minimum speed must the arrow have? (b) Does it matter where between the axle and rim of the wheel you aim? If so, where is the best location?

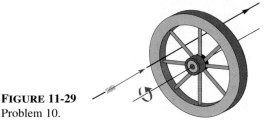

FIGURE 11-29
Problem 10.

SECTION 11-4 Rotation with Constant Angular Acceleration

11E. The angular speed of an automobile engine is increased from 1200 rev/min to 3000 rev/min in 12 s. (a) What is its angular acceleration in rev/min², assuming it to be uniform? (b) How many revolutions does the engine make during this time?

12E. A phonograph turntable rotating at $33\frac{1}{3}$ rev/min slows down and stops in 30 s after the motor is turned off. (a) Find its (uniform) angular acceleration in units of rev/min². (b) How many revolutions did it make in this time?

13E. A disk, initially rotating at 120 rad/s, is slowed down with a constant angular acceleration of magnitude 4.0 rad/s². (a) How much time elapses before the disk stops? (b) Through what angle does the disk rotate in coming to rest?

14E. A pulley wheel that is 8.0 cm in diameter has a 5.6 m long cord wrapped around its periphery. Starting from rest, the wheel is given a constant angular acceleration of 1.5 rad/s². (a) Through what angle must the wheel turn for the cord to unwind? (b) How long does the unwinding take?

15E. A heavy flywheel rotating on its central axis is slowing down because of friction in its bearings. At the end of the first minute of slowing, its angular velocity is 0.90 of its initial angular velocity of 250 rev/min. Assuming constant frictional forces, find its angular velocity at the end of the second minute.

16E. The flywheel of an engine is rotating at 25.0 rad/s. When the engine is turned off, the flywheel decelerates at a constant rate and comes to rest after 20.0 s. Calculate (a) the angular acceleration (in rad/s^2) of the flywheel, (b) the angle (in rad) through which the flywheel rotates in coming to rest, and (c) the number of revolutions made by the flywheel in coming to rest.

17E. Starting from rest, a disk rotates about its central axis with constant angular acceleration. In 5.0 s, it has rotated 25 rad. (a) What was the angular acceleration during this time? (b) What was the average angular velocity? (c) What is the instantaneous angular velocity of the disk at the end of the 5.0 s? (d) Assuming that the acceleration does not change, through what additional angle will the disk turn during the next 5.0 s?

18P. Starting from rest at $t = 0$, a wheel undergoes a constant angular acceleration. When $t = 2.0$ s, the angular velocity of the wheel is 5.0 rad/s. The acceleration continues until $t = 20$ s, when it abruptly ceases. Through what angle does the wheel rotate in the interval $t = 0$ to $t = 40$ s?

19P. A wheel, starting from rest, rotates with a constant angular acceleration of 2.00 rad/s^2. During a certain 3.00 s interval, it turns through 90.0 rad. (a) How long had the wheel been turning before the start of the 3.00 s interval? (b) What was the angular velocity of the wheel at the start of the 3.00 s interval?

20P. A wheel has a constant angular acceleration of 3.0 rad/s^2. In a 4.0 s interval, it turns through an angle of 120 rad. Assuming that the wheel started from rest, how long had it been in motion at the start of this 4.0 s interval?

21P. A flywheel completes 40 rev as it slows from an angular speed of 1.5 rad/s to a complete stop. (a) Assuming uniform acceleration, what is the time required for it to come to rest? (b) What is the angular acceleration? (c) How much time is required for it to complete the first 20 of the 40 revolutions?

22P. At $t = 0$, a flywheel has an angular velocity of 4.7 rad/s, an angular acceleration of -0.25 rad/s^2, and a reference line at $\theta_0 = 0$. (a) Through what maximum angle θ_{max} will the reference line turn in the positive direction? At what times t will the line be at (b) $\theta = \frac{1}{2}\theta_{max}$ and (c) $\theta = -10.5$ rad (consider both positive and negative values of t)? (d) Graph θ versus t, and indicate the answers to (a), (b), and (c) on the graph.

23P. A disk rotates about its central axis starting from rest and accelerates with constant angular acceleration. At one time it is rotating at 10 rev/s. After 60 more complete revolutions, its angular speed is 15 rev/s. Calculate (a) the angular acceleration, (b) the time required to complete the 60 revolutions mentioned, (c) the time required to attain the 10 rev/s angular speed, and (d) the number of revolutions from rest until the time the disk attained the 10 rev/s angular speed.

24P. A wheel turns through 90 revolutions in 15 s, its angular speed at the end of the period being 10 rev/s. (a) What was its angular speed at the beginning of the 15 s interval, assuming constant angular acceleration? (b) How much time had elapsed between the time the wheel was at rest and the beginning of the 15 s interval?

SECTION 11-5 Relating The Linear and Angular Variables

25E. What is the acceleration of a point on the rim of a 12 in. ($= 30$ cm) diameter record rotating at $33\frac{1}{3}$ rev/min?

26E. A phonograph record on a turntable rotates at $33\frac{1}{3}$ rev/min. (a) What is the angular speed in rad/s? What is the linear speed of a point on the record at the needle at (b) the beginning and (c) the end of the recording? The distances of the needle from the turntable axis are 5.9 in. and 2.9 in. at these two positions.

27E. What is the angular speed of a car traveling at 50 km/h and rounding a circular turn of radius 110 m?

28E. A flywheel with a diameter of 1.20 m is rotating at 200 rev/min. (a) What is the angular velocity of the flywheel in rad/s? (b) What is the linear velocity of a point on the rim of the flywheel? (c) What constant angular acceleration (in rev/min^2) will increase the wheel's angular speed to 1000 rev/min in 60 s? (d) How many revolutions does the wheel make during this 60 s acceleration?

29E. A point on the rim of a 0.75 m diameter grinding wheel changes speed uniformly from 12 m/s to 25 m/s in 6.2 s. What is the average angular acceleration of the wheel during this interval?

30E. The Earth's orbit about the Sun is almost a circle. With respect to the Sun, what are the Earth's (a) angular speed, (b) linear speed, and (c) acceleration?

31E. At 7:14 A.M. on June 30, 1908, a huge explosion occurred above remote central Siberia at latitude 61° N and longitude 102° E; the fireball thus created was the brightest flash seen by anyone before the advent of nuclear weapons. The *Tunguska Event,* which according to one chance witness "covered an enormous part of the sky," was probably the explosion of a *stony asteroid* about 140 m wide. (a) Considering only the Earth's rotation, determine how much later the asteroid would have had to arrive to put the explosion above Helsinki at longitude 25° E? This would have obliterated the city. (b) If the asteroid had, instead, been a *metallic asteroid,* it could have reached the Earth's surface. How much later would such an asteroid have had to arrive to put the impact in the Atlantic Ocean at longitude 20° W? (The resulting tsunamis would have wiped out coastal civilization on both sides of the Atlantic.)

32E. An astronaut is being tested in a centrifuge. The centrifuge has a radius of 10 m and, in starting, rotates according to $\theta = 0.30t^2$, where t in seconds gives θ in radians. When $t = 5.0$ s, what are the astronaut's (a) angular velocity, (b) linear speed, (c) tangential acceleration (magnitude only), and (d) radial acceleration (magnitude only)?

33E. What are (a) the angular speed, (b) the radial acceleration, and (c) the tangential acceleration of a spaceship negotiating a circular turn of radius 3220 km at a speed of 29,000 km/h?

34E. A coin of mass M is placed a distance R from the center of a phonograph turntable. The coefficient of static friction is μ_s. The angular speed of the turntable is slowly increased to a value ω_0, at which time the coin slides off. (a) Find ω_0 in terms of the quantities M, R, g, and μ_s. (b) Make a sketch showing the approximate path of the coin as it flies off the turntable.

35P. The flywheel of a steam engine runs with a constant angular speed of 150 rev/min. When steam is shut off, the friction of the bearings and of the air brings the wheel to rest in 2.2 h. (a) What is the constant angular acceleration, in rev/min², of the wheel during the slowdown? (b) How many rotations does the wheel make before coming to rest? (c) At the instant the flywheel is turning at 75 rev/min, what is the tangential component of the linear acceleration of a flywheel particle that is 50 cm from the axis of rotation? (d) What is the magnitude of the net linear acceleration of the particle in (c)?

36P. A gyroscope flywheel of radius 2.83 cm is accelerated from rest at 14.2 rad/s² until its angular speed is 2760 rev/min. (a) What is the tangential acceleration of a point on the rim of the flywheel? (b) What is the radial acceleration of this point when the flywheel is spinning at full speed? (c) Through what distance does a point on the rim move during the acceleration?

37P. If an airplane propeller rotates at 2000 rev/min while the airplane flies at a speed of 480 km/h relative to the ground, what is the speed of a point on the tip of the propeller, at radius 1.5 m, as seen by (a) the pilot and (b) an observer on the ground? The plane's velocity is parallel to the propeller's axis of rotation.

38P. An early method of measuring the speed of light makes use of a rotating slotted wheel. A beam of light passes through a slot at the outside edge of the wheel, as in Fig. 11-30, travels to a distant mirror, and returns to the wheel just in time to pass through the next slot in the wheel. One such slotted wheel has a radius of 5.0 cm and 500 slots at its edge. Measurements taken when the mirror was $l = 500$ m from the wheel indicated a speed of light of 3.0×10^5 km/s. (a) What was the (constant) angular speed of the wheel? (b) What was the linear speed of a point on the edge of the wheel?

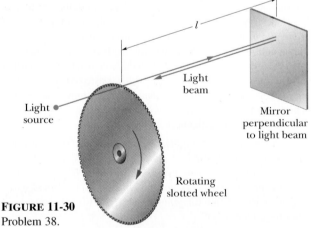

FIGURE 11-30
Problem 38.

39P. A car starts from rest and moves around a circular track of radius 30.0 m. Its speed increases at the constant rate of 0.500 m/s². (a) What is the magnitude of its *net* linear acceleration 15.0 s later? (b) What angle does this net acceleration vector make with the car's velocity at this time?

40P. (a) What is the angular speed about the polar axis of a point on the Earth's surface at a latitude of 40° N? (b) What is the linear speed of the point? (c) What are the corresponding values for a point at the equator?

41P. In Fig. 11-31, wheel A of radius $r_A = 10$ cm is coupled by belt B to wheel C of radius $r_C = 25$ cm. Wheel A increases its angular speed from rest at a uniform rate of 1.6 rad/s². Find the time for wheel C to reach a rotational speed of 100 rev/min, assuming the belt does not slip. (*Hint:* If the belt does not slip, the linear speeds at the rims of the two wheels must be equal.)

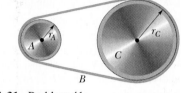

FIGURE 11-31 Problem 41.

42P. In Fig. 11-32, four pulleys are connected by two belts. Pulley A (radius 15 cm) is the drive pulley, and it rotates at 10 rad/s. Pulley B (radius 10 cm) is connected by belt 1 to pulley A. Pulley B' (radius 5 cm) is concentric with pulley B and is rigidly attached to it. Pulley C (radius 25 cm) is connected by belt 2 to pulley B'. Calculate (a) the linear speed of a point on belt 1, (b) the angular speed of pulley B, (c) the angular speed of pulley B', (d) the linear speed of a point on belt 2, and (e) the angular speed of pulley C. (See hint to Problem 41.)

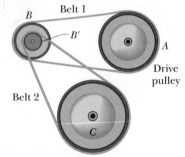
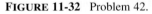

FIGURE 11-32 Problem 42.

43P. A phonograph turntable is rotating at $33\frac{1}{3}$ rev/min. A small object is on the turntable 6.0 cm from the axis of rotation. (a) Calculate the acceleration of the object, assuming that the object does not slip. (b) What is the minimum value of the coefficient of static friction between the object and the turntable? (c) Suppose that the turntable achieved its angular speed by starting from rest and undergoing a constant angular acceleration for 0.25 s. Calculate the minimum coefficient of static friction required for the object not to slip during the acceleration period.

44P. A pulsar is a rapidly rotating neutron star that emits radio pulses with precise synchronization, one such pulse for each rotation of the star. The *period T* of rotation is found by measuring the time between pulses. At present, the pulsar in the central region of the Crab nebula (see Fig. 11-33) has a period of rotation of $T = 0.033$ s, and this period is observed to be increasing at the rate of 1.26×10^{-5} s/y. (a) What is the value of the angular acceleration in rad/s²? (b) If its angular acceleration is constant, how many years from now will the pulsar stop rotating? (c) The pulsar originated in a supernova explosion seen in the year A.D. 1054. What was T for the pulsar when it was born? (Assume constant angular acceleration since then.)

FIGURE 11-33 Problem 44. The Crab nebula resulted from a star whose explosion was seen in A.D. 1054. In addition to the gaseous debris seen here, the explosion left a spinning neutron star at its center. The star has a diameter of only 30 km.

SECTION 11-6 Kinetic Energy of Rotation

45E. Calculate the rotational inertia of a wheel that has a kinetic energy of 24,400 J when rotating at 602 rev/min.

46P. The oxygen molecule, O_2, has a total mass of 5.30×10^{-26} kg and a rotational inertia of 1.94×10^{-46} kg·m² about an axis through the center of the line joining the atoms and perpendicular to that line. Suppose that such a molecule in a gas has a speed of 500 m/s and that its rotational kinetic energy is two-thirds of its translational kinetic energy. Find its angular velocity.

SECTION 11-7 Calculating the Rotational Inertia

47E. Calculate the kinetic energies of two uniform solid cylinders, each rotating about its central axis. They have the same mass, 1.25 kg, and rotate with the same angular velocity, 235 rad/s, but the first has a radius of 0.25 m and the second a radius of 0.75 m.

48E. A molecule has a rotational inertia of 14,000 u·pm² and is spinning at an angular speed of 4.3×10^{12} rad/s. (a) Express the rotational inertia in kg·m². (b) Calculate the rotational kinetic energy in electron-volts.

49E. A communications satellite is a solid cylinder with mass 1210 kg, diameter 1.21 m, and length 1.75 m. Prior to launching from the shuttle cargo bay, it is set spinning at 1.52 rev/s about the cylinder axis (Fig. 11-34). Calculate the satellite's (a) rotational inertia about the rotation axis and (b) rotational kinetic energy.

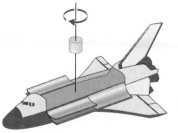

FIGURE 11-34 Exercise 49.

50E. Two particles, each with mass m, are fastened to each other, and to a rotation axis at O, by two thin rods, each with length l and mass M as shown in Fig. 11-35. The combination rotates around the rotation axis with angular velocity ω. Obtain algebraic expressions for (a) the rotational inertia of the combination about O and (b) the kinetic energy of rotation about O.

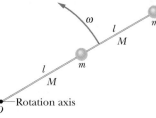

FIGURE 11-35 Exercise 50.

51E. Each of the three helicopter rotor blades shown in Fig 11-36 is 5.20 m long and has a mass of 240 kg. The rotor is rotating at 350 rev/min. (a) What is the rotational inertia of the rotor assembly about the axis of rotation? (Each blade can be considered to be a thin rod.) (b) What is the kinetic energy of rotation?

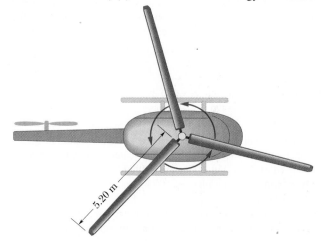

FIGURE 11-36 Exercise 51 and Problem 85.

52E. Assume the Earth to be a sphere of uniform density. Calculate (a) its rotational inertia and (b) its rotational kinetic energy. (c) Suppose that this energy could be harnessed for our use. How long could the Earth supply 1.0 kW of power to each of the 6.4×10^9 persons on Earth?

53E. Calculate the rotational inertia of a meter stick, with mass 0.56 kg, about an axis perpendicular to the stick and located at the 20 cm mark. (Treat the stick as a thin rod.)

54P. Show that the axis about which a given rigid body has its smallest rotational inertia must pass through its center of mass.

55P. Derive the expression for the rotational inertia of a hoop of mass M and radius R about its central axis; see Table 11-2(a).

56P. Figure 11-37 shows a uniform solid block of mass M and edge lengths a, b, and c. Calculate its rotational inertia about an axis through one corner and perpendicular to the large faces.

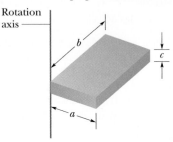

FIGURE 11-37 Problem 56.

57P. The masses and coordinates of four particles are as follows: 50 g, $x = 2.0$ cm, $y = 2.0$ cm; 25 g, $x = 0$, $y = 4.0$ cm; 25 g, $x = -3.0$ cm, $y = -3.0$ cm; 30 g, $x = -2.0$ cm, $y = 4.0$ cm. What is the rotational inertia of this collection with respect to the (a) x, (b) y, and (c) z axes? (d) If the answers to (a) and (b) are A and B, respectively, then what is the answer to (c) in terms of A and B?

58P. (a) Show that the rotational inertia of a solid cylinder of mass M and radius R about its central axis is equal to the rotational inertia of a thin hoop of mass M and radius $R/\sqrt{2}$ about its central axis. (b) Show that the rotational inertia I of any given body of mass M about any given axis is equal to the rotational inertia of an *equivalent hoop* about that axis, if the hoop has the same mass M and a radius k given by

$$k = \sqrt{\frac{I}{M}}.$$

The radius k of the equivalent hoop is called the *radius of gyration* of the given body.

59P. Delivery trucks that operate by making use of energy stored in a rotating flywheel have been used in Europe. The trucks are charged by using an electric motor to get the flywheel up to its top speed of 200π rad/s. One such flywheel is a solid, homogeneous cylinder with a mass of 500 kg and a radius of 1.0 m. (a) What is the kinetic energy of the flywheel after charging? (b) If the truck operates with an average power requirement of 8.0 kW, for how many minutes can it operate between chargings?

SECTION 11-8 Torque

60E. The length of a bicycle pedal arm is 0.152 m, and a downward force of 111 N is applied by the foot. What is the magnitude of the torque about the pivot point when the arm makes an angle of (a) 30°, (b) 90°, and (c) 180° with the vertical?

61E. A small 0.75 kg ball is attached to one end of a 1.25 m long massless rod, and the other end of the rod is hung from a pivot. When the resulting pendulum is 30° from the vertical, what is the magnitude of the torque about the pivot?

62E. A bicyclist of mass 70 kg puts all his weight on each downward-moving pedal as he pedals up a steep road. Take the diameter of the circle in which the pedals rotate to be 0.40 m, and determine the magnitude of the maximum torque he exerts.

63P. The body in Fig. 11-38 is pivoted at O, and two forces act on it as shown. (a) Find an expression for the magnitude of the net torque on the body about the pivot. (b) If $r_1 = 1.30$ m, $r_2 = 2.15$ m, $F_1 = 4.20$ N, $F_2 = 4.90$ N, $\theta_1 = 75.0°$, and $\theta_2 = 60.0°$, what is the net torque about the pivot?

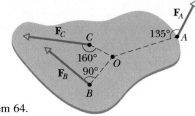

FIGURE 11-38 Problem 63.

64P. The body in Fig. 11-39 is pivoted at O. Three forces act on it in the directions shown on the figure: $F_A = 10$ N at point A, 8.0 m from O; $F_B = 16$ N at point B, 4.0 m from O; and $F_C = 19$ N at point C, 3.0 m from O. What is the net torque about O?

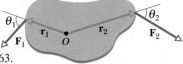

FIGURE 11-39 Problem 64.

SECTION 11-9 Newton's Second Law for Rotation

65E. When a torque of 32.0 N·m is applied to a certain wheel, the wheel acquires an angular acceleration of 25.0 rad/s². What is the rotational inertia of the wheel?

66E. Launching herself from a board, a diver changed her angular velocity from zero to 6.20 rad/s in 220 ms. Her rotational inertia is 12.0 kg·m². (a) What was her angular acceleration during the launch? (b) What external torque acted on the diver during the launch?

67E. A cylinder having a mass of 2.0 kg can rotate about its central axis through point O. Forces are applied as in Fig. 11-40:

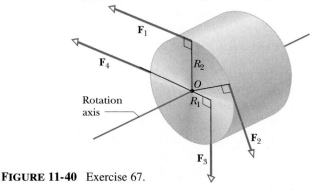

FIGURE 11-40 Exercise 67.

$F_1 = 6.0$ N, $F_2 = 4.0$ N, $F_3 = 2.0$ N, and $F_4 = 5.0$ N. Also, $R_1 = 5.0$ cm and $R_2 = 12$ cm. Find the magnitude and direction of the angular acceleration of the cylinder. (During the rotation, the forces maintain their same angles relative to the cylinder).

68E. A small object with mass 1.30 kg is mounted on one end of a rod 0.780 m long and of negligible mass. The system rotates in a horizontal circle about the other end of the rod at 5010 rev/min. (a) Calculate the rotational inertia of the system about the axis of rotation. (b) There is an air drag of 2.30×10^{-2} N on the object, directed opposite its motion. What torque must be applied to the system to keep it rotating at constant speed?

69E. A thin spherical shell has a radius of 1.90 m. An applied torque of 960 N·m imparts to the shell an angular acceleration equal to 6.20 rad/s² about an axis through the center of the shell. (a) What is the rotational inertia of the shell about the axis of rotation? (b) Calculate the mass of the shell.

70P. Figure 11-41 shows the massive shield door at a neutron test facility at Lawrence Livermore Laboratory; this is the world's heaviest hinged door. The door has a mass of 44,000 kg, a rotational inertia about an axis through its hinges of 8.7×10^4 kg·m², and a (front) face width of 2.4 m. Neglecting friction, what steady force, applied at its outer edge and perpendicular to the plane of the door, can move it from rest through an angle of 90° in 30 s? Assume no friction acts on the hinges.

FIGURE 11-41   Problem 70.

71P. A pulley, with a rotational inertia of 1.0×10^{-3} kg·m² about its axle and a radius of 10 cm, is acted on by a force applied tangentially at its rim. The force magnitude varies in time as $F = 0.50t + 0.30t^2$, where F is in newtons if t is given in seconds. The pulley is initially at rest. At $t = 3.0$ s what are (a) its angular acceleration and (b) its angular velocity?

72P. A wheel of radius 0.20 m is mounted on a frictionless horizontal axis. A massless cord is wrapped around the wheel and attached to a 2.0 kg object that slides on a frictionless surface inclined at an angle of 20° with the horizontal, as shown in Fig. 11-42. The object accelerates down the incline at 2.0 m/s². What is the rotational inertia of the wheel about its axis of rotation?

FIGURE 11-42   Problem 72.

73P. Two uniform solid spheres have the same mass, 1.65 kg, but one has a radius of 0.226 m while the other has a radius of 0.854 m. (a) For each of the spheres, find the torque required to bring the sphere from rest to an angular velocity of 317 rad/s in 15.5 s. Each sphere rotates about an axis through its center. (b) For each sphere, what force applied tangentially at the equator would provide the needed torque?

74P. In an Atwood's machine (Fig. 5-23), one block has a mass of 500 g, and the other a mass of 460 g. The pulley, which is mounted in horizontal frictionless bearings, has a radius of 5.00 cm. When released from rest, the heavier block is observed to fall 75.0 cm in 5.00 s (without the cord slipping on the pulley). (a) What is the acceleration of each block? What is the tension in the part of the cord that supports (b) the heavier block and (c) the lighter block? (d) What is the angular acceleration of the pulley? (e) What is its rotational inertia?

75P. Figure 11-43 shows two blocks, each of mass m, suspended from the ends of a rigid weightless rod of length $l_1 + l_2$, with $l_1 = 20$ cm and $l_2 = 80$ cm. The rod is held horizontally on the fulcrum shown in the figure and then released. Calculate the accelerations of the two blocks as they start to move.

FIGURE 11-43   Problem 75.

76P. Two identical blocks, each of mass M, are connected by a massless string over a pulley of radius R and rotational inertia I (Fig. 11-44). The string does not slip on the pulley; it is not known whether there is friction between the table and the sliding block; the pulley's axis is frictionless. When this system is released, it is found that the pulley turns through an angle θ in time t

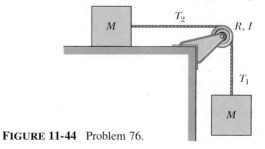

FIGURE 11-44   Problem 76.

and the acceleration of the blocks is constant. (a) What is the angular acceleration of the pulley? (b) What is the acceleration of the two blocks? (c) What are the tensions in the upper and lower sections of the string? Express the answers in terms of M, I, R, θ, g, and t.

SECTION 11-10 Work and Rotational Kinetic Energy

77E. (a) If $R = 12$ cm, $M = 400$ g, and $m = 50$ g in Fig. 11-18, find the speed of the block after it has descended 50 cm starting from rest. Solve the problem using energy conservation principles. (b) Repeat (a) with $R = 5.0$ cm.

78E. An automobile crankshaft develops 100 hp ($= 74.6$ kW) when rotating at a speed of 1800 rev/min. What torque does the crankshaft deliver?

79E. A 32.0 kg wheel, essentially a thin hoop with radius 1.20 m, is rotating at 280 rev/min. It must be brought to a stop in 15.0 s. (a) How much work must be done to stop it? (b) What is the required power?

80E. A thin rod of length l and mass m is suspended freely from one end. It is pulled to one side and then allowed to swing like a pendulum, passing through its lowest position with an angular speed ω. (a) Calculate its kinetic energy as it passes through its lowest position. (b) How high does its center of mass rise above its lowest position? Neglect friction and air resistance.

81P. Calculate (a) the torque, (b) the energy, and (c) the average power required to accelerate the Earth in 1 day from rest to its present angular speed about its axis.

82P. A meter stick is held vertically with one end on the floor and is then allowed to fall. Find the speed of the other end when it hits the floor, assuming that the end on the floor does not slip. (*Hint:* Consider the stick to be a thin rod and use the conservation of energy.)

83P. A rigid body is made of three identical thin rods, each with length l, fastened together in the form of a letter H (Fig. 11-45). The body is free to rotate about a horizontal axis that runs along the length of one of the legs of the H. The body is allowed to fall from rest from a position in which the plane of the H is horizontal. What is the angular speed of the body when the plane of the H is vertical?

FIGURE 11-45 Problem 83.

84P. A uniform cylinder of radius 10 cm and mass 20 kg is mounted so as to rotate freely about a horizontal axis that is parallel to and 5.0 cm from the central longitudinal axis of the cylinder. (a) What is the rotational inertia of the cylinder about the axis of rotation? (b) If the cylinder is released from rest with its central longitudinal axis at the same height as the axis about which the cylinder rotates, what is the angular speed of the cylinder as it passes through its lowest position? (*Hint:* Use the principle of conservation of energy.)

85P. A uniform helicopter rotor blade (see Fig. 11-36) is 7.80 m

long and has a mass of 110 kg. (a) What force is exerted on the bolt attaching the blade to the rotor axle when the rotor is turning at 320 rev/min? (*Hint:* For this calculation the blade can be considered to be a point mass at the center of mass. Why?) (b) Calculate the torque that must be applied to the rotor to bring it to full speed from rest in 6.7 s. Ignore air resistance. (The blade cannot be considered to be a point mass for this calculation. Why not? Assume the mass distribution of a uniform, thin rod.) (c) How much work did the torque do on the blade in order for the blade to reach a speed of 320 rev/min?

86P. A uniform spherical shell of mass M and radius R rotates about a vertical axis on frictionless bearings (Fig. 11-46). A massless cord passes around the equator of the shell, over a pulley of rotational inertia I and radius r, and is attached to a small object of mass m. There is no friction on the pulley's axle; the cord does not slip on the pulley. What is the speed of the object after it has fallen a distance h from rest? Use work–energy considerations.

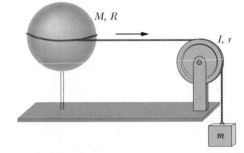

FIGURE 11-46 Problem 86.

87P. A tall, cylinder-shaped chimney falls over when its base is ruptured. Treating the chimney as a thin rod with height h, express the (a) radial and (b) tangential components of the linear acceleration of the top of the chimney as a function of the angle θ made by the chimney with the vertical. (c) At what angle θ does the linear acceleration equal g?

88P. Attached to each end of a thin steel rod of length 1.20 m and mass 6.40 kg is a small ball of mass 1.06 kg. The rod is constrained to rotate in a horizontal plane about a vertical axis through its midpoint. At a certain instant, it is observed to be rotating with an angular velocity of 39.0 rev/s. Because of friction, it comes to rest 32.0 s later. Assuming a constant frictional torque, compute (a) the angular acceleration, (b) the retarding torque exerted by the friction, (c) the total mechanical energy dissipated by the friction, and (d) the number of revolutions executed during the 32.0 s. (e) Now suppose that the frictional torque is known not to be constant. Which, if any, of the quantities (a), (b), (c), or (d) can still be computed without additional information? If such a quantity exists, give its value.

89P. A uniform rod of mass 1.5 kg is 2.0 m long (Fig. 11-47). The rod can pivot about a horizontal, frictionless pin through one end. It is released from rest at an angle of 40° above the horizontal. (a) What is the angular acceleration of the rod at the instant it is released? (The rotational inertia of the rod about the pin is 2.0 kg·m².) (b) Use the principle of conservation of energy to determine the angular speed of the rod as it passes through the horizontal position.

FIGURE 11-47 Problem 89.

90P. The rigid body shown in Fig. 11-48 consists of three particles connected by massless rods. It is to be rotated about an axis perpendicular to its plane through point P. If $M = 0.40$ kg, $a = 30$ cm, and $b = 50$ cm, how much work is required to take the body from rest to an angular speed of 5.0 rad/s?

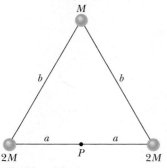

FIGURE 11-48 Problem 90.

91P*. A car is fitted with an energy-conserving flywheel, which in operation is geared to the driveshaft so that the flywheel rotates at 240 rev/s when the car is traveling at 80 km/h. The total mass of the car is 800 kg; the flywheel weighs 200 N and is a uniform disk 1.1 m in diameter. The car descends a 1500-m-long, 5° slope, starting from rest, with the flywheel engaged and no power generated from the engine. Neglecting friction and the rotational inertia of the wheels, find (a) the speed of the car at the bottom of the slope, (b) the angular acceleration of the flywheel at the bottom of the slope, and (c) the rate at which energy is stored in the rotation of the flywheel by the driveshaft as the car reaches the bottom of the slope.

ELECTRONIC COMPUTATION

92. In a certain machine two disks are free to rotate independently on parallel axles. The first has a radius of 7.0 cm and a rotational inertia of 1.5 kg·m². The second has a radius of 15 cm and a rotational inertia of 3.5 kg·m². While the first disk is at rest and the second is rotating at 175 rad/s, the axles are moved closer together so that the rims of the disks touch. Each disk exerts a normal force of 150 N on the other. (a) Take the coefficient of kinetic friction between the rims to be 0.25 and generate a table listing the values of the angular velocities for every 2 s during the first 80 s after the disks touch. Plot these values as functions of time. Also plot the total rotational kinetic energy of the disks. Is it conserved? (b) What influence does the frictional force have on the motions of the disks? Repeat the calculations for a coefficient of kinetic friction of 0.50. Is the time the disks take to reach their final angular velocities for the first μ_k value the same as for the second value? Are the final angular velocities the same for the two values? Is the final rotational kinetic energy the same?

93. Five particle-like objects, positioned in the xy plane according to the following table, form a rigidly connected body. What is the rotational inertia of the body about (a) the x axis, (b) the y axis, and (c) the z axis? (d) What is the center of mass of the body?

Object	1	2	3	4	5
Mass (grams)	500	400	300	600	450
x (cm)	15	-13	17	-4.0	-5.0
y (cm)	20	13	-6.0	-7.0	9.0

94. In Fig. 11-49, two blocks, of mass $m_1 = 400$ g and $m_2 = 600$ g, are connected by a massless cord that is wrapped around a uniform disk of mass $M = 500$ g and radius $R = 12.0$ cm. The disk can rotate without friction about a fixed horizontal axis through its center; the cord cannot slip on the disk. The system is released from rest. Find (a) the magnitude of the acceleration of the blocks, (b) the tension T_1 in the cord at the left, and (c) the tension T_2 in the cord at the right.

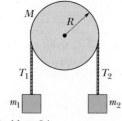

FIGURE 11-49 Problem 94.

12
Rolling, Torque, and Angular Momentum

In 1897, a European "aerialist" made the first triple somersault during the flight from a swinging trapeze to the hands of a partner. For the next 85 years aerialists attempted to complete a <u>quadruple</u> somersault, but not until 1982 was it done before an audience: Miguel Vazquez of the Ringling Bros. and Barnum & Bailey Circus rotated his body in four complete circles in midair before his brother Juan caught him. Both were stunned by their success. Why was the feat so difficult, and what feature of physics made it (finally) possible?

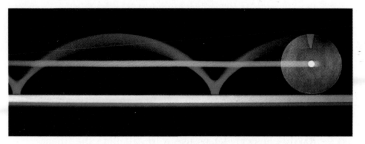

FIGURE 12-1 A time exposure photograph of a rolling disk. Small lights have been attached to the disk, one at its center and one at its edge. The latter traces out a curve called a *cycloid*.

12-1 ROLLING

When a bicycle moves along a straight track, the center of each wheel moves forward in pure translation. A point on the rim of the wheel, however, traces out a more complex path, as Fig. 12-1 shows. In what follows, we analyze the motion of a rolling wheel first by viewing it as a combination of pure translation and pure rotation, and then by viewing it as rotation alone.

Rolling as Rotation and Translation Combined

Imagine that you are watching the wheel of a bicycle, which passes you at constant speed while rolling smoothly, without slipping, along a street. As shown in Fig. 12-2, the center of mass O of the wheel moves forward at constant speed v_{cm}. The point P at which the wheel contacts the street also moves forward at speed v_{cm}, so that it always remains directly below O.

During a time interval t, you see both O and P move forward by a distance s. The bicycle rider sees the wheel rotate through an angle θ about the center of the wheel, with the point of the wheel that was touching the street at the beginning of t moving through arc length s. Equation 11-15 relates the arc length s to the rotation angle θ:

$$s = R\theta, \tag{12-1}$$

where R is the radius of the wheel. The linear speed v_{cm} of the center of the wheel (the center of mass of this uniform wheel) is ds/dt, and the angular speed ω of the wheel about its center is $d\theta/dt$. So, differentiating Eq. 12-1 with respect to time gives us

$$v_{cm} = \omega R \quad \text{(rolling motion).} \tag{12-2}$$

(Note that Eq. 12-2 holds only if the wheel rolls *smoothly*; that is, it does not slip over the street.)

Figure 12-3 shows that the rolling motion of a wheel is a combination of purely translational and purely rotational motions. Figure 12-3a shows the purely rotational motion (as if the rotation axis through the center were stationary): every point on the wheel rotates about the center with angular speed ω. (This is the type of motion we considered in Chapter 11.) Every point on the outside edge of the wheel has linear speed v_{cm} given by Eq. 12-2. Figure 12-3b shows the purely translational motion (as if the wheel did not rotate at all): every point on the wheel moves to the right with speed v_{cm}.

The combination of Figs. 12-3a and 12-3b yields the actual rolling motion of the wheel, Fig. 12-3c. Note that in this combination of motions, the portion of the wheel at the bottom (at point P) is stationary and the portion at the top (at point T) is moving at speed $2v_{cm}$, faster than any other portion of the wheel. These results are demonstrated in Fig. 12-4, which is a time exposure of a rolling bicycle wheel. The blurring of the spokes near the top of the wheel com-

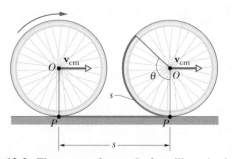

FIGURE 12-2 The center of mass O of a rolling wheel moves a distance s at velocity $\mathbf{v}_{cm}$, while the wheel rotates through angle θ. The contact point P between the wheel and the surface over which the wheel rolls also moves a distance s.

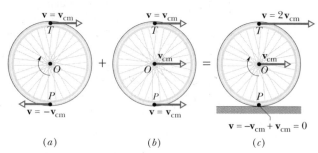

FIGURE 12-3 Rolling motion of a wheel as a combination of purely rotational motion and purely translational motion. (a) The purely rotational motion: all points on the wheel move with the same angular speed ω. Points on the outside edge of the wheel all move with the same linear speed $v = v_{cm}$. The linear velocities $\mathbf{v}$ of two such points, at top (T) and bottom (P) of the wheel, are shown. (b) The purely translational motion: all points on the wheel move to the right with the same linear velocity $\mathbf{v}_{cm}$ as the center of the wheel. (c) The rolling motion of the wheel is the combination of (a) and (b).

FIGURE 12-4 A photograph of a rolling bicycle wheel. The spokes near the top of the wheel are more blurred than those near the bottom of the wheel because they are moving faster, as Fig. 12-3c shows.

pared with the sharper images of the spokes near the bottom of the wheel shows that the wheel is moving faster near its top than near its bottom.

The motion of any round body rolling smoothly over a surface can be separated into purely rotational and purely translational motions, as in Figs. 12-3a and 12-3b.

Rolling as Pure Rotation

Figure 12-5 suggests another way to look at the rolling motion of a wheel, namely, as pure rotation about an axis that always extends through the point where the wheel contacts the street as the wheel moves. That is, we consider the rolling motion to be pure rotation about an axis passing through point P in Fig. 12-3c and perpendicular to the

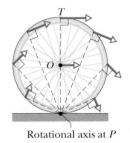

Rotational axis at P

FIGURE 12-5 Rolling can be viewed as pure rotation, with angular speed ω, about an axis that always extends through P. The vectors show the instantaneous linear velocities of selected points on the rolling wheel. You can obtain the vectors by combining the translational and rotational motions as in Fig. 12-3.

plane of the figure. The vectors in Fig. 12-5 then represent the instantaneous velocities of points on the rolling wheel.

Question. What angular speed about this new axis will a stationary observer assign to a rolling bicycle wheel? **Answer.** The same angular speed ω that the rider assigns to the wheel as she or he observes it in pure rotation about an axis through its center of mass.

To verify this answer, let us use it to calculate the linear speed of the top of the rolling wheel from the point of view of a stationary observer. If we call the wheel's radius R, the top is a distance 2R from the axis through P in Fig. 12-5, so the linear speed at the top should be (using Eq. 12-2)

$$v_{\text{top}} = (\omega)(2R) = 2(\omega R) = 2v_{\text{cm}},$$

in exact agreement with Fig. 12-3c. You can similarly verify the linear speeds shown for the portion of the wheel at points O and P in Fig. 12-3c.

CHECKPOINT **1:** The rear wheel on a clown's bicycle has twice the radius of the front wheel. (a) Is the linear speed at the very top of the rear wheel greater than, less than, or the same as that of the front wheel? (b) Is the angular speed of the rear wheel greater than, less than, or the same as that of the front wheel?

The Kinetic Energy

Let us now calculate the kinetic energy of the rolling wheel as measured by the stationary observer. If we view the

As this hoop rolls, does the dog walk as fast as the pony?

rolling as pure rotation about an axis through P in Fig. 12-5, we have

$$K = \tfrac{1}{2}I_P\omega^2, \qquad (12\text{-}3)$$

in which ω is the angular speed of the wheel and I_P is the rotational inertia of the wheel about the axis through P. From the parallel-axis theorem (Eq. 11-27) we have

$$I_P = I_{cm} + MR^2, \qquad (12\text{-}4)$$

in which M is the mass of the wheel and I_{cm} is its rotational inertia about an axis through its center of mass. Substituting Eq. 12-4 into Eq. 12-3, we obtain

$$K = \tfrac{1}{2}I_{cm}\omega^2 + \tfrac{1}{2}MR^2\omega^2,$$

and using the relation $v_{cm} = \omega R$ (Eq. 12-2) yields

$$K = \tfrac{1}{2}I_{cm}\omega^2 + \tfrac{1}{2}Mv_{cm}^2. \qquad (12\text{-}5)$$

We can interpret the first of these terms ($\tfrac{1}{2}I_{cm}\omega^2$) as the kinetic energy associated with the rotation of the wheel about an axis through its center of mass (Fig. 12-3a), and the second term ($\tfrac{1}{2}Mv_{cm}^2$) as the kinetic energy associated with the translational motion of the wheel (Fig. 12-3b).

Friction and Rolling

If the wheel rolls at constant speed, as in Fig. 12-2, it has no tendency to slide at the point of contact P, and thus there is no frictional force acting on the wheel there. However, if a force acts on the wheel, changing the speed v_{cm} of the center of the wheel or the angular speed ω about the center, then there is a tendency for the wheel to slide at P, and a frictional force acts on the wheel at P to oppose that tendency. Until the wheel actually begins to slide, the frictional force is a *static* frictional force $\mathbf{f}_s$. If the wheel begins to slide, then the force is a *kinetic* frictional force $\mathbf{f}_k$.

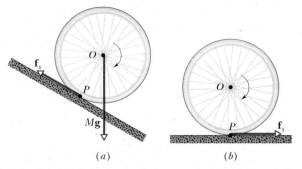

FIGURE 12-6 (a) A wheel rolls down an incline without sliding. A static frictional force $\mathbf{f}_s$ acts on the wheel at P, opposing the wheel's tendency to slide because of its weight $M\mathbf{g}$. (b) A wheel rolls horizontally without sliding while its angular speed is increased. A static frictional force $\mathbf{f}_s$ acts on the wheel at P, opposing the tendency to slide. If in (a) or (b) the wheel slides, the frictional force is a kinetic frictional force $\mathbf{f}_k$.

In Fig. 12-6a, a wheel rolls down an incline. Its weight $M\mathbf{g}$ acts on the wheel at the wheel's center of mass. Since that force $M\mathbf{g}$ lacks any moment arm relative to the center of the wheel, it cannot produce a torque about the center and cause the wheel to rotate about the center. But since $M\mathbf{g}$ tends to slide the wheel down the incline, a frictional force acts on the wheel at contact point P to oppose the tendency to slide. This force, which points up along the incline, *does* have a moment arm relative to the center; the moment arm is the radius of the wheel. Thus this frictional force produces a torque about the center and causes the wheel to rotate about the center.

In Fig. 12-6b, a wheel is being made to rotate faster while rolling along a flat surface, as on an accelerating bicycle. The increase in ω tends to slide the bottom of the wheel toward the left. A frictional force acts on the wheel toward the right at P to oppose the tendency to slide. (This frictional force is the external force acting on an accelerating bicycle–rider system that causes the acceleration.)

SAMPLE PROBLEM 12-1

A uniform solid cylindrical disk, whose mass M is 1.4 kg and whose radius R is 8.5 cm, rolls across a horizontal table at a speed v of 15 cm/s.

(a) What is the speed of the top of the rolling disk?

SOLUTION: When we speak of the speed of a rolling object, we always mean the speed of its center of mass. From Fig. 12-3c we see that the speed of the top of the disk is just twice this, or

$$v_{top} = 2v_{cm} = (2)(15 \text{ cm/s}) = 30 \text{ cm/s.} \quad \text{(Answer)}$$

(b) What is the angular speed ω of the rolling disk?

SOLUTION: From Eq. 12-2 we have

$$\omega = \frac{v_{cm}}{R} = \frac{15 \text{ cm/s}}{8.5 \text{ cm}}$$

$$= 1.8 \text{ rad/s} = 0.28 \text{ rev/s.} \quad \text{(Answer)}$$

This value applies whether the axis of rotation is taken to be an axis through point P in Fig. 12-5 or an axis through the center of mass.

(c) What is the kinetic energy K of the rolling disk?

SOLUTION: From Eq. 12-5 we have, putting $I_{cm} = \tfrac{1}{2}MR^2$ and using the relation $v_{cm} = \omega R$,

$$K = \tfrac{1}{2}I_{cm}\omega^2 + \tfrac{1}{2}Mv_{cm}^2$$

$$= (\tfrac{1}{2})(\tfrac{1}{2}MR^2)(v_{cm}/R)^2 + \tfrac{1}{2}Mv_{cm}^2 = \tfrac{3}{4}Mv_{cm}^2$$

$$= \tfrac{3}{4}(1.4 \text{ kg})(0.15 \text{ m/s})^2$$

$$= 0.024 \text{ J} = 24 \text{ mJ.} \quad \text{(Answer)}$$

TABLE 12-1 THE RELATIVE SPLITS BETWEEN ROTATIONAL AND TRANSLATIONAL ENERGIES FOR ROLLING OBJECTS

OBJECT	ROTATIONAL INERTIA I_{cm}	PERCENTAGE OF ENERGY IN	
		TRANSLATION	ROTATION
Hoop	$1MR^2$	50%	50%
Disk	$\frac{1}{2}MR^2$	67%	33%
Sphere	$\frac{2}{5}MR^2$	71%	29%
General^a	βMR^2	$100\dfrac{1}{1+\beta}$ %	$100\dfrac{\beta}{1+\beta}$ %

$^a\beta$ may be computed for any rolling object as I_{cm}/MR^2.

(d) What fraction of the kinetic energy is associated with the motion of translation and what fraction with the motion of rotation about an axis through the center of mass?

SOLUTION: The kinetic energy associated with translation is the second term of Eq. 12-5, or $\frac{1}{2}Mv_{cm}^2$. The fraction we seek is then, using the expression derived in (c),

$$\text{frac} = \frac{\frac{1}{2}Mv_{cm}^2}{\frac{3}{4}Mv_{cm}^2} = \frac{2}{3} \quad \text{or} \quad 67\%. \quad \text{(Answer)}$$

The remaining 33% is associated with rotation about an axis through the center of mass.

   The relative split between translational and rotational energy depends on the rotational inertia of the rolling object. As Table 12-1 shows, the rolling object (the hoop) that has its mass farthest from the central axis of rotation (and so has the largest rotational inertia) has the largest share of its kinetic energy in rotational motion. The object (the sphere) that has its mass closest to the central axis of rotation (and so has the smallest rotational inertia) has the smallest share in that form.

   The formulas at the bottom of Table 12-1 apply to the generic rolling object with **rotational inertia parameter** β. This parameter has the value 1 for a hoop, $\frac{1}{2}$ for a disk, and $\frac{2}{5}$ for a sphere.

SAMPLE PROBLEM 12-2

A bowling ball, whose radius R is 11 cm and whose mass M is 7.2 kg, rolls from rest down a ramp whose length L is 2.1 m. The ramp is inclined at an angle θ of 34° to the horizontal; see the sphere in Fig. 12-7. How fast is the ball moving when it reaches the bottom of the ramp? Assume the ball is uniform in density.

SOLUTION: The center of the ball falls a vertical distance $h = L \sin \theta$; so the decrease in gravitational potential energy is $MgL \sin \theta$. This loss of potential energy equals the gain in kinetic energy. Thus we can write (see Eq. 12-5)

$$MgL \sin \theta = \tfrac{1}{2}I_{cm}\omega^2 + \tfrac{1}{2}Mv_{cm}^2. \quad (12\text{-}6)$$

From Table 11-2(g) we see that, for a solid sphere, $I_{cm} = \frac{2}{5}MR^2$. We can also replace ω with its equal, v_{cm}/R. Substituting both these quantities into Eq. 12-6 yields

$$MgL \sin \theta = (\tfrac{1}{2})(\tfrac{2}{5})(MR^2)(v_{cm}/R)^2 + \tfrac{1}{2}Mv_{cm}^2.$$

Solving for v_{cm} yields

$$v_{cm} = \sqrt{(\tfrac{10}{7})\, gL \sin \theta}$$
$$= \sqrt{(\tfrac{10}{7})(9.8 \text{ m/s}^2)(2.1 \text{ m})(\sin 34°)}$$
$$= 4.1 \text{ m/s.} \quad \text{(Answer)}$$

Note that the answer does not depend on the mass or radius of the ball.

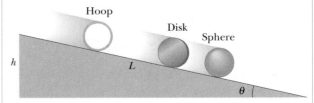
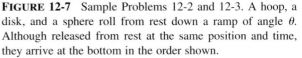

FIGURE 12-7 Sample Problems 12-2 and 12-3. A hoop, a disk, and a sphere roll from rest down a ramp of angle θ. Although released from rest at the same position and time, they arrive at the bottom in the order shown.

SAMPLE PROBLEM 12-3

Here we generalize the result of Sample Problem 12-2. A uniform hoop, disk, and sphere, having the same mass M and the same radius R, are released simultaneously from rest at the top of a ramp whose length L is 2.5 m and whose ramp angle θ is 12° (Fig. 12-7).

(a) Which object reaches the bottom first?

SOLUTION: Table 12-1 gives us the answer. The sphere puts the largest share of its kinetic energy (71%) into translational motion, so it wins the race. Next comes the disk and then the hoop.

(b) How fast are the objects moving at the ramp's bottom?

SOLUTION: The center of mass of each object rolling down the ramp falls the same vertical distance h. Like a body in free fall, the object loses potential energy in amount Mgh and thus gains this amount of kinetic energy. At the bottom of the ramp then, the total kinetic energies of all three objects are the same. How these kinetic energies are divided between the translational and rotational forms depends on each object's distribution of mass.

   From Eq. 12-5 we can write (putting $\omega = v_{cm}/R$)

$$Mgh = \tfrac{1}{2}I_{cm}\omega^2 + \tfrac{1}{2}Mv_{cm}^2$$
$$= \tfrac{1}{2}I_{cm}(v_{cm}^2/R^2) + \tfrac{1}{2}Mv_{cm}^2$$
$$= \tfrac{1}{2}(I_{cm}/R^2)v_{cm}^2 + \tfrac{1}{2}Mv_{cm}^2. \quad (12\text{-}7)$$

Putting $h = L \sin \theta$ and solving for v_{cm}, we obtain

$$v_{cm} = \sqrt{\frac{2gL \sin \theta}{1 + I_{cm}/MR^2}}, \quad \text{(Answer)} \quad (12\text{-}8)$$

which is the symbolic answer to the question.

Note that the speed depends not on the mass or the radius of the rolling object, but only on the distribution of its mass about its central axis, which enters through the term I_{cm}/MR^2. A marble and a bowling ball will have the same speed at the bottom of the ramp and will thus roll down the ramp in the same time. A bowling ball will beat a disk of any mass or radius, and almost anything that rolls will beat a hoop.

For the rolling hoop (see the hoop listing in Table 12-1) we have $I_{cm}/MR^2 = 1$, so Eq. 12-8 yields

$$\begin{aligned} v_{cm} &= \sqrt{\frac{2gL \sin \theta}{1 + I_{cm}/MR^2}} \\ &= \sqrt{\frac{(2)(9.8 \text{ m/s}^2)(2.5 \text{ m})(\sin 12°)}{1 + 1}} \\ &= 2.3 \text{ m/s}. \quad \text{(Answer)} \end{aligned}$$

From a similar calculation, we obtain $v_{cm} = 2.6$ m/s for the disk ($I_{cm}/MR^2 = \frac{1}{2}$) and 2.7 m/s for the sphere ($I_{cm}/MR^2 = \frac{2}{5}$). This supports our prediction of (a) that the win, place, and show sequence in this race will be sphere, disk, and hoop.

SAMPLE PROBLEM 12-4

Figure 12-8 shows a round uniform body of mass M and radius R rolling down a ramp at an angle θ. This time let us analyze the motion directly from Newton's laws, rather than by energy methods as we did in Sample Problem 12-3.

(a) What is the linear acceleration of the rolling body?

SOLUTION: Figure 12-8 also shows the forces that act on the body: its weight $M\mathbf{g}$, a normal force $\mathbf{N}$, and a static frictional force $\mathbf{f}_s$. The weight can be considered to act at the center of mass, which is at the center of this uniform body. The normal force and the frictional force act on the portion of the body in contact with the ramp at point P. The weight and normal force have zero moment arms about an axis through the center of the body. So they cannot cause the body to rotate about that center. Clockwise rotation of the body results from a negative torque due to the frictional force; that force has a moment arm of R about the center of the body.

We now apply the linear form of Newton's second law ($\Sigma F = Ma$) along the ramp, taking the positive direction to be up the ramp. We obtain

$$\sum F = f_s - Mg \sin \theta = Ma. \quad (12\text{-}9)$$

This equation has two unknowns, f_s and a. To obtain another equation in the same two unknowns, we next apply the angular form of Newton's second law ($\Sigma \tau = I\alpha$) about the rotation axis through the center of mass. (Although we derived the

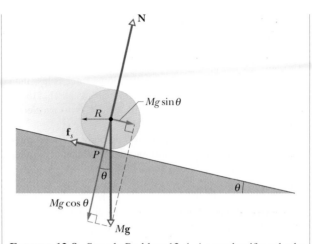

FIGURE 12-8 Sample Problem 12-4. A round uniform body of radius R rolls down a ramp. The forces that act on it are its weight $M\mathbf{g}$, a normal force $\mathbf{N}$, and a frictional force $\mathbf{f}_s$ pointing up the ramp. (For clarity, $\mathbf{N}$ has been shifted along its line of action until its tail is at the center of the body.)

relation $\Sigma \tau = I\alpha$ in Chapter 11 for an axis fixed in an inertial frame, it holds also for a rotation axis through the center of mass of this or another accelerating body, provided the axis does not change direction.) We get

$$\sum \tau = -f_s R = I_{cm}\alpha = \frac{I_{cm}a}{R}, \quad (12\text{-}10)$$

where we have used the relation $\alpha = a/R$ (Eq. 11-20).

Solving Eq. 12-10 for the frictional force f_s gives

$$f_s = -\frac{I_{cm}a}{R^2}, \quad (12\text{-}11)$$

where the minus sign reminds us that the frictional force $\mathbf{f}_s$ acts in the direction opposite that of the acceleration $\mathbf{a}$. Substituting Eq. 12-11 into Eq. 12-9 and solving for a, we have

$$a = -\frac{g \sin \theta}{1 + I_{cm}/MR^2}. \quad \text{(Answer)} \quad (12\text{-}12)$$

We could have, instead, found a second equation by summing torques and applying Newton's law in angular form about an axis through the *point of contact P*. This time, $\Sigma \tau$ would consist only of the torque due to the force component $Mg \sin \theta$, acting at the center of the body with moment arm R:

$$\sum \tau = -(Mg \sin \theta)(R) = I_P\alpha = \frac{I_P a}{R}, \quad (12\text{-}13)$$

where I_P is the rotational inertia about an axis through P. To find I_P, we would use the parallel-axis theorem:

$$I_P = I_{cm} + MR^2. \quad (12\text{-}14)$$

Substituting I_P from Eq. 12-14 in Eq. 12-13 and solving for a, we would again obtain Eq. 12-12.

(b) What is the frictional force f_s?

SOLUTION: Substituting Eq. 12-12 into Eq. 12-11 yields

$$f_s = Mg \frac{\sin \theta}{1 + MR^2/I_{cm}}. \quad \text{(Answer)} \quad (12\text{-}15)$$

Study of Eq. 12-15 shows that the frictional force is less than $Mg \sin \theta$, the component of the weight that acts parallel to the ramp. This is necessarily true if the object is to accelerate down the ramp.

Table 12-1 shows that if the rolling body is a solid disk, $I_{cm}/MR^2 = \frac{1}{2}$. The acceleration and the frictional force then follow from Eqs. 12-12 and 12-15 and are

$$a = -\tfrac{2}{3} g \sin \theta \quad \text{and} \quad f_s = \tfrac{1}{3} Mg \sin \theta.$$

(c) What is the speed of the rolling body at the bottom of the ramp if the ramp has length L?

SOLUTION: The motion is one of constant acceleration, so we can use the relation

$$v^2 = v_0^2 + 2a(x - x_0). \quad (12\text{-}16)$$

Putting $x - x_0 = -L$ and $v_0 = 0$ and introducing a from Eq. 12-12, we obtain Eq. 12-8—the result that we derived by energy methods.

CHECKPOINT 2: Disks A and B are identical and roll across a floor with equal speeds. Then disk A rolls up an incline, reaching a maximum height h, and disk B moves up an incline that is identical except that it is frictionless. Is the maximum height reached by disk B greater than, less than, or equal to h?

12-2 THE YO-YO

A yo-yo is a physics lab that you can fit in your pocket. If a yo-yo rolls down its string for a distance h, it loses potential energy in amount mgh but gains kinetic energy in both translational ($\frac{1}{2}mv_{cm}^2$) and rotational ($\frac{1}{2}I_{cm}\omega^2$) forms. When it is climbing back up, it loses kinetic energy and regains potential energy.

In a modern yo-yo the string is not tied to the axle but is looped around it. When the yo-yo "hits" the bottom of its string, an upward force on the axle from the string stops the descent, removing the small remaining translational kinetic energy. The yo-yo then spins, axle inside loop, with only rotational kinetic energy. The yo-yo keeps spinning ("sleeping") until you "wake it" by jerking on the string, causing the string to catch on the axle and the yo-yo to climb back up. The rotational kinetic energy of the yo-yo at the bottom of its string (and thus the sleeping time) can be considerably increased by throwing the yo-yo downward so that it starts down the string with initial speeds v_{cm} and ω instead of rolling down from rest.

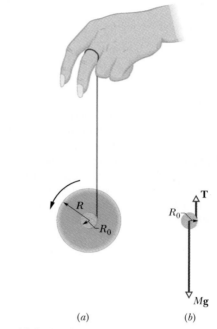

FIGURE 12-9 (a) A yo-yo, shown in cross section. The string, of assumed negligible thickness, is wound around an axle of radius R_0. (b) A free-body diagram for the falling yo-yo. Only the axle is shown. In a real yo-yo, the thickness of the string cannot be neglected. It changes the effective radius of the yo-yo axle, which then varies with the amount of wound-up string.

Let us analyze the motion of the yo-yo directly with Newton's second law. Figure 12-9a shows an idealized yo-yo, in which the thickness of the string can be neglected. Figure 12-9b is a free-body diagram, in which only the yo-yo axle is shown. We seek an expression for the linear acceleration a of the yo-yo. Applying Newton's second law in its linear form ($\Sigma F = ma$) yields

$$\Sigma F = T - Mg = Ma. \quad (12\text{-}17)$$

Here M is the mass of the yo-yo, and T is the tension in the yo-yo's string.

Applying Newton's second law in angular form ($\Sigma \tau = I\alpha$) about an axis through the center of mass yields

$$\Sigma \tau = TR_0 = I\alpha, \quad (12\text{-}18)$$

where R_0 is the radius of the yo-yo axle and I is the rotational inertia of the yo-yo about its central axis. The linear acceleration a of the yo-yo is downward (and thus *negative*). From the perspective of Fig. 12-9, the angular acceleration α of the yo-yo is counterclockwise (and thus *positive*) because from that perspective the torque given by Eq. 12-18 is counterclockwise. So we relate α to a with the relation $a = -\alpha R_0$. Solving this relation for α ($= -a/R_0$) and substituting into Eq. 12-18, we find

$$TR_0 = -\frac{Ia}{R_0}.$$

After eliminating T between this and Eq. 12-17, we solve for a, obtaining

$$a = -g \frac{1}{1 + I/MR_0^2}. \qquad (12\text{-}19)$$

Thus an ideal yo-yo rolls down its string with constant acceleration. For a small acceleration, you need a light yo-yo with a large rotational inertia and a small axle radius.

SAMPLE PROBLEM 12-5

A yo-yo is constructed of two brass disks whose thickness b is 8.5 mm and whose radius R is 3.5 cm, joined by a short axle whose radius R_0 is 3.2 mm.

(a) What is the yo-yo's rotational inertia about its central axis? Neglect the rotational inertia of the axle. The density ρ of brass is 8400 kg/m³.

SOLUTION: The rotational inertia I of a disk about its central axis is $\frac{1}{2}MR^2$. In this problem, we can treat the two disks together, as a single disk. We first find its mass from its density and its volume V:

$$M = V\rho = (2)(\pi R^2)(b)(\rho)$$
$$= (2)(\pi)(0.035 \text{ m})^2(0.0085 \text{ m})(8400 \text{ kg/m}^3)$$
$$= 0.550 \text{ kg}.$$

The rotational inertia is then

$$I = \tfrac{1}{2}MR^2 = (\tfrac{1}{2})(0.550 \text{ kg})(0.035 \text{ m})^2$$
$$= 3.4 \times 10^{-4} \text{ kg}\cdot\text{m}^2. \qquad \text{(Answer)}$$

(b) A string of length $l = 1.1$ m and of negligible thickness is wound on the axle. What is the linear acceleration of the yo-yo as it rolls down the string from rest?

SOLUTION: From Eq. 12-19,

$$a = -g \frac{1}{1 + I/MR_0^2}$$
$$= -\frac{9.8 \text{ m/s}^2}{1 + \dfrac{3.4 \times 10^{-4} \text{ kg}\cdot\text{m}^2}{(0.550 \text{ kg})(0.0032 \text{ m})^2}}$$
$$= -0.16 \text{ m/s}^2. \qquad \text{(Answer)}$$

The acceleration points downward and has this value whether the yo-yo is rolling down the string or climbing it.

Note that the quantity I/MR_0^2 in Eq. 12-19 is simply the rotational inertia parameter β that was introduced in Table 12-1. For this yo-yo, we have $\beta = 60$, a much greater value than that for any of the objects listed in that table. The acceleration of our yo-yo is small, corresponding to that of a hoop rolling down a 1.9° ramp.

(c) What is the tension in the string of the yo-yo?

SOLUTION: We can find this by substituting a from Eq. 12-19 into Eq. 12-17 and solving for T. We find

$$T = \frac{Mg}{1 + MR_0^2/I}, \qquad (12\text{-}20)$$

which tells us, as it must, that the tension in the string is smaller than the weight of the yo-yo. Numerically, we have

$$T = \frac{(0.550 \text{ kg})(9.8 \text{ m/s}^2)}{1 + (0.550 \text{ kg})(0.0032 \text{ m})^2/(3.4 \times 10^{-4} \text{ kg}\cdot\text{m}^2)}$$
$$= 5.3 \text{ N}. \qquad \text{(Answer)}$$

This is the tension whether the yo-yo is falling down or climbing up the string.

12-3 TORQUE REVISITED

In Chapter 11 we defined torque τ for a rigid body that can rotate around a fixed axis, with each particle in the body forced to move in a path that is a circle about that axis. We now expand the definition of torque to apply it to an individual particle that moves along any path relative to a fixed *point* (rather than a fixed axis). The path need no longer be a circle.

Figure 12-10a shows such a particle at point P in the xy plane. A single force $\mathbf{F}$ in that plane acts on the particle, and the particle's position relative to the origin O is given by position vector $\mathbf{r}$. The torque τ acting on the particle relative to the fixed point O is a vector quantity defined as

$$\boldsymbol{\tau} = \mathbf{r} \times \mathbf{F} \qquad \text{(torque defined)}. \qquad (12\text{-}21)$$

We can evaluate the vector (or cross) product in this definition of τ by using the rules for such products given in Section 3-7. To find the direction of τ, we slide the vector $\mathbf{F}$ (without changing its direction) until its tail is at origin O, so that the two vectors in the vector product are tail to tail as in Fig. 12-10b. We then use the right-hand rule for vector products in Fig. 3-20a, sweeping the fingers of the right hand from $\mathbf{r}$ (the first vector in the product) into $\mathbf{F}$ (the second vector). The outstretched right thumb then gives the direction of τ. In Fig. 12-10b, τ points along the positive direction of the z axis.

To find the magnitude of τ, we use the general result of Eq. 3-20 ($c = ab \sin \phi$), finding

$$\tau = rF \sin \phi, \qquad (12\text{-}22)$$

where ϕ is the angle between $\mathbf{r}$ and $\mathbf{F}$ when the vectors are tail to tail. From Fig. 12-10b, we see that Eq. 12-22 can be rewritten as

$$\tau = rF_\perp, \qquad (12\text{-}23)$$

where $F_\perp$ ($= F \sin \phi$) is the component of $\mathbf{F}$ perpendicular to $\mathbf{r}$. From Fig. 12-10c, we see that Eq. 12-22 can also be

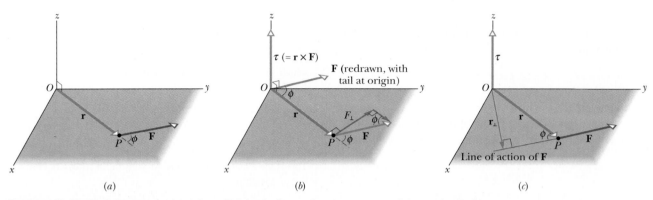

FIGURE 12-10 Defining torque. (*a*) A force **F**, lying in the *xy* plane, acts on a particle at point *P*. (*b*) This force produces a torque $\boldsymbol{\tau}$ ($= \mathbf{r} \times \mathbf{F}$) on the particle with respect to the origin *O*. By the right-hand rule for vector (cross) products, the torque vector points in the direction of increasing *z*. Its magnitude is given by $rF_\perp$ in (*b*) and by $r_\perp F$ in (*c*).

rewritten as

$$\tau = r_\perp F, \qquad (12\text{-}24)$$

where $r_\perp$ ($= r \sin \phi$) is the moment arm of **F** (the perpendicular distance between *O* and the line of action of **F**).

SAMPLE PROBLEM 12-6

In Fig. 12-11*a*, three forces, each of magnitude 2.0 N, act on a particle. The particle is in the *xz* plane at point *P* given by position vector **r**, where $r = 3.0$ m and $\theta = 30°$. Force $\mathbf{F}_1$ is parallel to the *x* axis, force $\mathbf{F}_2$ is parallel to the *z* axis, and force $\mathbf{F}_3$ is parallel to the *y* axis. What is the torque, with respect to the origin *O*, due to each force?

SOLUTION: Figures 12-11*b* and 12-11*c* are direct views of the *xz* plane, redrawn with vectors $\mathbf{F}_1$ and $\mathbf{F}_2$ shifted so their tails are at the origin, to better show the angles between those vectors and vector **r**. The angle between **r** and $\mathbf{F}_3$ is 90°. Applying Eq. 12-22 for each force, we find the magnitudes of the torques to be

$$\tau_1 = rF_1 \sin \phi_1 = (3.0 \text{ m})(2.0 \text{ N})(\sin 150°)$$
$$= 3.0 \text{ N} \cdot \text{m},$$

$$\tau_2 = rF_2 \sin \phi_2 = (3.0 \text{ m})(2.0 \text{ N})(\sin 120°)$$
$$= 5.2 \text{ N} \cdot \text{m},$$

and $\quad \tau_3 = rF_3 \sin \phi_3 = (3.0 \text{ m})(2.0 \text{ N})(\sin 90°)$
$$= 6.0 \text{ N} \cdot \text{m}. \qquad \text{(Answer)}$$

To find the directions of these torques, we use the right-hand rule, placing the fingers of the right hand so as to rotate **r** into **F** through the *smaller* of the two angles they make. In Fig. 12-11*d*, we represent $\mathbf{F}_3$ with a circled cross $\otimes$, suggesting the tail of an arrow. (Were it in the opposite direction, $\mathbf{F}_3$ would be represented by a circled dot $\odot$, suggesting the tip of an arrow.) Rotating **r** into $\mathbf{F}_3$ with the fingers of the right hand yields the direction of $\boldsymbol{\tau}_3$ (in the direction of the thumb). All three of the torque vectors are shown in Fig. 12-11*e*.

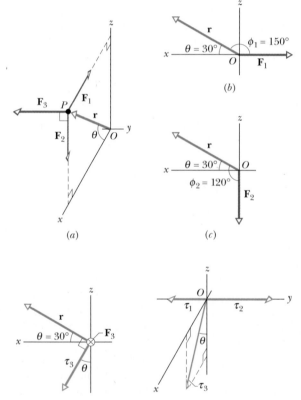

FIGURE 12-11 Sample Problem 12-6. (*a*) A particle at point *P* is acted on by three forces, each parallel to a coordinate axis. The angle ϕ (used in finding torque) is shown (*b*) for $\mathbf{F}_1$ and (*c*) for $\mathbf{F}_2$. (*d*) Torque $\boldsymbol{\tau}_3$ is perpendicular to both **r** and $\mathbf{F}_3$ (force $\mathbf{F}_3$ is directed into the plane of the figure). (*e*) The torques (relative to the origin *O*) acting on the particle.

PROBLEM SOLVING TACTICS

TACTIC 1: *Vector Products and Torques*
Equation 12-21 for torques is our first application of the vector

(or cross) product. You might want to review Section 3-7, where the rules for the vector prouct are given. In that section, Problem Solving Tactic 5 lists many common errors in finding the direction of the vector product.

Keep in mind that a torque is calculated *with respect to* (or *about*) a point, which must be known if the value of the torque is to be meaningful. Changing the point can change the torque in both magnitude and direction. For example, in Sample Problem 12-6, the torques due to the three forces are calculated about the origin O. You can show that the torques due to the same three forces are each zero if they are calculated about point P (at the position of the particle).

C HECKPOINT **3:** The position vector **r** of a particle points along the positive direction of the z axis. If the torque on the particle is (a) zero, (b) in the negative direction of x, and (c) in the negative direction of y, in what direction is the force producing the torque?

12-4 ANGULAR MOMENTUM

Like all other linear quantities, linear momentum has its angular counterpart. Figure 12-12 shows a particle with linear momentum **p** ($= m$**v**) located at point P in the xy plane. The **angular momentum** ℓ of this particle with respect to the origin O is a vector quantity defined as

$$\ell = \mathbf{r} \times \mathbf{p} = m(\mathbf{r} \times \mathbf{v}) \quad \text{(angular momentum defined),}$$
$$(12\text{-}25)$$

where **r** is the position vector of the particle with respect to O. As the particle moves relative to O, in the direction of its momentum **p** ($= m$**v**), position vector **r** rotates around O. Note carefully that to have angular momentum about O, the particle does *not* itself have to rotate around O. Comparison of Eqs. 12-21 and 12-25 shows that angular momentum bears the same relation to linear momentum that torque does to force. The SI unit of angular momentum is the kilogram-meter-squared per second (kg·m²/s), equivalent to the joule-second (J·s).

To find the direction of the angular momentum vector ℓ in Fig. 12-12, we slide the vector **p** until its tail is at origin O. Then we use the right-hand rule for vector products, sweeping the fingers from **r** into **p**. The outstretched thumb then shows that vector ℓ points in the positive direction of the z axis in Fig. 12-12. This positive direction is consistent with the counterclockwise rotation of the particle's position vector **r** about the z axis, as the particle continues to move. (A negative direction of ℓ would be consistent with a clockwise rotation of **r** about the z axis.)

To find the magnitude of ℓ, we use the general result of Eq. 3-20, to write

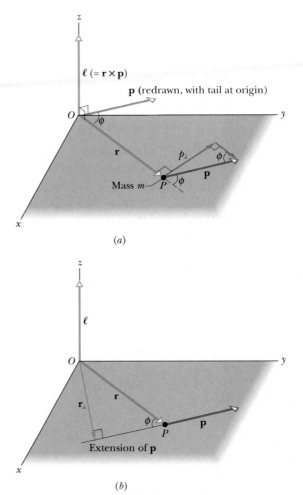

FIGURE 12-12 Defining angular momentum. A particle of mass m at point P has linear momentum **p** ($= m$**v**), assumed to lie in the xy plane. The particle has angular momentum ℓ ($= \mathbf{r} \times \mathbf{p}$) with respect to the origin O. By the right-hand rule, the angular momentum vector points in the direction of increasing z. (a) The magnitude of ℓ is given by $\ell = rp_\perp = rmv_\perp$. (b) The magnitude of ℓ is also given by $\ell = r_\perp p = r_\perp mv$.

$$\ell = rmv \sin \phi, \quad (12\text{-}26)$$

where ϕ is the angle between **r** and **p** when these two vectors are tail to tail. From Fig. 12-12a, we see that Eq. 12-26 can be rewritten as

$$\ell = rp_\perp = rmv_\perp, \quad (12\text{-}27)$$

where $p_\perp$ is the component of **p** perpendicular to **r** and $p_\perp = mv_\perp$. From Fig. 12-12b, we see that Eq. 12-26 can also be rewritten as

$$\ell = r_\perp p = r_\perp mv, \quad (12\text{-}28)$$

where $r_\perp$ is the perpendicular distance between O and an extension of **p**.

Just as is true for torque, angular momentum has

meaning only with respect to a specified origin. Moreover, if the particle in Fig. 12-12 did not lie in the xy plane, or if the linear momentum **p** of the particle did not also lie in that plane, the angular momentum ℓ would not be parallel to the z axis. The direction of the angular momentum vector is always perpendicular to the plane formed by the vectors **r** and **p**.

CHECKPOINT 4: In part a of the figure, particles 1 and 2 move around point O in opposite directions, in circles with radii 2 m and 4 m. In part b, particles 3 and 4 travel along straight lines, in the same direction, with their closest distances to point O being 2 m and 4 m. Particle 5 moves directly away from O. All five particles have the same mass and the same constant speed. (a) Rank the particles according to the magnitudes of their angular momentum about point O, greatest first. (b) Which particles have negative angular momentum about point O?

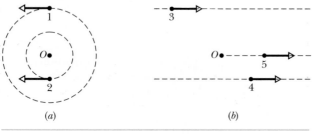

(a)   (b)

SAMPLE PROBLEM 12-7

Figure 12-13 shows an overhead view of two particles moving at constant momentum along horizontal paths. Particle 1, with momentum magnitude $p_1 = 5.0$ kg·m/s, has position vector $\mathbf{r}_1$ and will pass 2.0 m from point O. Particle 2, with momentum magnitude $p_2 = 2.0$ kg·m/s, has position vector $\mathbf{r}_2$ and will pass 4.0 m from point O. What is the net angular momentum **L** about point O of the two-particle system?

SOLUTION: To find **L**, we first must find the individual angular momentum vectors ℓ_1 and ℓ_2 of the two particles. To find the magnitude of ℓ_1, we use Eq. 12-28, substituting $r_{\perp1} = 2.0$ m and $p_1 = 5.0$ kg·m/s:

$$\ell_1 = r_{\perp1}p_1 = (2.0 \text{ m})(5.0 \text{ kg·m/s})$$
$$= 10 \text{ kg·m}^2/\text{s}.$$

To find the direction of vector ℓ_1, we use Eq. 12-25 and the right-hand rule for vector products. For $\mathbf{r}_1 \times \mathbf{p}_1$, the vector product is out of the page, perpendicular to the plane of Fig. 12-13. This is a positive direction, consistent with the counterclockwise rotation of the particle's position vector $\mathbf{r}_1$ around O as particle 1 moves. Thus the angular momentum vector for particle 1 is

$$\ell_1 = +10 \text{ kg·m}^2/\text{s}.$$

Similarly, the magnitude of ℓ_2 is

$$\ell_2 = r_{\perp2}p_2 = (4.0 \text{ m})(2.0 \text{ kg·m/s})$$
$$= 8.0 \text{ kg·m}^2/\text{s},$$

and the vector product $\mathbf{r}_2 \times \mathbf{p}_2$ is into the page, which is a negative direction, consistent with the clockwise rotation of $\mathbf{r}_2$ around O as particle 2 moves. Thus, the angular momentum vector for particle 2 is

$$\ell_2 = -8.0 \text{ kg·m}^2/\text{s}.$$

The net angular momentum for the two-particle system is then

$$L = \ell_1 + \ell_2$$
$$= +10 \text{ kg·m}^2/\text{s} + (-8.0 \text{ kg·m}^2/\text{s})$$
$$= +2.0 \text{ kg·m}^2/\text{s}. \tag{Answer}$$

The plus sign means that the system's net angular momentum about point O is out of the page.

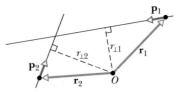

FIGURE 12-13   Sample Problem 12-7. Two particles pass near point O.

12-5 NEWTON'S SECOND LAW IN ANGULAR FORM

Newton's second law written in the form

$$\sum \mathbf{F} = \frac{d\mathbf{p}}{dt} \quad \text{(single particle)} \tag{12-29}$$

expresses the close relation between force and linear momentum for a single particle. We have seen enough of the parallelism between linear and angular quantities to be pretty sure that there is also a close relation between torque and angular momentum. Guided by Eq. 12-29, we can even guess that it must be

$$\sum \boldsymbol{\tau} = \frac{d\ell}{dt} \quad \text{(single particle)} \tag{12-30}$$

Equation 12-30 is indeed an angular form of Newton's second law for a single particle:

> The (vector) sum of all the torques acting on a particle is equal to the time rate of change of the angular momentum of that particle.

Equation 12-30 has no meaning unless the torques $\boldsymbol{\tau}$ and the angular momentum $\boldsymbol{\ell}$ are defined with respect to the same origin.

Proof of Equation 12-30

We start with Eq. 12-25, the definition of the angular momentum of a particle:

$$\boldsymbol{\ell} = m(\mathbf{r} \times \mathbf{v}),$$

where $\mathbf{r}$ is the position vector of the particle and $\mathbf{v}$ is the velocity of the particle. Differentiating* each side with respect to time t yields

$$\frac{d\boldsymbol{\ell}}{dt} = m\left(\mathbf{r} \times \frac{d\mathbf{v}}{dt} + \frac{d\mathbf{r}}{dt} \times \mathbf{v}\right). \qquad (12\text{-}31)$$

But $d\mathbf{v}/dt$ is the acceleration $\mathbf{a}$ of the particle, and $d\mathbf{r}/dt$ is its velocity $\mathbf{v}$. Thus we can rewrite Eq. 12-31 as

$$\frac{d\boldsymbol{\ell}}{dt} = m(\mathbf{r} \times \mathbf{a} + \mathbf{v} \times \mathbf{v}).$$

Now $\mathbf{v} \times \mathbf{v} = 0$ (the vector product of any vector with itself is zero because the angle between the two vectors is necessarily zero). This leads to

$$\frac{d\boldsymbol{\ell}}{dt} = m(\mathbf{r} \times \mathbf{a}) = \mathbf{r} \times m\mathbf{a}.$$

We now use Newton's second law ($\Sigma\mathbf{F} = m\mathbf{a}$) to replace $m\mathbf{a}$ with its equal, the vector sum of the forces that act on the particle, obtaining

$$\frac{d\boldsymbol{\ell}}{dt} = \mathbf{r} \times \left(\sum \mathbf{F}\right) = \sum(\mathbf{r} \times \mathbf{F}). \qquad (12\text{-}32)$$

Finally, Eq. 12-21 shows us that $\mathbf{r} \times \mathbf{F}$ is the torque associated with the force $\mathbf{F}$, so Eq. 12-32 becomes

$$\sum \boldsymbol{\tau} = \frac{d\boldsymbol{\ell}}{dt}.$$

This is Eq. 12-30, the relation that we set out to prove.

$\mathbb{C}$HECKPOINT 5: The figure shows the position vector $\mathbf{r}$ of a particle at a certain instant, and four choices for the direction of a force that is to accelerate the particle. All four choices are in the xy plane. (a) Rank the choices according to the magnitude of the time rate of change ($d\boldsymbol{\ell}/dt$) they produce in the angular momentum of the particle about point O, greatest first. (b) Which

choice results in a negative rate of change about O?

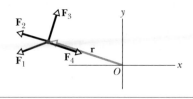

SAMPLE PROBLEM 12-8

A penguin of mass m falls from rest at point A, a horizontal distance d from the origin O in Fig. 12-14.

(a) Find an expression for the angular momentum of the falling penguin about O.

SOLUTION: The angular momentum is given by Eq. 12-25 ($\boldsymbol{\ell} = \mathbf{r} \times \mathbf{p}$); its magnitude is (from Eq. 12-26)

$$\ell = rmv \sin \phi.$$

Here, $r \sin \phi = d$ no matter how far the penguin falls, and $v = gt$. Thus $\boldsymbol{\ell}$ has magnitude

$$\ell = mgtd. \qquad \text{(Answer)} \qquad (12\text{-}33)$$

The right-hand rule shows that the angular momentum vector $\boldsymbol{\ell}$ is directed into the plane of Fig. 12-14, in the direction of decreasing z. We represent $\boldsymbol{\ell}$ with a circled cross $\otimes$ at the origin. The vector $\boldsymbol{\ell}$ changes with time in magnitude only; its direction remains unchanged.

(b) What torque does the weight $m\mathbf{g}$ acting on the penguin exert about the origin O?

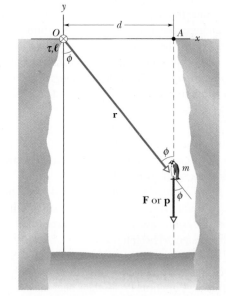

FIGURE 12-14 Sample Problem 12-8. A penguin of mass m falls vertically from point A. The torque $\boldsymbol{\tau}$ and the angular momentum $\boldsymbol{\ell}$ of the falling penguin with respect to the origin O are directed into the plane of the figure at O.

---

*In differentiating a vector product, be sure not to change the order of the two quantities (here $\mathbf{r}$ and $\mathbf{v}$) that form that product. (See Eq. 3-21.)

SOLUTION: The torque is given by Eq. 12-21 ($\boldsymbol{\tau} = \mathbf{r} \times \mathbf{F}$); its magnitude is (from Eq. 12-22)

$$\tau = rF \sin \phi.$$

Again $r \sin \phi = d$, and $F = mg$. Thus

$$\tau = mgd = \text{a constant.} \quad \text{(Answer)} \quad (12\text{-}34)$$

Note that the magnitude of the torque is simply the product of the force magnitude mg and the moment arm d. The right-hand rule shows that the torque vector $\boldsymbol{\tau}$ is directed into the plane of Fig. 12-14, in the direction of decreasing z, hence parallel to $\boldsymbol{\ell}$. (Note that we can also derive Eq. 12-34 by differentiating Eq. 12-33 with respect to t and then substituting the result into Eq. 12-30.)

We see that $\boldsymbol{\tau}$ and $\boldsymbol{\ell}$ depend very much (through d) on the location of the origin. If the penguin falls from the origin, we have $d = 0$ and thus no torque or angular momentum.

12-6 THE ANGULAR MOMENTUM OF A SYSTEM OF PARTICLES

Now we turn our attention to the motion of a system of particles with respect to an origin. Note that "a system of particles" includes a rigid body as a special case. The total angular momentum $\mathbf{L}$ of a system of particles is the (vector) sum of the angular momenta $\boldsymbol{\ell}$ of the particles:

$$\mathbf{L} = \boldsymbol{\ell}_1 + \boldsymbol{\ell}_2 + \boldsymbol{\ell}_3 + \cdots + \boldsymbol{\ell}_n = \sum_{i=1}^{n} \boldsymbol{\ell}_i, \quad (12\text{-}35)$$

in which i (= 1, 2, 3, . . .) labels the particles.

With time, the angular momenta of individual particles may change, either because of interactions within the system (between the individual particles) or because of influences that may act on the system from the outside. We can find the change in $\mathbf{L}$ as these changes take place by taking the time derivative of Eq. 12-35. Thus

$$\frac{d\mathbf{L}}{dt} = \sum_{i=1}^{n} \frac{d\boldsymbol{\ell}_i}{dt}. \quad (12\text{-}36)$$

From Eq. 12-30, $d\boldsymbol{\ell}_i/dt$ is just $\Sigma\boldsymbol{\tau}_i$, the (vector) sum of the torques that act on the ith particle.

Some torques are *internal,* associated with forces that the particles within the system exert on one another; other torques are *external,* associated with forces that act from outside the system. The internal forces, because of Newton's law of action and reaction, cancel in pairs. So, to add the torques, we need consider only those associated with external forces. Equation 12-36 then becomes

$$\sum \boldsymbol{\tau}_{\text{ext}} = \frac{d\mathbf{L}}{dt} \quad \text{(system of particles).} \quad (12\text{-}37)$$

Equation 12-37 is Newton's second law for rotation in angular form, expressed for a system of particles; it is analogous to $\Sigma\mathbf{F}_{\text{ext}} = d\mathbf{P}/dt$ (Eq. 9-28). In words, Eq. 12-37 tells us that the (vector) sum of the *external torques* acting on a system of particles is equal to the time rate of change of the *angular momentum* of that system. Equation 12-37 has meaning only if the torque and angular momentum vectors are referred to the same origin. In an inertial reference frame, Eq. 12-37 can be applied with respect to any point. In an accelerating frame (such as a wheel rolling down a ramp), Eq. 12-37 can be applied *only* with respect to the *center of mass* of the system.

12-7 THE ANGULAR MOMENTUM OF A RIGID BODY ROTATING ABOUT A FIXED AXIS

We next evaluate the angular momentum of a system of particles that form a rigid body which rotates about a fixed axis. Figure 12-15a shows such a body. The fixed axis of rotation is the z axis, and the body rotates about it with constant angular speed ω. We wish to find the angular momentum of the body about the axis of rotation.

We can find the angular momentum by summing the z components of the angular momenta of the mass elements in the body. In Fig. 12-15a, a typical mass element Δm_i of the body moves around the z axis in a circular path. The position of the mass element is located relative to the origin

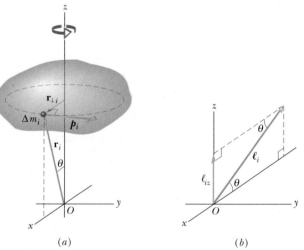

FIGURE 12-15 (a) A rigid body rotates about the z axis with angular speed ω. A mass element Δm_i within the body moves about the z axis in a circle with radius $r_{\perp i}$. The mass element has linear momentum $\mathbf{p}_i$, and it is located relative to the origin O by position vector $\mathbf{r}_i$. Here Δm_i is shown when $r_{\perp i}$ is parallel to the x axis. (b) The angular momentum $\boldsymbol{\ell}_i$, with respect to O, of the mass element in (a). The z component $\boldsymbol{\ell}_{iz}$ of $\boldsymbol{\ell}_i$ is also shown.

O by position vector $\mathbf{r}_i$. The radius of the mass element's circular path is $r_{\perp i}$, the perpendicular distance between the element and the z axis.

The magnitude of the angular momentum ℓ_i of this mass element, with respect to O, is given by Eq. 12-26:

$$\ell_i = (r_i)(p_i)(\sin 90°) = (r_i)(\Delta m_i \, v_i),$$

where p_i and v_i are the linear momentum and linear speed of the mass element, and $90°$ is the angle between $\mathbf{r}_i$ and $\mathbf{p}_i$. The angular momentum vector ℓ_i for the mass element in Fig. 12-15a is shown in Fig. 12-15b; its direction must be perpendicular to those of $\mathbf{r}_i$ and $\mathbf{p}_i$.

We are interested in the component of ℓ_i that is parallel to the rotation axis, here the z axis. That z component is

$$\ell_{iz} = \ell_i \sin \theta = (r_i \sin \theta)(\Delta m_i \, v_i) = r_{\perp i} \, \Delta m_i \, v_i.$$

The z component of the angular momentum for the rotating rigid body as a whole is found by adding up the contributions of all the mass elements that make up the body. Thus, because $v = \omega r_\perp$, we may write

$$L_z = \sum_{i=1}^{n} \ell_{iz} = \sum_{i=1}^{n} \Delta m_i \, v_i r_{\perp i} = \sum_{i=1}^{n} \Delta m_i(\omega r_{\perp i})r_{\perp i}$$

$$= \omega \left(\sum_{i=1}^{n} \Delta m_i \, r_{\perp i}^2 \right). \qquad (12\text{-}38)$$

We can remove ω from the summation here because it is a constant: it has the same value for all points of the rotating rigid body.

The quantity $\Sigma \Delta m_i \, r_{\perp i}^2$ in Eq. 12-38 is the rotational inertia I of the body about the fixed axis (see Eq. 11-24). Thus Eq. 12-38 reduces to

$$L = I\omega \qquad \text{(rigid body, fixed axis).} \qquad (12\text{-}39)$$

We have dropped the subscript z, but you must remember that the angular momentum defined by Eq. 12-39 is the

angular momentum about the rotation axis. Also, I in that equation is the rotational inertia about that same axis.

Table 12-2, which supplements Table 11-3, extends our list of corresponding linear and angular relations.

SAMPLE PROBLEM 12-9

A tomahawk expert knows how to throw the instrument so that it completes an integer number of full revolutions about its center of mass during its flight, to bury its edge in the target (Fig. 12-16). Suppose that for a flight of horizontal distance $d = 5.90$ m, with a horizontal component of velocity $v_x = 20.0$ m/s, a tomahawk rotates 1.00 rev. Suppose also that the rotational inertia I of the tomahawk about its center of mass is 1.95×10^{-3} kg·m².

(a) What is the magnitude of the tomahawk's angular momentum about the center of mass during the tomahawk's flight?

SOLUTION: The tomahawk rotates at a constant angular speed ω about an axis that is through its center of mass. So, we can calculate the angular momentum of the tomahawk about that axis with Eq. 12-39 ($L = I\omega$) if we first find ω. From Eq. 11-5, we know that ω is related to an angle of rotation $\Delta \theta$ and the time interval Δt_1 for the rotation by

$$\omega = \frac{\Delta \theta}{\Delta t_1}.$$

Replacing Δt_1 with d/v_x and substituting $\Delta \theta = 1.00$ rev $= 2\pi$ rad, we have

$$\omega = \frac{v_x \, \Delta \theta}{d} = \frac{(20.0 \text{ m/s})(2\pi \text{ rad})}{5.90 \text{ m}} = 21.3 \text{ rad/s}.$$

Substituting this and the given value of I in Eq. 12-39, we find

$$L = I\omega = (1.95 \times 10^{-3} \text{ kg·m}^2)(21.3 \text{ rad/s})$$

$$= 4.15 \times 10^{-2} \text{ kg·m}^2\text{/s}. \qquad \text{(Answer)}$$

(b) The launch required a time interval $\Delta t_2 = 0.150$ s. About

---

TABLE 12-2 MORE CORRESPONDING RELATIONS FOR TRANSLATIONAL AND ROTATIONAL MOTION[a]

TRANSLATIONAL		ROTATIONAL	
Force	$\mathbf{F}$	Torque	$\boldsymbol{\tau} \, (= \mathbf{r} \times \mathbf{F})$
Linear momentum	$\mathbf{p}$	Angular momentum	$\boldsymbol{\ell} \, (= \mathbf{r} \times \mathbf{p})$
Linear momentum[b]	$\mathbf{P} \, (= \Sigma \mathbf{p}_i)$	Angular momentum[b]	$\mathbf{L} \, (= \Sigma \boldsymbol{\ell}_i)$
Linear momentum[b]	$\mathbf{P} = M\mathbf{v}_{cm}$	Angular momentum[c]	$L = I\omega$
Newton's second law[b]	$\Sigma \mathbf{F}_{ext} = \dfrac{d\mathbf{P}}{dt}$	Newton's second law[b]	$\Sigma \boldsymbol{\tau}_{ext} = \dfrac{d\mathbf{L}}{dt}$
Conservation law[d]	$\mathbf{P} = $ a constant	Conservation law[d]	$\mathbf{L} = $ a constant

[a] See also Table 11-3.   [b] For systems of particles, including rigid bodies.

[c] For a rigid body about a fixed axis, with L being the component along that axis.

[d] For an isolated system.

the tomahawk's center of mass and from the perspective of Fig. 12-16, what was the average torque applied by the expert to the tomahawk during the launch?

SOLUTION: From Eq. 12-37, we can relate the average torque $\bar{\tau}$ to the change in angular momentum ΔL during the launching time interval Δt_2 with

$$\bar{\tau} = \frac{\Delta L}{\Delta t_2} = \frac{L_f - L_i}{\Delta t_2}.$$

Here, the initial and final angular momenta for the launching period are $L_i = 0$ and $L_f = -4.15 \times 10^{-2}$ kg·m²/s (negative because the tomahawk rotates clockwise in Fig. 12-16). Substituting these values and the given value for Δt_2 yields

$$\bar{\tau} = \frac{-4.15 \times 10^{-2} \text{ kg·m}^2/\text{s}}{0.150 \text{ s}} = -0.277 \text{ N·m}. \quad \text{(Answer)}$$

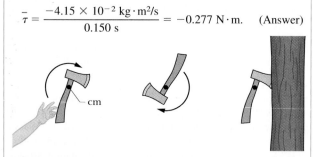

FIGURE 12-16 Sample Problem 12-9. Once thrown, a tomahawk rotates around an axis through its center of mass. (The parabolic path of the center of mass is not shown.)

CHECKPOINT 6: In the figure, a disk, a hoop, and a solid sphere are made to spin about fixed central axes (like a top) by means of strings wrapped around them, with the strings producing the same constant tangential force **F** on all three objects. The three objects have the same mass and radius, and they are initially stationary. Rank the objects according to (a) their angular momentum about their central axes and (b) their angular speed, greatest first, when the strings have been pulled for a certain time t.

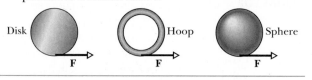

Disk   Hoop   Sphere
F   F   F

12-8 CONSERVATION OF ANGULAR MOMENTUM

So far we have discussed two powerful conservation laws, the conservation of energy and the conservation of linear momentum. Now we meet a third law of this type, the conservation of angular momentum. We start from Eq. 12-37 ($\Sigma\boldsymbol{\tau}_{\text{ext}} = d\mathbf{L}/dt$), which is Newton's second law in an-

gular form. If no net external torque acts on the system, this equation becomes $d\mathbf{L}/dt = 0$, or

$$\mathbf{L} = \text{a constant} \quad \text{(isolated system).} \quad (12\text{-}40)$$

This result, called the **law of conservation of angular momentum,** can also be written as

$$\mathbf{L}_i = \mathbf{L}_f \quad \text{(isolated system),} \quad (12\text{-}41)$$

where the subscripts refer to the values of **L** at some initial time i and later time f. Equations 12-40 and 12-41 tell us:

> If the net external torque acting on a system is zero, the angular momentum **L** of the system remains constant, no matter what changes take place within the system.

Equations 12-40 and 12-41 are vector equations; as such, they are equivalent to three scalar equations corresponding to the conservation of angular momentum in three mutually perpendicular directions. Depending on the torques acting on a system, the angular momentum of the system might be conserved in only one or two directions but not in all directions:

> If any component of the net *external* torque on a system is zero, then that component of the angular momentum of the system along that axis cannot change, no matter what changes take place within the system.

We can apply this law to the isolated body in Fig. 12-15, which rotates around the z axis. Suppose that the initially rigid body somehow redistributes its mass relative to that rotation axis, changing its rotational inertia. Equations 12-40 and 12-41 state that the angular momentum of the body cannot change. Substituting Eq. 12-39 (for the angular momentum along the rotational axis) into Eq. 12-41, we write this conservation law as

$$I_i \omega_i = I_f \omega_f. \quad (12\text{-}42)$$

Here the subscripts refer to the values of the rotational inertia I and angular speed ω before and after the redistribution of mass.

Like the other two conservation laws that we have discussed, Eqs. 12-40 and 12-41 hold beyond the limitations of Newtonian mechanics. They hold for particles whose speeds approach that of light (where the theory of relativity reigns), and they remain true in the world of subatomic particles (where quantum mechanics reigns). No exceptions to the law of conservation of angular momentum have ever been found.

We now discuss four examples involving this law.

1. *The spinning volunteer.* Figure 12-17 shows a student seated on a stool that can rotate freely about a vertical axis. The student, who has been set into rotation at a modest initial angular speed ω_i, holds two dumbbells in his outstretched hands. His angular momentum vector **L** lies along the vertical axis, pointing upward.

The instructor now asks the student to pull in his arms; this action reduces his rotational inertia from its initial value I_i to a smaller value I_f, because he moves mass closer to the rotation axis. His rate of rotation increases markedly, from ω_i to ω_f. If the student wishes to slow down, he has only to extend his arms once more.

No net external torque acts on the system consisting of the student, stool, and dumbbells. Thus the angular momentum of that system about the rotation axis must remain constant, no matter how the student maneuvers the weights. In Fig. 12-17a, the student's angular speed ω_i is relatively low and his rotational inertia I_i relatively high. According to Eq. 12-42, his angular speed in Fig. 12-17b must be greater to compensate for the decreased rotational inertia due to mass being closer to the rotational axis.

2. *The springboard diver.* Figure 12-18 shows a diver doing a forward one-and-a-half-somersault dive. As you should expect, her center of mass follows a parabolic path. She leaves the springboard with a definite angular momentum **L** about an axis through her center of mass, represented by a vector pointing into the plane of Fig. 12-18, perpendicular to the page. When she is in the air, the diver

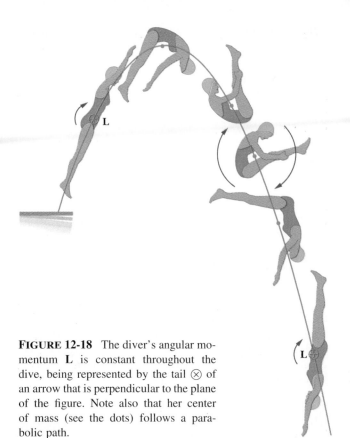

FIGURE 12-18 The diver's angular momentum **L** is constant throughout the dive, being represented by the tail $\otimes$ of an arrow that is perpendicular to the plane of the figure. Note also that her center of mass (see the dots) follows a parabolic path.

forms an isolated system and her angular momentum cannot change. By pulling her arms and legs into the closed *tuck position,* she can considerably reduce her rotational inertia about the same axis and thus, according to Eq. 12-42, considerably increase her angular speed. Pulling out of the tuck position (into the *open layout position*) at the end of the dive increases her rotational inertia and thus slows her rotation rate so she can enter the water with little splash. Even in a more complicated dive involving both twisting and somersaulting, the angular momentum of the diver must be conserved, in both magnitude *and* direction, throughout the dive.

3. *Spacecraft orientation.* Figure 12-19, which represents a spacecraft with a rigidly mounted flywheel, suggests a scheme (albeit crude) for orientation control. The *spacecraft + flywheel* form an isolated system. So, if the total angular momentum **L** of the system is zero because neither spacecraft nor flywheel is turning, it must remain zero (as long as the system remains isolated).

To change the orientation of the spacecraft, the flywheel is started up (Fig. 12-19a). The spacecraft will start to rotate in the opposite sense to maintain the system's angular momentum at zero. When the flywheel is then brought to rest, the spacecraft will also stop rotating but will have changed its orientation (Fig. 12-19b). Throughout, the angular momentum of the system *spacecraft + flywheel* never differs from zero.

Interestingly, the spacecraft *Voyager 2*, on its 1986

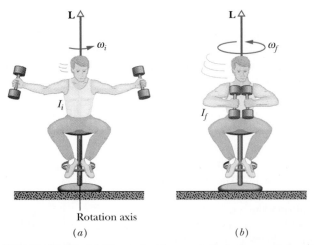

FIGURE 12-17 (*a*) The student has a relatively large rotational inertia and a relatively small angular speed. (*b*) By decreasing his rotational inertia, the student automatically increases his angular speed. The angular momentum **L** of the rotating system remains unchanged.

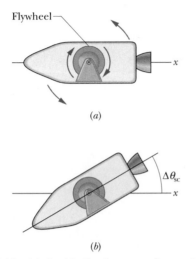

FIGURE 12-19 (*a*) An idealized spacecraft containing a flywheel. If the flywheel is made to rotate clockwise as shown, the spacecraft itself will rotate counterclockwise because the total angular momentum must remain zero. (*b*) When the flywheel is braked to rest, the spacecraft will also stop rotating but will have reoriented its axis by the angle $\Delta\theta_{sc}$.

flyby of the planet Uranus, was set into unwanted rotation by this flywheel effect every time its tape recorder was turned on at high speed. The ground staff at the Jet Propulsion Laboratory had to program the on-board computer to turn on counteracting thruster jets every time the tape recorder was turned on or off.

4. *The incredible shrinking star.* When the nuclear fire in the core of a star burns low, the star may eventually begin to collapse, building up pressure in its interior. The collapse may go so far as to reduce the radius of the star from something like that of the Sun to the incredibly small value of a few kilometers. The star then becomes a *neutron star*—its material has been compressed to an incredibly dense gas of neutrons.

During this shrinking process, the star is an isolated system and its angular momentum **L** cannot change. Because its rotational inertia is greatly reduced, its angular speed is correspondingly greatly increased, to as much as 600–800 revolutions per *second*. For comparison, the Sun, a typical star, rotates at about one revolution per month.

CHECKPOINT 7: A rhinoceros beetle rides the rim of a small disk that rotates like a merry-go-round. If the beetle crawls toward the center of the disk, do the following (each relative to the central axis) increase, decrease, or remain the same: (a) the rotational inertia of the beetle–disk system, (b) the angular momentum of the system, and (c) the angular speed of the beetle and disk?

SAMPLE PROBLEM 12-10

Figure 12-20*a* shows a student, again sitting on a stool that can rotate freely about a vertical axis. The student, initially at rest, is holding a bicycle wheel whose rim is loaded with lead and whose rotational inertia I about its central axis is $1.2 \text{ kg} \cdot \text{m}^2$. The wheel is rotating at an angular speed ω_i of 3.9 rev/s; as seen from overhead, the rotation is counterclockwise. The axis of the wheel is vertical, and the angular momentum $\mathbf{L}_i$ of the wheel points vertically upward. The student now inverts the wheel (Fig. 12-20*b*); as a result, the student and stool rotate about the stool axis. With what angular speed and direction does the student then rotate? (The rotational inertia I_0 of the *student + stool + wheel* system about the stool axis is $6.8 \text{ kg} \cdot \text{m}^2$.)

SOLUTION: There is no net torque acting on the *student + stool + wheel* to change the angular momentum of that system about any vertical axis. The initial angular momentum $\mathbf{L}_i$ of the system is that of the bicycle wheel alone. After the wheel has been inverted, the system must still have a net angular momentum of the same magnitude *and* direction.

After the inversion, the angular momentum of the wheel is $-\mathbf{L}_i$. In addition, the *student + stool* must acquire some angular momentum; call it **L**. Then, as shown in Fig. 12-20*c*, we have

$$L_i = L + (-L_i),$$

or

$$L = 2L_i = I_0\omega,$$

in which ω is the angular speed acquired by the student after the wheel's inversion. This yields

$$\omega = \frac{2L_i}{I_0} = \frac{2I\omega_i}{I_0}$$

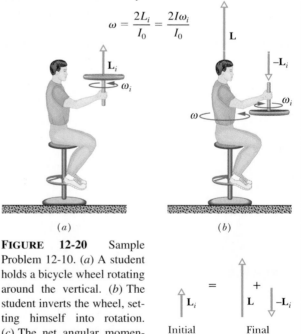

FIGURE 12-20 Sample Problem 12-10. (*a*) A student holds a bicycle wheel rotating around the vertical. (*b*) The student inverts the wheel, setting himself into rotation. (*c*) The net angular momentum of the system must remain the same in spite of the inversion.

$$= \frac{(2)(1.2 \text{ kg} \cdot \text{m}^2)(3.9 \text{ rev/s})}{6.8 \text{ kg} \cdot \text{m}^2}$$

$$= 1.4 \text{ rev/s.} \qquad \text{(Answer)}$$

This positive result tells us that the student rotates counterclockwise about the stool axis as seen from overhead. If the student wishes to stop rotating, he has only to invert the wheel once more.

In inverting the wheel, the student will become well aware of the need to apply a torque. However, this torque is internal to the *student + stool + wheel* system and so cannot change the total angular momentum of this system.

We could, however, decide to adopt as our system *student + stool* alone; the wheel then would be external to our system. From this point of view, as the student exerts a torque on the wheel, the wheel exerts a reaction torque on him and this torque is now an external torque. It is the action of this external torque that changes the angular momentum of the *student + stool* system, setting it spinning. Whether a torque is internal or external depends only on how you choose to define your system.

SAMPLE PROBLEM 12-11

During a jump to his partner, an aerialist is to make a triple somersault lasting a time $t = 1.87$ s. For the first and last quarter-revolution, he is in the extended orientation shown in Fig. 12-21, with rotational inertia $I_1 = 19.9$ kg·m² around his center of mass. During the rest of the flight he is in a moderate tuck, with rotational inertia $I_2 = 5.50$ kg·m².

(a) What must be his initial angular speed ω_1 around his center of mass?

SOLUTION: He rotates in the extended position for two quarter-turns, or a total angle $\theta_1 = 0.500$ rev, in total time t_1; he is in the tuck for an angle θ_2 during a time t_2. These two times are given by

$$t_1 = \frac{\theta_1}{\omega_1}, \qquad \text{and} \qquad t_2 = \frac{\theta_2}{\omega_2}, \qquad (12\text{-}43)$$

where ω_2 is his angular speed in the tuck. We can find an expression for ω_2 by noting that his angular momentum is conserved throughout the flight. Then, from Eq. 12-42,

$$I_2\omega_2 = I_1\omega_1,$$

from which

$$\omega_2 = \frac{I_1}{I_2}\omega_1. \qquad (12\text{-}44)$$

His total flight time is

$$t = t_1 + t_2,$$

which, with substitutions from Eqs. 12-43 and 12-44, becomes

$$t = \frac{\theta_1}{\omega_1} + \frac{\theta_2 I_2}{\omega_1 I_1} = \frac{1}{\omega_1}\left(\theta_1 + \theta_2\frac{I_2}{I_1}\right). \qquad (12\text{-}45)$$

Inserting the given data, we obtain

$$1.87 \text{ s} = \frac{1}{\omega_1}\left(0.500 \text{ rev} + 2.50 \text{ rev}\frac{5.50 \text{ kg} \cdot \text{m}^2}{19.9 \text{ kg} \cdot \text{m}^2}\right),$$

where, for the triple somersault, he must spend $\theta = 2.5$ revolutions in a tuck. From this equation we find

$$\omega_1 = 0.6369 \text{ rev/s} \approx 0.637 \text{ rev/s.} \qquad \text{(Answer)}$$

(b) If he now attempts a quadruple somersault, with the same ω_1 and t, by using a tighter tuck, what must his rotational inertia I_2 be during the tuck?

SOLUTION: The angle of rotation θ_2 during the tuck is now 3.50 rev, and Eq. 12-45 becomes

$$1.87 \text{ s} = \frac{1}{0.6369 \text{ rev/s}}$$

$$\times \left(0.500 \text{ rev} + 3.50 \text{ rev}\frac{I_2}{19.9 \text{ kg} \cdot \text{m}^2}\right),$$

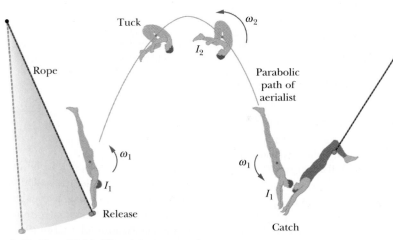

FIGURE 12-21 Sample Problem 12-11. The triple somersault.

from which we find

$$I_2 = 3.929 \text{ kg} \cdot \text{m}^2 \approx 3.93 \text{ kg} \cdot \text{m}^2. \quad \text{(Answer)}$$

This smaller value for I_2 allows faster turning in the tuck and is about the smallest value possible for an aerialist. To make a four-and-a-half somersault, an aerialist would have to increase either the time of flight or the initial angular speed, but either change would make the catch by his partner more difficult. (Can you see why?)

(c) For the quadruple somersault, what is the rotation period T during the tuck?

SOLUTION: We first find the angular speed ω_2 during the tuck from Eq. 12-44:

$$\omega_2 = \frac{I_1}{I_2}\omega_1 = \frac{19.9 \text{ kg} \cdot \text{m}^2}{3.929 \text{ kg} \cdot \text{m}^2} 0.6369 \text{ rev/s}$$

$$= 3.226 \text{ rev/s}.$$

Then we find the time T for one rotation from

$$T = \frac{1 \text{ rev}}{\omega_2} = \frac{1 \text{ rev}}{3.226 \text{ rev/s}} = 0.310 \text{ s}. \quad \text{(Answer)}$$

One reason the quadruple somersault is so difficult is that rotation occurs too quickly for the aerialist to see his surroundings clearly or to "fine-tune" the angular speed by adjusting his rotational inertia during flight.

SAMPLE PROBLEM 12-12

Four thin rods, each with mass M and length $d = 1.0$ m, are rigidly connected in the form of a plus sign; the entire assembly rotates in a horizontal plane around a vertical axle at the center, with initial (clockwise) angular velocity $\omega_i = -2.0$ rad/s (see Fig. 12-22). A mud ball with mass m and initial speed $v_i = 12$ m/s is thrown at, and sticks to, the end of one rod. Let $M = 3m$. What is the final angular velocity ω_f of the *plus sign + mud ball* system if the initial path of the mud ball is each of the four paths shown in Fig. 12-22: path 1 (contact is made when the ball's velocity is perpendicular to the rod), path 2 (radial contact), path 3 (perpendicular contact), and path 4 (contact is made at 60° to the perpendicular)?

SOLUTION: The total angular momentum L of the system about the axle is conserved during the collision:

$$L_f = L_i, \quad \text{(12-46)}$$

where the subscripts f and i represent final and initial values. Let I_+ represent the rotational inertia of the plus sign about the axle. From Table 11-2(f), we have, for the four rods,

$$I_+ = 4\left(\frac{Md^2}{3}\right).$$

The rotational inertia of the mud ball about the axle as the ball rotates on the plus sign is $I_{\text{mb}} = md^2$. Let ℓ_i represent the

initial (before contact) angular momentum of the mud ball about the axle, and ω_f represent the final angular velocity of the system. Using $L = I\omega$, we may rewrite Eq. 12-46 as

$$I_+\omega_f + I_{\text{mb}}\omega_f = I_+\omega_i + \ell_i,$$

and then as

$$\left(\tfrac{4}{3}Md^2\right)\omega_f + (md^2)\omega_f = \left(\tfrac{4}{3}Md^2\right)\omega_i + \ell_i. \quad \text{(12-47)}$$

Substituting $M = 3m$ and $\omega_i = -2.0$ rad/s and solving for ω_f, we find that

$$\omega_f = \frac{1}{5md^2}\left(4md^2\,(-2 \text{ rad/s}) + \ell_i\right). \quad \text{(12-48)}$$

We evaluate the magnitude of ℓ_i for paths 1 and 3 with Eq. 12-28, where $r_\perp = d$ and $v = v_i$. For path 2, we use Eq. 12-28 with $r_\perp = 0$. For path 4 we use Eq. 12-27, where $r = d$ and $v_\perp = v_i \cos 60°$. To determine the sign of ℓ_i for an approaching mud ball, we draw a position vector from the axle of the plus sign to the mud ball. Then we note how the position vector rotates around the axle as the mud ball continues to move toward the plus sign. If the position vector rotates clockwise, ℓ_i is negative; if it rotates counterclockwise, ℓ_i is positive. The results are:

path 1: $\ell_i = -mdv_i$;   path 2: $\ell_i = 0$;

path 3: $\ell_i = mdv_i$;   path 4: $\ell_i = mdv_i \cos 60°$.

The speed v_i is given as 12 m/s. Substituting these values in turn into Eq. 12-48, we find ω_f to be:

path 1: -4.0 rad/s;   path 2: -1.6 rad/s;

path 3: 0.80 rad/s;   path 4: -0.40 rad/s.   (Answer)

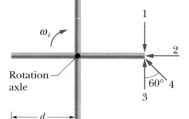

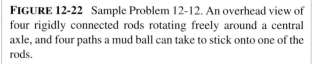

FIGURE 12-22 Sample Problem 12-12. An overhead view of four rigidly connected rods rotating freely around a central axle, and four paths a mud ball can take to stick onto one of the rods.

12-9 QUANTIZED ANGULAR MOMENTUM

A physical quantity is said to be **quantized** if it can have only certain discrete values, and all intermediate values are prohibited. We have met two examples so far, both on a microscopic level: the quantization of mass (Section 2-9) and the quantization of energy (Section 8-9). Angular momentum, on a microscopic level, is a third example.

Fundamental particles such as electrons and protons always have a certain intrinsic angular momentum called **spin angular momentum** S, as if they always spin like a top at a certain rate. (However, they do not spin. Spin angular momentum is a more abstract quantity than the angular momentum of, say, a merry-go-round, which involves actual rotation.) Spin angular momentum S is quantized; its value (along any given direction) is given by

$$S = m_s \frac{h}{2\pi}, \qquad (12\text{-}49)$$

where h is Planck's constant (6.63×10^{-34} J·s) and m_s is a quantum number. For electrons, protons, positrons, and antiprotons, the values of m_s can be only $+\frac{1}{2}$ and $-\frac{1}{2}$. If $m_s = +\frac{1}{2}$, the particle is said to be *spin up* and it has spin $S = +\frac{1}{2}(h/2\pi)$. If $m_s = -\frac{1}{2}$, the particle is said to be *spin down* and it has spin $S = -\frac{1}{2}(h/2\pi)$.

When fundamental particles collide or transform into other fundamental particles, the angular momentum of the system of those particles must be conserved. For example, it must be conserved when a proton (p) and an antiproton ($\bar{\text{p}}$) collide and *annihilate* each other. The result can be the appearance of four positive pions (π^+) and four negative pions (π^-), which are also fundamental particles:

$$\text{p} + \bar{\text{p}} \rightarrow 4\pi^+ + 4\pi^-.$$

For any pion, $m_s = 0$. So, by Eq. 12-49, for any pion, $S = 0$. Thus, after the p–$\bar{\text{p}}$ annihilation, the system has a total spin angular momentum of zero. By conservation of angular momentum, the total spin angular momentum of the colliding proton and antiproton must also be zero if the annihilation is to occur by the process above. That is, either the proton or the antiproton must be spin up and the other must be spin down.

REVIEW & SUMMARY

Rolling Bodies

For a wheel of radius R that rolls without slipping,

$$v_{cm} = \omega R, \qquad (12\text{-}2)$$

where v_{cm} is the linear speed of the wheel's center and ω is the angular speed of the wheel about its center. The wheel may also be viewed as rotating instantaneously about the point P of the "road" that is in contact with the wheel. The angular speed of the wheel about this point is the same as the angular speed of the wheel about its center. The rolling wheel has kinetic energy

$$K = \tfrac{1}{2}I_{cm}\omega^2 + \tfrac{1}{2}Mv_{cm}^2, \qquad (12\text{-}5)$$

where I_{cm} is the rotational moment of the wheel about its center.

Torque as a Vector

In three dimensions, *torque* $\boldsymbol{\tau}$ is a vector quantity defined relative to a fixed point (usually an origin); it is

$$\boldsymbol{\tau} = \mathbf{r} \times \mathbf{F}, \qquad (12\text{-}21)$$

where $\mathbf{F}$ is a force applied to a particle and $\mathbf{r}$ is a position vector locating the particle relative to the fixed point (or origin). The magnitude of $\boldsymbol{\tau}$ is given by

$$\tau = rF\sin\phi = rF_\perp = r_\perp F, \qquad (12\text{-}22, 12\text{-}23, 12\text{-}24)$$

where ϕ is the angle between $\mathbf{F}$ and $\mathbf{r}$, $F_\perp$ is the component of $\mathbf{F}$ perpendicular to $\mathbf{r}$, and $r_\perp$ is the moment arm of $\mathbf{F}$. The direction of $\boldsymbol{\tau}$ is given by the right-hand rule for cross products.

Angular Momentum of a Particle

The **angular momentum** $\boldsymbol{\ell}$ of a particle with linear momentum $\mathbf{p}$, mass m, and linear velocity $\mathbf{v}$ is a vector quantity defined relative to a fixed point (usually an origin); it is

$$\boldsymbol{\ell} = \mathbf{r} \times \mathbf{p} = m(\mathbf{r} \times \mathbf{v}). \qquad (12\text{-}25)$$

The magnitude of $\boldsymbol{\ell}$ is given by

$$\ell = rmv\sin\phi \qquad (12\text{-}26)$$
$$= rp_\perp = rmv_\perp \qquad (12\text{-}27)$$
$$= r_\perp p = r_\perp mv, \qquad (12\text{-}28)$$

where ϕ is the angle between $\mathbf{r}$ and $\mathbf{p}$, $p_\perp$ and $v_\perp$ are the components of $\mathbf{p}$ and $\mathbf{v}$ perpendicular to $\mathbf{r}$, and $r_\perp$ is the perpendicular distance between the fixed point and an extension of $\mathbf{p}$. The direction of $\boldsymbol{\ell}$ is given by the right-hand rule.

Newton's Second Law in Angular Form

Newton's second law for a particle can be written in vector angular form as

$$\sum \boldsymbol{\tau} = \frac{d\boldsymbol{\ell}}{dt}, \qquad (12\text{-}30)$$

where $\sum\boldsymbol{\tau}$ is the net torque acting on the particle, and $\boldsymbol{\ell}$ is the angular momentum of the particle.

Angular Momentum of a System of Particles

The angular momentum $\mathbf{L}$ of a system of particles is the vector sum of the angular momenta of the individual particles:

$$\mathbf{L} = \boldsymbol{\ell}_1 + \boldsymbol{\ell}_2 + \cdots + \boldsymbol{\ell}_n = \sum_{i=1}^{n} \boldsymbol{\ell}_i. \qquad (12\text{-}35)$$

The time rate of change of this angular momentum is equal to the sum of the external torques on the system (the torques due to interactions of the particles of the system with particles external to the system):

$$\sum \boldsymbol{\tau}_{ext} = \frac{d\mathbf{L}}{dt} \qquad \text{(system of particles)}. \qquad (12\text{-}37)$$

Angular Momentum of a Rigid Body

For a rigid body rotating about a fixed axis, the component of its angular momentum parallel to the rotation axis is

$$L = I\omega \qquad \text{(rigid body, fixed axis).} \qquad (12\text{-}39)$$

Conservation of Angular Momentum

The angular momentum **L** of a system remains constant if the net external torque acting on the system is zero:

$$\mathbf{L} = \text{a constant} \qquad \text{(isolated system)} \qquad (12\text{-}40)$$

or

$$\mathbf{L}_i = \mathbf{L}_f \qquad \text{(isolated system).} \qquad (12\text{-}41)$$

This is the **law of conservation of angular momentum.** It is one of the fundamental conservation laws of nature, having been veri-

fied even in situations (involving high-speed particles or subatomic dimensions) in which Newton's laws are not applicable.

Quantized Angular Momentum

The spin angular momentum S of fundamental particles is quantized; its value (along any given direction) is

$$S = m_s \frac{h}{2\pi}, \qquad (12\text{-}49)$$

where h is Planck's constant (6.63×10^{-34} J·s) and m_s is a quantum number. *Spin up* and *spin down* correspond to $m_s = +\frac{1}{2}$ and $m_s = -\frac{1}{2}$, respectively. The angular momentum of a system of fundamental particles is conserved during a collision or a transformation.

QUESTIONS

1. In Fig. 12-23, a block slides down a frictionless ramp and a sphere rolls without sliding down a ramp of the same angle θ. The block and sphere have the same mass, start from rest at point A, and descend through point B. (a) In that descent, is the work done by the block's weight on the block greater than, less than, or the same as the work done by the sphere's weight on the sphere? At B, which object has more (b) translational kinetic energy and (c) speed down the ramp?

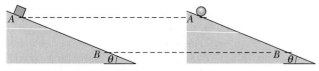

FIGURE 12-23 Question 1.

2. A cannonball and a marble roll from rest down an incline without sliding. (a) Does the cannonball take more, less, or the same time as the marble to reach the bottom? (b) Is the fraction of the cannonball's kinetic energy that is associated with translation more than, less than, or the same as that of the marble?

3. A cannonball rolls down an incline without sliding. If the roll is now repeated with an incline that is less steep but of the same height as the first incline, are (a) the ball's time to reach the bottom and (b) its translational kinetic energy at the bottom greater than, less than, or the same as previously?

4. A solid brass cylinder and a solid wood cylinder have the same radius and mass (the wood cylinder is longer). Released together from rest, they roll down an incline. (a) Which cylinder reaches the bottom first, or do they tie? (b) The wood cylinder is then shortened to match the length of the brass cylinder, and the brass cylinder is drilled out along its long axis to match the mass of the wood cylinder. Which cylinder now wins the race?

5. In Fig. 12-24, a woman rolls a cylindrical drum, by means of

a board on top, through the distance $L/2$, which is half the board's length. The drum rolls smoothly, and the board does not slide over the drum. (a) What length of board has rolled over the top of the drum? (b) How far has the woman walked?

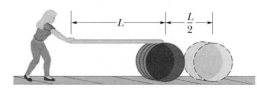

FIGURE 12-24 Question 5.

6. The position vector **r** of a particle relative to a certain point has a magnitude of 3 m, and the force **F** on the particle has a magnitude of 4 N. What is the angle between the directions of **r** and **F** if the magnitude of the associated torque equals (a) zero and (b) 12 N·m?

7. Figure 12-25 shows a particle moving at constant velocity **v** and five points with their xy coordinates. Rank the points according to the magnitude of the angular momentum of the particle measured about them, greatest first.

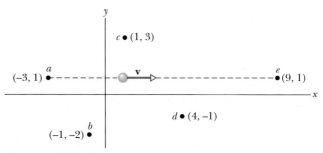

FIGURE 12-25 Question 7.

8. (a) In Checkpoint 4, what is the torque on particles 1 and 2 about point O due to the centripetal forces that cause those particles to circle at constant speed? (b) As particles 3, 4, and 5 move from the left to the right of point O, do their individual angular momenta increase, decrease, or stay the same?

9. Figure 12-26 shows three particles of the same mass and the same constant speed moving as indicated by the velocity vectors. Points a, b, c, and d form a square, with point e at the center. Rank the points according to the magnitude of the net angular momentum of the three-particle system when measured about the points, greatest first.

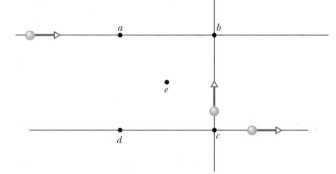

FIGURE 12-26 Question 9.

10. The following gives the angular momentum $\ell(t)$ of a particle in four situations: (1) $\ell = 3t + 4$; (2) $\ell = -6t^2$; (3) $\ell = 2$; (4) $\ell = 4/t$. In which situation is the net torque on the particle (a) zero, (b) positive and constant, (c) negative and increasing in magnitude $(t > 0)$, and (d) negative and decreasing in magnitude $(t > 0)$?

11. A bola, which consists of three heavy balls connected to a common point by identical lengths of sturdy string, is readied for launch by holding one of the balls overhead and rotating the wrist, causing the other two balls to rotate in a horizontal circle about the hand. The bola is then released, and its configuration rapidly changes from that shown in the overhead view of Fig. 12-27a to that of Fig. 12-27b. During that change do its (a) angular momentum about its center and (b) angular speed increase, decrease, or stay the same?

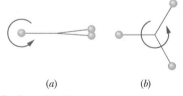

(a)   (b)

FIGURE 12-27 Question 11.

12. A rhinoceros beetle rides the rim of a disk rotating like a merry-go-round counterclockwise. If it then walks along the rim in the direction of the rotation, will the following increase, decrease, or remain the same: (a) the angular momentum of the beetle–disk system, (b) the angular momentum and angular velocity of the beetle, and (c) the angular momentum and angular

velocity of the disk? (d) What are your answers if the beetle walks in the direction opposite the rotation?

13. In Fig. 12-28, a disk is spinning freely at the bottom of an axle and with an angular momentum (about its center) of 50 units, counterclockwise. Four more spinning disks are to be dropped down the axle to land on and couple (via friction) to the first disk. Their angular momenta are: (1) 20 units clockwise, (2) 10 units counterclockwise, (3) 10 units clockwise, and (4) 60 units clockwise. (a) What is the final angular momentum of the system? (b) In what order should the five disks be added so that at one stage the disks that are then coupled on the axle are stationary?

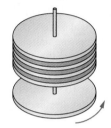

FIGURE 12-28 Question 13.

14. Figure 12-29 shows an overhead view of a rectangular slab that can spin like a merry-go-round about its center at O. Also shown are seven paths along which wads of bubble gum can be thrown (all with the same speed and mass) to stick onto the stationary slab. (a) Rank the paths according to the angular speed that the slab (and gum) will have after the gum sticks, greatest first. (b) For which paths will the angular momentum of the slab (and gum) about O be negative?

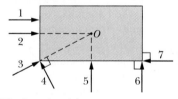

FIGURE 12-29 Question 14.

15. What happens to the initially stationary yo-yo in Fig. 12-30 if you pull it via its string with (a) force $\mathbf{F}_2$ (the line of action passes through the point of contact on the table, as indicated), (b) force $\mathbf{F}_1$ (the line of action passes above the point of contact), and (c) force $\mathbf{F}_3$ (the line of action passes to the right of the point of contact)?

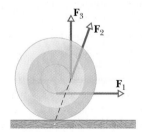

FIGURE 12-30 Question 15.

EXERCISES & PROBLEMS

SECTION 12-1 Rolling

1E. A thin-walled pipe rolls along the floor. What is the ratio of its translational kinetic energy to its rotational kinetic energy about an axis parallel to its length and through its center of mass?

2E. A 140 kg hoop rolls along a horizontal floor so that its center of mass has a speed of 0.150 m/s. How much work must be done on the hoop to stop it?

3E. An automobile traveling 80.0 km/h has tires of 75.0 cm diameter. (a) What is the angular speed of the tires about the axle? (b) If the car is brought to a stop uniformly in 30.0 turns of the tires (without skidding), what is the angular acceleration of the wheels? (c) How far does the car move during the braking?

4E. A 1000 kg car has four 10 kg wheels. When the car is moving, what fraction of the total kinetic energy of the car is due to rotation of the wheels about their axles? Assume that the wheels have the same rotational inertia as uniform disks of the same mass and size. Why do you not need the radius of the wheels?

5E. A wheel of radius 0.250 m, which is moving initially at 43.0 m/s, rolls to a stop in 225 m. Calculate (a) its linear acceleration and (b) its angular acceleration. (c) The wheel's rotational inertia is 0.155 kg·m² about its central axis. Calculate the torque exerted by the friction on the wheel, about the central axis.

6E. An automobile has a total mass of 1700 kg. It accelerates from rest to 40 km/h in 10 s. Assume each wheel is a uniform 32 kg disk. Find, for the end of the 10 s interval, (a) the rotational kinetic energy of each wheel about its axle, (b) the total kinetic energy of each wheel, and (c) the total kinetic energy of the automobile.

7E. A uniform sphere rolls down an incline. (a) What must be the incline angle if the linear acceleration of the center of the sphere is to be 0.10g? (b) For this angle, what would be the acceleration of a frictionless block sliding down the incline?

8E. A solid sphere of weight 8.00 lb rolls up an incline with an inclination angle of 30.0°. At the bottom of the incline the center of mass of the sphere has a translational speed of 16.0 ft/s. (a) What is the kinetic energy of the sphere at the bottom of the incline? (b) How far does the sphere travel up the incline? (c) Does the answer to (b) depend on the weight of the sphere?

9P. A constant horizontal force of 10 N is applied to a wheel of mass 10 kg and radius 0.30 m as shown in Fig. 12-31. The wheel

rolls without slipping on the horizontal surface, and the acceleration of its center of mass is 0.60 m/s². (a) What are the magnitude and direction of the frictional force on the wheel? (b) What is the rotational inertia of the wheel about an axis through its center of mass and perpendicular to the plane of the wheel?

10P. Consider a 66 cm diameter tire on a car traveling at 80 km/h on a level road in the direction of increasing x. As seen by a passenger in the car, what are the linear velocity and the magnitude of the linear acceleration of (a) the center of the wheel, (b) a point at the top of the tire, and (c) a point at the bottom of the tire? (d) Repeat (a) to (c), in the same order, for a stationary observer alongside the road.

11P. A body of radius R and mass m is rolling smoothly with speed v on a horizontal surface. It then rolls up a hill to a maximum height h. (a) If $h = 3v^2/4g$, what is the body's rotational inertia about the rotational axis through its center of mass? (b) What might the body be?

12P. A homogeneous sphere starts from rest at the upper end of the track shown in Fig. 12-32 and rolls without slipping until it rolls off the right-hand end. If $H = 6.0$ m and $h = 2.0$ m and the track is horizontal at the right-hand end, how far horizontally from point A does the sphere land on the floor?

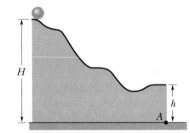

FIGURE 12-32 Problem 12.

13P. A small sphere, with radius r and mass m, rolls without slipping on the inside of a large fixed hemisphere with radius R and a vertical axis of symmetry. It starts at the top from rest. (a) What is its kinetic energy at the bottom? (b) What fraction of its kinetic energy at the bottom is associated with rotation about an axis through its center of mass? (c) What normal force does the small sphere exert on the hemisphere at the bottom if $r \ll R$?

14P. A solid cylinder of radius 10 cm and mass 12 kg starts from rest and rolls without slipping a distance of 6.0 m down a house roof that is inclined at 30°. (See Fig. 12-33.) (a) What is the angular speed of the cylinder about its center as it leaves the house roof? (b) The outside wall of the house is 5.0 m high. How far from the edge of the roof does the cylinder hit the level ground?

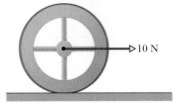

FIGURE 12-31 Problem 9.

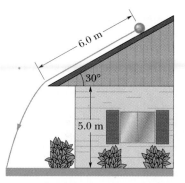

FIGURE 12-33 Problem 14.

15P. A small solid marble of mass m and radius r will roll without slipping along the loop-the-loop track shown in Fig. 12-34 if it is released from rest somewhere on the straight section of track. (a) From what minimum height h above the bottom of the track must the marble be released to ensure that it does not leave the track at the top of the loop? (The radius of the loop-the-loop is R; assume $R \gg r$.) (b) If the marble is released from height $6R$ above the bottom of the track, what is the horizontal component of the force acting on it at point Q?

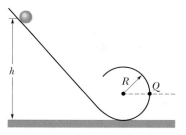

FIGURE 12-34 Problem 15.

16P. A bowler throws a bowling ball of radius $R = 11$ cm down a lane. The ball slides on the lane, with initial speed $v_{cm,0} = 8.5$ m/s and initial angular speed $\omega_0 = 0$. The coefficient of kinetic friction between the ball and the lane is 0.21. The kinetic frictional force $\mathbf{f}_k$ acting on the ball (Fig. 12-35) causes a linear acceleration of the ball while producing a torque that causes an angular acceleration of the ball. When speed v_{cm} has decreased enough and angular speed ω has increased enough, the ball stops sliding and then rolls smoothly. (a) What then is v_{cm} in terms of ω? During the sliding, what are the ball's (b) linear acceleration and (c) angular acceleration? (d) How long does the ball slide? (e) How far does the ball slide? (f) What is the speed of the ball when smooth rolling begins?

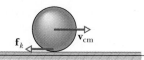

FIGURE 12-35 Problem 16.

SECTION 12-2 The Yo-Yo

17E. A yo-yo has a rotational inertia of 950 g·cm² and a mass of 120 g. Its axle radius is 3.2 mm, and its string is 120 cm long. The yo-yo rolls from rest down to the end of the string. (a) What is its linear acceleration? (b) How long does it take to reach the end of the string? As it reaches the end of the string, what are its (c) linear speed, (d) translational kinetic energy, (e) rotational kinetic energy, and (f) angular speed?

18P. Suppose that the yo-yo in Exercise 17, instead of rolling from rest, is thrown so that its initial speed down the string is 1.3 m/s. (a) How long does the yo-yo take to reach the end of the string? As it reaches the end of the string, what are its (b) total kinetic energy, (c) linear speed, (d) translational kinetic energy, (e) angular speed, and (f) rotational kinetic energy?

SECTION 12-3 Torque Revisited

19E. Given that $\mathbf{r} = x\mathbf{i} + y\mathbf{j} + z\mathbf{k}$ and $\mathbf{F} = F_x\mathbf{i} + F_y\mathbf{j} + F_z\mathbf{k}$, show that the torque $\boldsymbol{\tau} = \mathbf{r} \times \mathbf{F}$ is given by

$$\boldsymbol{\tau} = (yF_z - zF_y)\mathbf{i} + (zF_x - xF_z)\mathbf{j} + (xF_y - yF_x)\mathbf{k}.$$

20E. Show that, if $\mathbf{r}$ and $\mathbf{F}$ lie in a given plane, the torque $\boldsymbol{\tau} = \mathbf{r} \times \mathbf{F}$ has no component in that plane.

21E. What are the magnitude and direction of the torque about the origin on a plum located at coordinates $(-2.0$ m, 0, 4.0 m$)$ due to force $\mathbf{F}$ whose only component is (a) $F_x = 6.0$ N, (b) $F_x = -6.0$ N, (c) $F_z = 6.0$ N, and (d) $F_z = -6.0$ N?

22E. What are the magnitude and direction of the torque about the origin on a particle located at coordinates $(0, -4.0$ m, 3.0 m$)$ due to (a) force $\mathbf{F}_1$ with components $F_{1x} = 2.0$ N and $F_{1y} = F_{1z} = 0$, and (b) force $\mathbf{F}_2$ with components $F_{2x} = 0$, $F_{2y} = 2.0$ N, and $F_{2z} = 4.0$ N?

23P. Force $\mathbf{F} = (2.0$ N$)\mathbf{i} - (3.0$ N$)\mathbf{k}$ acts on a pebble with position vector $\mathbf{r} = (0.50$ m$)\mathbf{j} - (2.0$ m$)\mathbf{k}$, relative to the origin. What is the resulting torque acting on the pebble about (a) the origin and (b) a point with coordinates $(2.0$ m, 0, -3.0 m$)$?

24P. What is the torque about the origin on a jar of jalapeño peppers located at coordinates $(3.0$ m, -2.0 m, 4.0 m$)$ due to (a) force $\mathbf{F}_1 = (3.0$ N$)\mathbf{i} - (4.0$ N$)\mathbf{j} + (5.0$ N$)\mathbf{k}$, (b) force $\mathbf{F}_2 = (-3.0$ N$)\mathbf{i} - (4.0$ N$)\mathbf{j} - (5.0$ N$)\mathbf{k}$, and (c) the vector sum of $\mathbf{F}_1$ and $\mathbf{F}_2$? (d) Repeat (c) for a point with coordinates $(3.0$ m, 2.0 m, 4.0 m$)$ instead of the origin.

25P. What is the net torque about the origin on a flea located at coordinates $(0, -4.0$ m, 5.0 m$)$ when forces $\mathbf{F}_1 = (3.0$ N$)\mathbf{k}$ and $\mathbf{F}_2 = (-2.0$ N$)\mathbf{j}$ act on the flea?

26P. Force $\mathbf{F} = (-8.0$ N$)\mathbf{i} + (6.0$ N$)\mathbf{j}$ acts on a particle with position vector $\mathbf{r} = (3.0$ m$)\mathbf{i} + (4.0$ m$)\mathbf{j}$. What are (a) the torque on the particle about the origin and (b) the angle between the directions of $\mathbf{r}$ and $\mathbf{F}$?

SECTION 12-4 Angular Momentum

27E. Two objects are moving as shown in Fig. 12-36. What is their total angular momentum about point O?

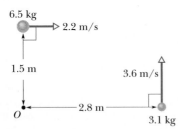

FIGURE 12-36 Exercise 27.

28E. A 1200 kg airplane is flying in a straight line at 80 m/s, 1.3 km above the ground. What is the magnitude of its angular momentum with respect to a point on the ground directly under the path of the plane?

29E. A particle P with mass 2.0 kg has position vector $\mathbf{r}$ ($r = 3.0$ m) and velocity $\mathbf{v}$ ($v = 4.0$ m/s) as shown in Fig. 12-37. It is

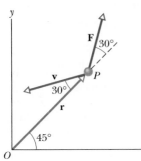

FIGURE 12-37
Exercise 29.

acted on by force $\mathbf{F}$ ($F = 2.0$ N). All three vectors lie in the xy plane. About the origin, what are (a) the angular momentum of the particle and (b) the torque acting on the particle?

30E. If we are given r, p, and ϕ, we can calculate the angular momentum of a particle from Eq. 12-26. Sometimes, however, we are given the components (x, y, z) of $\mathbf{r}$ and (v_x, v_y, v_z) of $\mathbf{v}$ instead. (a) Show that the components of ℓ along the x, y, and z axes are then given by $\ell_x = m(yv_z - zv_y)$, $\ell_y = m(zv_x - xv_z)$, and $\ell_z = m(xv_y - yv_x)$. (b) Show that if the particle moves only in the xy plane, the angular momentum vector has only a z component.

31E. At a certain time, the position vector in meters of a 0.25 kg object is $\mathbf{r} = 2.0\mathbf{i} - 2.0\mathbf{k}$. At that instant, its velocity in meters per second is $\mathbf{v} = -5.0\mathbf{i} + 5.0\mathbf{k}$, and the force in newtons acting on it is $\mathbf{F} = 4.0\mathbf{j}$. (a) What is the angular momentum of the object about the origin? (b) What torque acts on it? (*Hint:* See Exercises 19 and 30.)

32P. What is the magnitude of the angular momentum, about the Earth's center, of an 84 kg person on the equator due to the rotation of the Earth?

33P. Two particles, each of mass m and speed v, travel in opposite directions along parallel lines separated by a distance d. (a) In terms of m, v, and d, find an expression for the magnitude L of the angular momentum of the two-particle system around a point

midway between the two lines. (b) Does the expression change if we change the point about which L is calculated? (c) Now reverse the direction of travel for one of the particles and repeat (a) and (b).

34P. A 2.0 kg object moves in a plane with velocity components $v_x = 30$ m/s and $v_y = 60$ m/s as it passes through the point $(x, y) = (3.0, -4.0)$ m. (a) What is its angular momentum relative to the origin at this moment? (b) What is its angular momentum relative to the point $(-2.0, -2.0)$ m at this same moment?

35P. (a) Use the data given in the appendices to compute the total of the magnitudes of the angular momenta of all the planets owing to their revolution about the Sun. (b) What fraction of this total is associated with the planet Jupiter?

SECTION 12-5 Newton's Second Law in Angular Form

36E. A 3.0 kg particle is at $x = 3.0$ m, $y = 8.0$ m with a velocity of $\mathbf{v} = (5.0$ m/s$)\mathbf{i} - (6.0$ m/s$)\mathbf{j}$. It is acted on by a 7.0 N force in the negative x direction. (a) What is the angular momentum of the particle about the origin? (b) What torque about the origin acts on the particle? (c) At what rate is the angular momentum of the particle changing with time?

37E. A particle is acted on by two torques about the origin: $\boldsymbol{\tau}_1$ has a magnitude of 2.0 N·m and points in the direction of increasing x, and $\boldsymbol{\tau}_2$ has a magnitude of 4.0 N·m and points in the direction of decreasing y. What are the magnitude and direction of $d\boldsymbol{\ell}/dt$, where $\boldsymbol{\ell}$ is the angular momentum of the particle about the origin?

38E. What torque about the origin acts on a particle moving in the xy plane if the particle has the following values of angular momentum about the origin:

$$(a) -4.0 \text{ kg·m}^2/\text{s}, \quad (c) -4.0\sqrt{t} \text{ kg·m}^2/\text{s},$$
$$(b) -4.0t^2 \text{ kg·m}^2/\text{s}, \quad (d) -4.0/t^2 \text{ kg·m}^2/\text{s}?$$

39E. A 3.0 kg toy car on the x axis has velocity $\mathbf{v} = -2.0t^3$ m/s along that axis. About the origin and for $t > 0$, what are (a) the car's angular momentum and (b) the torque acting on the car? (c) Repeat (a) and (b) for a point with coordinates (2.0 m, 5.0 m, 0) instead of the origin. (d) Repeat (a) and (b) for a point with coordinates (2.0 m, −5.0 m, 0) instead of the origin.

40P. At $t = 0$, a 2.0 kg particle has position vector $\mathbf{r} = (4.0$ m$)\mathbf{i} - (2.0$ m$)\mathbf{j}$ relative to the origin. Its velocity is given by $\mathbf{v} = (-6.0t^4$ m/s$)\mathbf{i} + (3.0$ m/s$)\mathbf{j}$. About the origin and for $t > 0$, what are (a) the particle's angular momentum and (b) the torque acting on the particle? (c) Repeat (a) and (b) for a point with coordinates $(-2.0$ m, -3.0 m, 0) instead of the origin.

41P. A projectile of mass m is fired from the ground with an initial speed v_0 and an initial angle θ_0 above the horizontal. (a) Find an expression for the magnitude of its angular momentum about the firing point as a function of time. (b) Find the rate at which the angular momentum changes with time. (c) Evaluate the magnitude of $\mathbf{r} \times \mathbf{F}$ directly and compare the result with (b). Why should the results be identical?

SECTION 12-7 The Angular Momentum of a Rigid Body Rotating About a Fixed Axis

42E. A sanding disk with rotational inertia 1.2×10^{-3} kg·m² is attached to an electric drill whose motor delivers a torque of 16 N·m. Find (a) the angular momentum of the disk about its central axis and (b) the angular speed of the disk 33 ms after the motor is turned on.

43E. The angular momentum of a flywheel having a rotational inertia of 0.140 kg·m² about its axis decreases from 3.00 to 0.800 kg·m²/s in 1.50 s. (a) What is the average torque acting on the flywheel about its central axis during this period? (b) Assuming a uniform angular acceleration, through what angle will the flywheel have turned? (c) How much work was done on the wheel? (d) What is the average power of the flywheel?

44E. Three particles, each of mass m, are fastened to each other and to a rotation axis at O by three massless strings, each with length l as shown in Fig. 12-38. The combination rotates around the rotational axis with angular velocity ω in such a way that the particles remain in a straight line. In terms of m, l, and ω, and relative to point O, what are (a) the rotational inertia of the combination, (b) the angular momentum of the middle particle, and (c) the total angular momentum of the three particles?

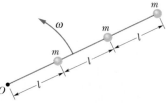

FIGURE 12-38 Exercise 44.

45E. A uniform rod rotates in a horizontal plane about a vertical axis through one end. The rod is 6.00 m long, weighs 10.0 N, and rotates at 240 rev/min clockwise when seen from above. Calculate (a) the rotational inertia of the rod about the axis of rotation and (b) the angular momentum of the rod about that axis.

46P. Figure 12-39 shows a rigid structure consisting of a circular hoop, of radius R and mass m, and a square made of four thin bars, each of length R and mass m. The rigid structure rotates at a constant speed about a vertical axis with a period of rotation of 2.5 s. Assuming $R = 0.50$ m and $m = 2.0$ kg, calculate (a) the structure's rotational inertia about the axis of rotation and (b) its angular momentum about that axis.

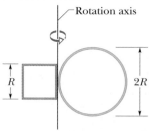

FIGURE 12-39 Problem 46.

47P. Wheels A and B in Fig. 12-40 are connected by a belt that does not slip. The radius of wheel B is three times the radius of wheel A. What would be the ratio of the rotational inertias I_A/I_B if both wheels had (a) the same angular momenta about their central axes and (b) the same rotational kinetic energies?

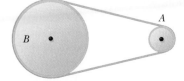

FIGURE 12-40 Problem 47.

48P. An impulsive force $F(t)$ acts for a short time Δt on a rotating rigid body with rotational inertia I. Show that

$$\int \tau \, dt = \overline{F} R \, \Delta t = I(\omega_f - \omega_i),$$

where R is the moment arm of the force, $\overline{F}$ is the average value of the force during the time it acts on the body, and ω_i and ω_f are the angular velocities of the body just before and just after the force acts. (The quantity $\int \tau \, dt = \overline{F} R \, \Delta t$ is called the *angular impulse,* in analogy with $\overline{F} \, \Delta t$, the linear impulse.)

49P*. Two cylinders having radii R_1 and R_2 and rotational inertias I_1 and I_2 about the central axis are supported by axes perpendicular to the plane of Fig. 12-41. The large cylinder is initially

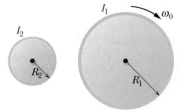

FIGURE 12-41 Problem 49.

rotating with angular velocity ω_0. The small cylinder is moved to the right until it touches the large cylinder and is caused to rotate by the frictional force between the two. Eventually, slipping ceases, and the two cylinders rotate at constant rates in opposite directions. Find the final angular velocity ω_2 of the small cylinder in terms of I_1, I_2, R_1, R_2, and ω_0. (*Hint:* Neither angular momentum nor kinetic energy is conserved. Apply the angular impulse equation of Problem 48.)

SECTION 12-8 Conservation of Angular Momentum

50E. The rotor of an electric motor has rotational inertia $I_m = 2.0 \times 10^{-3}$ kg·m² about its central axis. The motor is used to change the orientation of the space probe in which it is mounted. The motor axis is mounted parallel to the axis of the probe, which has rotational inertia $I_p = 12$ kg·m² about its axis. Calculate the number of revolutions of the rotor required to turn the probe through 30° about its axis.

51E. A man stands on a platform that is rotating (without friction) with an angular speed of 1.2 rev/s; his arms are outstretched and he holds a weight in each hand. The rotational inertia of the system of man, weights, and platform about the central axis is 6.0 kg·m². If by moving the weights the man decreases the rotational inertia of the system to 2.0 kg·m², (a) what is the resulting angular speed of the platform and (b) what is the ratio of the new kinetic energy of the system to the original kinetic energy? (c) What provided the added kinetic energy?

52E. Two disks are mounted on low-friction bearings on the same axle and can be brought together so that they couple and rotate as one unit. (a) The first disk, with rotational inertia 3.3 kg·m² about its central axis, is set spinning at 450 rev/min. The second disk, with rotational inertia 6.6 kg·m² about its central axis, is set spinning at 900 rev/min in the same direction as the first. They then couple together. What is their angular speed after coupling? (b) If instead the second disk is set spinning at 900 rev/min in the direction opposite the first disk's rotation, what is the angular speed after coupling?

53E. A wheel is rotating freely with an angular speed of 800 rev/min on a shaft whose rotational inertia is negligible. A second wheel, initially at rest and with twice the rotational inertia of the first, is suddenly coupled to the same shaft. (a) What is the angular speed of the resultant combination of the shaft and two wheels? (b) What fraction of the original rotational kinetic energy is lost?

54E. The rotational inertia of a collapsing spinning star changes to one-third its initial value. What is the ratio of the new rotational kinetic energy to the initial rotational kinetic energy?

55E. Suppose that the Sun runs out of nuclear fuel and suddenly collapses to form a white dwarf star, with a diameter equal to that of the Earth. Assuming no mass loss, what would then be the Sun's new rotation period, which currently is about 25 days? Assume that the Sun and the white dwarf are uniform, solid spheres.

56E. In a playground, there is a small merry-go-round of radius 1.20 m and mass 180 kg. Its radius of gyration (see Problem 58 of Chapter 11) is 91.0 cm. A child of mass 44.0 kg runs at a speed of 3.00 m/s along a path that is tangent to the rim of the initially stationary merry-go-round and then jumps on. Neglect friction between the bearings and the shaft of the merry-go-round. Calculate (a) the rotational inertia of the merry-go-round about its axis of rotation, (b) the angular momentum of the child, while running, about the axis of rotation of the merry-go-round, and (c) the angular speed of the merry-go-round and child after the child has jumped on.

57E. A horizontal platform in the shape of a circular disk rotates on a frictionless bearing about a vertical axle through the center of the disk. The platform has a mass of 150 kg, a radius of 2.0 m, and a rotational inertia of 300 kg·m² about the axis of rotation. A 60 kg student walks slowly from the rim of the platform toward the center. If the angular speed of the system is 1.5 rad/s when the student starts at the rim, what is the angular speed when she is 0.50 m from the center?

58E. With center and spokes of negligible mass, a certain bicycle wheel has a thin rim of radius 1.14 ft and weight 8.36 lb; it can turn on its axle with negligible friction. A man holds the wheel above his head with the axle vertical while he stands on a turntable free to rotate without friction; the wheel rotates clockwise, as seen from above, with an angular speed of 57.7 rad/s, and the

turntable is initially at rest. The rotational inertia of *wheel + man + turntable* about the common axis of rotation is 1.54 slug·ft². The man's free hand suddenly stops the rotation of the wheel (relative to the turntable). Determine the resulting angular velocity (magnitude and direction) of the system.

59P. Two skaters, each of mass 50 kg, approach each other along parallel paths separated by 3.0 m. They have equal and opposite velocities of 1.4 m/s. The first skater carries one end of a long pole with negligible mass, and the second skater grabs the other end of it as she passes; see Fig. 12-42. Assume frictionless ice. (a) Describe quantitatively the motion of the skaters after they have become connected by the pole. (b) By pulling on the pole, the skaters reduce their separation to 1.0 m. What is their angular speed then? (c) Calculate the kinetic energy of the system in (a) and (b). (d) What is the source of the added kinetic energy?

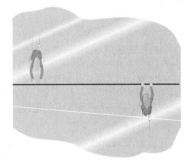

FIGURE 12-42   Problem 59.

60P. Two children, each with mass M, sit on opposite ends of a narrow board with length L and mass M (the same as the mass of each child). The board is pivoted at its center and is free to rotate in a horizontal circle without friction. (Treat it as a thin rod.) (a) What is the rotational inertia of the board plus the children about a vertical axis through the center of the board? (b) What is the angular momentum of the system if it is rotating with angular speed ω_0 in a clockwise direction as seen from above? What is the direction of the angular momentum? (c) While the system is rotating, the children pull themselves toward the center of the board until they are half as far from the center as before. What is the resulting angular speed in terms of ω_0? (d) What is the change in kinetic energy of the system as a result of the children changing their positions? (What is the source of the added kinetic energy?)

61P. A toy train track is mounted on a large wheel that is free to turn with negligible friction about a vertical axis (Fig. 12-43). A toy train of mass m is placed on the track and, with the system initially at rest, the electrical power is turned on. The train reaches a steady speed v with respect to the track. What is the angular

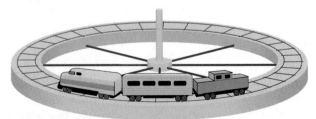

FIGURE 12-43   Problem 61.

velocity ω of the wheel, if its mass is M and its radius R? (Treat the wheel as a hoop, and neglect the mass of the spokes and hub.)

62P. A cockroach of mass m runs counterclockwise around the rim of a lazy Susan (a circular dish mounted on a vertical axle) of radius R and rotational inertia I and having frictionless bearings. The cockroach's speed (relative to the Earth) is v, whereas the lazy Susan turns clockwise with angular speed ω_0. The cockroach finds a bread crumb on the rim and, of course, stops. (a) What is the angular speed of the lazy Susan after the cockroach stops? (b) Is mechanical energy conserved?

63P. A girl of mass M stands on the rim of a frictionless merry-go-round of radius R and rotational inertia I that is not moving. She throws a rock of mass m horizontally in a direction that is tangent to the outer edge of the merry-go-round. The speed of the rock, relative to the ground, is v. Afterwards what are (a) the angular speed of the merry-go-round and (b) the linear speed of the girl?

64P. A phonograph record of mass 0.10 kg and radius 0.10 m rotates about a vertical axis through its center with an angular speed of 4.7 rad/s. The rotational inertia of the record about its axis of rotation is 5.0×10^{-4} kg·m². A wad of putty of mass 0.020 kg drops vertically onto the record from above and sticks to the edge of the record. What is the angular speed of the record immediately after the putty sticks to it?

65P. A uniform thin rod of length 0.50 m and mass 4.0 kg can rotate in a horizontal plane about a vertical axis through its center. The rod is at rest when a 3.0-g bullet traveling in the horizontal plane of the rod is fired into one end of the rod. As viewed from above, the direction of the bullet's velocity makes an angle of 60° with the rod (Fig. 12-44). If the bullet lodges in the rod and the angular velocity of the rod is 10 rad/s immediately after the collision, what is the magnitude of the bullet's velocity just before impact?

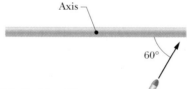

FIGURE 12-44 Problem 65.

66P. A cockroach of mass m lies on the rim of a uniform disk of mass $10.0m$ that can rotate freely about its center like a merry-go-round. Initially the cockroach and disk rotate together with an angular velocity of ω_0. Then the cockroach walks half way to the center of the disk. (a) What is the change $\Delta\omega$ in the angular velocity of the cockroach-disk system? (b) What is the ratio K/K_0 of the new kinetic energy of the system to its initial kinetic energy? (c) What accounts for the change in the kinetic energy?

67P. A uniform disk of mass $10m$ and radius $3.0r$ can rotate freely about its fixed center like a merry-go-round. A smaller uniform disk of mass m and radius r lies on top of the larger disk, concentric with it. Initially the two disks rotate together with an angular velocity of 20 rad/s. Then a slight disturbance causes the smaller disk to slide outward across the larger disk, until the outer edge of the smaller disk catches on the outer edge of the larger

disk. Afterwards, the two disks again rotate together (without further sliding). (a) What then is their angular velocity about the center of the larger disk? (b) What is the ratio K/K_0 of the new kinetic energy of the two-disk system to the system's initial kinetic energy?

68P. If the Earth's polar ice caps melted and the water returned to the oceans, the oceans would be made deeper by about 30 m. What effect would this have on the Earth's rotation? Make an estimate of the resulting change in the length of the day. (Concern has been expressed that warming of the atmosphere resulting from industrial pollution could cause the ice caps to melt.)

69P*. The particle of mass m in Fig. 12-45 slides down the frictionless surface and collides with the uniform vertical rod, sticking to it. The rod pivots about O through the angle θ before momentarily coming to rest. Find θ in terms of the other parameters given in the figure.

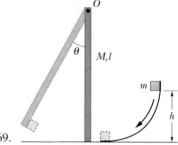

FIGURE 12-45 Problem 69.

70P*. Two 2.00 kg balls are attached to the ends of a thin rod of negligible mass, 50.0 cm long. The rod is free to rotate in a vertical plane without friction about a horizontal axis through its center. While the rod is horizontal (Fig. 12-46), a 50.0 g putty wad drops onto one of the balls with a speed of 3.00 m/s and sticks to it. (a) What is the angular speed of the system just after the putty wad hits? (b) What is the ratio of the kinetic energy of the entire system after the collision to that of the putty wad just before? (c) Through what angle will the system rotate until it momentarily stops?

FIGURE 12-46 Problem 70.

SECTION 12-9 Quantized Angular Momentum

71E. What is the value of spin angular momentum S (along any given direction) for an electron?

72E. A positive pion (π^+) can spontaneously decay to (suddenly transform to) a positive muon (μ^+) and a neutrino (ν):

$$\pi^+ \rightarrow \mu^+ + \nu.$$

If the neutrino has $m_s = +\frac{1}{2}$, what are the values of (a) m_s and (b) S for the positive muon?

73E. Two protons colliding at high speeds can transform to three protons and one antiproton:

$$p + p \rightarrow p + p + p + \bar{p}.$$

If the two colliding protons are both spin up, what are the spin orientations (up or down) of the transformation products?

ELECTRONIC COMPUTATION

74. When a bowling ball is bowled, it usually skids for a short distance, then rolls without slipping. While it is skidding, a force of kinetic friction acts to slow the translational motion and increase the angular speed. Skidding stops when $\omega R = v$, where ω is the angular speed, R is the radius, and v is the speed of the center of mass. The net force acting on the ball has magnitude $f = \mu_k N$, where μ_k is the coefficient of kinetic friction and N is the magnitude of the normal force. Since the alley is horizontal, $N = mg$, where m is the mass of the ball. The net torque on the ball has magnitude $\tau = fR$, where R is the radius of the ball. A bowling ball has a mass of 7.25 kg, a radius of 10.9 cm, and a rotational inertia that is nearly the same as that of a uniform sphere. Set up a program to compute the angular speed and translational speed every 0.1 s from the time the ball is bowled to the time it stops skidding. Take the coefficient of kinetic friction to be 0.35 and the initial translational speed to be 20 m/s. For each of the initial conditions given below, plot a graph of the speed as a function of time. Find the time when the ball stops skidding and also the linear speed of the ball at that time. Except for a slight decrease due to air drag, this is the speed with which the ball hits the pins. (a) The ball is not rotating initially. (b) The ball is released with an angular speed of 150 rad/s, spinning in the same direction as when it stops skidding. (c) The ball is released with

the same angular speed but in the "wrong" direction. (d) How should a bowling ball be released to have the greatest linear speed as it hits the pins?

75. A projectile with a mass of 0.15 kg is fired with an initial speed of 50 m/s at an angle of 35° above the horizontal. (a) Neglect air resistance and use a numerical integration program to calculate its angular momentum about the firing point every 0.1 s from the time it is fired until it again reaches the height from which it was fired. Use $\tau = \mathbf{r} \times \mathbf{F}$, where $\mathbf{r}$ is the position vector of the projectile and $\mathbf{F}$ is the gravitational force on the projectile, to calculate the torque τ on the projectile (about the firing point) at each of those times. Plot the magnitude of the angular momentum as a function of the square of the time and verify that the graph is a straight line. Plot the magnitude of the torque as a function of time and verify that the graph is a straight line. Are these graphs consistent with Newton's second law for rotation, $\tau = d\mathbf{L}/dt$? (b) Now include air resistance. Take the terminal speed to be $v_t = 75$ m/s. The acceleration can be written $\mathbf{a} = -g\mathbf{j} - bv\mathbf{v}$, where $b = g/v_t^2$. Does the magnitude of the angular momentum increase in proportion to the square of the time? Is the magnitude of the torque proportional to the time? At any position of the projectile, is the magnitude of the torque greater or less than it is in the absence of drag? Give an argument based on the directions of the forces acting on the projectile to show that your numerical result is plausible.

76. At the instant the displacement of a 2.00 kg object relative to the origin is $\mathbf{d} = (2.00 \text{ m})\mathbf{i} + (4.00 \text{ m})\mathbf{j} - (3.00 \text{ m})\mathbf{k}$, its velocity is $\mathbf{v} = -(6.00 \text{ m/s})\mathbf{i} + (3.00 \text{ m/s})\mathbf{j} + (3.00 \text{ m/s})\mathbf{k}$, and it is subject to a force $\mathbf{F} = (6.00 \text{ N})\mathbf{i} - (8.00 \text{ N})\mathbf{j} + (4.0 \text{ N})\mathbf{k}$. Find (a) the acceleration of the object, (b) the angular momentum of the object about the origin, (c) the torque about the origin acting on the object, and (d) the angle between the velocity of the object and the force acting on the object.

Appendix A
The International System of Units (SI)*

1. THE SI BASE UNITS

QUANTITY	NAME	SYMBOL	DEFINITION
length	meter	m	". . . the length of the path traveled by light in vacuum in 1/299,792,458 of a second." (1983)
mass	kilogram	kg	". . . this prototype [a certain platinum–iridium cylinder] shall henceforth be considered to be the unit of mass." (1889)
time	second	s	". . . the duration of 9,192,631,770 periods of the radiation corresponding to the transition between the two hyperfine levels of the ground state of the cesium-133 atom." (1967)
electric current	ampere	A	". . . that constant current which, if maintained in two straight parallel conductors of infinite length, of negligible circular cross section, and placed 1 meter apart in vacuum, would produce between these conductors a force equal to 2×10^{-7} newton per meter of length." (1946)
thermodynamic temperature	kelvin	K	". . . the fraction 1/273.16 of the thermodynamic temperature of the triple point of water." (1967)
amount of substance	mole	mol	". . . the amount of substance of a system which contains as many elementary entities as there are atoms in 0.012 kilogram of carbon-12." (1971)
luminous intensity	candela	cd	". . . the luminous intensity, in the perpendicular direction, of a surface of 1/600,000 square meter of a blackbody at the temperature of freezing platinum under a pressure of 101.325 newtons per square meter." (1967)

*Adapted from "The International System of Units (SI)," National Bureau of Standards Special Publication 330, 1972 edition. The definitions above were adopted by the General Conference of Weights and Measures, an international body, on the dates shown. In this book we do not use the candela.

2. SOME SI DERIVED UNITS

QUANTITY	NAME OF UNIT	SYMBOL	
area	square meter	m^2	
volume	cubic meter	m^3	
frequency	hertz	Hz	s^{-1}
mass density (density)	kilogram per cubic meter	kg/m^3	
speed, velocity	meter per second	m/s	
angular velocity	radian per second	rad/s	
acceleration	meter per second per second	m/s^2	
angular acceleration	radian per second per second	rad/s^2	
force	newton	N	$kg \cdot m/s^2$
pressure	pascal	Pa	N/m^2
work, energy, quantity of heat	joule	J	$N \cdot m$
power	watt	W	J/s
quantity of electric charge	coulomb	C	$A \cdot s$
potential difference, electromotive force	volt	V	W/A
electric field strength	volt per meter (or newton per coulomb)	V/m	N/C
electric resistance	ohm	Ω	V/A
capacitance	farad	F	$A \cdot s/V$
magnetic flux	weber	Wb	$V \cdot s$
inductance	henry	H	$V \cdot s/A$
magnetic flux density	tesla	T	Wb/m^2
magnetic field strength	ampere per meter	A/m	
entropy	joule per kelvin	J/K	
specific heat	joule per kilogram kelvin	$J/(kg \cdot K)$	
thermal conductivity	watt per meter kelvin	$W/(m \cdot K)$	
radiant intensity	watt per steradian	W/sr	

3. THE SI SUPPLEMENTARY UNITS

QUANTITY	NAME OF UNIT	SYMBOL
plane angle	radian	rad
solid angle	steradian	sr

Appendix **B**
*Some Fundamental Constants of Physics**

CONSTANT	SYMBOL	COMPUTATIONAL VALUE	BEST (1986) VALUE	
			VALUE[a]	UNCERTAINTY[b]
Speed of light in a vacuum	c	3.00×10^8 m/s	2.99792458	exact
Elementary charge	e	1.60×10^{-19} C	1.60217733	0.30
Gravitational constant	G	6.67×10^{-11} m³/s²·kg	6.67259	128
Universal gas constant	R	8.31 J/mol·K	8.314510	8.4
Avogadro constant	N_A	6.02×10^{23} mol⁻¹	6.0221367	0.59
Boltzmann constant	k	1.38×10^{-23} J/K	1.380658	8.5
Stefan-Boltzmann constant	σ	5.67×10^{-8} W/m²·K⁴	5.67051	34
Molar volume of ideal gas at STP[d]	V_m	2.24×10^{-2} m³/mol	2.241409	8.4
Permittivity constant	ϵ_0	8.85×10^{-12} F/m	8.85418781762	exact
Permeability constant	μ_0	1.26×10^{-6} H/m	1.25663706143	exact
Planck constant	h	6.63×10^{-34} J·s	6.6260755	0.60
Electron mass[c]	m_e	9.11×10^{-31} kg	9.1093897	0.59
		5.49×10^{-4} u	5.48579903	0.023
Proton mass[c]	m_p	1.67×10^{-27} kg	1.6726231	0.59
		1.0073 u	1.0072764660	0.005
Ratio of proton mass to electron mass	m_p/m_e	1840	1836.152701	0.020
Electron charge-to-mass ratio	e/m_e	1.76×10^{11} C/kg	1.75881961	0.30
Neutron mass[c]	m_n	1.68×10^{-27} kg	1.6749286	0.59
		1.0087 u	1.0086649235	0.0023
Hydrogen atom mass[c]	m_{1_H}	1.0078 u	1.0078250316	0.0005
Deuterium atom mass[c]	m_{2_H}	2.0141 u	2.0141017779	0.0005
Helium atom mass[c]	$m_{4_{He}}$	4.0026 u	4.0026032	0.067
Muon mass	m_μ	1.88×10^{-28} kg	1.8835326	0.61
Electron magnetic moment	μ_e	9.28×10^{-24} J/T	9.2847701	0.34
Proton magnetic moment	μ_p	1.41×10^{-26} J/T	1.41060761	0.34
Bohr magneton	μ_B	9.27×10^{-24} J/T	9.2740154	0.34
Nuclear magneton	μ_N	5.05×10^{-27} J/T	5.0507866	0.34
Bohr radius	r_B	5.29×10^{-11} m	5.29177249	0.045
Rydberg constant	R	1.10×10^7 m⁻¹	1.0973731534	0.0012
Electron Compton wavelength	λ_C	2.43×10^{-12} m	2.42631058	0.089

[a]Values given in this column should be given the same unit and power of 10 as the computational value.   [b]Parts per million.   [c]Masses given in u are in unified atomic mass units, where 1 u = $1.6605402 \times 10^{-27}$ kg.   [d]STP means standard temperature and pressure: 0°C and 1.0 atm (0.1 MPa).

*The values in this table were largely selected from a longer list in *Symbols, Units and Nomenclature in Physics* (IUPAP), prepared by E. Richard Cohen and Pierre Giacomo, 1986.

Appendix C
Some Astronomical Data

SOME DISTANCES FROM THE EARTH

To the moon*	3.82×10^8 m
To the sun*	1.50×10^{11} m
To the nearest star (Proxima Centauri)	4.04×10^{16} m
To the center of our galaxy	2.2×10^{20} m
To the Andromeda Galaxy	2.1×10^{22} m
To the edge of the observable universe	$\sim 10^{26}$ m

*Mean distance.

THE SUN, THE EARTH, AND THE MOON

PROPERTY	UNIT	SUN	EARTH	MOON
Mass	kg	1.99×10^{30}	5.98×10^{24}	7.36×10^{22}
Mean radius	m	6.96×10^8	6.37×10^6	1.74×10^6
Mean density	kg/m^3	1410	5520	3340
Free-fall acceleration at the surface	m/s^2	274	9.81	1.67
Escape velocity	km/s	618	11.2	2.38
Period of rotation[a]	—	37 d at poles[b] 26 d at equator[b]	23 h 56 min	27.3 d
Radiation power[c]	W	3.90×10^{26}		

[a]Measured with respect to the distant stars.

[b]The sun, a ball of gas, does not rotate as a rigid body.

[c]Just outside the Earth's atmosphere solar energy is received, assuming normal incidence, at the rate of 1340 W/m^2.

SOME PROPERTIES OF THE PLANETS

	MERCURY	VENUS	EARTH	MARS	JUPITER	SATURN	URANUS	NEPTUNE	PLUTO
Mean distance from sun, 10^6 km	57.9	108	150	228	778	1430	2870	4500	5900
Period of revolution, y	0.241	0.615	1.00	1.88	11.9	29.5	84.0	165	248
Period of rotation,[a] d	58.7	-243^b	0.997	1.03	0.409	0.426	-0.451^b	0.658	6.39
Orbital speed, km/s	47.9	35.0	29.8	24.1	13.1	9.64	6.81	5.43	4.74
Inclination of axis to orbit	<28°	≈3°	23.4°	25.0°	3.08°	26.7°	97.9°	29.6°	57.5°
Inclination of orbit to Earth's orbit	7.00°	3.39°		1.85°	1.30°	2.49°	0.77°	1.77°	17.2°
Eccentricity of orbit	0.206	0.0068	0.0167	0.0934	0.0485	0.0556	0.0472	0.0086	0.250
Equatorial diameter, km	4880	12,100	12,800	6790	143,000	120,000	51,800	49,500	2300
Mass (Earth = 1)	0.0558	0.815	1.000	0.107	318	95.1	14.5	17.2	0.002
Density (water = 1)	5.60	5.20	5.52	3.95	1.31	0.704	1.21	1.67	2.03
Surface value of g,[c] m/s^2	3.78	8.60	9.78	3.72	22.9	9.05	7.77	11.0	0.5
Escape velocity,[c] km/s	4.3	10.3	11.2	5.0	59.5	35.6	21.2	23.6	1.1
Known satellites	0	0	1	2	16 + ring	18 + rings	15 + rings	8 + rings	1

[a]Measured with respect to the distant stars.

[b]Venus and Uranus rotate opposite their orbital motion.

[c]Gravitational acceleration measured at the planet's equator.

Appendix D
Conversion Factors

Conversion factors may be read directly from these tables. For example, 1 degree = 2.778×10^{-3} revolutions, so $16.7° = 16.7 \times 2.778 \times 10^{-3}$ rev. The SI quantities are fully capitalized.

Adapted in part from G. Shortley and D. Williams, *Elements of Physics*, Prentice-Hall, Englewood Cliffs, NJ, 1971.

PLANE ANGLE

	°	′	″	RADIAN	rev
1 degree =	1	60	3600	1.745×10^{-2}	2.778×10^{-3}
1 minute =	1.667×10^{-2}	1	60	2.909×10^{-4}	4.630×10^{-5}
1 second =	2.778×10^{-4}	1.667×10^{-2}	1	4.848×10^{-6}	7.716×10^{-7}
1 RADIAN =	57.30	3438	2.063×10^{5}	1	0.1592
1 revolution =	360	2.16×10^{4}	1.296×10^{6}	6.283	1

SOLID ANGLE

1 sphere = 4π steradians = 12.57 steradians

LENGTH

	cm	METER	km	in.	ft	mi
1 centimeter =	1	10^{-2}	10^{-5}	0.3937	3.281×10^{-2}	6.214×10^{-6}
1 METER =	100	1	10^{-3}	39.37	3.281	6.214×10^{-4}
1 kilometer =	10^{5}	1000	1	3.937×10^{4}	3281	0.6214
1 inch =	2.540	2.540×10^{-2}	2.540×10^{-5}	1	8.333×10^{-2}	1.578×10^{-5}
1 foot =	30.48	0.3048	3.048×10^{-4}	12	1	1.894×10^{-4}
1 mile =	1.609×10^{5}	1609	1.609	6.336×10^{4}	5280	1

1 angström = 10^{-10} m
1 nautical mile = 1852 m
  = 1.151 miles = 6076 ft

1 fermi = 10^{-15} m
1 light-year = 9.460×10^{12} km
1 parsec = 3.084×10^{13} km

1 fathom = 6 ft
1 Bohr radius = 5.292×10^{-11} m
1 yard = 3 ft

1 rod = 16.5 ft
1 mil = 10^{-3} in.
1 nm = 10^{-9} m

AREA

	METER^2	cm^2	ft^2	in.^{2}
1 SQUARE METER =	1	10^{4}	10.76	1550
1 square centimeter =	10^{-4}	1	1.076×10^{-3}	0.1550
1 square foot =	9.290×10^{-2}	929.0	1	144
1 square inch =	6.452×10^{-4}	6.452	6.944×10^{-3}	1

1 square mile = 2.788×10^7 ft^2
  = 640 acres

1 barn = 10^{-28} m^2

1 acre = 43,560 ft^2
1 hectare = 10^4 m^2 = 2.471 acres

VOLUME

	METER^3	cm^3	L	ft^3	in.^{3}
1 CUBIC METER = 1	10^6	1000	35.31	6.102 × 10^4
1 cubic centimeter = 10^{-6}	1	1.000 × 10^{-3}	3.531 × 10^{-5}	6.102 × 10^{-2}
1 liter = 1.000 × 10^{-3}	1000	1	3.531 × 10^{-2}	61.02
1 cubic foot = 2.832 × 10^{-2}	2.832 × 10^4	28.32	1	1728
1 cubic inch = 1.639 × 10^{-5}	16.39	1.639 × 10^{-2}	5.787 × 10^{-4}	1

1 U.S. fluid gallon = 4 U.S. fluid quarts = 8 U.S. pints = 128 U.S. fluid ounces = 231 in.^{3}

1 British imperial gallon = 277.4 in.^{3} = 1.201 U.S. fluid gallons

MASS

Quantities in the colored areas are not mass units but are often used as such. When we write, for example, 1 kg "=" 2.205 lb, this means that a kilogram is a *mass* that *weighs* 2.205 pounds at a location where g has the standard value of 9.80665 m/s^2.

	g	KILOGRAM	slug	u	oz	lb	ton
1 gram = 1	0.001	6.852 × 10^{-5}	6.022 × 10^{23}	3.527 × 10^{-2}	2.205 × 10^{-3}	1.102 × 10^{-6}
1 KILOGRAM = 1000	1	6.852 × 10^{-2}	6.022 × 10^{26}	35.27	2.205	1.102 × 10^{-3}
1 slug = 1.459 × 10^4	14.59	1	8.786 × 10^{27}	514.8	32.17	1.609 × 10^{-2}
1 atomic mass unit = 1.661 × 10^{-24}	1.661 × 10^{-27}	1.138 × 10^{-28}	1	5.857 × 10^{-26}	3.662 × 10^{-27}	1.830 × 10^{-30}
1 ounce = 28.35	2.835 × 10^{-2}	1.943 × 10^{-3}	1.718 × 10^{25}	1	6.250 × 10^{-2}	3.125 × 10^{-5}
1 pound = 453.6	0.4536	3.108 × 10^{-2}	2.732 × 10^{26}	16	1	0.0005
1 ton = 9.072 × 10^5	907.2	62.16	5.463 × 10^{29}	3.2 × 10^4	2000	1

1 metric ton = 1000 kg

DENSITY

Quantities in the colored areas are weight densities and, as such, are dimensionally different from mass densities. See note for mass table.

	slug/ft^3	KILOGRAM/ METER^3	g/cm^3	lb/ft^3	lb/in.^{3}
1 slug per foot^3 = 1	515.4	0.5154	32.17	1.862 × 10^{-2}
1 KILOGRAM per METER^3 = 1.940 × 10^{-3}	1	0.001	6.243 × 10^{-2}	3.613 × 10^{-5}
1 gram per centimeter^3 = 1.940	1000	1	62.43	3.613 × 10^{-2}
1 pound per foot^3 = 3.108 × 10^{-2}	16.02	1.602 × 10^{-2}	1	5.787 × 10^{-4}
1 pound per inch^3 = 53.71	2.768 × 10^4	27.68	1728	1

TIME

	y	d	h	min	SECOND
1 year = 1	365.25	8.766 × 10^3	5.259 × 10^5	3.156 × 10^7
1 day = 2.738 × 10^{-3}	1	24	1440	8.640 × 10^4
1 hour = 1.141 × 10^{-4}	4.167 × 10^{-2}	1	60	3600
1 minute = 1.901 × 10^{-6}	6.944 × 10^{-4}	1.667 × 10^{-2}	1	60
1 SECOND = 3.169 × 10^{-8}	1.157 × 10^{-5}	2.778 × 10^{-4}	1.667 × 10^{-2}	1

SPEED

	ft/s	km/h	METER/SECOND	mi/h	cm/s
1 foot per second = 1		1.097	0.3048	0.6818	30.48
1 kilometer per hour = 0.9113		1	0.2778	0.6214	27.78
1 METER per SECOND = 3.281		3.6	1	2.237	100
1 mile per hour = 1.467		1.609	0.4470	1	44.70
1 centimeter per second = 3.281×10^{-2}		3.6×10^{-2}	0.01	2.237×10^{-2}	1

1 knot = 1 nautical mi/h = 1.688 ft/s   1 mi/min = 88.00 ft/s = 60.00 mi/h

FORCE

Force units in the colored areas are now little used. To clarify: 1 gram-force (= 1 gf) is the force of gravity that would act on an object whose mass is 1 gram at a location where g has the standard value of 9.80665 m/s².

	dyne	NEWTON	lb	pdl	gf	kgf
1 dyne = 1		10^{-5}	2.248×10^{-6}	7.233×10^{-5}	1.020×10^{-3}	1.020×10^{-6}
1 NEWTON = 10^5		1	0.2248	7.233	102.0	0.1020
1 pound = 4.448×10^5		4.448	1	32.17	453.6	0.4536
1 poundal = 1.383×10^4		0.1383	3.108×10^{-2}	1	14.10	1.410×10^{-2}
1 gram-force = 980.7		9.807×10^{-3}	2.205×10^{-3}	7.093×10^{-2}	1	0.001
1 kilogram-force = 9.807×10^5		9.807	2.205	70.93	1000	1

PRESSURE

	atm	dyne/cm²	inch of water	cm Hg	PASCAL	lb/in.²	lb/ft²
1 atmosphere = 1		1.013×10^6	406.8	76	1.013×10^5	14.70	2116
1 dyne per centimeter² = 9.869×10^{-7}		1	4.015×10^{-4}	7.501×10^{-5}	0.1	1.405×10^{-5}	2.089×10^{-3}
1 inch of water[a] at 4°C = 2.458×10^{-3}		2491	1	0.1868	249.1	3.613×10^{-2}	5.202
1 centimeter of mercury[a] at 0°C = 1.316×10^{-2}		1.333×10^4	5.353	1	1333	0.1934	27.85
1 PASCAL = 9.869×10^{-6}		10	4.015×10^{-3}	7.501×10^{-4}	1	1.450×10^{-4}	2.089×10^{-2}
1 pound per inch² = 6.805×10^{-2}		6.895×10^4	27.68	5.171	6.895×10^3	1	144
1 pound per foot² = 4.725×10^{-4}		478.8	0.1922	3.591×10^{-2}	47.88	6.944×10^{-3}	1

[a] Where the acceleration of gravity has the standard value of 9.80665 m/s².

1 bar = 10^6 dyne/cm² = 0.1 MPa   1 millibar = 10^3 dyne/cm² = 10^2 Pa   1 torr = 1 mm Hg

ENERGY, WORK, HEAT

Quantities in the colored areas are not properly energy units but are included for convenience. They arise from the relativistic mass–energy equivalence formula $E = mc^2$ and represent the energy released if a kilogram or unified atomic mass unit (u) is completely converted to energy (bottom two rows) or the mass that would be completely converted to one unit of energy (rightmost two columns).

	Btu	erg	ft·lb	hp·h	JOULE	cal	kW·h	eV	MeV	kg	u
1 British thermal unit =	1	1.055×10^{10}	777.9	3.929×10^{-4}	1055	252.0	2.930×10^{-4}	6.585×10^{21}	6.585×10^{15}	1.174×10^{-14}	7.070×10^{12}
1 erg =	9.481×10^{-11}	1	7.376×10^{-8}	3.725×10^{-14}	10^{-7}	2.389×10^{-8}	2.778×10^{-14}	6.242×10^{11}	6.242×10^{5}	1.113×10^{-24}	670.2
1 foot-pound =	1.285×10^{-3}	1.356×10^{7}	1	5.051×10^{-7}	1.356	0.3238	3.766×10^{-7}	8.464×10^{18}	8.464×10^{12}	1.509×10^{-17}	9.037×10^{9}
1 horsepower-hour =	2545	2.685×10^{13}	1.980×10^{6}	1	2.685×10^{6}	6.413×10^{5}	0.7457	1.676×10^{25}	1.676×10^{19}	2.988×10^{-11}	1.799×10^{16}
1 JOULE =	9.481×10^{-4}	10^{7}	0.7376	3.725×10^{-7}	1	0.2389	2.778×10^{-7}	6.242×10^{18}	6.242×10^{12}	1.113×10^{-17}	6.702×10^{9}
1 calorie =	3.969×10^{-3}	4.186×10^{7}	3.088	1.560×10^{-6}	4.186	1	1.163×10^{-6}	2.613×10^{19}	2.613×10^{13}	4.660×10^{-17}	2.806×10^{10}
1 kilowatt-hour =	3413	3.600×10^{13}	2.655×10^{6}	1.341	3.600×10^{6}	8.600×10^{5}	1	2.247×10^{25}	2.247×10^{19}	4.007×10^{-11}	2.413×10^{16}
1 electron-volt =	1.519×10^{-22}	1.602×10^{-12}	1.182×10^{-19}	5.967×10^{-26}	1.602×10^{-19}	3.827×10^{-20}	4.450×10^{-26}	1	10^{-6}	1.783×10^{-36}	1.074×10^{-9}
1 million electron-volts =	1.519×10^{-16}	1.602×10^{-6}	1.182×10^{-13}	5.967×10^{-20}	1.602×10^{-13}	3.827×10^{-14}	4.450×10^{-20}	10^{-6}	1	1.783×10^{-30}	1.074×10^{-3}
1 kilogram =	8.521×10^{13}	8.987×10^{23}	6.629×10^{16}	3.348×10^{10}	8.987×10^{16}	2.146×10^{16}	2.497×10^{10}	5.610×10^{35}	5.610×10^{29}	1	6.022×10^{26}
1 unified atomic mass unit =	1.415×10^{-13}	1.492×10^{-3}	1.101×10^{-10}	5.559×10^{-17}	1.492×10^{-10}	3.564×10^{-11}	4.146×10^{-17}	9.320×10^{8}	932.0	1.661×10^{-27}	1

POWER

	Btu/h	ft·lb/s	hp	cal/s	kW	WATT
1 British thermal unit per hour =	1	0.2161	3.929×10^{-4}	6.998×10^{-2}	2.930×10^{-4}	0.2930
1 foot-pound per second =	4.628	1	1.818×10^{-3}	0.3239	1.356×10^{-3}	1.356
1 horsepower =	2545	550	1	178.1	0.7457	745.7
1 calorie per second =	14.29	3.088	5.615×10^{-3}	1	4.186×10^{-3}	4.186
1 kilowatt =	3413	737.6	1.341	238.9	1	1000
1 WATT =	3.413	0.7376	1.341×10^{-3}	0.2389	0.001	1

MAGNETIC FIELD

	gauss	TESLA	milligauss
1 gauss =	1	10^{-4}	1000
1 TESLA =	10^{4}	1	10^{7}
1 milligauss =	0.001	10^{-7}	1

1 tesla = 1 weber/meter^2

MAGNETIC FLUX

	maxwell	WEBER
1 maxwell =	1	10^{-8}
1 WEBER =	10^{8}	1

Appendix E
Mathematical Formulas

GEOMETRY

Circle of radius r: circumference $= 2\pi r$; area $= \pi r^2$.

Sphere of radius r: area $= 4\pi r^2$; volume $= \frac{4}{3}\pi r^3$.

Right circular cylinder of radius r and height h:
area $= 2\pi r^2 + 2\pi rh$; volume $= \pi r^2 h$.

Triangle of base a and altitude h: area $= \frac{1}{2}ah$.

QUADRATIC FORMULA

If $ax^2 + bx + c = 0$, then $x = \dfrac{-b \pm \sqrt{b^2 - 4ac}}{2a}$.

TRIGONOMETRIC FUNCTIONS OF ANGLE θ

$\sin\theta = \dfrac{y}{r}$ $\quad \cos\theta = \dfrac{x}{r}$

$\tan\theta = \dfrac{y}{x}$ $\quad \cot\theta = \dfrac{x}{y}$

$\sec\theta = \dfrac{r}{x}$ $\quad \csc\theta = \dfrac{r}{y}$

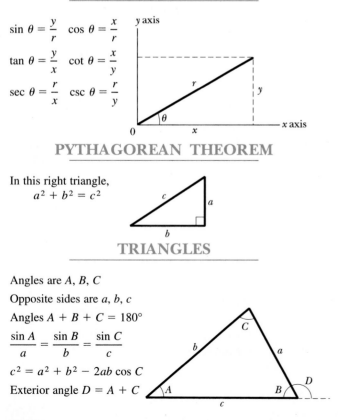

PYTHAGOREAN THEOREM

In this right triangle,
$$a^2 + b^2 = c^2$$

TRIANGLES

Angles are A, B, C

Opposite sides are a, b, c

Angles $A + B + C = 180°$

$\dfrac{\sin A}{a} = \dfrac{\sin B}{b} = \dfrac{\sin C}{c}$

$c^2 = a^2 + b^2 - 2ab\cos C$

Exterior angle $D = A + C$

MATHEMATICAL SIGNS AND SYMBOLS

$=$ equals

$\approx$ equals approximately

$\sim$ is the order of magnitude of

$\neq$ is not equal to

$\equiv$ is identical to, is defined as

$>$ is greater than ($\gg$ is much greater than)

$<$ is less than ($\ll$ is much less than)

$\geq$ is greater than or equal to (or, is no less than)

$\leq$ is less than or equal to (or, is no more than)

$\pm$ plus or minus

$\propto$ is proportional to

Σ the sum of

$\bar{x}$ the average value of x

TRIGONOMETRIC IDENTITIES

$\sin(90° - \theta) = \cos\theta$

$\cos(90° - \theta) = \sin\theta$

$\sin\theta/\cos\theta = \tan\theta$

$\sin^2\theta + \cos^2\theta = 1$

$\sec^2\theta - \tan^2\theta = 1$

$\csc^2\theta - \cot^2\theta = 1$

$\sin 2\theta = 2\sin\theta\cos\theta$

$\cos 2\theta = \cos^2\theta - \sin^2\theta = 2\cos^2\theta - 1 = 1 - 2\sin^2\theta$

$\sin(\alpha \pm \beta) = \sin\alpha\cos\beta \pm \cos\alpha\sin\beta$

$\cos(\alpha \pm \beta) = \cos\alpha\cos\beta \mp \sin\alpha\sin\beta$

$\tan(\alpha \pm \beta) = \dfrac{\tan\alpha \pm \tan\beta}{1 \mp \tan\alpha\tan\beta}$

$\sin\alpha \pm \sin\beta = 2\sin\frac{1}{2}(\alpha \pm \beta)\cos\frac{1}{2}(\alpha \mp \beta)$

$\cos\alpha + \cos\beta = 2\cos\frac{1}{2}(\alpha + \beta)\cos\frac{1}{2}(\alpha - \beta)$

$\cos\alpha - \cos\beta = -2\sin\frac{1}{2}(\alpha + \beta)\sin\frac{1}{2}(\alpha - \beta)$

BINOMIAL THEOREM

$$(1 + x)^n = 1 + \frac{nx}{1!} + \frac{n(n-1)x^2}{2!} + \cdots \qquad (x^2 < 1)$$

EXPONENTIAL EXPANSION

$$e^x = 1 + x + \frac{x^2}{2!} + \frac{x^3}{3!} + \cdots$$

LOGARITHMIC EXPANSION

$$\ln(1 + x) = x - \tfrac{1}{2}x^2 + \tfrac{1}{3}x^3 - \cdots \qquad (|x| < 1)$$

TRIGONOMETRIC EXPANSIONS
(θ in radians)

$$\sin \theta = \theta - \frac{\theta^3}{3!} + \frac{\theta^5}{5!} - \cdots$$

$$\cos \theta = 1 - \frac{\theta^2}{2!} + \frac{\theta^4}{4!} - \cdots$$

$$\tan \theta = \theta + \frac{\theta^3}{3} + \frac{2\theta^5}{15} + \cdots$$

CRAMER'S RULE

Two simultaneous equations in unknowns x and y,

$$a_1 x + b_1 y = c_1 \qquad \text{and} \qquad a_2 x + b_2 y = c_2,$$

have the solutions

$$x = \frac{\begin{vmatrix} c_1 & b_1 \\ c_2 & b_2 \end{vmatrix}}{\begin{vmatrix} a_1 & b_1 \\ a_2 & b_2 \end{vmatrix}} = \frac{c_1 b_2 - c_2 b_1}{a_1 b_2 - a_2 b_1}$$

and

$$y = \frac{\begin{vmatrix} a_1 & c_1 \\ a_2 & c_2 \end{vmatrix}}{\begin{vmatrix} a_1 & b_1 \\ a_2 & b_2 \end{vmatrix}} = \frac{a_1 c_2 - a_2 c_1}{a_1 b_2 - a_2 b_1}.$$

PRODUCTS OF VECTORS

Let $\mathbf{i}$, $\mathbf{j}$, and $\mathbf{k}$ be unit vectors in the x, y, and z directions. Then

$$\mathbf{i} \cdot \mathbf{i} = \mathbf{j} \cdot \mathbf{j} = \mathbf{k} \cdot \mathbf{k} = 1, \qquad \mathbf{i} \cdot \mathbf{j} = \mathbf{j} \cdot \mathbf{k} = \mathbf{k} \cdot \mathbf{i} = 0,$$

$$\mathbf{i} \times \mathbf{i} = \mathbf{j} \times \mathbf{j} = \mathbf{k} \times \mathbf{k} = 0,$$

$$\mathbf{i} \times \mathbf{j} = \mathbf{k}, \qquad \mathbf{j} \times \mathbf{k} = \mathbf{i}, \qquad \mathbf{k} \times \mathbf{i} = \mathbf{j},$$

Any vector $\mathbf{a}$ with components a_x, a_y, and a_z along the x, y, and z axes can be written

$$\mathbf{a} = a_x \mathbf{i} + a_y \mathbf{j} + a_z \mathbf{k}.$$

Let $\mathbf{a}$, $\mathbf{b}$, and $\mathbf{c}$ be arbitrary vectors with magnitudes a, b, and c. Then

$$\mathbf{a} \times (\mathbf{b} + \mathbf{c}) = (\mathbf{a} \times \mathbf{b}) + (\mathbf{a} \times \mathbf{c})$$

$$(s\mathbf{a}) \times \mathbf{b} = \mathbf{a} \times (s\mathbf{b}) = s(\mathbf{a} \times \mathbf{b}) \qquad (s = \text{a scalar}).$$

Let θ be the smaller of the two angles between $\mathbf{a}$ and $\mathbf{b}$. Then

$$\mathbf{a} \cdot \mathbf{b} = \mathbf{b} \cdot \mathbf{a} = a_x b_x + a_y b_y + a_z b_z = ab \cos \theta$$

$$\mathbf{a} \times \mathbf{b} = -\mathbf{b} \times \mathbf{a} = \begin{vmatrix} \mathbf{i} & \mathbf{j} & \mathbf{k} \\ a_x & a_y & a_z \\ b_x & b_y & b_z \end{vmatrix}$$

$$= \mathbf{i} \begin{vmatrix} a_y & a_z \\ b_y & b_z \end{vmatrix} - \mathbf{j} \begin{vmatrix} a_x & a_z \\ b_x & b_z \end{vmatrix} + \mathbf{k} \begin{vmatrix} a_x & a_y \\ b_x & b_y \end{vmatrix}$$

$$= (a_y b_z - b_y a_z)\mathbf{i}$$

$$+ (a_z b_x - b_z a_x)\mathbf{j} + (a_x b_y - b_x a_y)\mathbf{k}$$

$$|\mathbf{a} \times \mathbf{b}| = ab \sin \theta$$

$$\mathbf{a} \cdot (\mathbf{b} \times \mathbf{c}) = \mathbf{b} \cdot (\mathbf{c} \times \mathbf{a}) = \mathbf{c} \cdot (\mathbf{a} \times \mathbf{b})$$

$$\mathbf{a} \times (\mathbf{b} \times \mathbf{c}) = (\mathbf{a} \cdot \mathbf{c})\mathbf{b} - (\mathbf{a} \cdot \mathbf{b})\mathbf{c}$$

DERIVATIVES AND INTEGRALS

In what follows, the letters u and v stand for any functions of x, and a and m are constants. To each of the indefinite integrals should be added an arbitrary constant of integration. The *Handbook of Chemistry and Physics* (CRC Press Inc.) gives a more extensive tabulation.

1. $\dfrac{dx}{dx} = 1$

2. $\dfrac{d}{dx}(au) = a\dfrac{du}{dx}$

3. $\dfrac{d}{dx}(u + v) = \dfrac{du}{dx} + \dfrac{dv}{dx}$

4. $\dfrac{d}{dx}x^m = mx^{m-1}$

5. $\dfrac{d}{dx}\ln x = \dfrac{1}{x}$

6. $\dfrac{d}{dx}(uv) = u\dfrac{dv}{dx} + v\dfrac{du}{dx}$

7. $\dfrac{d}{dx}e^x = e^x$

8. $\dfrac{d}{dx}\sin x = \cos x$

9. $\dfrac{d}{dx}\cos x = -\sin x$

10. $\dfrac{d}{dx}\tan x = \sec^2 x$

11. $\dfrac{d}{dx}\cot x = -\csc^2 x$

12. $\dfrac{d}{dx}\sec x = \tan x \sec x$

13. $\dfrac{d}{dx}\csc x = -\cot x \csc x$

14. $\dfrac{d}{dx}e^u = e^u\dfrac{du}{dx}$

15. $\dfrac{d}{dx}\sin u = \cos u\dfrac{du}{dx}$

16. $\dfrac{d}{dx}\cos u = -\sin u\dfrac{du}{dx}$

1. $\displaystyle\int dx = x$

2. $\displaystyle\int au\,dx = a\int u\,dx$

3. $\displaystyle\int (u + v)\,dx = \int u\,dx + \int v\,dx$

4. $\displaystyle\int x^m\,dx = \dfrac{x^{m+1}}{m + 1} \quad (m \neq -1)$

5. $\displaystyle\int \dfrac{dx}{x} = \ln |x|$

6. $\displaystyle\int u\dfrac{dv}{dx}\,dx = uv - \int v\dfrac{du}{dx}\,dx$

7. $\displaystyle\int e^x\,dx = e^x$

8. $\displaystyle\int \sin x\,dx = -\cos x$

9. $\displaystyle\int \cos x\,dx = \sin x$

10. $\displaystyle\int \tan x\,dx = \ln |\sec x|$

11. $\displaystyle\int \sin^2 x\,dx = \tfrac{1}{2}x - \tfrac{1}{4}\sin 2x$

12. $\displaystyle\int e^{-ax}\,dx = -\dfrac{1}{a}e^{-ax}$

13. $\displaystyle\int xe^{-ax}\,dx = -\dfrac{1}{a^2}(ax + 1)e^{-ax}$

14. $\displaystyle\int x^2e^{-ax}\,dx = -\dfrac{1}{a^3}(a^2x^2 + 2ax + 2)e^{-ax}$

15. $\displaystyle\int_0^\infty x^ne^{-ax}\,dx = \dfrac{n!}{a^{n+1}}$

16. $\displaystyle\int_0^\infty x^{2n}e^{-ax^2}\,dx = \dfrac{1 \cdot 3 \cdot 5 \cdots (2n - 1)}{2^{n+1}a^n}\sqrt{\dfrac{\pi}{a}}$

17. $\displaystyle\int \dfrac{dx}{\sqrt{x^2 + a^2}} = \ln(x + \sqrt{x^2 + a^2})$

18. $\displaystyle\int \dfrac{x\,dx}{(x^2 + a^2)^{3/2}} = -\dfrac{1}{(x^2 + a^2)^{1/2}}$

19. $\displaystyle\int \dfrac{dx}{(x^2 + a^2)^{3/2}} = \dfrac{x}{a^2(x^2 + a^2)^{1/2}}$

Appendix F
Properties of the Elements

All physical properties are for a pressure of 1 atm unless otherwise specified.

ELEMENT	SYMBOL	ATOMIC NUMBER, Z	MOLAR MASS, g/mol	DENSITY, g/cm³ AT 20°C	MELTING POINT, °C	BOILING POINT, °C	SPECIFIC HEAT, J/(g·°C) AT 25°C
Actinium	Ac	89	(227)	10.06	1323	(3473)	0.092
Aluminum	Al	13	26.9815	2.699	660	2450	0.900
Americium	Am	95	(243)	13.67	1541	—	—
Antimony	Sb	51	121.75	6.691	630.5	1380	0.205
Argon	Ar	18	39.948	1.6626×10^{-3}	−189.4	−185.8	0.523
Arsenic	As	33	74.9216	5.78	817 (28 atm)	613	0.331
Astatine	At	85	(210)	—	(302)	—	—
Barium	Ba	56	137.34	3.594	729	1640	0.205
Berkelium	Bk	97	(247)	14.79	—	—	—
Beryllium	Be	4	9.0122	1.848	1287	2770	1.83
Bismuth	Bi	83	208.980	9.747	271.37	1560	0.122
Boron	B	5	10.811	2.34	2030	—	1.11
Bromine	Br	35	79.909	3.12 (liquid)	−7.2	58	0.293
Cadmium	Cd	48	112.40	8.65	321.03	765	0.226
Calcium	Ca	20	40.08	1.55	838	1440	0.624
Californium	Cf	98	(251)	—	—	—	—
Carbon	C	6	12.01115	2.26	3727	4830	0.691
Cerium	Ce	58	140.12	6.768	804	3470	0.188
Cesium	Cs	55	132.905	1.873	28.40	690	0.243
Chlorine	Cl	17	35.453	3.214×10^{-3} (0°C)	−101	−34.7	0.486
Chromium	Cr	24	51.996	7.19	1857	2665	0.448
Cobalt	Co	27	58.9332	8.85	1495	2900	0.423
Copper	Cu	29	63.54	8.96	1083.40	2595	0.385
Curium	Cm	96	(247)	13.3	—	—	—
Dysprosium	Dy	66	162.50	8.55	1409	2330	0.172
Einsteinium	Es	99	(254)	—	—	—	—
Erbium	Er	68	167.26	9.15	1522	2630	0.167
Europium	Eu	63	151.96	5.243	817	1490	0.163
Fermium	Fm	100	(237)	—	—	—	—
Fluorine	F	9	18.9984	1.696×10^{-3} (0°C)	−219.6	−188.2	0.753
Francium	Fr	87	(223)	—	(27)	—	—
Gadolinium	Gd	64	157.25	7.90	1312	2730	0.234
Gallium	Ga	31	69.72	5.907	29.75	2237	0.377
Germanium	Ge	32	72.59	5.323	937.25	2830	0.322
Gold	Au	79	196.967	19.32	1064.43	2970	0.131
Hafnium	Hf	72	178.49	13.31	2227	5400	0.144
Hahnium	Ha	105	—	—	—	—	—
Hassium	Hs	108	—	—	—	—	—

continued on next page

ELEMENT	SYMBOL	ATOMIC NUMBER, Z	MOLAR MASS, g/mol	DENSITY, g/cm³ AT 20°C	MELTING POINT, °C	BOILING POINT, °C	SPECIFIC HEAT, J/(g·°C) AT 25°C
Helium	He	2	4.0026	0.1664×10^{-3}	−269.7	−268.9	5.23
Holmium	Ho	67	164.930	8.79	1470	2330	0.165
Hydrogen	H	1	1.00797	0.08375×10^{-3}	−259.19	−252.7	14.4
Indium	In	49	114.82	7.31	156.634	2000	0.233
Iodine	I	53	126.9044	4.93	113.7	183	0.218
Iridium	Ir	77	192.2	22.5	2447	(5300)	0.130
Iron	Fe	26	55.847	7.874	1536.5	3000	0.447
Krypton	Kr	36	83.80	3.488×10^{-3}	−157.37	−152	0.247
Lanthanum	La	57	138.91	6.189	920	3470	0.195
Lawrencium	Lr	103	(257)		—	—	
Lead	Pb	82	207.19	11.35	327.45	1725	0.129
Lithium	Li	3	6.939	0.534	180.55	1300	3.58
Lutetium	Lu	71	174.97	9.849	1663	1930	0.155
Magnesium	Mg	12	24.312	1.738	650	1107	1.03
Manganese	Mn	25	54.9380	7.44	1244	2150	0.481
Meitnerium	Mt	109	—	—	—	—	—
Mendelevium	Md	101	(256)	—	—	—	—
Mercury	Hg	80	200.59	13.55	−38.87	357	0.138
Molybdenum	Mo	42	95.94	10.22	2617	5560	0.251
Neodymium	Nd	60	144.24	7.007	1016	3180	0.188
Neon	Ne	10	20.183	0.8387×10^{-3}	−248.597	−246.0	1.03
Neptunium	Np	93	(237)	20.25	637	—	1.26
Nickel	Ni	28	58.71	8.902	1453	2730	0.444
Nielsbohrium	Ns	107	—	—	—	—	—
Niobium	Nb	41	92.906	8.57	2468	4927	0.264
Nitrogen	N	7	14.0067	1.1649×10^{-3}	−210	−195.8	1.03
Nobelium	No	102	(255)	—	—	—	—
Osmium	Os	76	190.2	22.59	3027	5500	0.130
Oxygen	O	8	15.9994	1.3318×10^{-3}	−218.80	−183.0	0.913
Palladium	Pd	46	106.4	12.02	1552	3980	0.243
Phosphorus	P	15	30.9738	1.83	44.25	280	0.741
Platinum	Pt	78	195.09	21.45	1769	4530	0.134
Plutonium	Pu	94	(244)	19.8	640	3235	0.130
Polonium	Po	84	(210)	9.32	254	—	—
Potassium	K	19	39.102	0.862	63.20	760	0.758
Praseodymium	Pr	59	140.907	6.773	931	3020	0.197
Promethium	Pm	61	(145)	7.22	(1027)	—	—
Protactinium	Pa	91	(231)	15.37 (estimated)	(1230)	—	—
Radium	Ra	88	(226)	5.0	700	—	—
Radon	Rn	86	(222)	9.96×10^{-3} (0°C)	(−71)	−61.8	0.092
Rhenium	Re	75	186.2	21.02	3180	5900	0.134
Rhodium	Rh	45	102.905	12.41	1963	4500	0.243
Rubidium	Rb	37	85.47	1.532	39.49	688	0.364
Ruthenium	Ru	44	101.107	12.37	2250	4900	0.239
Rutherfordium	Rf	104	—	—	—	—	—
Samarium	Sm	62	150.35	7.52	1072	1630	0.197

continued on next page

ELEMENT	SYMBOL	ATOMIC NUMBER, Z	MOLAR MASS, g/mol	DENSITY, g/cm³ AT 20°C	MELTING POINT, °C	BOILING POINT, °C	SPECIFIC HEAT, J/(g·°C) AT 25°C
Scandium	Sc	21	44.956	2.99	1539	2730	0.569
Seaborgium	Sg	106	—	—	—	—	—
Selenium	Se	34	78.96	4.79	221	685	0.318
Silicon	Si	14	28.086	2.33	1412	2680	0.712
Silver	Ag	47	107.870	10.49	960.8	2210	0.234
Sodium	Na	11	22.9898	0.9712	97.85	892	1.23
Strontium	Sr	38	87.62	2.54	768	1380	0.737
Sulfur	S	16	32.064	2.07	119.0	444.6	0.707
Tantalum	Ta	73	180.948	16.6	3014	5425	0.138
Technetium	Tc	43	(99)	11.46	2200	—	0.209
Tellurium	Te	52	127.60	6.24	449.5	990	0.201
Terbium	Tb	65	158.924	8.229	1357	2530	0.180
Thallium	Tl	81	204.37	11.85	304	1457	0.130
Thorium	Th	90	(232)	11.72	1755	(3850)	0.117
Thulium	Tm	69	168.934	9.32	1545	1720	0.159
Tin	Sn	50	118.69	7.2984	231.868	2270	0.226
Titanium	Ti	22	47.90	4.54	1670	3260	0.523
Tungsten	W	74	183.85	19.3	3380	5930	0.134
Uranium	U	92	(238)	18.95	1132	3818	0.117
Vanadium	V	23	50.942	6.11	1902	3400	0.490
Xenon	Xe	54	131.30	5.495×10^{-3}	−111.79	−108	0.159
Ytterbium	Yb	70	173.04	6.965	824	1530	0.155
Yttrium	Y	39	88.905	4.469	1526	3030	0.297
Zinc	Zn	30	65.37	7.133	419.58	906	0.389
Zirconium	Zr	40	91.22	6.506	1852	3580	0.276

The values in parentheses in the column of molar masses are the mass numbers of the longest-lived isotopes of those elements that are radioactive. Melting points and boiling points in parentheses are uncertain.

The data for gases are valid only when these are in their usual molecular state, such as H_2, He, O_2, Ne, etc. The specific heats of the gases are the values at constant pressure.

Source: Adapted from Wehr, Richards, Adair, *Physics of the Atom,* 4th ed., Addison-Wesley, Reading, MA, 1984, and from J. Emsley, *The Elements,* 2nd ed., Clarendon Press, Oxford, 1991.

	Metals
	Metalloids
	Nonmetals

Alkali metals IA		Transition metals														Noble gases 0

THE HORIZONTAL PERIODS

Period	IA	IIA	IIIB	IVB	VB	VIB	VIIB	VIIIB			IB	IIB	IIIA	IVA	VA	VIA	VIIA	0
1	1 H																	2 He
2	3 Li	4 Be											5 B	6 C	7 N	8 O	9 F	10 Ne
3	11 Na	12 Mg											13 Al	14 Si	15 P	16 S	17 Cl	18 Ar
4	19 K	20 Ca	21 Sc	22 Ti	23 V	24 Cr	25 Mn	26 Fe	27 Co	28 Ni	29 Cu	30 Zn	31 Ga	32 Ge	33 As	34 Se	35 Br	36 Kr
5	37 Rb	38 Sr	39 Y	40 Zr	41 Nb	42 Mo	43 Tc	44 Ru	45 Rh	46 Pd	47 Ag	48 Cd	49 In	50 Sn	51 Sb	52 Te	53 I	54 Xe
6	55 Cs	56 Ba	57-71 *	72 Hf	73 Ta	74 W	75 Re	76 Os	77 Ir	78 Pt	79 Au	80 Hg	81 Tl	82 Pb	83 Bi	84 Po	85 At	86 Rn
7	87 Fr	88 Ra	89-103 †	104 Rf	105 Ha	106 Sg	107 Ns	108 Hs	109 Mt	110	111	112						

Inner transition metals

Lanthanide series *	57 La	58 Ce	59 Pr	60 Nd	61 Pm	62 Sm	63 Eu	64 Gd	65 Tb	66 Dy	67 Ho	68 Er	69 Tm	70 Yb	71 Lu
Actinide series †	89 Ac	90 Th	91 Pa	92 U	93 Np	94 Pu	95 Am	96 Cm	97 Bk	98 Cf	99 Es	100 Fm	101 Md	102 No	103 Lr

The names for elements 104–109 (Rutherfordium, Hahnium, Seaborgium, Nielsbohrium, Hassium, and Meitnerium, respectively) are those recommended by the American Chemical Society Nomenclature Committee. As of 1996, the names and symbols for elements 104–108 have not yet been approved by the appropriate international body. Elements 110, 111 and 112 have been discovered but, as of 1996, have not been provisionally named.

Appendix H
Winners of the Nobel Prize in Physics*

1901 Wilhelm Konrad Röntgen *(1845–1923)* for the discovery of x rays

1902 Hendrik Antoon Lorentz *(1853–1928)* and Pieter Zeeman *(1865–1943)* for their researches into the influence of magnetism upon radiation phenomena

1903 Antoine Henri Becquerel *(1852–1908)* for his discovery of spontaneous radioactivity

Pierre Curie *(1859–1906)* and Marie Sklowdowska-Curie *(1867–1934)* for their joint researches on the radiation phenomena discovered by Becquerel

1904 Lord Rayleigh (John William Strutt) *(1842–1919)* for his investigations of the densities of the most important gases and for his discovery of argon

1905 Philipp Eduard Anton von Lenard *(1862–1947)* for his work on cathode rays

1906 Joseph John Thomson *(1856–1940)* for his theoretical and experimental investigations on the conduction of electricity by gases

1907 Albert Abraham Michelson *(1852–1931)* for his optical precision instruments and metrological investigations carried out with their aid

1908 Gabriel Lippmann *(1845–1921)* for his method of reproducing colors photographically based on the phenomena of interference

1909 Guglielmo Marconi *(1874–1937)* and Carl Ferdinand Braun *(1850–1918)* for their contributions to the development of wireless telegraphy

1910 Johannes Diderik van der Waals *(1837–1923)* for his work on the equation of state for gases and liquids

1911 Wilhelm Wien *(1864–1928)* for his discoveries regarding the laws governing the radiation of heat

1912 Nils Gustaf Dalén *(1869–1937)* for his invention of automatic regulators for use in conjunction with gas accumulators for illuminating lighthouses and buoys

1913 Heike Kamerlingh Onnes *(1853–1926)* for his investigations of the properties of matter at low temperatures which led, among other things, to the production of liquid helium

1914 Max von Laue *(1879–1960)* for his discovery of the diffraction of Röntgen rays by crystals

1915 William Henry Bragg *(1862–1942)* and William Lawrence Bragg *(1890–1971)* for their services in the analysis of crystal structure by means of x rays

1917 Charles Glover Barkla *(1877–1944)* for his discovery of the characteristic x rays of the elements

1918 Max Planck *(1858–1947)* for his discovery of energy quanta

1919 Johannes Stark *(1874–1957)* for his discovery of the Doppler effect in canal rays and the splitting of spectral lines in electric fields

1920 Charles-Édouard Guillaume *(1861–1938)* for the service he rendered to precision measurements in physics by his discovery of anomalies in nickel steel alloys

1921 Albert Einstein *(1879–1955)* for his services to theoretical physics, and especially for his discovery of the law of the photoelectric effect

1922 Niels Bohr *(1885–1962)* for the investigation of the structure of atoms, and of the radiation emanating from them

1923 Robert Andrews Millikan *(1868–1953)* for his work on the elementary charge of electricity and on the photoelectric effect

1924 Karl Manne Georg Siegbahn *(1886–1978)* for his discoveries and research in the field of x-ray spectroscopy

1925 James Franck *(1882–1964)* and Gustav Hertz *(1887–1975)* for their discovery of the laws governing the impact of an electron upon an atom

1926 Jean Baptiste Perrin *(1870–1942)* for his work on the discontinuous structure of matter, and especially for his discovery of sedimentation equilibrium

1927 Arthur Holly Compton *(1892–1962)* for his discovery of the effect named after him

Charles Thomson Rees Wilson (1869–1959) for his method of making the paths of electrically charged particles visible by condensation of vapor

1928 Owen Willans Richardson *(1879–1959)* for his work on the thermionic phenomenon and especially for the discovery of the law named after him

1929 Prince Louis Victor de Broglie *(1892–1987)* for his discovery of the wave nature of electrons

*See *Nobel Lectures, Physics,* 1901–1970, Elsevier Publishing Company, for biographies of the awardees and for lectures given by them on receiving the prize.

1930 Sir Chandrasekhara Venkata Raman *(1888–1970)* for his work on the scattering of light and for the discovery of the effect named after him

1932 Werner Heisenberg *(1901–1976)* for the creation of quantum mechanics, the application of which has, among other things, led to the discovery of the allotropic forms of hydrogen

1933 Erwin Schrödinger *(1887–1961)* and Paul Adrien Maurice Dirac *(1902–1984)* for the discovery of new productive forms of atomic theory

1935 James Chadwick *(1891–1974)* for his discovery of the neutron

1936 Victor Franz Hess *(1883–1964)* for the discovery of cosmic radiation

Carl David Anderson *(1905–1991)* for his discovery of the positron

1937 Clinton Joseph Davisson *(1881–1958)* and George Paget Thomson *(1892–1975)* for their experimental discovery of the diffraction of electrons by crystals

1938 Enrico Fermi *(1901–1954)* for his demonstrations of the existence of new radioactive elements produced by neutron irradiation, and for his related discovery of nuclear reactions brought about by slow neutrons

1939 Ernest Orlando Lawrence *(1901–1958)* for the invention and development of the cyclotron and for results obtained with it, especially for artificial radioactive elements

1943 Otto Stern *(1888–1969)* for his contribution to the development of the molecular-ray method and his discovery of the magnetic moment of the proton

1944 Isidor Isaac Rabi *(1898–1988)* for his resonance method for recording the magnetic properties of atomic nuclei

1945 Wolfgang Pauli *(1900–1958)* for the discovery of the Exclusion Principle (also called Pauli Principle)

1946 Percy Williams Bridgman *(1882–1961)* for the invention of an apparatus to produce extremely high pressures and for the discoveries he made therewith in the field of high-pressure physics

1947 Sir Edward Victor Appleton *(1892–1965)* for his investigations of the physics of the upper atmosphere, especially for the discovery of the so-called Appleton layer

1948 Patrick Maynard Stuart Blackett *(1897–1974)* for his development of the Wilson cloud-chamber method, and his discoveries therewith in nuclear physics and cosmic radiation

1949 Hideki Yukawa *(1907–1981)* for his prediction of the existence of mesons on the basis of theoretical work on nuclear forces

1950 Cecil Frank Powell *(1903–1969)* for his development of the photographic method of studying nuclear processes and his discoveries regarding mesons made with this method

1951 Sir John Douglas Cockcroft *(1897–1967)* and Ernest Thomas Sinton Walton *(1903–1995)* for their pioneer work on the transmutation of atomic nuclei by artificially accelerated atomic particles

1952 Felix Bloch *(1905–1983)* and Edward Mills Purcell *(1912–   )* for their development of new nuclear-magnetic precision methods and discoveries in connection therewith

1953 Frits Zernike *(1888–1966)* for his demonstration of the phase-contrast method, especially for his invention of the phase-contrast microscope

1954 Max Born *(1882–1970)* for his fundamental research in quantum mechanics, especially for his statistical interpretation of the wave function

Walther Bothe *(1891–1957)* for the coincidence method and his discoveries made therewith

1955 Willis Eugene Lamb *(1913–   )* for his discoveries concerning the fine structure of the hydrogen spectrum

Polykarp Kusch *(1911–1993)* for his precision determination of the magnetic moment of the electron

1956 William Shockley *(1910–1989)*, John Bardeen *(1908–1991)* and Walter Houser Brattain *(1902–1987)* for their researches on semiconductors and their discovery of the transistor effect

1957 Chen Ning Yang *(1922–   )* and Tsung Dao Lee *(1926–   )* for their penetrating investigation of the parity laws which has led to important discoveries regarding the elementary particles

1958 Pavel Aleksejevič Čerenkov *(1904–   )*, Il' ja Michajlovič Frank *(1908–1990)* and Igor' Evgen' evič Tamm *(1895–1971)* for the discovery and interpretation of the Cerenkov effect

1959 Emilio Gino Segrè *(1905–1989)* and Owen Chamberlain *(1920–   )* for their discovery of the antiproton

1960 Donald Arthur Glaser *(1926–   )* for the invention of the bubble chamber

1961 Robert Hofstadter *(1915–1990)* for his pioneering studies of electron scattering in atomic nuclei and for his thereby achieved discoveries concerning the structure of the nucleons

Rudolf Ludwig Mössbauer *(1929–   )* for his researches concerning the resonance absorption of γ rays and his discovery in this connection of the effect which bears his name

1962 Lev Davidovič Landau *(1908–1968)* for his pioneering theories of condensed matter, especially liquid helium

1963 Eugene P. Wigner *(1902–1995)* for his contributions to the theory of the atomic nucleus and the elementary particles, particularly through the discovery and application of fundamental symmetry principles

Maria Goeppert Mayer *(1906–1972)* and J. Hans D. Jensen *(1907–1973)* for their discoveries concerning nuclear shell structure

1964 Charles H. Townes *(1915–   )*, Nikolai G. Basov *(1922–   )* and Alexander M. Prochorov *(1916–   )* for fundamental work in the field of quantum electronics which has led to the construction of oscillators and amplifiers based on the maser–laser principle

1965 Sin-itiro Tomonaga *(1906–1979)*, Julian Schwinger *(1918–1994)* and Richard P. Feynman *(1918–1988)* for their fundamental work in quantum electrodynamics, with deep-ploughing consequences for the physics of elementary particles

1966 Alfred Kastler *(1902–1984)* for the discovery and development of optical methods for studying Hertzian resonance in atoms

1967 Hans Albrecht Bethe *(1906–   )* for his contributions to the theory of nuclear reactions, especially his discoveries concerning the energy production in stars

1968 Luis W. Alvarez *(1911–1988)* for his decisive contribution to elementary particle physics, in particular the discovery of a large number of resonance states, made possible through his development of the techniques of using the hydrogen bubble chamber and its data analysis

1969 Murray Gell-Mann *(1929–   )* for his contributions and discoveries concerning the classification of elementary particles and their interactions

1970 Hannes Alfvén *(1908–1995)* for fundamental work and discoveries in magneto-hydrodynamics with fruitful applications in different parts of plasma physics

Louis Néel *(1904–   )* for fundamental work and discoveries concerning antiferromagnetism and ferrimagnetism which have led to important applications in solid state physics

1971 Dennis Gabor *(1900–1979)* for his discovery of the principles of holography

1972 John Bardeen *(1908–1991)*, Leon N. Cooper *(1930–   )* and J. Robert Schrieffer *(1931–   )* for their development of a theory of superconductivity

1973 Leo Esaki *(1925–   )* for his discovery of tunneling in semiconductors

Ivar Giaever *(1929–   )* for his discovery of tunneling in superconductors

Brian D. Josephson *(1940–   )* for his theoretical prediction of the properties of a supercurrent through a tunnel barrier

1974 Antony Hewish *(1924–   )* for the discovery of pulsars

Sir Martin Ryle *(1918–1984)* for his pioneering work in radioastronomy

1975 Aage Bohr *(1922–   )*, Ben Mottelson *(1926–   )* and James Rainwater *(1917–1986)* for the discovery of the connection between collective motion and particle motion and the development of the theory of the structure of the atomic nucleus based on this connection

1976 Burton Richter *(1931–   )* and Samuel Chao Chung Ting *(1936–   )* for their (independent) discovery of an important fundamental particle

1977 Philip Warren Anderson *(1923–   )*, Nevill Francis Mott *(1905–1996)* and John Hasbrouck Van Vleck *(1899–1980)* for their fundamental theoretical investigations of the electronic structure of magnetic and disordered systems

1978 Peter L. Kapitza *(1894–1984)* for his basic inventions and discoveries in low-temperature physics

Arno A. Penzias *(1933–   )* and Robert Woodrow Wilson *(1936–   )* for their discovery of cosmic microwave background radiation

1979 Sheldon Lee Glashow *(1932–   )*, Abdus Salam *(1926–1996)*, and Steven Weinberg *(1933–   )* for their unified model of the action of the weak and electromagnetic forces and for their prediction of the existence of neutral currents

1980 James W. Cronin *(1931–   )* and Val L. Fitch *(1923–   )* for the discovery of violations of fundamental symmetry principles in the decay of neutral K mesons

1981 Nicolaas Bloembergen *(1920–   )* and Arthur Leonard Schawlow *(1921–   )* for their contribution to the development of laser spectroscopy

Kai M. Siegbahn *(1918–   )* for his contribution to high-resolution electron spectroscopy

1982 Kenneth Geddes Wilson *(1936–   )* for his method of analyzing the critical phenomena inherent in the changes of matter under the influence of pressure and temperature

1983 Subrehmanyan Chandrasekhar *(1910–1995)* for his theoretical studies of the structure and evolution of stars

William A. Fowler *(1911–1995)* for his studies of the formation of the chemical elements in the universe

1984 Carlo Rubbia *(1934–   )* and Simon van der Meer *(1925–   )* for their decisive contributions to the Large Project, which led to the discovery of the field particles *W* and *Z*, communicators of the weak interaction

1985 Klaus von Klitzing *(1943–   )* for his discovery of the quantized Hall resistance

1986 Ernst Ruska *(1906–1988)* for his invention of the electron microscope

Gerd Binnig *(1947–   )*, Heinrich Rohrer *(1933–   )* for their invention of the scanning tunneling microscope

1987 Karl Alex Müller *(1927–   )* and J. George Bednorz *(1950–   )* for their discovery of a new class of superconductors

1988 Leon M. Lederman *(1922–   )*, Melvin Schwartz *(1932–   )* and Jack Steinberger *(1921–   )* for the first use of a neutrino beam and the discovery of the muon neutrino

1989 Norman Ramsey *(1915–   )*, Hans Dehmelt *(1922–   )* and Wolfgang Paul *(1913–1993)* for their work that led to the development of atomic clocks and precision timing

1990 Jerome I. Friedman *(1930–   )*, Henry W. Kendall *(1926–   )* and Richard E. Taylor *(1929–   )* for demonstrating that protons and neutrons consist of quarks

1991 Pierre de Gennes *(1932–   )* for studies of order phenomena, such as in liquid crystals and polymers

1992 George Charpak *(1924–   )* for his invention of fast electronic detectors for high-energy particles

1993 Joseph H. Taylor *(1941–)* and Russell A. Hulse *(1950–)* for the discovery and interpretation of the first binary pulsar.

1994 Bertram N. Brockhouse *(1918–)* and Clifford G. Shull *(1915–)* for the development of neutron scattering techniques

1995 Martin L. Perl *(1927–)* for the discovery of the tau lepton

Frederick Reines *(1918–)* for the detection of the neutrino

1996 David M. Lee *(1931–)*, Robert C. Richardson *(1937–)* and Douglas D. Oscheroff *(1944–)* for the discovery of superfluidity in helium-3

Answers to Checkpoints, Odd-Numbered Questions, Exercises, and Problems

Chapter 1

EP 3. (a) 186 mi; (b) 3.0×10^8 mm **5.** (a) 10^9; (b) 10^{-4}; (c) 9.1×10^5 **7.** 32.2 km **9.** 0.020 km^3 **11.** (a) 250 ft^2; (b) 23.3 m^2; (c) 3060 ft^3; (d) 86.6 m^3 **13.** 8×10^2 km **15.** (a) 11.3 m^2/L; (b) 1.13×10^4 m^{-1}; (c) 2.17×10^{-3} gal/ft^2 **17.** (a) $d_{Sun}/d_{Moon} = 400$; (b) $V_{Sun}/V_{Moon} = 6.4 \times 10^7$; (c) 3.5×10^3 km **19.** (a) 0.98 ft/ns; (b) 0.30 mm/ps **21.** 3.156×10^7 s **23.** 5.79×10^{12} days **25.** (a) 0.013; (b) 0.54; (c) 10.3; (d) 31 m/s **27.** 15° **29.** 3.3 ft **31.** 2 days 5 hours **33.** (a) 2.99×10^{-26} kg; (b) 4.68×10^{46} **35.** 1.3×10^9 kg **37.** (a) 10^3 kg/m^3; (b) 158 kg/s **39.** (a) 1.18×10^{-29} m^3; (b) 0.282 nm

Chapter 2

CP 1. b and c **2.** zero **3.** (a) 1 and 4; (b) 2 and 3; (c) 3 **4.** (a) plus; (b) minus; (c) minus; (d) plus **5.** 1 and 4 **6.** (a) plus; (b) minus; (c) $a = -g = -9.8$ m/s^2 **Q 1.** (a) yes; (b) no; (c) yes; (d) yes **3.** (a) 2, 3; (b) 1, 3; (c) 4 **5.** all tie (see Eq. 2-16) **7.** (a) $-g$; (b) 2 m/s upward **9.** same **11.** $x = t^2$ and $x = 8(t - 2) + (1.5)(t - 2)^2$ **13.** increase **EP 1.** (a) Lewis: 10.0 m/s, Rodgers: 5.41 m/s; (b) 1 h 10 min **3.** 309 ft **5.** 2 cm/y **7.** 6.71×10^8 mi/h, 9.84×10^8 ft/s, 1.00 ly/y **9.** (a) 5.7 ft/s; (b) 7.0 ft/s **11.** (a) 45 mi/h (72 km/h); (b) 43 mi/h (69 km/h); (c) 44 mi/h (71 km/h); (d) 0 **13.** (a) 28.5 cm/s; (b) 18.0 cm/s; (c) 40.5 cm/s; (d) 28.1 cm/s; (e) 30.3 cm/s **15.** (a) mathematically, an infinite number; (b) 60 km **17.** (a) 4 s $> t > 2$ s; (b) 3 s $> t > 0$; (c) 7 s $> t > 3$ s; (d) $t = 3$ s **19.** 100 m **23.** (a) The signs of v and a are: AB: +, −; BC: 0, 0; CD: +, +; DE: +, 0; (b) no; (c) no **25.** (e) situations (a), (b), and (d) **27.** (a) 80 m/s; (b) 110 m/s; (c) 20 m/s^2 **29.** (a) 1.10 m/s, 6.11 mm/s^2; (b) 1.47 m/s, 6.11 mm/s^2 **31.** (a) 2.00 s; (b) 12 cm from left edge of screen; (c) 9.00 cm/s^2, to the left; (d) to the right; (e) to the left; (f) 3.46 s **33.** 0.556 s **35.** each, 0.28 m/s^2 **37.** 2.8 m/s^2 **39.** 1.62×10^{15} m/s^2 **41.** 21g **43.** (a) 25g; (b) 400 m **45.** 90 m **47.** (a) 5.0 m/s^2; (b) 4.0 s; (c) 6.0 s; (d) 90 m **49.** (a) 5.00 m/s; (b) 1.67 m/s^2; (c) 7.50 m **51.** (a) 0.74 s; (b) -20 ft/s^2 **53.** (a) 0.75 s; (b) 50 m **55.** (a) 34.7 ft; (b) 41.6 s **57.** (a) 3.26 ft/s^2 **61.** (a) 31 m/s; (b) 6.4 s **63.** (a) 48.5 m/s; (b) 4.95 s; (c) 34.3 m/s; (d) 3.50 s **65.** (a) 5.44 s; (b) 53.3 m/s; (d) 5.80 m **67.** (a) 3.2 s; (b) 1.3 s **69.** 4.0 m/s **71.** (a) 350 ms; (b) 82 ms (each is for ascent and descent through the 15 cm) **73.** 857 m/s^2, upward **75.** (a) 1.23 cm; (b) 4 times, 9 times, 16 times, 25 times **77.** (a) 8.85 m/s; (b) 1.00 m **79.** 22 cm and 89 cm below the nozzle **81.** (a) 3.41 s; (b) 57 m **83.** (a) 40.0 ft/s **85.** 1.5 s **87.** (a) 5.4 s; (b) 41 m/s **89.** 20.4 m

91. (a) $d = v_i^2/2a' + T_R v_i$; (b) 9.0 m/s^2; (c) 0.66 s. **93.** (a) $v_j^2 = 2a'd_0(j - 1) + v_1^2$; (c) 7.0 m/s^2; (d) 14 m.

Chapter 3

CP 1. (a) 7 m; (b) 1 m **2.** c, d, f **3.** (a) +, +; (b) +, −; (c) +, + **4.** (a) 90°; (b) 0 (vectors are parallel); (c) 180° (vectors are antiparallel) **5.** (a) 0° or 180°; (b) 90° **Q 1.** A and B **3.** No, but **a** and $-\mathbf{b}$ are commutative: $\mathbf{a} + (-\mathbf{b}) = (-\mathbf{b}) + \mathbf{a}$. **5.** (a) **a** and **b** are parallel; (b) $\mathbf{b} = 0$; (c) **a** and **b** are perpendicular **7.** (a)–(c) yes (example: 5**i** and $-2\mathbf{i}$) **9.** all but e **11.** (a) minus, minus; (b) minus, minus **13.** (a) **B** and **C**, **D** and **E**; (b) **D** and **E** **15.** no (their orientations can differ) **17.** (a) 0 (vectors are parallel); (b) 0 (vectors are antiparallel) **EP 1.** The displacements should be (a) parallel, (b) antiparallel, (c) perpendicular **3.** (b) 3.2 km, 41° south of west **5.** $\mathbf{a} + \mathbf{b}$: 4.2, 40° east of north; $\mathbf{b} - \mathbf{a}$: 8.0, 24° north of west **7.** (a) 38 units at 320°; (b) 130 units at 1.2°; (c) 62 units at 130° **9.** $a_x = -2.5$, $a_y = -6.9$ **11.** $r_x = 13$ m, $r_y = 7.5$ m **13.** (a) 14 cm, 45° left of straight down; (b) 20 cm, vertically up; (c) zero **15.** 4.74 km **17.** 168 cm, 32.5° above the floor **19.** $r_x = 12$, $r_y = -5.8$, $r_z = -2.8$ **21.** (a) $8\mathbf{i} + 2\mathbf{j}$, 8.2, 14°; (b) $2\mathbf{i} - 6\mathbf{j}$, 6.3, $-72°$ relative to **i** **23.** (a) 5.0, $-37°$; (b) 10, 53°; (c) 11, 27°; (d) 11, 80°; (e) 11, 260°; the angles are relative to $+x$, the last two vectors are in opposite directions **25.** 4.1 **27.** (a) $r_x = 1.59$, $r_y = 12.1$; (b) 12.2; (c) 82.5° **29.** 3390 ft, horizontally **31.** (a) -2.83 m, -2.83 m, $+5.00$ m, 0 m, 3.00 m, 5.20 m; (b) 5.17 m, 2.37 m; (c) 5.69 m, 24.6° north of east; (d) 5.69 m, 24.6° south of west **35.** (a) $a_x = 9.51$ m, $a_y = 14.1$ m; (b) $a_x' = 13.4$ m, $a_y' = 10.5$ m **37.** (a) $+y$; (b) $-y$; (c) 0; (d) 0; (e) $+z$; (f) $-z$; (g) ab, both; (h) ab/d, $+z$ **39.** yes **41.** (a) up, unit magnitude; (b) zero; (c) south, unit magnitude; (d) 1.00; (e) 0 **43.** (a) -18.8; (b) 26.9, $+z$ direction **45.** (a) 12, out of page; (b) 12, into page; (c) 12, out of page **47.** (a) $11\mathbf{i} + 5\mathbf{j} - 7\mathbf{k}$; (b) 120° **51.** (a) 57°; (b) $c_x = \pm 2.2$, $c_y = \mp 4.5$ **53.** (a) -21; (b) -9; (c) $5\mathbf{i} - 11\mathbf{j} - 9\mathbf{k}$

Chapter 4

CP 1. (a) $(8\mathbf{i} - 6\mathbf{j})$ m; (b) yes, the xy plane **2.** (a) first; (b) third **3.** (1) and (3) a_x and a_y are both constant and thus **a** is constant; (2) and (4) a_y is constant but a_x is not, thus **a** is not **4.** 4 m/s^3, -2 m/s, 3 m **5.** (a) v_x constant; (b) v_y initially positive, decreases to zero, and then becomes progressively more negative; (c) $a_x = 0$ throughout; (d) $a_y = -g$ throughout **6.** (a) $-(4$ m/s$)\mathbf{i}$; (b) $-(8$ m/s$^2)\mathbf{j}$ **7.** (1) 0, distance not changing; (2) $+70$ km/h, distance increasing; (3) $+80$ km/h, distance decreasing **Q 1.** (1) and (3) a_y is constant but a_x is not and thus **a** is not; (2) a_x is constant but a_y

is not and thus **a** is not; (4) a_x and a_y are both constant and thus **a** is constant; -2 m/s², 3 m/s **3.** (a) highest point; (b) lowest point **5.** (a) all tie; (b) 1 and 2 tie (the rocket is shot upward), then 3 and 4 tie (it is shot into the ground!) **7.** $(2\mathbf{i} - 4\mathbf{j})$ m/s **9.** (a) all tie; (b) all tie; (c) c, b, a; (d) c, b, a **11.** (a) no; (b) same **13.** (a) in your hands; (b) behind you; (c) in front of you **15.** (a) straight down; (b) curved; (c) more curved **17.** (a) 3; (b) 4. **EP** **1.** (a) $(-5.0\mathbf{i} + 8.0\mathbf{j})$ m; (b) 9.4 m, 122° from $+x$; (d) $(8\mathbf{i} - 8\mathbf{j})$ m; (e) 11 m, $-45°$ from $+x$ **3.** (a) $(-7.0\mathbf{i} + 12\mathbf{j})$ m; (b) xy plane **5.** (a) 671 mi, 63.4° south of east; (b) 298 mi/h, 63.4° south of east; (c) 400 mi/h **7.** (a) 6.79 km/h; (b) 6.96° **9.** (a) $(3\mathbf{i} - 8t\mathbf{j})$ m/s; (b) $(3\mathbf{i} - 16\mathbf{j})$ m/s; (c) 16 m/s, $-79°$ to $+x$ **11.** (a) $(8t\mathbf{j} + \mathbf{k})$ m/s; (b) $8\mathbf{j}$ m/s² **13.** $(-2.10\mathbf{i} + 2.81\mathbf{j})$ m/s² **15.** (a) $-1.5\mathbf{j}$ m/s; (b) $(4.5\mathbf{i} - 2.25\mathbf{j})$ m **17.** 60.0° **19.** (a) 63 ms; (b) 1.6×10^3 ft/s **21.** (a) 2.0 ns; (b) 2.0 mm; (c) $(1.0 \times 10^9\mathbf{i} - 2.0 \times 10^8\mathbf{j})$ cm/s **23.** (a) 3.03 s; (b) 758 m; (c) 29.7 m/s **25.** (a) 16 m/s, 23° above the horizontal; (b) 27 m/s, 57° below the horizontal **27.** (a) 32.4 m; (b) -37.7 m **29.** (b) 76° **31.** (a) 51.8 m; (b) 27.4 m/s; (c) 67.5 m **33.** (a) 194 m/s; (b) 38° **35.** 1.9 in. **37.** (a) 11 m; (b) 23 m; (c) 17 m/s, 63° below horizontal **41.** (a) 73 ft; (b) 7.6°; (c) 1.0 s **43.** 23 ft/s **45.** (a) 11 m; (b) 45 m/s **47.** 30 m above the release point **49.** 19 ft/s **51.** (a) 202 m/s; (b) 806 m; (c) 161 m/s, -171 m/s **53.** (a) 20 cm; (b) no, the ball hits the net only 4.4 cm above the ground **55.** yes; its center passes about 4.1 ft above the fence **57.** (a) 9.00×10^{22} m/s², toward the center; (b) 1.52×10^{-16} s **59.** (a) 6.7×10^6 m/s; (b) 1.4×10^{-7} s **61.** (a) 7.49 km/s; (b) 8.00 m/s² **63.** (a) 0.94 m; (b) 19 m/s; (c) 2400 m/s², toward center; (d) 0.05 s **65.** (a) 1.3×10^5 m/s; (b) 7.9×10^5 m/s² or $(8.0 \times 10^4)g$, toward the center; (c) both answers increase **67.** (a) 0.034 m/s²; (b) 84 min **69.** 2.58 cm/s² **71.** 160 m/s² **73.** 36 s, no **75.** 0.018 mi/s² from either frame **77.** 130° **79.** 60° **81.** (a) 5.8 m/s; (b) 16.7 m; (c) 67° **83.** 185 km/h, 22° south of west **85.** (a) from 75° east of south; (b) 30° east of north; substitute west for east to get second solution **87.** (a) 30° upstream; (b) 69 min; (c) 80 min; (d) 80 min; (e) perpendicular to the current, the shortest possible time is 60 min **89.** 0.83c **91.** (a) 0.35c; (b) 0.62c **93.** For launch angles from 5° to 70°, it always moves away from the launch site. For a 75° launch angle, it moves toward the site from 11.5 s to 18.5 s after launch. For an 80° launch angle, it moves toward the site from 10.5 s to 20.5 s after launch. For an 85° launch angle, it moves toward the site from 10.5 s to 20.5 s after launch. For a 90° launch angle, it moves toward the site from 10 s to 20.5 s after launch. **95.** (a) 1.6 s; (b) no; (c) 14 m/s; (d) yes **97.** (a) $\Delta\mathbf{D} = (1.0$ m$)\mathbf{i} - (2.0$ m$)\mathbf{j} + (1.0$ m$)\mathbf{k}$; (b) 2.4 m; (c) $\bar{\mathbf{v}} = (0.025$ m/s$)\mathbf{i} - (0.050$ m/s$)\mathbf{j} + (0.025$ m/s$)\mathbf{k}$ (d) cannot be determined without additional information

Chapter 5

CP **1.** c, d, and e **2.** (a) and (b) 2 N, leftward (acceleration is zero in each situation) **3.** (a) and (b) 1, 4, 3, 2 **4.** (a) equal; (b) greater (acceleration is upward, thus net force on body must be upward) **5.** (a) equal; (b) greater; (c) less **6.** (a) increase; (b) yes; (c) same; (d) yes **7.** (a) $F \sin\theta$; (b) increase **8.** 0 **Q** **1.** (a) yes; (b) yes; (c) yes; (d) yes **3.** (a) 2 and 4; (b) 2 and 4 **5.** (a) 50 N, upward; (b) 150 N, upward **7.** (a) less; (b) greater **9.** (a) no; (b) no; (c) no **11.** (a) increases; (b) increases; (c) decreases; (d) decreases **13.** (a) 20 kg; (b) 18 kg; (c) 10 kg; (d) all tie; (e) 3, 2, 1 **15.** d, c, a, b **EP** **1.** (a) $F_x = 1.88$ N, $F_y = 0.684$ N; (b) $(1.88\mathbf{i} + 0.684\mathbf{j})$ N **3.** (a) $(-6.26\mathbf{i} - 3.23\mathbf{j})$ N; (b) 7.0 N, 207° relative to $+x$ **5.** $(-2\mathbf{i} + 6\mathbf{j})$ N **7.** (a) 0; (b) $+20$ N; (c) -20 N; (d) -40 N; (e) -60 N **9.** (a) $(1\mathbf{i} - 1.3\mathbf{j})$ m/s²; (b) 1.6 m/s² at $-50°$ from $+x$ **11.** (a) $\mathbf{F}_2$ and $\mathbf{F}_3$ are in the $-x$ direction, $\mathbf{a} = 0$; (b) $\mathbf{F}_2$ and $\mathbf{F}_3$ are in the $-x$ direction, $\mathbf{a}$ is on the x axis, $a = 0.83$ m/s²; (c) $\mathbf{F}_2$ and $\mathbf{F}_3$ are at 34° from $-x$ direction; $\mathbf{a} = 0$ **13.** (a) 22 N, 2.3 kg; (b) 1100 N, 110 kg; (c) 1.6×10^4 N, 1.6×10^3 kg **15.** (a) 11 N, 2.2 kg; (b) 0, 2.2 kg **17.** (a) 44 N; (b) 78 N; (c) 54 N; (d) 152 N **19.** 1.18×10^4 N **21.** 1.2×10^5 N **23.** 16 N **25.** (a) 13 ft/s²; (b) 190 lb **27.** (a) 42 N; (b) 72 N; (c) 4.9 m/s² **29.** (a) 0.02 m/s²; (b) 8×10^4 km; (c) 2×10^3 m/s **31.** (a) 1.1×10^{-15} N; (b) 8.9×10^{-30} N **33.** (a) 5500 N; (b) 2.7 s; (c) 4 times as far; (d) twice the time **35.** (a) 4.9×10^5 N; (b) 1.5×10^6 N **37.** (a) 110 lb, up; (b) 110 lb, down **39.** (a) 0.74 m/s²; (b) 7.3 m/s² **41.** (a) $\cos\theta$; (b) $\sqrt{\cos\theta}$ **43.** 1.8×10^4 N **45.** (a) 4.6×10^3 N; (b) 5.8×10^3 N **47.** (a) 250 m/s²; (b) 2.0×10^4 N **49.** 23 kg **51.** (a) 620 N; (b) 580 N **53.** 1.9×10^5 lb **55.** (a) rope breaks; (b) 1.6 m/s² **57.** 4.6 N **59.** (a) allow a downward acceleration with magnitude ≥ 4.2 ft/s²; (b) 13 ft/s or greater **61.** 195 N, up **63.** (a) 566 N; (b) 1130 N **65.** 18,000 N **67.** (a) 1.4×10^4 N; (b) 1.1×10^4 N; (c) 2700 N, toward the counterweight **69.** 6800 N, at 21° to the line of motion of the barge **71.** (a) 4.6 m/s²; (b) 2.6 m/s² **73.** (b) $Fl/(m + M)$; (c) $MF/(m + M)$; (d) $F(m + 2M)/2(m + M)$ **75.** $T_1 = 13$ N, $T_2 = 20$ N, $a = 3.2$ m/s²

Chapter 6

CP **1.** (a) zero (because there is no attempt at sliding); (b) 5 N; (c) no; (d) yes **2.** (a) same (10 N); (b) decreases; (c) decreases **3.** greater **4.** (a) **a** downward; **N** upward; (b) **a** and **N** upward **5.** (a) $4R_1$; (b) $4R_1$ **6.** (a) same; (b) increases; (c) increases **Q** **1.** They slide at the same angle for all orders. **3.** (a) upward; (b) horizontal, toward you; (c) no change; (d) increases; (e) increases **5.** The frictional force $\mathbf{f}_s$ is initially directed up the ramp, decreases in magnitude to zero, and then is directed down the ramp, increasing in magnitude until the magnitude reaches $f_{s,max}$; thereafter, the magnitude of the frictional force is f_k, which is a constant smaller value. **7.** (a) decreases; (b) decreases; (c) increases; (d) increases **9.** (a) zero; (b) infinite **11.** 4, 3; then 1, 2, and 5 tie **13.** (a) less; (b) greater **EP** **1.** (a) 200 N; (b) 120 N **3.** 2° **5.** 440 N

7. (a) 110 N; (b) 130 N; (c) no; (d) 46 N; (e) 17 N
9. (a) 90 N; (b) 70 N; (c) 0.89 m/s²   **11.** (a) no; (b) ($-12\mathbf{i}$ + $5\mathbf{j}$) N   **13.** 20°   **15.** (a) 0.13 N; (b) 0.12   **17.** $\mu_s = 0.58$, $\mu_k = 0.54$ **19.** (a) 0.11 m/s², 0.23 m/s²; (b) 0.041, 0.029
21. 36 m   **23.** (a) 300 N; (b) 1.3 m/s²   **25.** (a) 66 N; (b) 2.3 m/s²   **27.** (a) $\mu_k mg/(\sin\theta - \mu_k\cos\theta)$; (b) $\theta_0 = \tan^{-1}\mu_s$
29. (b) 3.0×10^7 N   **31.** 100 N   **33.** 3.3 kg   **35.** (a) 11 ft/s²; (b) 0.46 lb; (c) blocks move independently
37. (a) 27 N; (b) 3.0 m/s²   **39.** (a) 6.1 m/s², leftward; (b) 0.98 m/s², leftward   **41.** (a) 3.0×10^5 N; (b) 1.2°
43. 9.9 s   **45.** 3.75   **47.** 12 cm   **49.** 68 ft
51. (a) 3210 N; (b) yes   **53.** 0.078   **55.** (a) 0.72 m/s; (b) 2.1 m/s²; (c) 0.50 N   **57.** $\sqrt{Mgr/m}$ **59.** (a) 30 cm/s; (b) 180 cm/s², radially inward; (c) 3.6×10^{-3} N, radially inward; (d) 0.37   **61.** (a) 275 N; (b) 877 N   **63.** 874 N
65. (a) at the bottom of the circle; (b) 31 ft/s   **67.** (a) 9.5 m/s; (b) 20 m   **69.** 13°   **71.** (a) 0.0338 N; (b) 9.77 N

Chapter 7

CP   1. (a) decrease; (b) same; (c) negative, zero   **2.** d, c, b, a **3.** (a) same; (b) smaller   **4.** (a) positive; (b) negative; (c) zero   **5.** zero   **Q   1.** all tie   **3.** (a) increasing; (b) same; (c) same; (d) increasing   **5.** (a) positive; (b) negative; (c) negative   **7.** (a) positive; (b) zero; (c) negative; (d) negative; (e) zero; (f) positive   **9.** all tie   **11.** c, d, a and b tie; then f, e.   **13.** (a) 3 m; (b) 3 m; (c) 0 and 6 m; (d) negative direction of x **15.** (a) A; (b) B
17. twice   **EP   1.** 1.8×10^{13} J;   **3.** (a) 3610 J; (b) 1900 J; (c) 1.1×10^{10} J   **5.** (a) 1×10^5 megatons TNT; (b) 1×10^7 bombs   **7.** father, 2.4 m/s; son, 4.8 m/s
9. (a) 200 N; (b) 700 m; (c) -1.4×10^5 J; (d) 400 N, 350 m, -1.4×10^5 J   **11.** 5000 J   **13.** 47 keV   **15.** 7.9 J
17. 530 J   **19.** -37 J   **21.** (a) 314 J; (b) -155 J; (c) 0; (d) 158 J   **23.** (a) 98 N; (b) 4.0 cm; (c) 3.9 J; (d) -3.9 J
25. (a) $-3Mgd/4$; (b) Mgd; (c) $Mgd/4$ (d) $\sqrt{gd/2}$ **27.** 25 J
31. -6 J   **33.** (a) 12 J; (b) 4.0 m; (c) 18 J
35. (a) -0.043 J; (b) -0.13 J   **37.** (a) 6.6 m/s; (b) 4.7 m
39. (a) up; (b) 5.0 cm; (c) 5.0 J   **41.** 270 kW
43. 235 kW   **45.** 490 W   **47.** (a) 100 J; (b) 67 W; (c) 33 W   **49.** 0.99 hp   **51.** (a) 0; (b) -350 W
53. (a) 79.4 keV; (b) 3.12 MeV; (c) 10.9 MeV   **55.** (a) 32 J; (b) 8 W; (c) 78°

Chapter 8

CP   1. no   **2.** 3, 1, 2   **3.** (a) all tie; (b) all tie   **4.** (a) CD, AB, BC (zero); (b) positive direction of x **5.** 2, 1, 3
6. decrease   **7.** (a) seventh excited state, with energy E_7; (b) 1.3 eV   **Q   1.** -40 J   **3.** (c) and (d) tie; then (a) and (b) tie   **5.** (a) all tie; (b) all tie   **7.** (a) 3, 2, 1; (b) 1, 2, 3
9. less than (smaller decrease in potential energy)
11. (a) $E < 3$ J, $K < 2$ J; (b) $E < 5$ J, $K < 4$ J
13. (a) increasing; (b) decreasing; (c) decreasing; (d) constant in AB and BC, decreasing in CD **EP   1.** 15 J
3. (a) 167 J; (b) -167 J; (c) 196 J; (d) 29 J   **5.** (a) 0; (b) $mgh/2$; (c) mgh; (d) $mgh/2$; (e) mgh **7.** (a) -0.80 J; (b) -0.80 J; (c) $+1.1$ J   **9.** (a) $mgL(1 - \cos\theta)$; (b) $-mgL(1 - \cos\theta)$; (c) $mgL(1 - \cos\theta)$ **11.** (a) 18 J;

(b) 0; (c) 30 J; (d) 0; (e) parts b and d   **13.** (a) 2.08 m/s; (b) 2.08 m/s   **15.** (a) $\sqrt{2gL}$; (b) $2\sqrt{gL}$; (c) $\sqrt{2gL}$
17. 830 ft   **19.** (a) 6.75 J; (b) -6.75 J; (c) 6.75 J; (d) 6.75 J; (e) -6.75 J; (f) 0.459 m   **21.** (a) 21.0 m/s; (b) 21.0 m/s
23. (a) 0.98 J; (b) -0.98 J; (c) 3.1 N/cm   **25.** (a) 39.2 J; (b) 39.2 J; (c) 4.00 m   **27.** (a) 54 m/s; (b) 52 m/s; (c) 76 m, below   **29.** (a) 39 ft/s; (b) 4.3 in.   **31.** (a) 300 J; (b) 93.8 J; (c) 6.38 m   **33.** (a) 4.8 m/s; (b) 2.4 m/s
35. (a) $[v_0^2 + 2gL(1 - \cos\theta_0)]^{1/2}$; (b) $(2gL\cos\theta_0)^{1/2}$; (c) $[gL(3 + 2\cos\theta_0)]^{1/2}$ **37.** (a) $U(x) = -Gm_1m_2/x$; (b) $Gm_1m_2d/x_1(x_1 + d)$ **39.** (a) $8mg$ leftward and mg downward; (b) $2.5R$ **43.** $mgL/32$ **47.** (a) $1.12(A/B)^{1/6}$; (b) repulsive; (c) attractive   **49.** (a) turning point on left, none on right; molecule breaks apart; (b) turning points on both left and right; molecule does not break apart; (c) -1.2×10^{-19} J; (d) 2.2×10^{-19} J; (e) $\approx 1 \times 10^{-9}$ on each, directed toward the other; (f) $r < 0.2$ nm; (g) $r > 0.2$ nm; (h) $r = 0.2$ nm
51. -25 J   **53.** (a) 2200 J; (b) -1500 J; (c) 700 J
55. 17 kW   **57.** (a) -0.74 J; (b) -0.53 J   **59.** -12 J
61. 54%   **63.** 880 MW   **65.** (a) 39 kW; (b) 39 kW
67. (a) 1.5 MJ; (b) 0.51 MJ; (c) 1.0 MJ; (d) 63 m/s
69. (a) 67 J; (b) 67 J; (c) 46 cm   **71.** Your force on the cabbage does work.   **73.** (a) -0.90 J; (b) 0.46 J; (c) 1.0 m/s
75. (a) 18 ft/s; (b) 18 ft   **77.** 4.3 m   **79.** (a) 31.0 J; (b) 5.35 m/s; (c) conservative   **81.** 1.2 m   **85.** in the center of the flat part   **87.** (a) 24 ft/s; (b) 3.0 ft; (c) 9.0 ft; (d) 49 ft   **89.** (a) 216 J; (b) 1180 N; (c) 432 J; (d) motor also supplies thermal energy to crate and belt   **91.** (a) 1.1×10^{17} J; (b) 1.2 kg   **93.** 7.28 MeV   **95.** (a) release; (b) 17.6 MeV   **97.** (a) 5.3 eV; (b) 0.9 eV   **99.** (a) 7.2 J; (b) -7.2 J; (c) 86 cm; (d) 26 cm

Chapter 9

CP   1. (a) origin; (b) fourth quadrant; (c) on y axis below origin; (d) origin; (e) third quadrant; (f) origin   **2.** (a) to (c) at the center of mass, still at the origin (their forces are internal to the system and cannot move the center of mass)   **3.** (a) 1, 3, and then 2 and 4 tie (zero force); (b) 3   **4.** (a) 0; (b) no; (c) negative x **5.** (a) 500 km/h; (b) 2600 km/h; (c) 1600 km/h   **6.** (a) yes; (b) no   **Q   1.** point 4   **3.** (a) at the center of the sled; (b) $L/4$, to the right; (c) not at all (no net external force); (d) $L/4$, to the left; (e) L; (f) $L/2$; (g) $L/2$
5. (a) ac, cd, and bc; (b) bc; (c) bd and ad **7.** (a) 2 N, rightward; (b) 2 N, rightward; (c) greater than 2 N, rightward
9. b, c, a **11.** (a) yes; (b) 6 kg·m/s in $-x$ direction; (c) can't tell   **EP   1.** (a) 4600 km; (b) $0.73R_e$ **3.** (a) $x_{cm} = 1.1$ m, $y_{cm} = 1.3$ m; (b) shifts toward topmost particle   **5.** $x_{cm} = -0.25$ m, $y_{cm} = 0$ **7.** in the iron, at midheight and midwidth, 2.7 cm from midlength   **9.** $x_{cm} = y_{cm} = 20$ cm, $z_{cm} = 16$ cm   **11.** (a) $H/2$; (b) $H/2$; (c) descends to lowest point and then ascends to $H/2$; (d) $(HM/m)(\sqrt{1 + m/M} - 1)$
13. 72 km/h   **15.** (a) center of mass does not move; (b) 0.75 m   **17.** 4.8 m/s   **19.** (a) 22 m; (b) 9.3 m/s
21. 53 m   **23.** 13.6 ft   **25.** (a) 52.0 km/h; (b) 28.8 km/h
27. a proton   **29.** (a) 30°; (b) $-0.572\mathbf{j}$ kg·m/s
31. (a) $(-4.0 \times 10^4\,\mathbf{i})$ kg·m/s; (b) west; (c) 0   **33.** $0.707c$
35. 0.57 m/s, toward center of mass   **37.** it increases by

4.4 m/s   **39.** (a) rocket case: 7290 m/s, payload: 8200 m/s; (b) before: 1.271×10^{10} J, after: 1.275×10^{10} J   **41.** (a) -1; (b) 1830; (c) 1830; (d) same   **43.** 14 m/s, 135° from the other pieces   **45.** 190 m/s   **47.** (a) $0.200v_{rel}$; (b) $0.210v_{rel}$; (c) $0.209v_{rel}$ **49.** (a) 1.57×10^6 N; (b) 1.35×10^5 kg; (c) 2.08 km/s   **51.** 108 m/s   **53.** 2.2×10^{-3} **57.** fast barge: 46 N more; slow barge: no change   **59.** (a) 7.8 MJ; (b) 6.2   **61.** 690 W   **63.** 5.5×10^6 N   **65.** 24 W   **67.** 100 m   **69.** (a) 860 N; (b) 2.4 m/s   **71.** (a) 3.0×10^5 J; (b) 10 kW; (c) 20 kW   **73.** (a) 2.1×10^6 kg; (b) $\sqrt{100 + 1.5t}$ m/s; (c) $(1.5 \times 10^6)/\sqrt{100 + 1.5t}$ N; (d) 6.7 km   **75.** $t = (3d/2)^{2/3}(m/2P)^{1/3}$

Chapter 10

CP   **1.** (a) unchanged; (b) unchanged; (c) decreased   **2.** (a) zero; (b) positive; (c) positive direction of y **3.** (a) 4 kg·m/s; (b) 8 kg·m/s; (c) 3 J   **4.** (a) 0; (b) 4 kg·m/s   **5.** (a) 10 kg·m/s; (b) 14 kg·m/s; (c) 6 kg·m/s   **6.** (a) 2 kg·m/s; (b) 3 kg·m/s   **7.** (a) increases; (b) increases   Q   **1.** all tie   **3.** b and c   **5.** (a) one stationary; (b) 2; (c) 5; (d) equal (pool player's result)   **7.** (a) 1 and 4 tie; then 2 and 3 tie; (b) 1; 3 and 4 tie; then 2   **9.** (a) rightward; (b) rightward; (c) smaller   **11.** positive direction of x axis   EP   **1.** (a) 750 N; (b) 6.0 m/s   **3.** 6.2×10^4 N   **5.** 3000 N ($= 660$ lb)   **7.** 1.1 m   **9.** (a) 42 N·s; (b) 2100 N   **11.** (a) $(7.4 \times 10^3\, \mathbf{i} - 7.4 \times 10^3\, \mathbf{j})$ N·s; (b) $(-7.4 \times 10^3\, \mathbf{i})$ N·s; (c) 2.3×10^3 N; (d) 2.1×10^4 N; (e) $-45°$ **13.** (a) 1.0 kg·m/s; (b) 250 J; (c) 10 N; (d) 1700 N   **15.** 5 N   **17.** $2\mu v$ **19.** 990 N   **21.** (a) 1.8 N·s, upward; (b) 180 N, downward   **25.** 8 m/s   **27.** 38 km/s   **29.** 4.2 m/s   **31.** (a) 99 g; (b) 1.9 m/s; (c) 0.93 m/s   **33.** (a) 1.2 kg; (b) 2.5 m/s   **35.** 7.8 kg   **37.** (a) 1/3; (b) $4h$ **39.** 35 cm   **41.** 3.0 m/s   **43.** (a) $(10\mathbf{i} + 15\mathbf{j})$ m/s; (b) 500 J lost   **45.** (a) 2.7 m/s; (b) 1400 m/s   **47.** (a) A: 4.6 m/s, B: 3.9 m/s; (b) 7.5 m/s   **49.** 20 J for the heavy particle, 40 J for the light particle   **51.** $mv^2/6$ **53.** 13 tons   **55.** 25 cm   **57.** 0.975 m/s, 0.841 m/s   **59.** (a) 4.1 ft/s; (b) 1700 ft·lb; (c) $v_{24} = 5.3$ ft/s, $v_{32} = 3.3$ ft/s   **61.** (a) 30° from the incoming proton's direction; (b) 250 m/s and 430 m/s   **63.** (a) 41°; (b) 4.76 m/s; (c) no   **65.** $v = V/4$ **67.** (a) 117° from the final direction of B; (b) no   **69.** 120°   **71.** (a) 1.9 m/s, 30° to initial direction; (b) no   **73.** (a) 3.4 m/s, deflected by 17° to the right; (b) 0.95 MJ   **75.** (a) 117 MeV; (b) equal and opposite momenta; (c) π^- **77.** (a) 4.94 MeV; (b) 0; (c) 4.85 MeV; (d) 0.09 MeV

Chapter 11

CP   **1.** (b) and (c)   **2.** (a) and (d)   **3.** (a) yes; (b) no; (c) yes; (d) yes   **4.** all tie   **5.** 1, 2, 4, 3   **6.** (a) 1 and 3 tie, 4; then 2 and 5 tie (zero)   **7.** (a) downward in the figure; (b) less   Q   **1.** (a) positive; (b) zero; (c) negative; (d) negative   **3.** (a) 2 and 3; (b) 1 and 3; (c) 4   **5.** (a) and (c)   **7.** (a) all tie; (b) 2, 3; then 1 and 4 tie   **9.** b, c, a **11.** less   **13.** 90°; then 70° and 110° tie   **15.** Finite angular

displacements are not commutative.   EP   **1.** (a) 1.50 rad; (b) 85.9°; (c) 1.49 m   **3.** (a) 0.105 rad/s; (b) 1.75×10^{-3} rad/s; (c) 1.45×10^{-4} rad/s   **5.** (a) $\omega(2) = 4.0$ rad/s, $\omega(4) = 28$ rad/s; (b) 12 rad/s²; (c) $\alpha(2) = 6.0$ rad/s², $\alpha(4) = 18$ rad/s²   **7.** (a) $\omega_0 + at^4 - bt^3$; (b) $\theta_0 + \omega_0 t + at^5/5 - bt^4/4$ **9.** 11 rad/s   **11.** (a) 9000 rev/min²; (b) 420 rev   **13.** (a) 30 s; (b) 1800 rad   **15.** 200 rev/min   **17.** (a) 2.0 rad/s²; (b) 5.0 rad/s; (c) 10 rad/s; (d) 75 rad   **19.** (a) 13.5 s; (b) 27.0 rad/s   **21.** (a) 340 s; (b) -4.5×10^{-3} rad/s²; (c) 98 s   **23.** (a) 1.0 rev/s²; (b) 4.8 s; (c) 9.6 s; (d) 48 rev   **25.** 6.1 ft/s² (1.8 m/s²), toward the center   **27.** 0.13 rad/s   **29.** 5.6 rad/s²   **31.** (a) 5.1 h; (b) 8.1 h   **33.** (a) 2.50×10^{-3} rad/s; (b) 20.2 m/s²; (c) 0   **35.** (a) -1.1 rev/min²; (b) 9900 rev; (c) -0.99 mm/s²; (d) 31 m/s²   **37.** (a) 310 m/s; (b) 340 m/s   **39.** (a) 1.94 m/s²; (b) 75.1°, toward the center of the track   **41.** 16 s   **43.** (a) 73 cm/s²; (b) 0.075; (c) 0.11   **45.** 12.3 kg·m²   **47.** first cylinder: 1100 J; second cylinder: 9700 J   **49.** (a) 221 kg·m²; (b) 1.10×10^4 J   **51.** (a) 6490 kg·m²; (b) 4.36 MJ   **53.** 0.097 kg·m²   **57.** (a) 1300 g·cm²; (b) 550 g·cm²; (c) 1900 g·cm²; (d) $A + B$ **59.** (a) 49 MJ; (b) 100 min   **61.** 4.6 N·m   **63.** (a) $r_1 F_1 \sin \theta_1 - r_2 F_2 \sin \theta_2$; (b) -3.8 N·m   **65.** 1.28 kg·m²   **67.** 9.7 rad/s², counterclockwise   **69.** (a) 155 kg·m²; (b) 64.4 kg   **71.** (a) 420 rad/s²; (b) 500 rad/s   **73.** small sphere: (a) 0.689 N·m and (b) 3.05 N; large sphere: (a) 9.84 N·m and (b) 11.5 N   **75.** 1.73 m/s²; 6.92 m/s²   **77.** (a) 1.4 m/s; (b) 1.4 m/s   **79.** (a) 19.8 kJ; (b) 1.32 kW   **81.** (a) 8.2×10^{28} N·m; (b) 2.6×10^{29} J; (c) 3.0×10^{21} kW   **83.** $\sqrt{9g/4\ell}$ **85.** (a) 4.8×10^5 N; (b) 1.1×10^4 N·m; (c) 1.3×10^6 J   **87.** (a) $3g(1 - \cos \theta)$; (b) $\frac{3}{2}g \sin \theta$; (c) 41.8°   **89.** (a) 5.6 rad/s²; (b) 3.1 rad/s   **91.** (a) 42.1 km/h; (b) 3.09 rad/s²; (c) 7.57 kW   **93.** (a) 3.4×10^5 g·cm²; (b) 2.9×10^5 g·cm²; (c) 6.3×10^5 g·cm²; (d) (1.2 cm) $\mathbf{i}$ + (5.9 cm) $\mathbf{j}$

Chapter 12

CP   **1.** (a) same; (b) less   **2.** less   **3.** (a) $\pm z$; (b) $+y$; (c) $-x$ **4.** (a) 1 and 3 tie, then 2 and 4 tie, then 5 (zero); (b) 2 and 3   **5.** (a) 3, 1; then 2 and 4 tie (zero); (b) 3   **6.** (a) all tie (same τ, same t, thus same ΔL); (b) sphere, disk, hoop (reverse order of I)   **7.** (a) decreases; (b) same; (c) increases   Q   **1.** (a) same; (b) block; (c) block   **3.** (a) greater; (b) same   **5.** (a) L; (b) $1.5L$ **7.** b, then c and d tie; then a and e tie (zero)   **9.** a, then b and c tie; then e, d (zero)   **11.** (a) same; (b) increases, because of decrease in rotational inertia   **13.** (a) 30 units clockwise; (b) 2 then 4, then the others; or 4 then 2, then the others   **15.** (a) spins in place; (b) rolls toward you; (c) rolls away from you   EP   **1.** 1.00   **3.** (a) 59.3 rad/s; (b) -9.31 rad/s²; (c) 70.7 m   **5.** (a) -4.11 m/s²; (b) -16.4 rad/s²; (c) -2.54 N·m   **7.** (a) 8.0°; (b) $0.14g$ **9.** (a) 4.0 N, to the left; (b) 0.60 kg·m²   **11.** (a) $\frac{1}{2}mR^2$; (b) a solid circular cylinder   **13.** (a) $mg(R - r)$; (b) 2/7; (c) $(17/7)mg$ **15.** (a) $2.7R$; (b) $(50/7)mg$ **17.** (a) 13 cm/s²; (b) 4.4 s; (c) 55 cm/s; (d) 1.8×10^{-2} J; (e) 1.4 J; (f) 27 rev/s   **21.** (a) 24 N·m, in $+y$ direction; (b) 24 N·m, $-y$; (c) 12 N·m, $+y$;

(d) 12 N·m, $-y$ **23.** (a) $(-1.5\mathbf{i} - 4.0\mathbf{j} - \mathbf{k})$ N·m;
(b) $(-1.5\mathbf{i} - 4.0\mathbf{j} - \mathbf{k})$ N·m   **25.** $-2.0\mathbf{i}$ N·m   **27.** 9.8
kg·m²/s   **29.** (a) 12 kg·m²/s, out of page; (b) 3.0 N·m,
out of page   **31.** (a) 0; (b) $(8.0\mathbf{i} + 8.0\mathbf{k})$ N·m   **33.** (a) mvd;
(b) no; (c) 0, yes   **35.** (a) 3.15×10^{43} kg·m²/s; (b) 0.616
37. 4.5 N·m, parallel to xy plane at $-63°$ from $+x$
39. (a) 0; (b) 0; (c) $30t^3$ kg·m²/s, $90t^2$ N·m, both in $-z$
direction; (d) $30t^3$ kg·m²/s, $90t^2$ N·m, both in $+z$ direction
41. (a) $\frac{1}{2}mgt^2v_0 \cos\theta_0$; (b) $mgtv_0 \cos\theta_0$; (c) $mgtv_0 \cos\theta_0$
43. (a) -1.47 N·m; (b) 20.4 rad; (c) -29.9 J; (d) 19.9 W
45. (a) 12.2 kg·m²; (b) 308 kg·m²/s, down   **47.** (a) 1/3;
(b) 1/9   **49.** $\omega_0 R_1 R_2 I_1/(I_1 R_2^2 + I_2 R_1^2)$ **51.** (a) 3.6 rev/s;
(b) 3.0; (c) work done by man in moving weights inward
53. (a) 267 rev/min; (b) 2/3   **55.** 3.0 min   **57.** 2.6 rad/s
59. (a) they revolve in a circle of 1.5 m radius at 0.93 rad/s;
(b) 8.4 rad/s; (c) $K_a = 98$ J, $K_b = 880$ J; (d) from the work
done in pulling inward   **61.** $m/(M + m)(v/R)$
63. (a) $mvR/(I + MR^2)$; (b) $mvR^2/(I + MR^2)$ **65.** 1300 m/s
67. (a) 18 rad/s; (b) 0.92
69. $\theta = \cos^{-1}\left[1 - \dfrac{6m^2h}{\ell(2m + M)(3m + M)}\right]$
71. 5.28×10^{-35} J·s   **73.** Any three are spin up; the other is
spin down.   **75.** (a) The magnitude of the angular momentum
increases in proportion to t^2 and the magnitude of the torque
increases in proportion to t, in agreement with the second law
for rotation. (b) The magnitudes of the angular momentum and
torque again increase with time. But the change in the
magnitude of the angular momentum in any interval is less than
is predicted by proportionality to t^2 law and the change in the
torque is less than is predicted by proportionality to t. At any
position of the projectile the torque is less when drag is present
than when it is not.

Chapter 13

CP   **1.** c, e, f **2.** (a) no; (b) at site of $\mathbf{F}_1$, perpendicular to
plane of figure; (c) 45 N   **3.** (a) at C (to eliminate forces there
from a torque equation); (b) plus; (c) minus; (d) equal   **4.** d
5. (a) equal; (b) B; (c) B **Q**   **1.** (a) yes; (b) yes; (c) yes;
(d) no   **3.** b **5.** (a) yes; (b) no; (c) no (it could balance the
torques but the forces would then be unbalanced)   **7.** (a) a,
then b and c tie, then d **9.** (a) 20 N (the key is the pulley
with the 20 N weight); (b) 25 N   **11.** (a) sin θ; (b) same;
(c) larger   **13.** tie of A and B, then C **EP**   **1.** (a) two;
(b) seven   **3.** (a) 2.5 m; (b) 7.3°   **5.** 120°   **7.** 7920 N
9. (a) 840 N; (b) 530 N   **11.** 0.536 m   **13.** (a) 2770 N;
(b) 3890 N   **15.** (a) 1160 N, down; (b) 1740 N, up; (c) left,
stretched; (d) right, compressed   **17.** (a) 280 N; (b) 880 N,
71° above the horizontal   **19.** bars BC, CD, and DA are under
tension due to forces T, diagonals AC and BD are compressed
due to forces $\sqrt{2}T$ **21.** (a) 1800 lb; (b) 822 lb; (c) 1270 lb
23. (a) 49 N; (b) 28 N; (c) 57 N; (d) 29°   **25.** (a) 1900 N, up;
(b) 2100 N, down   **27.** (a) 340 N; (b) 0.88 m; (c) increases,
decreases   **29.** $W\sqrt{2rh - h^2}/(r - h)$ **31.** (a) $L/2$; (b) $L/4$;
(c) $L/6$; (d) $L/8$; (e) $25L/24$ **33.** (a) 6630 N; (b) $F_h = 5740$ N;
(c) $F_v = 5960$ N   **35.** 2.20 m   **37.** (a) 1.50 m; (b) 433 N;

(c) 250 N   **39.** (a) $a_1 = L/2$, $a_2 = 5L/8$, $h = 9L/8$;
(b) $b_1 = 2L/3$, $b_2 = L/2$, $h = 7L/6$ **41.** (a) 47 lb; (b) 120 lb;
(c) 72 lb   **43.** (a) 445 N; (b) 0.50; (c) 315 N
45. (a) 3.9 m/s²; (b) 2000 N on each rear wheel, 3500 N on
each front wheel; (c) 790 N on each rear wheel, 1410 N on
each front wheel   **47.** (a) 1.9×10^{-3}; (b) 1.3×10^7 N/m²;
(c) 6.9×10^9 N/m²   **49.** 3.1 cm   **51.** 2.4×10^9 N/m²
53. (a) 1.8×10^7 N; (b) 1.4×10^7 N; (c) 16   **55.** (a) 867 N;
(b) 143 N; (c) 0.165

Chapter 14

CP   **1.** all tie   **2.** (a) 1, tie of 2 and 4, then 3; (b) line d
3. negative y direction   **4.** (a) increase; (b) negative
5. (a) 2; (b) 1   **6.** (a) path 1 (decreased E (more negative)
gives decreased a); (b) less than (decreased a gives decreased T)
Q   **1.** (a) between, closer to less massive particle; (b) no;
(c) no (other than infinity)   **3.** $3GM/d^2$, leftward   **5.** b, tie
of a and c, then d **7.** b, a, c **9.** (a) negative; (b) negative;
(c) postive; (d) all tie   **11.** (a) all tie; (b) all tie
13. (a) same; (b) greater   **EP**   **1.** 19 m   **3.** 2.16
5. 1/2   **7.** 3.4×10^5 km   **9.** (a) 3.7×10^{-5} N, increasing y **11.** $M = m$ **13.** 3.2×10^{-7} N   **15.** $(GmM/d^2) \times$
$\left[1 - \dfrac{1}{8(1 - R/2d)^2}\right]$ **17.** 2.6×10^6 m   **19.** (a) $1.3 \times$
10^{12} m/s²; (b) 1.6×10^6 m/s   **21.** (a) 17 N; (b) 2.5
23. (b) 1.9 h   **27.** (a) $a_g = (3.03 \times 10^{43}$ kg·m/s²$)/M_h$;
(b) decrease; (c) 9.82 m/s²; (d) 7.30×10^{-15} m/s²; (e) no
29. 7.91 km/s   **31.** (a) $(3.0 \times 10^{-7}m)$ N; (b) $(3.3 \times 10^{-7}m)$
N; (c) $(6.7 \times 10^{-7}mr)$ N   **33.** (a) 9.83 m/s²; (b) 9.84 m/s²;
(c) 9.79 m/s²   **35.** (a) -1.4×10^{-4} J; (b) less; (c) positive;
(d) negative   **37.** (a) 0.74; (b) 3.7 m/s²; (c) 5.0 km/s
39. (a) 0.0451; (b) 28.5   **41.** $-Gm(M_E/R + M_M/r)$
43. (a) 5.0×10^{-11} J; (b) -5.0×10^{-11} J   **45.** (a) 1700 m/s;
(b) 250 km; (c) 1400 m/s   **47.** (a) 2.2×10^{-7} rad/s;
(b) 90 km/s   **51.** (a) -1.67×10^{-8} J; (b) 0.56×10^{-8} J
55. 6.5×10^{23} kg   **57.** 5×10^{10} **59.** (a) 7.82 km/s;
(b) 87.5 min   **61.** (a) 6640 km; (b) 0.0136   **63.** (a) 39.5
AU³/M_S·y²; (b) $T^2 = r^3/M$ **65.** (a) 1.9×10^{13} m;
(b) $3.5R_P$ **67.** south, at 35.4° above the horizon
71. $2\pi r^{3/2}/\sqrt{G(M + m/4)}$ **73.** $\sqrt{GM/L}$ **75.** (a) 2.8 y;
(b) 1.0×10^{-4} **77.** (a) 1/2; (b) 1/2; (c) B, by 1.1×10^8 J
79. (a) 54 km/s; (b) 960 m/s; (c) $R_p/R_a = v_a/v_p$ **81.** (a) $4.6 \times$
10^5 J; (b) 260   **83.** (a) 7.5 km/s; (b) 97 min; (c) 410 km;
(d) 7.7 km/s; (e) 92 min; (f) 3.2×10^{-3} N; (g) if the satellite–
Earth system is considered isolated, its $\mathbf{L}$ is conserved
85. (a) 5540 s; (b) 7.68 km/s; (c) 7.60 km/s; (d) 5.78×10^{10} J;
(e) -11.8×10^{10} J; (f) -6.02×10^{10} J; (g) $6.63 \times$
10^6 m; (h) 170 s, new orbit   **87.** (a) $(-7.0$ mm$)\mathbf{i} +$
$(3.0$ cm$)\mathbf{j}$; (b) $(-0.19$ m/s$)\mathbf{i} + (0.40$ m/s$)\mathbf{j}$ **89.** (a) $1.98 \times$
10^{30} kg; (b) 1.96×10^{30} kg

Chapter 15

CP   **1.** all tie   **2.** (a) all tie; (b) $0.95\rho_0$, ρ_0, $1.1\rho_0$
3. 13 cm³/s, outward   **4.** (a) all tie; (b) 1, then 2 and 3 tie, 4;

(c) 4, 3, 2, 1 **Q 1.** e, then b and d tie, then a and c tie
3. (a) 1, 3, 2; (b) all tie; (c) no (you must consider the weight exerted on the scale via the walls) **5.** 3, 4, 1, 2
7. (a) downward; (b) downward; (c) same **9.** (a) same; (b) same; (c) lower; (d) higher **11.** (a) block 1, counterclockwise; block 2, clockwise; (b) block 1, tip more; block 2, right itself **EP 1.** 1000 kg/m³ **3.** 1.1×10^5 Pa or 1.1 atm **5.** 2.9×10^4 N **7.** 6.0 lb/in.² **9.** 1.90×10^4 Pa **11.** 5.4×10^4 Pa **13.** 0.52 m **15.** (a) 6.06×10^9 N; (b) 20 atm **17.** 0.412 cm **19.** $\frac{1}{4}\rho g A(h_2 - h_1)^2$
21. 44 km **23.** (a) $\rho g W D^2/2$; (b) $\rho g W D^3/6$; (c) $D/3$
25. (a) 2.2; (b) 2.4 **27.** -3.9×10^{-3} atm **29.** (a) fA/a; (b) 20 lb **31.** 1070 g **33.** 1.5 g/cm³ **35.** 600 kg/m³
37. (a) 670 kg/m³; (b) 740 kg/m³ **39.** 390 kg
41. (a) 1.2 kg; (b) 1300 kg/m³ **43.** 0.126 m³ **45.** five
47. (a) 1.80 m³; (b) 4.75 m³ **49.** 2.79 g/cm³
51. (a) 9.4 N; (b) 1.6 N **53.** 4.0 m **55.** 28 ft/s **57.** 43 cm/s **59.** (a) 2.40 m/s; (b) 245 Pa **61.** (a) 12 ft/s; (b) 13 lb/in.² **63.** 0.72 ft·lb/ft³ **65.** (a) 2; (b) $R_1/R_2 = \frac{1}{2}$; (c) drain it until $h_2 = h_1/4$ **67.** 116 m/s **69.** (a) 6.4 m³; (b) 5.4 m/s; (c) 9.8×10^4 Pa **71.** (a) 560 Pa; (b) 5.0×10^4 N
73. 40 m/s **75.** (b) $H - h$; (c) $H/2$ **77.** (b) 0.69 ft³/s
79. (b) 63.3 m/s

Chapter 16

CP 1. (a) $-x_m$; (b) $+x_m$; (c) 0 **2.** a **3.** (a) 5 J; (b) 2 J; (c) 5 J **4.** all tie (in Eq. 16-32, m is included in I)
5. 1, 2, 3 (the ratio m/b matters; k does not) **Q 1.** c
3. (a) 0; (b) between 0 and $+x_m$; (c) between $-x_m$ and 0; (d) between $-x_m$ and 0 **5.** (a) toward $-x_m$; (b) toward $+x_m$; (c) between $-x_m$ and 0; (d) between $-x_m$ and 0; (e) decreasing; (f) increasing **7.** (a) 3, 2, 1; (b) all tie **9.** 3, 2, 1
11. system with spring A **13.** b (infinite period; does not oscillate), c, a **15.** (a) same; (b) same; (c) same; (d) smaller; (e) smaller; (f) and (g) larger ($T = \infty$) **EP 1.** (a) 0.50 s; (b) 2.0 Hz; (c) 18 cm **3.** (a) 245 N/m; (b) 0.284 s
5. 708 N/m **7.** $f > 500$ Hz **9.** (a) 100 N/m; (b) 0.45 s
11. (a) 6.28×10^5 rad/s; (b) 1.59 mm **13.** (a) 1.0 mm; (b) 0.75 m/s; (c) 570 m/s² **15.** (a) 1.29×10^5 N/m; (b) 2.68 Hz **17.** (a) 4.0 s; (b) $\pi/2$ rad/s; (c) 0.37 cm; (d) (0.37 cm) $\cos \frac{\pi}{4}t$; (e) (-0.58 cm/s) $\sin \frac{\pi}{4}t$; (f) 0.58 cm/s; (g) 0.91 cm/s²; (h) 0; (i) 0.58 cm/s **19.** (b) 12.47 kg; (c) 54.43 kg **21.** 1.6 kg **23.** (a) 1.6 Hz; (b) 1.0 m/s, 0; (c) 10 m/s², ±10 cm; (d) $(-10 \text{ N/m})x$ **25.** 22 cm
27. (a) 25 cm; (b) 2.2 Hz **29.** (a) 0.500 m; (b) -0.251 m; (c) 3.06 m/s **31.** (a) $0.183A$; (b) same direction
37. (a) $k_1 = (n + 1)k/n$, $k_2 = (n + 1)k$; (b) $f_1 = \sqrt{(n + 1)/n}f$, $f_2 = \sqrt{n + 1}f$ **39.** (b) 42 min **41.** (a) 200 N/m; (b) 1.39 kg; (c) 1.91 Hz **43.** (a) 130 N/m; (b) 0.62 s; (c) 1.6 Hz; (d) 5.0 cm; (e) 0.51 m/s **45.** (a) 3/4; (b) 1/4; (c) $x_m/\sqrt{2}$
47. (a) 3.5 m; (b) 0.75 s **49.** (a) 0.21 m; (b) 1.6 Hz; (c) 0.10 m **51.** (a) 0.0625 J; (b) 0.03125 J **53.** 12 s
55. (a) 39.5 rad/s; (b) 34.2 rad/s; (c) 124 rad/s² **57.** (a) 8.3 s; (b) no **59.** 9.47 m/s² **61.** 8.77 s **63.** 5.6 cm
65. $2\pi\sqrt{(R^2 + 2d^2)/2gd}$ **67.** (a) 0.205 kg·m²; (b) 47.7 cm; (c) 1.50 s **71.** (a) $2\pi\sqrt{(L^2 + 12x^2)/12gx}$; (b) 0.289 m

73. 9.78 m/s² **75.** $2\pi\sqrt{m/3k}$ **77.** $(1/2\pi)(\sqrt{g^2 + v^4/R^2}/L)^{1/2}$
79. (b) smaller **81.** (a) 2.0 s; (b) 18.5 N·m/rad
83. $0.29L$ **85.** 0.39 **87.** (a) 0.102 kg/s; (b) 0.137 J
89. $k = 490$ N/cm, $b = 1100$ kg/s **91.** 1.9 in.
93. (a) $y_m = 8.8 \times 10^{-4}$ m, $T = 0.18$ s, $\omega = 35$ rad/s; (b) $y_m = 5.6 \times 10^{-2}$ m, $T = 0.48$ s, $\omega = 13$ rad/s; (c) $y_m = 3.3 \times 10^{-2}$ m, $T = 0.31$ s, $\omega = 20$ rad/s

Chapter 17

CP 1. a, 2; b, 3; c, 1 **2.** (a) 2, 3, 1; (b) 3, then 1 and 2 tie
3. a **4.** 0.20 and 0.80 tie, then 0.60, 0.45 **5.** (a) 1; (b) 3; (c) 2 **6.** (a) 75 Hz; (b) 525 Hz **Q 1.** $7d$ **3.** tie of A and B, then C, D **5.** intermediate (closer to fully destructive interference) **7.** a and d tie, then b and c tie **9.** (a) 8; (b) antinode; (c) longer; (d) lower **11.** (a) integer multiples of 3; (b) node; (c) node **13.** string A **15.** decrease
EP 1. (a) 75 Hz; (b) 13 ms **3.** (a) 7.5×10^{14} to 4.3×10^{14} Hz; (b) 1.0 to 200 m; (c) 6.0×10^{16} to 3.0×10^{19} Hz
5. $y = 0.010 \sin \pi(3.33x + 1100t)$, with x and y in m and t in s **11.** (a) $z = 3.0 \sin(60y - 10\pi t)$, with z in mm, y in cm, and t in s; (b) 9.4 cm/s **13.** (a) $y = 2.0 \sin 2\pi(0.10x - 400t)$, with x and y in cm and t in s; (b) 50 m/s; (c) 40 m/s
15. (b) 2.0 cm/s; (c) $y = (4.0 \text{ cm}) \sin(\pi x/10 - \pi t/5 + \pi)$, where x is in cm and t is in s; (d) -2.5 cm/s **17.** 129 m/s
19. 135 N **23.** (a) 15 m/s; (b) 0.036 N **25.** $y = 0.12 \sin(141x + 628t)$, with y in mm, x in m, and t in s
27. (a) 5.0 cm; (b) 40 cm; (c) 12 m/s; (d) 0.033 s; (e) 9.4 m/s; (f) $5.0 \sin(16x + 190t + 0.79)$, with x in m, y in cm, and t in s **29.** (a) $v_1 = 28.6$ m/s, $v_2 = 22.1$ m/s; (b) $M_1 = 188$ g, $M_2 = 313$ g **31.** (a) $\sqrt{k(\Delta l)(l + \Delta l)/m}$ **33.** (a) $P_2 = 2P_1$; (b) $P_2 = P_1/4$ **35.** (a) 3.77 m/s; (b) 12.3 N; (c) zero; (d) 46.3 W; (e) zero; (f) zero; (g) ±0.50 cm **37.** 82.8°, 1.45 rad, 0.23 wavelength **39.** 5.0 cm **41.** (a) 4.4 mm; (b) 112°
43. (a) $0.83y_1$; (b) 37° **45.** (a) $2f_3$; (b) λ_3 **47.** 10 cm
49. (a) 82.0 m/s; (b) 16.8 m; (c) 4.88 Hz **51.** 240 cm, 120 cm, 80 cm **53.** 7.91 Hz, 15.8 Hz, 23.7 Hz **55.** $f_{1A} = f_{4B}$, $f_{2A} = f_{8B}$ **57.** (a) 2.0 Hz, 200 cm, 400 cm/s; (b) $x = 50$ cm, 150 cm, 250 cm, etc.; (c) $x = 0$, 100 cm, 200 cm, etc.
63. (a) 1.3 m; (b) $y' = 0.002 \sin(9.4x) \cos(3800t)$, with x and y in m and t in s **67.** (b) in the positive x direction; interchange the amplitudes of the original two traveling waves; (c) largest at $x = \lambda/4 = 6.26$ cm; smallest at $x = 0$ and $x = \lambda/2 = 12.5$ cm; (d) the largest amplitude is 4.0 mm, which is the sum of the amplitudes of the original traveling waves; the smallest amplitude is 1.0 mm, which is the difference of the amplitudes of the original traveling waves

Chapter 18

CP 1. beginning to decrease (example: mentally move the curves of Fig. 18-6 rightward past the point at $x = 42$ m)
2. (a) fully constructive, 0; (b) fully constructive, 4 **3.** (a) 1 and 2 tie, then 3; (b) 3, then 1 and 2 tie **4.** second
5. loosen **6.** a, greater; b, less; c, can't tell; d, can't tell; e, greater; f, less **7.** (a) 222 m/s; (b) $+20$ m/s **Q 1.** pulse

along path 2 **3.** (a) 2.0 wavelengths; (b) 1.5 wavelengths; (c) fully constructive, fully destructive **5.** (a) exactly out of phase; (b) exactly out of phase **7.** 70 dB **9.** (a) two; (b) antinode **11.** all odd harmonics **13.** 501, 503, and 508 Hz; or 505, 507, and 508 Hz **EP 1.** (a) $\approx 6\%$ **3.** the radio listener by about 0.85 s **5.** 7.9×10^{10} Pa **7.** If only the length is uncertain, it must be known to within 10^{-4} m. If only the time is imprecise, the uncertainty must be no more than one part in 10^8. **9.** 43.5 m **11.** 40.7 m **13.** 100 kHz **15.** (a) 2.29, 0.229, 22.9 kHz; (b) 1.14, 0.114, 11.4 kHz **17.** (a) 6.0 m/s; (b) $y = 0.30 \sin(\pi x/12 + 50\pi t)$, with x and y in cm and t in s **19.** 4.12 rad **21.** (a) $343 \times (1 + 2m)$ Hz, with m being an integer from 0 to 28; (b) $686m$ Hz, with m being an integer from 1 to 29 **23.** (a) eight; (b) eight **25.** 64.7 Hz, 129 Hz **27.** (a) 0.080 W/m²; (b) 0.013 W/m² **29.** 36.8 nm **31.** (a) 1000; (b) 32 **33.** (a) 39.7 μW/m²; (b) 171 nm; (c) 0.893 Pa **35.** (a) 59.7; (b) 2.81×10^{-4} **37.** $s_m \propto r^{-1/2}$ **39.** (a) 5000; (b) 71; (c) 71 **41.** 171 m **43.** 3.16 km **45.** (a) 5200 Hz; (b) amplitude$_{SAD}$/amplitude$_{SBD}$ = 2 **47.** 20 kHz **49.** by a factor of 4 **51.** water filled to a height of $\frac{7}{8}, \frac{5}{8}, \frac{3}{8}, \frac{1}{8}$ m **53.** (a) 5.0 cm from one end; (b) 1.2; (c) 1.2 **55.** (a) 1130, 1500, and 1880 Hz **57.** (a) 230 Hz; (b) higher **59.** (a) node; (c) 22 s **61.** 387 Hz **63.** 0.02 **65.** 3.8 Hz **67.** (a) 380 mi/h, away from owner; (b) 77 mi/h, away from owner **69.** 15.1 ft/s **71.** 2.6×10^8 m/s **73.** (a) 77.6 Hz; (b) 77.0 Hz **75.** 33.0 km **79.** (a) 970 Hz; (b) 1030 Hz; (c) 60 Hz, no **81.** (a) 1.02 kHz; (b) 1.04 kHz **83.** 1540 m/s **85.** 41 kHz **87.** (a) 2.0 kHz; (b) 2.0 kHz **89.** (a) 485.8 Hz; (b) 500.0 Hz; (c) 486.2 Hz; (d) 500.0 Hz **91.** 1×10^6 m/s, receding **93.** 0.13c

Chapter 19

CP 1. (a) all tie; (b) 50°X, 50°Y, 50°W **2.** (a) 2 and 3 tie, then 1, then 4; (b) 3, 2, then 1 and 4 tie **3.** A **4.** c and e **5.** (a) zero; (b) negative **6.** b and d tie, then a, c **Q 1.** 25 S°, 25 U°, 25 R° **3.** c, then the rest tie **5.** B, then A and C tie **7.** (a) both clockwise; (b) both clockwise **9.** c, a, b **11.** upward (with liquid water on the exterior and at the bottom, $\Delta T = 0$ horizontally and downward) **13.** at the temperature of your fingers **15.** 3, 2, 1 **EP 1.** 2.71 K **3.** 0.05 kPa, nitrogen **5.** (a) 320°F; (b) −12.3°F **7.** (a) −96°F; (b) 56.7°C **9.** (a) −40°; (b) 575°; (c) Celsius and Kelvin cannot give the same reading **11.** (a) Dimensions are inverse time **13.** 4.4×10^{-3} cm **15.** 0.038 in. **17.** (a) 9.996 cm; (b) 68°C **19.** 170 km **21.** 0.32 cm² **23.** 29 cm³ **25.** 0.432 cm³ **27.** −157°C **29.** 360°C **35.** +0.68 s/h **37.** (b) use 39.3 cm of steel and 13.1 cm of brass **39.** (a) 523 J/kg·K; (b) 0.600; (c) 26.2 J/mol·K **41.** 94.6 L **43.** 109 g **45.** 1.30 MJ **47.** 1.9 times as great **49.** (a) 33.9 Btu; (b) 172 F° **51.** (a) 52 MJ; (b) 0°C **53.** (a) 411 g; (b) 3.1¢ **55.** 0.41 kJ/kg·K **57.** 3.0 min **59.** 73 kW **61.** 2.17 g **63.** 33 m³ **65.** 33 g **67.** (a) 0°C; (b) 2.5°C **69.** 2500 J/kg·K **71.** A: 120 J, B: 75 J, C: 30 J **73.** (a) −200 J; (b) −293 J;

(c) −93 J **75.** −5.0 J **77.** 33.3 kJ **79.** 766°C **81.** (a) 1.2 W/m·K, 0.70 Btu/ft·F°·h; (b) 0.030 ft²·F°·h/Btu **83.** 1660 J/s **87.** arrangement b **89.** (a) 2.0 MW; (b) 220 W **91.** (a) 17 kW/m²; (b) 18 W/m² **93.** −6.1 nW **95.** 0.40 cm/h **97.** Cu-Al, 84.3°C; Al-brass, 57.6°C

Chapter 20

CP 1. all but c **2.** (a) all tie; (b) 3, 2, 1 **3.** gas A **4.** 5 (greatest change in T), then tie of 1, 2, 3, and 4 **5.** 1, 2, 3 ($Q_3 = 0$, Q_2 goes into work W_2, but Q_1 goes into greater work W_1 and increases gas temperature) **Q 1.** increased but less than doubled **3.** a, c, b **5.** 1180 J **7.** d, tie of a and b, then c **9.** constant-volume process **11.** (a) same; (b) increases; (c) decreases; (d) increases **13.** −4 J **15.** (a) 1, polyatomic; 2, diatomic; 3, monatomic; (b) more **EP 1.** (a) 0.0127; (b) 7.65×10^{21} **3.** 6560 **5.** number of molecules in the ink $\approx 3 \times 10^{16}$; number of people $\approx 5 \times 10^{20}$; statement is wrong, by a factor of about 20,000 **7.** (a) 5.47×10^{-8} mol; (b) 3.29×10^{16} **9.** (a) 106; (b) 0.892 m³ **11.** 27.0 lb/in.² **13.** (a) 2.5×10^{25}; (b) 1.2 kg **15.** 5600 J **17.** 1/5 **19.** (a) −45 J; (b) 180 K **21.** 100 cm³ **23.** 198°F **25.** 2.0×10^5 Pa **27.** 180 m/s **29.** 9.53×10^6 m/s **31.** 313°C **33.** 1.9 kPa **35.** (a) 0.0353 eV, 0.0483 eV; (b) 3400 J, 4650 J **37.** 9.1×10^{-6} **39.** (a) 6.75×10^{-20} J; (b) 10.7 **41.** 0.32 nm **43.** 15 cm **45.** (a) 3.27×10^{10}; (b) 172 m **47.** (a) 22.5 L; (b) 2.25; (c) 8.4×10^{-5} cm; (d) same as (c) **51.** (a) 3.2 cm/s; (b) 3.4 cm/s; (c) 4.0 cm/s **53.** (a) v_P, v_{rms}, $\bar{v}$ (b) reverse ranking **55.** (a) 1.0×10^4 K, 1.6×10^5 K; (b) 440 K, 7000 K **57.** 4.7 **59.** (a) $2N/3v_0$; (b) $N/3$; (c) $1.22v_0$; (d) $1.31v_0$ **61.** $RT \ln(V_f/V_i)$ **63.** (a) 15.9 J; (b) 34.4 J/mol·K; (c) 26.1 J/mol·K **65.** $(n_1C_1 + n_2C_2 + n_3C_3)/(n_1 + n_2 + n_3)$ **67.** (a) −5.0 kJ; (b) 2.0 kJ; (c) 5.0 kJ **69.** (a) 0.375 mol; (b) 1090 J; (c) 0.714 **71.** (a) 14 atm; (b) 620 K **79.** 0.63 **81.** (a) monatomic; (b) 2.7×10^4 K; (c) 4.5×10^4 mol; (d) 3.4 kJ, 340 kJ; (e) 0.01 **83.** 5 m³ **85.** (a) in joules, in the order Q, ΔE_{int}, W: $1 \rightarrow 2$: 3740, 3740, 0; $2 \rightarrow 3$: 0, −1810, 1810; $3 \rightarrow 1$: −3220, −1930, −1290; cycle: 520, 0, 520; (b) $V_2 = 0.0246$ m³, $p_2 = 2.00$ atm, $V_3 = 0.0373$ m³, $p_3 = 1.00$ atm

Chapter 21

CP 1. a, b, c **2.** smaller **3.** c, b, a **4.** a, d, c, b **5.** b **Q 1.** increase **3.** (a) increase; (b) same **5.** equal **7.** lower the temperature of the low temperature reservoir **9.** (a) same; (b) increase; (c) decrease **11.** (a) same; (b) increase; (c) decrease **13.** more than the age of the universe **EP 1.** 1.86×10^4 J **3.** 2.75 mol **7.** (a) 5.79×10^4 J; (b) 173 J/K **9.** +3.59 J/K **11.** (a) 14.6 J/K; (b) 30.2 J/K **13.** (a) 4.45 J/K; (b) no **15.** (a) 4500 J; (b) −5000 J; (c) 9500 J **17.** (a) 57.0°C; (b) −22.1 J/K; (c) +24.9 J/K; (d) +2.8 J/K **19.** (a) −710 mJ/K; (b) +710 mJ/K; (c) +723 mJ/K;

(d) -723 mJ/K; (e) $+13$ mJ/K; (f) 0   **23.** (a) (I) constant T, $Q = pV \ln 2$; constant V, $Q = 4.5pV$; (II) constant T, $Q = -pV \ln 2$; constant p, $Q = 7.5pV$; (b) (I) constant T, $W = pV \ln 2$; constant V, $W = 0$; (II) constant T, $W = -pV \ln 2$; constant p, $W = 3pV$; (c) $4.5pV$ for either case; (d) $4R \ln 2$ for either case   **25.** 0.75 J/K   **27.** (a) -943 J/K; (b) $+943$ J/K; (c) yes   **29.** (a) $3p_0V_0$; (b) $6RT_0$, $(3/2)R \ln 2$; (c) both are zero   **33.** (a) 31%; (b) 16 kJ   **35.** engine A, first; engine B, first and second; engine C, second; engine D, neither   **37.** 97 K   **39.** 99.99995%   **41.** 7.2 J/cycle   **43.** (a) 7200 J; (b) 960 J; (c) 13%   **45.** (a) 2270 J; (b) 14,800 J; (c) 15.4%; (d) 75.0%, greater   **49.** (a) 78%; (b) 81 kg/s   **55.** (a) 49 kJ; (b) 7.4 kJ   **57.** 21 J   **59.** (a) 0.071 J; (b) 0.50 J; (c) 2.0 J; (d) 5.0 J   **61.** 1.08 MJ   **63.** $[1 - (T_2/T_1)] \div [1 - (T_4/T_3)]$ **67.** (a) 1.27×10^{30}; (b) 7.9%; (c) 7.3%; (d) 7.3%; (e) 1.1%; (f) 0.0023%   **69.** (a) $W = N!/(n_1! \, n_2! \, n_3!)$; (b) $[(N/2)! \, (N/2)!]/[(N/3)! \, (N/3)! \, (N/3)!]$; (c) 4.2×10^{16}

Chapter 22

CP   **1.** C and D attract; B and D attract   **2.** (a) leftward; (b) leftward; (c) leftward   **3.** (a) a, c, b; (b) less than   **4.** $-15e$ (net charge of $-30e$ is equally shared)
Q   **1.** No, only for charged particles, charged particle-like objects, and spherical shells (including solid spheres) of uniform charge   **3.** a and b **5.** two points: one to the left of the particles and one between the protons   **7.** $6q^2/4\pi\epsilon_0 d^2$, leftward   **9.** (a) same; (b) less than; (c) cancel; (d) add; (e) the adding components; (f) positive direction of y; (g) negative direction of y; (h) positive direction of x; (i) negative direction of x **11.** (a) A, B, and D; (b) all four; (c) Connect A and D; disconnect them; then connect one of them to B. (There are two more solutions.)   **13.** (a) possibly; (b) definitely   **15.** same   **17.** D **EP**   **1.** 0.50 C
3. 2.81 N on each   **5.** (a) 4.9×10^{-7} kg; (b) 7.1×10^{-11} C   **7.** $3F/8$ **9.** (a) 1.60 N; (b) 2.77 N   **11.** (a) $q_1 = 9q_2$; (b) $q_1 = -25q_2$ **13.** either -1.00 μC and $+3.00$ μC or $+1.00$ μC and -3.00 μC   **15.** (a) 36 N, $-10°$ from the x axis; (b) $x = -8.3$ cm, $y = +2.7$ cm   **17.** (a) 5.7×10^{13} C, no; (b) 6.0×10^5 kg   **19.** (a) $Q = -2\sqrt{2}q$; (b) no   **21.** 3.1 cm   **23.** 2.89×10^{-9} N   **25.** -1.32×10^{13} C   **27.** (a) 3.2×10^{-19} C; (b) two   **29.** (a) 8.99×10^{-19} N; (b) 625   **31.** 5.1 m below the electron   **33.** 1.3 days   **35.** (a) 0; (b) 1.9×10^{-9} N   **37.** 10^{18} N   **39.** (a) ^{9}B; (b) ^{13}N; (c) ^{12}C   **41.** (a) $F = (Q^2/4\pi\epsilon_0 d^2)\alpha(1 - \alpha)$; (c) 0.5; (d) 0.15 and 0.85

Chapter 23

CP   **1.** (a) rightward; (b) leftward; (c) leftward; (d) rightward (p and e have same charge magnitude, and p is farther)   **2.** all tie   **3.** (a) toward positive y; (b) toward positive x; (c) toward negative y **4.** (a) leftward; (b) leftward; (c) decrease   **5.** (a) all tie; (b) 1 and 3 tie, then 2 and 4 tie
Q   **1.** (a) toward positive x; (b) downward and to the right;

(c) A **3.** two points: one to the left of the particles, the other between the protons   **5.** (a) yes; (b) toward; (c) no (the field vectors are not along the same line); (d) cancel; (e) add; (f) adding components; (g) toward negative y **7.** (a) 3, then 1 and 2 tie (zero); (b) all tie; (c) 1 and 2 tie, then 3   **9.** (a) rightward; (b) $+q_1$ and $-q_3$, increase; q_2, decrease; n, same   **11.** a, b, c **13.** (a) 4, 3, 1, 2; (b) 3, then 1 and 4 tie, then 2   **EP**   **1.** (a) 6.4×10^{-18} N; (b) 20 N/C   **3.** to the right in the figure   **7.** 56 pC   **9.** 3.07×10^{21} N/C, radially outward   **13.** (a) $q/8\pi\epsilon_0 d^2$, to the left; $3q/\pi\epsilon_0 d^2$, to the right; $7q/16\pi\epsilon_0 d^2$, to the left   **15.** 0   **17.** 9:30   **19.** $E = q/\pi\epsilon_0 a^2$, along bisector, away from triangle   **21.** $7.4q/4\pi\epsilon_0 d^2$, $28°$ counterclockwise to $+x$ **23.** 6.88×10^{-28} C$\cdot$m   **25.** $(1/4\pi\epsilon_0)(p/r^3)$, antiparallel to $\mathbf{p}$ **29.** $R/\sqrt{2}$ **31.** $(1/4\pi\epsilon_0)(4q/\pi R^2)$, toward decreasing y **37.** (a) 0.10 μC; (b) 1.3×10^{17}; (c) 5.0×10^{-6} **39.** 3.51×10^{15} m/s^2 **41.** 6.6×10^{-15} N   **43.** 2.03×10^{-7} N/C, up   **45.** (a) -0.029 C; (b) repulsive forces would explode the sphere   **47.** (a) 1.92×10^{12} m/s^2; (b) 1.96×10^5 m/s   **49.** (a) 8.87×10^{-15} N; (b) 120   **51.** 1.64×10^{-19} C ($\approx 3\%$ high)   **53.** (a) 0.245 N, $11.3°$ clockwise from the $+x$ axis; (b) $x = 108$ m, $y = -21.6$ m   **55.** 27μm   **57.** (a) yes; (b) upper plate, 2.73 cm   **59.** (a) 0; (b) 8.5×10^{-22} N$\cdot$m; (c) 0   **61.** $(1/2\pi)\sqrt{pE/I}$ **63.** (a) $E = (2q/4\pi\epsilon_0 d^2)(\alpha/(1 + \alpha^2)^{3/2})$; (c) 0.707; (d) 0.21 and 1.9

Chapter 24

CP   **1.** (a) $+EA$; (b) $-EA$; (c) 0; (d) 0   **2.** (a) 2; (b) 3; (c) 1   **3.** (a) equal; (b) equal; (c) equal   **4.** (a) $+50e$; (b) $-150e$ **5.** 3 and 4 tie, then 2, 1   **Q**   **1.** (a) 8 N$\cdot$m^2/C; (b) 0   **3.** (a) all tie (zero); (b) all tie   **5.** $+13q/\epsilon_0$ **7.** all tie   **9.** all tie   **11.** 2σ, σ, 3σ; or 3σ, σ, 2σ **13.** (a) all tie ($E = 0$); (b) all tie   **15.** (a) same ($E = 0$); (b) decrease; (c) decrease (to zero); (d) same   **EP**   **1.** (a) 693 kg/s; (b) 693 kg/s; (c) 347 kg/s; (d) 347 kg/s; (e) 575 kg/s   **3.** (a) 0; (b) -3.92 N$\cdot$m^2/C; (c) 0; (d) 0 for each field   **5.** (a) enclose $2q$ and $-2q$, or enclose all four charges; (b) enclose $2q$ and q; (c) not possible   **7.** 2.0×10^5 N$\cdot$m^2/C   **9.** $q/6\epsilon_0$ **11.** (a) $-\pi R^2 E$; (b) $\pi R^2 E$ **13.** -4.2×10^{-10} C   **15.** 0 through each of the three faces meeting at q, $q/24\epsilon_0$ through each of the other faces   **17.** 2.0 μC/m^2 **19.** (a) 4.5×10^{-7} C/m^2; (b) 5.1×10^4 N/C   **21.** (a) -3.0×10^{-6} C; (b) $+1.3 \times 10^{-5}$ C   **23.** (a) 0.32 μC; (b) 0.14 μC   **27.** (a) $E = q/2\pi\epsilon_0 Lr$, radially inward; (b) $-q$ on both inner and outer surfaces; (c) $E = q/2\pi\epsilon_0 Lr$, radially outward   **29.** 3.6 nC   **31.** (b) $\rho R^2/2\epsilon_0 r$ **33.** (a) 5.3×10^7 N/C; (b) 60 N/C   **35.** 5.0 nC/m^2 **37.** 0.44 mm   **39.** (a) 4.9×10^{-22} C/m^2; (b) downward   **41.** (a) $\rho x/\epsilon_0$; (b) $\rho d/2\epsilon_0$ **43.** (a) -750 N$\cdot$m^2/C; (b) -6.64 nC   **45.** (a) 4.0×10^6 N/C; (b) 0   **47.** (a) 0; (b) $q_a/4\pi\epsilon_0 r^2$; (c) $(q_a + q_b)/4\pi\epsilon_0 r^2$ **51.** (a) $-q$; (b) $+q$; (c) $E = q/4\pi\epsilon_0 r^2$, radially outward; (d) $E = 0$; (e) $E = q/4\pi\epsilon_0 r^2$, radially outward; (f) 0; (g) $E = q/4\pi\epsilon_0 r^2$, radially outward; (h) yes, charge is induced; (i) no; (j) yes; (k) no; (l) no   **53.** (a) $E = (q/4\pi\epsilon_0 a^3)r$; (b) $E = q/4\pi\epsilon_0 r^2$; (c) 0; (d) 0; (e) inner, $-q$; outer, 0   **55.** $q/2\pi a^2$ **59.** $\alpha = 0.80$

Chapter 25

CP   1. (a) negative; (b) increase   **2.** (a) positive;
(b) higher   **3.** (a) rightward; (b) 1, 2, 3, 5: positive; 4,
negative; (c) 3, then 1, 2, and 5 tie, then 4   **4.** all tie   **5.** a,
c (zero), b **6.** (a) 2, then 1 and 3 tie; (b) 3; (c) accelerate
leftward   **7.** closer (half of 9.23 fm)   **Q   1.** (a) higher;
(b) positive; (c) negative; (d) all tie   **3.** (a) 1 and 2; (b) none;
(c) no; (d) 1 and 2 yes, 3 and 4 no   **5.** b, then a, c, and d tie
7. (a) negative; (b) zero   **9.** (a) 1, then 2 and 3 tie; (b) 3
11. left   **13.** a, b, c **15.** (a) c, b, a; (b) zero
17. (a) positive; (b) positive; (c) negative; (d) all tie
19. (a) no; (b) yes   **21.** no (a particle at the intersection
would have two different potential energies)   **23.** (a)–(b) all
tie; (c) C, B, A; (d) all tie   **EP   1.** 1.2 GeV   **3.** (a) $3.0 \times$
10^{10} J; (b) 7.7 km/s; (c) 9.0×10^4 kg   **7.** 2.90 kV   **9.** 8.8
mm   **11.** (a) 136 MV/m; (b) 8.82 kV/m   **13.** (b) because
$V = 0$ point is chosen differently; (c) $q/(8\pi\epsilon_0 R)$; (d) potential
differences are independent of the choice for the $V = 0$ point
15. (a) -4500 V; (b) -4500 V   **17.** 843 V   **19.** 2.8×10^5
21. $x = d/4$ and $x = -d/2$ **23.** none   **25.** (a) 3.3 nC;
(b) 12 nC/m^2 **27.** 6.4×10^8 V   **29.** 190 MV
31. (a) -4.8 nm; (b) 8.1 nm; (c) no   **33.** 16.3 μV
35. (a) $\dfrac{2\lambda}{4\pi\epsilon_0} \ln\left[\dfrac{L/2 + (L^2/4 + d^2)^{1/2}}{d}\right]$; (b) 0
37. (a) $-5Q/4\pi\epsilon_0 R$; (b) $-5Q/4\pi\epsilon_0(z^2 + R^2)^{1/2}$
39. $0.113\sigma R/\epsilon_0$ **41.** $(Q/4\pi\epsilon_0 L)\ln(1 + L/d)$ **43.** 670 V/m
45. $p/2\pi\epsilon_0 r^3$ **47.** 39 V/m, $-x$ direction
51. (a) $\dfrac{c}{4\pi\epsilon_0}[\sqrt{L^2 + y^2} - y]$; (b) $\dfrac{c}{4\pi\epsilon_0}\left[1 - \dfrac{y}{\sqrt{L^2 + y^2}}\right]$
53. (a) 2.5 MV; (b) 5.1 J; (c) 6.9 J   **55.** (a) 2.72×10^{-14} J;
(b) 3.02×10^{-31} kg, about $\frac{1}{3}$ of accepted value   **57.** (a) 0.484
MeV; (b) 0   **59.** 2.1 d   **61.** 0   **63.** (a) 27.2 V; (b) -27.2
eV; (c) 13.6 eV; (d) 13.6 eV   **65.** 1.8×10^{-10} J
67. 1.48×10^7 m/s   **69.** $qQ/4\pi\epsilon_0 K$ **71.** 0.32 km/s
73. 1.6×10^{-9} m   **77.** (a) $V_1 = V_2$; (b) $q_1 = q/3$, $q_2 =$
$2q/3$; (c) 2   **79.** (a) -0.12 V; (b) 1.8×10^{-8} N/C, radially
inward   **81.** (a) 12,000 N/C; (b) 1800 V; (c) 5.8 cm
83. (c) 4.24 V

Chapter 26

CP   1. (a) same; (b) same   **2.** (a) decreases; (b) increases;
(c) decreases   **3.** (a) V, $q/2$; (b) $V/2$, q **4.** (a) $q_0 = q_1 +$
q_{34}; (b) equal (C_3 and C_4 are in series)   **5.** (a) same;
(b)–(d) increase; (e) same (same potential difference across
same plate separation)   **6.** (a) same; (b) decrease;
(c) increase   **Q   1.** a, 2; b, 1; c, 3   **3.** (a) increase;
(b) same   **5.** (a) parallel; (b) series   **7.** (a) $C/3$; (b) $3C$;
(c) parallel   **9.** (a) equal; (b) less   **11.** (a)–(d) less
13. (a) 2; (b) 3; (c) 1   **15.** Increase plate separation d, but
also plate area A, keeping A/d constant.   **EP   1.** 7.5 pC
3. 3.0 mC   **5.** (a) 140 pF; (b) 17 nC   **7.** (a) 84.5 pF;
(b) 191 cm^2 **9.** (a) 11 cm^2; (b) 11 pF; (c) 1.2 V
13. (b) 4.6×10^{-5}/K   **15.** 7.33 μF   **17.** 315 mC
19. (a) 10.0 μF; (b) $q_2 = 0.800$ mC, $q_1 = 1.20$ mC; (c) 200 V
for both   **21.** (a) $d/3$; (b) $3d$ **25.** (a) five in series; (b) three

arrays as in (a) in parallel (and other possibilities)   **27.** 43 pF
29. (a) 50 V; (b) 5.0×10^{-5} C; (c) 1.5×10^{-4} C
31. (a) $q_1 = 9.0$ μC, $q_2 = 16$ μC, $q_3 = 9.0$ μC, $q_4 = 16$ μC;
(b) $q_1 = 8.4$ μC, $q_2 = 17$ μC, $q_3 = 11$ μC, $q_4 = 14$ μC
33. 99.6 nJ   **35.** 72 F   **37.** 4.9%   **39.** 0.27 J   **41.** 0.11
J/m^3 **43.** (a) 2.0 J   **45.** (a) $q_1 = 0.21$ mC, $q_2 = 0.11$ mC,
$q_3 = 0.32$ mC; (b) $V_1 = V_2 = 21$ V, $V_3 = 79$ V; (c) $U_1 = 2.2$
mJ, $U_2 = 1.1$ mJ, $U_3 = 13$ mJ   **47.** (a) $q_1 = q_2 = 0.33$ mC,
$q_3 = 0.40$ mC; (b) $V_1 = 33$ V, $V_2 = 67$ V, $V_3 = 100$ V;
(c) $U_1 = 5.6$ mJ, $U_2 = 11$ mJ, $U_3 = 20$ mJ   **53.** Pyrex
55. (a) 6.2 cm; (b) 280 pF   **57.** 0.63 m^2 **59.** (a) 2.85 m^3;
(b) 1.01×10^4 **61.** (a) $\epsilon_0 A/(d-b)$; (b) $d/(d-b)$;
(c) $-q^2 b/2\epsilon_0 A$, sucked in   **65.** $\dfrac{\epsilon_0 A}{4d}\left(\kappa_1 + \dfrac{2\kappa_2\kappa_3}{\kappa_2 + \kappa_3}\right)$
67. (a) 13.4 pF; (b) 1.15 nC; (c) 1.13×10^4 N/C; (d) $4.33 \times$
10^3 N/C   **69.** (a) 7.1; (b) 0.77 μC   **71.** (a) 0.606; (b) 0.394

Chapter 27

CP   1. 8 A, rightward   **2.** (a)–(c) rightward   **3.** a and c
tie, then b **4.** Device 2   **5.** (a) and (b) tie, then (d), then
(c)   **Q   1.** a, b, and c tie, then d (zero)   **3.** b, a, c **5.** tie
of A, B, and C, then a tie of $A + B$ and $B + C$, then $A +$
$B + C$ **7.** (a)–(c) 1 and 2 tie, then 3   **9.** C, A, B
11. b, a, c **13.** (a) conductors: 1 and 4; semiconductors: 2
and 3; (b) 2 and 3; (c) all four   **EP   1.** 1.25×10^{15} **3.** 6.7
μC/m^2 **5.** 14-gauge   **7.** (a) 2.4×10^{-5} A/m^2; (b) $1.8 \times$
10^{-15} m/s   **9.** 0.67 A, toward the negative terminal
11. (a) 0.654 μA/m^2; (b) 83.4 MA   **13.** 13 min
15. (a) $J_0 A/3$; (b) $2J_0 A/3$ **17.** 2.0×10^{-8} $\Omega \cdot$m
19. 100 V   **21.** (a) 1.53 kA; (b) 54.1 MA/m^2; (c) $10.6 \times$
10^{-8} $\Omega \cdot$m, platinum   **23.** (a) 253°C; (b) yes   **25.** (a) 0.38 mV;
(b) negative; (c) 3 min 58 s   **27.** 54 Ω **29.** 3.0
31. (a) 1.3 mΩ; (b) 4.6 mm   **33.** (a) 6.00 mA; (b) $1.59 \times$
10^{-8} V; (c) 21.2 nΩ **35.** 2000 K   **37.** (a) copper: $5.32 \times$
10^5 A/m^2, aluminum: 3.27×10^5 A/m^2; (b) copper: 1.01 kg/m,
aluminum: 0.495 kg/m   **39.** 0.40 Ω **41.** (a) $R = \rho L/\pi ab$
43. 14 kC   **45.** 11.1 Ω **47.** (a) 1.0 kW; (b) 25¢
49. 0.135 W   **51.** (a) 1.74 A; (b) 2.15 MA/m^2; (c) 36.3
mV/m; (d) 2.09 W   **53.** (a) 1.3×10^5 A/m^2; (b) 94 mV
55. (a) $4.46 for a 31-day month; (b) 144 Ω; (c) 0.833 A
57. 660 W   **59.** (a) 3.1×10^{11}; (b) 25 μA; (c) 1300 W,
25 MW   **61.** 27 cm/s   **63.** (a) 120 Ω; (b) 107 Ω; (c) $5.3 \times$
10^{-3}/C°; (d) 5.9×10^{-3}/C°; (e) 276 Ω

Chapter 28

CP   1. (a) rightward; (b) all tie; (c) b, then a and c tie; (d) b,
then a and c tie   **2.** (a) all tie; (b) R_1, R_2, R_3 **3.** (a) less;
(b) greater; (c) equal   **4.** (a) $V/2$, i; (b) V, $i/2$ **5.** (a) 1, 2, 4,
3; (b) 4, tie of 1 and 2; then 3   **Q   1.** 3, 4, 1, 2   **3.** (a) no;
(b) yes; (c) all tie (the circuits are the same)   **5.** parallel, R_2,
R_1, series   **7.** (a) equal; (b) more   **9.** (a) less; (b) less;
(c) more   **11.** C_1, 15 V; C_2, 35 V; C_3, 20 V; C_4, 20 V; C_5,
30 V   **13.** 60 μC   **15.** c, b, a **17.** (a) all tie; (b) 1, 3, 2
19. 1, 3, and 4 tie (8 V on each resistor), then 2 and 5 tie (4 V
on each resistor)   **EP   1.** (a) $320; (b) 4.8¢   **3.** 11 kJ

5. (a) counterclockwise; (b) battery 1; (c) B   **7.** (a) 80 J;
(b) 67 J; (c) 13 J converted to thermal energy within battery
9. (a) 14 V; (b) 100 W; (c) 600 W; (d) 10 V, 100 W
11. (a) 50 V; (b) 48 V; (c) B is connected to the negative
terminal   **13.** 2.5 V   **15.** (a) 6.9 km; (b) 20 Ω
17. 8.0 Ω **19.** 10^{-6} **21.** the cable   **23.** (a) 1000 Ω;
(b) 300 mV; (c) 2.3×10^{-3} **25.** (a) 3.41 A or 0.586 A;
(b) 0.293 V or 1.71 V   **27.** 5.56 A   **29.** 4.0 Ω and 12 Ω
31. 4.50 Ω **33.** 0.00, 2.00, 2.40, 2.86, 3.00, 3.60, 3.75,
3.94 A   **35.** $V_d - V_c = +0.25$ V, by all paths   **37.** three
39. (a) 2.50 Ω; (b) 3.13 Ω **41.** nine   **43.** (a) left branch:
0.67 A down; center branch: 0.33 A up; right branch: 0.33 A
up; (b) 3.3 V   **47.** (a) 120 Ω; (b) $i_1 = 51$ mA, $i_2 = i_3 =$
19 mA, $i_4 = 13$ mA   **49.** (a) 19.5 Ω; (b) 0; (c) ∞; (d) 82.3 W,
57.6 W   **51.** (a) Cu: 1.11 A, Al: 0.893 A; (b) 126 m
53. (a) 13.5 kΩ; (b) 1500 Ω; (c) 167 Ω; (d) 1480 Ω
55. 0.45 A   **57.** (a) 12.5 V; (b) 50 A   **59.** -0.9%
65. (a) 0.41τ; (b) 1.1τ **67.** 4.6   **69.** (a) 0.955 μC/s;
(b) 1.08 μW; (c) 2.74 μW; (d) 3.82 μW   **71.** 2.35 MΩ
73. 0.72 MΩ **75.** 24.8 Ω to 14.9 kΩ **77.** (a) at $t = 0$,
$i_1 = 1.1$ mA, $i_2 = i_3 = 0.55$ mA; at $t = \infty$, $i_1 = i_2 = 0.82$ mA,
$i_3 = 0$; (c) at $t = 0$, $V_2 = 400$ V; at $t = \infty$, $V_2 = 600$ V;
(d) after several time constants ($\tau = 7.1$ s) have elapsed
79. (a) $V_T = -ir + \mathcal{E}$; (b) 13.6 V; (c) 0.060 Ω
81. (a) 6.4 V; (b) 3.6 W; (c) 17 W; (d) -5.6 W; (e) a

Chapter 29

CP **1.** a, $+z$; b, $-x$; c, $F_B = 0$ **2.** 2, then tie of 1 and 3
(zero); (b) 4   **3.** (a) $+z$ and $-z$ tie, then $+y$ and $-y$ tie, then
$+x$ and $-x$ tie (zero); (b) $+y$ **4.** (a) electron;
(b) clockwise   **5.** $-y$ **6.** (a) all tie; (b) 1 and 4 tie, then 2
and 3 tie   **Q** **1.** (a) all tie; (b) 1 and 2 (charge is negative)
3. a, no, **v** and $\mathbf{F}_B$ must be perpendicular; b, yes; c, no, **B** and
$\mathbf{F}_B$ must be perpendicular   **5.** (a) $\mathbf{F}_E$; (b) $\mathbf{F}_B$ **7.** (a) right;
(b) right   **9.** (a) negative; (b) equal; (c) equal; (d) half a
circle   **11.** (a) $\mathbf{B}_1$; (b) $\mathbf{B}_1$ into page; $\mathbf{B}_2$ out of page; (c) less
13. all   **15.** all tie   **17.** (a) positive; (b) (1) and (2) tie, then
(3) which is zero   **EP** **1.** M/QT **3.** (a) 9.56×10^{-14} N, 0;
(b) 0.267°   **5.** (a) east; (b) 6.28×10^{14} m/s^2; (c) 2.98 mm
7. 0.75**k** T   **9.** (a) 3.4×10^{-4} T, horizontal and to the left
as viewed along $\mathbf{v}_0$; (b) yes, if velocity is the same as the
electron's velocity   **11.** $(-11.4\mathbf{i} - 6.00\mathbf{j} + 4.80\mathbf{k})$ V/m
13. 680 kV/m   **17.** (b) 2.84×10^{-3} **19.** (a) 1.11×10^7 m/s;
(b) 0.316 mm   **21.** (a) 0.34 mm; (b) 2.6 keV
23. (a) 2.05×10^7 m/s; (b) 467 μT; (c) 13.1 MHz; (d) 76.3 ns
25. (a) 2.60×10^6 m/s; (b) 0.109 μs; (c) 0.140 MeV;
(d) 70 kV   **29.** (a) 1.0 MeV; (b) 0.5 MeV   **31.** $R_d = \sqrt{2}R_p$;
$R_\alpha = R_p$.   **33.** (a) $B\sqrt{mq/2V}\,\Delta x$; (b) 8.2 mm   **37.** (a) $-q$;
(b) $\pi m/qB$ **39.** $B_{\min} = \sqrt{mV/2ed^2}$ **41.** (a) 22 cm;
(b) 21 MHz   **43.** neutron moves tangent to original path,
proton moves in a circular orbit of radius 25 cm   **45.** 28.2 N,
horizontally west   **47.** 20.1 N   **49.** $Bitd/m$, away from
generator   **51.** $-0.35\mathbf{k}$ N   **53.** 0.10 T, at 31° from the
vertical   **55.** 4.3×10^{-3} N·m, negative y **59.** $qvaB/2$
61. (a) 540 Ω, in series; (b) 2.52 Ω, in parallel
63. (a) 12.7 A; (b) 0.0805 N·m   **65.** (a) 0.184 A·m^2;

(b) 1.45 N·m   **67.** (a) 20 min; (b) 5.9×10^{-2} N·m
69. (a) $(8.0 \times 10^{-4}$ N·m$)(-1.2\mathbf{i} - 0.90\mathbf{j} + 1.0\mathbf{k})$;
(b) -6.0×10^{-4} J

Chapter 30

CP **1.** a, c, b **2.** b, c, a **3.** d, tie of a and c, then b
4. d, a, tie of b and c (zero)   **Q** **1.** c, d, then a and b tie
3. 2 and 4   **5.** a, b, c **7.** b, d, c, a (zero)   **9.** (a) 1, $+x$;
2, $-y$; (b) 1, $+y$; 2, $+x$ **11.** outward   **13.** c and
d tie, then b, a **15.** d, then tie of a and e, then b, c
17. 0 (dot product is zero)   **EP** **1.** 32.1 A
3. (a) 3.3 μT; (b) yes   **5.** (a) $(0.24\mathbf{i})$ nT; (b) 0; (c) $(-43\mathbf{k})$ pT;
(d) $(0.14\mathbf{k})$ nT   **7.** (a) 16 A; (b) west to east   **9.** 0
11. (a) 0; (b) $\mu_0 i/4R$, into the page; (c) same as (b)
13. $\mu_0 i\theta$ $(1/b - 1/a)/4\pi$, out of page   **15.** (a) 1.0 mT, out of
the figure; (b) 0.80 mT, out of the figure   **25.** 200 μT, into
page   **27.** (a) it is impossible to have other than $B = 0$
midway between them; (b) 30 A   **29.** 4.3 A, out of page
35. $0.338\mu_0 i^2/a$, toward the center of the square
37. (b) to the right   **39.** (b) 2.3 km/s   **41.** $+5\mu_0 i$
47. (a) $\mu_0 ir/2\pi c^2$; (b) $\mu_0 i/2\pi r$; (c) $\dfrac{\mu_0 i}{2\pi(a^2 - b^2)} \dfrac{a^2 - r^2}{r}$;
(d) 0   **49.** $3i/8$, into page   **53.** 0.30 mT   **55.** 108 m
61. 0.272 A   **63.** (a) 4; (b) 1/2   **65.** (a) 2.4 A·m^2;
(b) 46 cm   **67.** (a) $\mu_0 i(1/a + 1/b)/4$, into page; (b) $\frac{1}{2}i\pi(a^2 + b^2)$,
into page   **69.** (a) 79 μT; (b) 1.1×10^{-6} N·m
71. (b) $(0.060\,\mathbf{j})$ A·m^2; (c) $(9.6 \times 10^{-11}\mathbf{j})$ T, $(-4.8 \times$
$10^{-11}\mathbf{j})$ T   **73.** (a) B from sum: 7.069×10^{-5} T; $\mu_0 in =$
5.027×10^{-5} T; 40% difference; (b) B from sum: $1.043 \times$
10^{-4} T; $\mu_0 in = 1.005 \times 10^{-4}$ T; 4% difference; (c) B from
sum: 2.506×10^{-4} T; $\mu_0 in = 2.513 \times 10^{-4}$ T; 0.3%
difference   **75.** (a) $\mathbf{B} = (\mu_0/2\pi)\,[i_1/(x - a) + i_2/x]\mathbf{j}$;
(b) $\mathbf{B} = (\mu_0/2\pi)\,(i_1/a)\,(1 + b/2)\mathbf{j}$

Chapter 31

CP **1.** b, then d and e tie, and then a and c tie (zero)   **2.** a
and b tie, then c (zero)   **3.** c and d tie, then a and b tie
4. b, out; c, out; d, into; e, into   **5.** d and e   **6.** (a) 2, 3, 1
(zero); (b) 2, 3, 1   **7.** a and b tie, then c **Q** **1.** (a) all tie
(zero); (b) all tie (nonzero); (c) 3, then tie of 1 and 2 (zero)
3. out   **5.** (a) into; (b) counterclockwise; (c) larger
7. (a) leftward; (b) rightward   **9.** c, a, b **11.** (a) 1, 3, 2;
(b) 1 and 3 tie, then 2   **13.** a, tie of b and c   **15.** (a) more;
(b) same; (c) same; (d) same (zero)   **17.** a, 2; b, 4; c, 1; d, 3
EP **1.** 57 μWb   **3.** 1.5 mV   **5.** (a) 31 mV; (b) right to
left   **7.** (a) 0.40 V; (b) 20 A   **9.** (b) 58 mA   **11.** 1.2 mV
13. 1.15 μWb   **15.** 51 mV, clockwise when viewed along the
direction of **B**   **17.** (a) 1.26×10^{-4} T, 0, -1.26×10^{-4} T;
(b) 5.04×10^{-8} V   **19.** (b) no   **21.** 15.5 μC
23. (a) 24 μV; (b) from c to b **25.** (b) design it so that
$Nab = (5/2\pi)$ m^2 **27.** (a) 0.598 μV; (b) counterclockwise
29. (a) $\dfrac{\mu_0 ia}{2\pi}\left(\dfrac{2r + b}{2r - b}\right)$; (b) $2\mu_0 iabv/\pi R(4r^2 - b^2)$
31. $A^2B^2/R\Delta t$ **33.** (a) 48.1 mV; (b) 2.67 mA; (c) 0.128 mW
35. $v_t = mgR/B^2L^2$ **37.** 268 W   **39.** (a) 240 μV; (b) 0.600

mA; (c) 0.144 μW; (d) 2.88 $\times$ 10^{-8} N; (e) same as (c)
41. 1, -1.07 mV; 2, -2.40 mV; 3, 1.33 mV **43.** at a:
4.4 $\times$ 10^7 m/s^2, to the right; at b: 0; at c: 4.4 $\times$ 10^7 m/s^2, to the
left **45.** 0.10 μWb **47.** (a) 800; (b) 2.5 $\times$ 10^{-4} H/m
49. (a) $\mu_0 i/W$; (b) $\pi\mu_0 R^2/W$ **51.** (a) decreasing;
(b) 0.68 mH **53.** (a) 0.10 H; (b) 1.3 V **55.** (a) 16 kV;
(b) 3.1 kV; (c) 23 kV **57.** (b) so that the changing magnetic
field of one does not induce current in the other;
(c) $1/L_{eq} = \sum_{j=1}^{N} (1/L_j)$ **59.** 6.91 **61.** 1.54 s **63.** (a) 8.45
ns; (b) 7.37 mA **65.** $(42 + 20t)$ V **67.** 12.0 A/s
69. (a) $i_1 = i_2 = 3.33$ A; (b) $i_1 = 4.55$ A, $i_2 = 2.73$ A;
(c) $i_1 = 0$, $i_2 = 1.82$ A; (d) $i_1 = i_2 = 0$ **71.** $\mathscr{E}L_1/R(L_1 + L_2)$
73. (a) $i(1 - e^{-Rt/L})$ **75.** $1.23\tau_L$ **77.** (a) 240 W;
(b) 150 W; (c) 390 W **79.** (a) 97.9 H; (b) 0.196 mJ
81. (a) 10.5 mJ; (b) 14.1 mJ **83.** (a) 34.2 J/m^3; (b) 49.4
mJ **85.** 1.5 $\times$ 10^8 V/m **87.** $(\mu_0 l/2\pi)\ln(b/a)$ **89.** (a) 1.3
mT; (b) 0.63 J/m^3 **91.** (a) 1.0 J/m^3; (b) 4.8 $\times$ 10^{-15} J/m^3
93. (a) 1.67 mH; (b) 6.00 mWb **95.** 13 H **99.** magnetic
field exists only within the cross section of solenoid 1

Chapter 32

CP **1.** d, b, c, a (zero) **2.** (a) 2; (b) 1 **3.** (a) away;
(b) away; (c) less **4.** (a) toward; (b) toward; (c) less
5. a, c, b, d (zero) **6.** tie of b, c, and d, then a
Q **1.** (a) a, c, f; (b) bar gh **3.** supplied **5.** (a) all down;
(b) 1 up, 2 down, 3 zero **7.** (a) 1 up, 2 up, 3 down;
(b) 1 down, 2 up, 3 zero **9.** (a) rightward; (b) leftward
11. (a) decreasing; (b) decreasing **13.** (a) tie of a and b, then
c, d; (b) none (plate lacks circular symmetry, so **B** is not
tangent to a circular loop); (c) none **15.** 1/4 EP **1.** (b)
sign is minus; (c) no, compensating positive flux through open
end near magnet **3.** 47 μWb, inward **5.** 55 μT **7.** (a)
600 MA; (b) yes; (c) no **9.** (a) 31.0 μT, 0°; (b) 55.9 μT,
73.9°; (c) 62.0 μT, 90° **11.** 4.6 $\times$ 10^{-24} J **13.** (a) 5.3 $\times$
10^{11} V/m; (b) 20 mT; (c) 660 **15.** (a) 7; (b) 7; (c) $3h/2\pi$, 0;
(d) $3eh/4\pi m$, 0; (e) $3.5h/2\pi$; (f) 8 **17.** (b) in the direction
of the angular momentum vector **19.** $\Delta\mu = e^2r^2B/4m$
21. 20.8 mJ/T **23.** yes **25.** (a) 4 K; (b) 1 K
29. (a) 3.0 μT; (b) 5.6 $\times$ 10^{-10} eV **31.** (a) 8.9 A·m^2;
(b) 13 N·m **35.** (a) 0.14 A; (b) 79 μC **37.** 2.4 $\times$
10^{13} V/m·s **39.** 1.9 pT **41.** 7.5 $\times$ 10^5 V/s
43. 7.2 $\times$ 10^{12} V/m·s **45.** (a) 2.1 $\times$ 10^{-8} A, downward;
(b) clockwise **47.** (a) 0.63 μT; (b) 2.3 $\times$ 10^{12} V/m·s
49. (a) 2.0 A; (b) 2.3 $\times$ 10^{11} V/m·s; (c) 0.50 A;
(d) 0.63 μT·m **51.** (a) 7.60 μA; (b) 859 kV·m/s;
(c) 3.39 mm; (d) 5.16 pT

Chapter 33

CP **1.** (a) $T/2$, (b) T, (c) $T/2$, (d) $T/4$ **2.** (a) 5 V;
(b) 150 μJ **3.** (a) 1; (b) 2 **4.** (a) C, B, A; (b) 1, A;
2, B; 3, S; 4, C; (c) A **5.** (a) increases; (b) decreases
6. (a) 1, lags; 2, leads; 3, in phase; (b) 3 ($\omega_d = \omega$ when
$X_L = X_C$) **7.** (a) increase (circuit is mainly capacitive;
increase C to decrease X_C to be closer to resonance for

maximum P_{av}); (b) closer **8.** step-up Q **1.** (a) $T/4$,
(b) $T/4$, (c) $T/2$ (see Fig. 33-2), (d) $T/2$ (see Eq. 31-40)
3. b, a, c **5.** (a) 3, 1, 2; (b) 2, tie of 1 and 3
7. slower **9.** (a) 1 and 4; (b) 2 and 3 **11.** (a) 3,
then 1 and 2 tie; (b) 2, 1, 3 **13.** (a) negative; (b) lead
15. (a)–(c) rightward, increase EP **1.** 9.14 nF
3. 45.2 mA **5.** (a) 6.00 μs; (b) 167 kHz; (c) 3.00 μs
7. (a) 89 rad/s; (b) 70 ms; (c) 25 μF **9.** 38 μH
11. 7.0 $\times$ 10^{-4} s **15.** (a) 3.0 nC; (b) 1.7 mA; (c) 4.5 nJ
17. (a) 3.60 mH; (b) 1.33 kHz; (c) 0.188 ms **19.** 600, 710,
1100, 1300 Hz **21.** (a) $Q/\sqrt{3}$; (b) 0.152 **25.** (a) 1.98 μJ;
(b) 5.56 μC; (c) 12.6 mA; (d) -46.9°; (e) $+46.9$° **27.** (a) 0;
(b) $2i(t)$ **29.** (a) 356 μs; (b) 2.50 mH; (c) 3.20 mJ
31. 8.66 mΩ **33.** $(L/R)\ln 2$ **35.** (a) $\pi/2$ rad; (b) $q =$
$(I/\omega')\, e^{-Rt/2L} \sin \omega't$ **39.** (a) 0.0955 A; (b) 0.0119 A
41. (a) 4.60 kHz; (b) 26.6 nF; (c) $X_L = 2.60$ kΩ, $X_C =$
0.650 kΩ **43.** (a) 0.65 kHz; (b) 24 Ω **45.** (a) 39.1 mA;
(b) 0; (c) 33.9 mA **47.** (a) 6.73 ms; (b) 2.24 ms;
(c) capacitor; (d) 59.0 μF **49.** (a) $X_C = 0$, $X_L = 86.7$ Ω,
$Z = 182$ Ω, $I = 198$ mA, $\phi = 28.5$° **51.** (a) $X_C = 37.9$ Ω,
$X_L = 86.7$ Ω, $Z = 167$ Ω, $I = 216$ mA, $\phi = 17.1$°
53. (a) 2.35 mH; (b) they move away from 1.40 kHz
55. 1000V **57.** (a) 36.0 V; (b) 27.3 V; (c) 17.0 V;
(d) -8.34 V **59.** (a) 224 rad/s; (b) 6.00 A; (c) 228 rad/s,
219 rad/s; (d) 0.040 **61.** (a) 707 Ω; (b) 32.2 mH;
(c) 21.9 nF **63.** (a) resonance at $f = 1/2\pi\sqrt{LC} = 85.7$ Hz;
(b) 15.6 μF; (c) 225 mA **65.** (a) 796 Hz; (b) no change;
(c) decreased; (d) increased **69.** 141 V **71.** (a) taking;
(b) supplying **73.** 0, 9.00 W, 3.14 W, 1.82 W
75. 177 Ω, no **77.** 7.61 A **83.** (a) 117 μF; (b) 0;
(c) 90.0 W, 0; (d) 0°, 90°; (e) 1, 0 **85.** (a) 2.59 A;
(b) 38.8 V, 159 V, 224 V, 64.2 V, 75.0 V; (c) 100 W for R,
0 for L and C. **87.** (a) 2.4 V; (b) 3.2 mA, 0.16 A
89. (a) 1.9 V, 5.9 W; (b) 19 V, 590 W; (c) 0.19 kV, 59 kW
91. (a) $X_C = [(2\pi)(45 \times 10^{-6}$ F$)f]^{-1}$; (c) 17.7 Hz
93. (a) $X_L = (2\pi)(40 \times 10^{-3}$ H$)f$; (c) 796 Hz
95. (b) 61 Hz; (c) 90 Ω and 61 Hz

Chapter 34

CP **1.** (a) (Use Fig. 34-5.) On right side of rectangle, **E** is in
negative y direction; on left side, **E** + d**E** is greater and in
same direction; (b) **E** is downward. On right side, **B** is in
negative z direction; on left side, **B** + d**B** is greater and in
same direction. **2.** positive direction of x **3.** (a) same;
(b) decrease **4.** a, d, b, c (zero) **5.** a **6.** (a) yes; (b) no
Q **1.** (a) positive direction of z; (b) x **3.** (a) same;
(b) increase; (c) decrease **5.** both 20° clockwise from the y
axis **7.** two **9.** b, 30°; c, 60°; d, 60°; e, 30°; f, 60°
11. d, b, a, c **13.** (a) b; (b) blue; (c) c **15.** 1.5
EP **1.** (a) 4.7 $\times$ 10^{-3} Hz; (b) 3 min 32 s **3.** (a) 4.5 $\times$ 10^{24}
Hz; (b) 1.0 $\times$ 10^4 km or 1.6 Earth radii **7.** it would steadily
increase; (b) the summed discrepancies between the apparent
time of eclipse and those observed from x; the radius of Earth's
orbit **9.** 5.0 $\times$ 10^{-21} H **11.** 1.07 pT **17.** 4.8 $\times$ 10^{-29}
W/m^2 **19.** 4.51 $\times$ 10^{-10} **21.** 89 cm **23.** 1.2 MW/m^2
25. 820 m **27.** (a) 1.03 kV/m; 3.43 μT **29.** (a) 1.4 $\times$

10^{-22} W; (b) 1.1×10^{15} W **31.** (a) 87 mV/m; (b) 0.30 nT; (c) 13 kW **33.** 3.3×10^{-8} Pa **35.** (a) 4.7×10^{-6} Pa; (b) 2.1×10^{10} times smaller **37.** 5.9×10^{-8} Pa **39.** (a) 3.97 GW/m²; (b) 13.2 Pa; (c) 1.67×10^{-11} N; (d) 3.14×10^3 m/s² **41.** $I(2 - frac)/c$ **43.** $p_{r\perp} \cos^2 \theta$ **45.** 1.9 mm/s **47.** (b) 580 nm **49.** (a) 1.9 V/m; (b) 1.7×10^{-11} Pa **51.** 1/8 **53.** 3.1% **55.** 20° or 70° **57.** 19 W/m² **59.** (a) 2 sheets; (b) 5 sheets **61.** 180° **63.** 1.26 **65.** 1.07 m **69.** (a) 0; (b) 20°; (c) still 0 and 20° **73.** 1.41 **75.** 1.22 **77.** 182 cm **79.** (a) no; (b) yes; (c) about 43° **81.** (a) 35.6°; (b) 53.1° **83.** (b) 23.2° **85.** (a) 53°; (b) yes **87.** (a) 55.5°; (b) 55.8°

Chapter 35

CP Kaleidoscope answer: two mirrors that form a V with an angle of 60° **1.** 0.2d, 1.8d, 2.2d **2.** (a) real; (b) inverted; (c) same **3.** (a) e; (b) virtual, same **4.** virtual, same as object, diverging Q **1.** c **3.** (a) a; (b) c **5.** (a) no; (b) yes (fourth is off mirror ed) **7.** (a) from infinity to the focal point; (b) decrease continually **9.** d (infinite), tie of a and b, then c **11.** mirror, equal; lens, greater **13.** (a) all but variation 2; (b) for 1, 3, and 4: right, inverted; for 5 and 6: left, same **15.** (a) less; (b) less EP **1.** (a) virtual; (b) same; (c) same; (d) $D + L$ **3.** 40 cm **7.** (a) 7; (b) 5; (c) 1 to 3; (d) depends on the position of O and your perspective **11.** new illumination is 10/9 of the old **15.** 10.5 cm **19.** (a) 2.00; (b) none **23.** 1.14 **25.** (b) separate the lenses by a distance $f_2 - |f_1|$, where f_2 is the focal length of the converging lens **27.** 45 mm, 90 mm **29.** (a) $+40$ cm; (b) at infinity **33.** (a) 40 cm, real; (b) 80 cm, real; (c) 240 cm, real; (d) -40 cm, virtual; (e) -80 cm, virtual; (f) -240 cm, virtual **35.** same orientation, virtual, 30 cm to left of second lens, $m = 1$ **37.** (a) final image coincides in location with the object; it is real, inverted, and $m = -1.0$ **39.** (a) coincides in location with the original object and is enlarged 5.0 times; (c) virtual; (d) yes

45. $i = \dfrac{(2 - n)r}{2(n - 1)}$, to the right of the right side of the sphere

47. 2.1 mm **49.** (b) when image is at near point **51.** (b) farsighted **53.** -125

Chapter 36

CP **1.** b (least n), c, a **2.** (a) top; (b) bright intermediate illumination (phase difference is 2.1 wavelengths) **3.** (a) 3λ, 3; (b) 2.5λ, 2.5 **4.** a and d tie (amplitude of resultant wave is $4E_0$), then b and c tie (amplitude of resultant wave is $2E_0$) **5.** (a) 1 and 4; (b) 1 and 4 Q **1.** a, c, b **3.** (a) 300 nm; (b) exactly out of phase **5.** c **7.** (a) increase; (b) 1λ **9.** down **11.** (a) maximum; (b) minimum; (c) alternates **13.** d **15.** (a) 0.5 wavelength; (b) 1 wavelength **17.** bright **19.** all EP **1.** (a) 5.09×10^{14} Hz; (b) 388 nm; (c) 1.97×10^8 m/s **5.** 2.1×10^8 m/s **7.** the time is longer for the pipeline containing air, by about 1.55 ns **9.** 22°, refraction reduces θ **11.** (a) pulse 2; (b) $0.03L/c$ **13.** (a) 1.70 (or 0.70); (b) 1.70 (or 0.70); (c) 1.30

(or 0.30); (d) brightness is identical, close to fully destructive interference **15.** (a) 0.833; (b) intermediate, closer to fully constructive interference **17.** $(2m + 1)\pi$ **19.** 2.25 mm **21.** 648 nm **23.** 1.6 mm **25.** 16 **27.** 0.072 mm **29.** 8.75λ **31.** 0.03% **33.** 6.64 μm **35.** $y = 17 \sin(\omega t + 13°)$ **39.** (a) 1.17 m, 3.00 m, 7.50 m; (b) no **41.** $I = \frac{1}{9}I_m[1 + 8 \cos^2(\pi d \sin \theta/\lambda)]$, I_m = intensity of central maximum **43.** $L = (m + \frac{1}{2})\lambda/2$, for $m = 0, 1, 2, \ldots$ **45.** 0.117 μm, 0.352 μm **47.** λ/5 **49.** 70.0 nm **51.** none **53.** (a) 552 nm; (b) 442 nm **55.** 338 nm **59.** $2n_2 L/\cos \theta_r = (m + \frac{1}{2})\lambda$, for $m = 0, 1, 2, \ldots$, where $\theta_r = \sin^{-1}[(\sin \theta_i)/n_2]$ **61.** intensity is diminished by 88% at 450 nm and by 94% at 650 nm **63.** (a) dark; (b) blue end **65.** 1.89 μm **67.** 1.00025 **69.** (a) 34; (b) 46 **73.** 588 nm **75.** 1.0003 **77.** $I = I_m \cos^2(2\pi x/\lambda)$

Chapter 37

CP **1.** (a) expand; (b) expand **2.** (a) second side maximum; (b) 2.5 **3.** (a) red; (b) violet **4.** diminish **5.** (a) increase; (b) same **6.** (a) left; (b) less Q **1.** (a) contract; (b) contract **3.** with megaphone (larger opening, less diffraction) **5.** four **7.** (a) larger; (b) red **9.** (a) decrease; (b) same; (c) in place **11.** (a) A; (b) left; (c) left; (d) right EP **1.** 690 nm **3.** 60.4 μm **5.** (a) 2.5 mm; (b) 2.2×10^{-4} rad **7.** (a) 70 cm; (b) 1.0 mm **9.** 41.2 m from the central axis **11.** 160° **15.** (d) 53°, 10°, 5.1° **19.** (a) 1.3×10^{-4} rad; (b) 10 km **21.** 50 m **23.** 30.5 μm **25.** 1600 km **27.** (a) 17.1 m; (b) 1.37×10^{-10} **29.** 27 cm **31.** 4.7 cm **33.** (a) 0.347°; (b) 0.97° **35.** (a) red; (b) 130 μm **37.** five **41.** λD/d **43.** (a) 5.05 μm; (b) 20.2 μm **45.** (a) 3.33 μm; (b) 0, $\pm 10.2°$, $\pm 20.7°$, $\pm 32.0°$, $\pm 45.0°$, $\pm 62.2°$ **47.** all wavelengths shorter than 635 nm **49.** 13,600 **51.** 500 nm **53.** (a) three; (b) 0.051° **55.** 523 nm **61.** 470 nm to 560 nm **63.** 491 **65.** 3650 **67.** (a) 1.0×10^4 nm; (b) 3.3 mm **69.** (a) 0.032°/nm, 0.076°/nm, 0.24°/nm; (b) 40,000, 80,000, 120,000 **71.** (a) tan θ; (b) 0.89 **73.** 0.26 nm **75.** 6.8° **77.** (a) 170 pm; (b) 130 pm **81.** 0.570 nm **83.** 30.6°, 15.3° (clockwise); 3.08°, 37.8° (counterclockwise)

Chapter 38

CP **1.** (a) same (speed of light postulate); (b) no (the start and end of the flight are spatially separated); (c) no (again, because of the spatial separation) **2.** (a) Sally's; (b) Sally's **3.** a, negative; b, positive; c, negative **4.** (a) right; (b) more **5.** (a) equal; (b) less Q **1.** all tie (pulse speed is c) **3.** (a) C_1; (b) C_1 **5.** (a) 3, 2, 1; (b) 1 and 3 tie, then 2 **7.** (a) negative; (b) positive **9.** c, then b and d tie, then a **11.** (a) 3, tie of 1 and 2, then 4; (b) 4, tie of 1 and 2, then 3; (c) 1, 4, 2, 3 **13.** greater than f_1 EP **1.** (a) 3×10^{-18}; (b) 8.2×10^{-8}; (c) 1.1×10^{-6}; (d) 3.7×10^{-5}; (e) 0.10 **3.** 0.75c **5.** 0.99c **7.** 55 m **9.** 1.32 m **11.** 0.63 m **13.** 6.4 cm **15.** (a) 26 y; (b) 52 y; (c) 3.7 y **17.** (b) 0.999 999 15c **19.** (a) $x' = 0, t' = 2.29$ s; (b) $x' =$

6.55×10^8 m, $t' = 3.16$ s   **21.** (a) 25.8 μs; (b) small flash
23. (a) 1.25; (b) 0.800 μs   **25.** 2.40 μs   **27.** (a) 0.84c, in
the direction of increasing x; (b) 0.21c, in the direction of
increasing x; the classical predictions are 1.1c and 0.15c
29. (a) 0.35c; (b) 0.62c **31.** 1.2 μs   **33.** seven
35. 22.9 MHz   **37.** +2.97 nm   **39.** (a) $\tau_0/\sqrt{1 - v^2/c^2}$
41. (a) 0.134c; (b) 4.65 keV; (c) 1.1%   **43.** (a) 0.9988, 20.6;
(b) 0.145, 1.01; (c) 0.073, 1.0027   **45.** (a) 5.71 GeV,
6.65 GeV, 6.58 GeV/c; (b) 3.11 MeV, 3.62 MeV,
3.59 MeV/c **47.** 18 smu/y   **49.** (a) 0.943c; (b) 0.866c
51. (a) 256 kV; (b) 0.746c **53.** $\sqrt{8}mc$ **55.** 6.65×10^6 mi,
or 270 earth circumferences   **57.** 110 km   **59.** (a) 2.7 $\times$

10^{14} J; (b) 1.8×10^7 kg; (c) 6.0×10^6 **61.** 4.00 u, probably
a helium nucleus   **63.** 330 mT
65.

SIGNAL	TIME SENT (h)	TIME REPLY RECEIVED (h)	TIME REPORTED	DISTANCE (m)
1	6.0	400	11.8	2.10×10^{14}
2	12.0	800	23.6	4.19×10^{14}
3	18.0	1200	35.5	6.29×10^{14}
4	24.0	1600	47.3	8.38×10^{14}
5	30.0	2000	59.1	1.05×10^{15}

67. (a) $vt \sin \theta$; (b) $t[1 - (v/c) \cos \theta]$; (c) 3.24c

Index

Page references followed by lowercase roman t indicate material in tables. Page references followed by lowercase italic *n* indicate material in footnotes.

Photo Credits

Chapter 16
Page 372: Tom van Dyke/Sygma. Page 373: Kent Knudson/FPG International. Page 387: Bettmann Archive. Page 392: Courtesy NASA.

Chapter 17
Page 399: John Visser/Bruce Coleman, Inc. Page 415: Richard Megna/Fundamental Photographs. Page 416: Courtesy T.D. Rossing, Northern Illinois University.

Chapter 18
Page 425: Stephen Dalton/Animals Animals. Page 426: Howard Sochurak/The Stock Market. Page 433: Ben Rose/The Image Bank. Page 434: Bob Gruen/Star File. Page 435: John Eastcott/Yva Momativk/DRK Photo. Page 441: Philippe Plailly/Science Photo Library/Photo Researchers. Page 444: Courtesy NASA.

Chapter 19
Page 453: Tom Owen Edmunds/The Image Bank. Page 459: AP/Wide World Photos. Page 471: Peter Arnold/Peter Arnold, Inc. Page 472: Courtesy Daedalus Enterprises, Inc.

Chapter 20
Page 484: Bryan and Cherry Alexander Photography.

Chapter 21
Page 509: Steven Dalton/Photo Researchers. Page 517: Richard Ustinich/The Image Bank. Page 523 (left): Cary Wolinski/Stock, Boston. Page 523 (right): Courtesy of Professor Bernard Hallet, Quaternary Research Center, University of Washington, Seattle. Page 534: Jeff Werner.

Chapter 22
Page 537: Michael Watson. Page 538: Fundamental Photographs. Page 539: Courtesy Xerox. Page 540: Johann Gabriel Doppelmayr, Neuentdeckte Phaenomena von Bewünderswurdigen Würckungen der Natur, Nuremberg, 1744. Page 547: Courtesy Lawrence Berkeley Laboratory.

Chapter 23
Page 554: Quesada/Burke Studios. Page 565: Russ Kinne/Comstock, Inc. Page 566: Courtesy Environmental Elements Corporation.

Chapter 24
Page 579: E.R. Degginger/Bruce Coleman, Inc. Page 589 (top): Courtesy E. Philip Krider, Institute for Atmospheric Physics, University of Arizona, Tucson. Page 589 (bottom): C. Johnny Autery.

Chapter 25
Pages 601 and 605: Courtesy NOAA. Page 617: Courtesy Westinghouse Corporation.

Chapter 26
Page 628: Goivaux Communication/Phototake. Page 629: Paul Silvermann/Fundamental Photographs. Page 638: ©The Harold E. Edgerton 1992 Trust/Courtesy Palm Press, Inc. Page 639: Courtesy The Royal Institute, England.

Chapter 27
Page 651: UPI/Corbis-Bettmann. Page 657: The Image Works. Page 663: Laurie Rubie/Tony Stone Images/New York, Inc. Page 666 (left): Courtesy AT&T Bell Laboratories. Page 666 (right): Courtesy Shoji Tonaka/International Superconductivity Technology Center, Tokyo, Japan.

Chapter 28
Page 673: Norbert Wu. Page 674: Courtesy Southern California Edison Company.

Chapter 29
Page 700: Johnny Johnson/Earth Scenes/Animals Animals. Page 701: Schneps/The Image Bank. Page 703: Courtesy Lawrence Berkeley Laboratory, University of California. Page 704: Courtesy Dr. Richard Cannon, Southeast Missouri State University, Cape Girardeau. Page 708: Courtesy John Le P. Webb, Sussex University, England. Page 710: Courtesy Dr. L.A. Frank, University of Iowa. Page 712: Courtesy Fermi National Accelerator Laboratory.

Chapter 30
Page 728: Michael Brown/Florida Today/Gamma Liaison. Page 730: Courtesy Education Development Center.

Chapter 31
Page 752: Dan McCoy/Black Star. Page 756: Courtesy Fender Musical Instruments Corporation. Page 760: Courtesy Jenn-Air Co. Page 765: Courtesy The Royal Institute, England.

Chapter 32
Page 786: Bob Zehring. Page 787: Runk/Schoenberger/Grant Heilman Photography. Page 794: Peter Lerman. Page 797 (top): Courtesy Ralph W. DeBlois. Page 797 (bottom): R.E. Rosenweig, Research and Science Laboratory, courtesy Exxon Co. USA.

Chapter 33
Page 811: Courtesy Haverfield Helicopter Co. Page 814: Courtesy Hewlett Packard. Page 827 (left): Steve Kagan/Gamma Liaison. Page 827 (right): Ted Cowell/Black Star.

Chapter 34
Page 841: Courtesy Hansen Publications. Page 850: ©1992 Ben and Miriam Rose, from the collection of the Center for Creative Photography, Tucson. Page 854: Diane Schiumo/Fundamental Photographs. Page 855: PSSC Physics, 2nd edition; ©1975 D.C. Heath and Co. with Education Development Center, Newton, MA. Page 856: Courtesy Lockheed Advanced Development Company. Page 858 (top right): Tony Stone Images/New York, Inc. Page 858 (bottom left): Courtesy Bausch & Lomb. Page 861: Will and Deni McIntyre/Photo Researchers. Page 867: Cornell University.

Chapter 35
Page 872: Courtesy Courtauld Institute Galleries, London. Page 874 (left): Frans Lanting/Minden Pictures, Inc. Page 874 (right): Wayne Sorce. Page 882: Dr. Paul A. Zahl/Photo Researchers.

Page 884: Courtesy Matthew J. Wheeler. Page 894: Piergiorgio Scharandis/Black Star.

Chapter 36

Page 901: E.R. Degginger. Page 905: Runk Schoenberger/Grant Heilman Photography. Page 906: From *Atlas of Optical Phenomena* by M. Cagnet et al., Springer-Verlag, Prentice Hall, 1962. Page 914: Richard Megna/Fundamental Photographs. Page 917: Courtesy Dr. Helen Ghiradella, Department of Biological Sciences, SUNY, Albany. Page 926: Courtesy Bausch & Lomb.

Chapter 37

Page 929: Georges Seurat, French, 1859–1891, *A Sunday on La Grande Jatte,* 1884. Oil on canvas; 1884–86, 207.5 × 308 cm; Helen Birch Bartlett Memorial Collection, 1926. Photograph ©1996, The Art Institute of Chicago. All Rights Reserved. Page 930: Ken Kay/Fundamental Photographs. Pages 931, 937 (bottom left) and 940: From *Atlas of Optical Phenomena* by Cagnet, Francon, Thierr, Springer-Verlag, Berlin, 1962. Page 937 (bottom right): AP/Wide World Photos, Inc. Page 938: Cath Ellis/Science Photo Library/Photo Researchers. Page 945 (left): Department of Physics, Imperial College/Science Photo Library/Photo Researchers. Page 945 (top right): Peter L. Chapman/Stock, Boston. Page 953: Kjell B. Sandved/Bruce Coleman, Inc. Page 954: Courtesy Professor Robert Greenler, Physics Department, University of Wisconsin.

Chapter 38

Page 958: T. Tracy/FPG International. Page 959: Courtesy Hebrew University of Jerusalem, Israel.